Prof.
Ching Li

Theory and Practice of Recursive Identification

The MIT Press Series in Signal Processing, Optimization, and Control
Alan S. Willsky, editor

1. *Location on Networks: Theory and Algorithms*, Gabriel Y. Handler and Pitu B. Mirchandani, 1979

2. *Digital Signal Processing and Control and Estimation Theory: Points of Tangency, Areas of Intersection, and Parallel Directions*, Alan S. Willsky, 1979

3. *Stability and Robustness of Multivariable Feedback Systems*, Michael George Safonov, 1980

4. *Theory and Practice of Recursive Identification*, Lennart Ljung and Torsten Söderström, 1983

Theory and Practice of Recursive Identification

Lennart Ljung
Torsten Söderström

The MIT Press
Cambridge, Massachusetts
London, England

Third printing, 1986

This book was set in Times New Roman by Asco Trade Typesetting Ltd., Hong Kong and printed and bound by Halliday Lithograph in the United States of America.

Library of Congress Cataloging in Publication Data

Ljung, Lennart.
   Theory and practice of recursive identification.
   (The MIT Press series in signal processing, optimization, and control; 4)
   Includes bibliographies and index.
   1. System identification.   I. Söderström, Torsten.   II. Title.   III. Series.
QA402.L6 1983   003   82-17928
ISBN 0-262-12095-X

Contents

7 Applications of Recursive Identification

Series Foreword

The fields of signal processing, optimization, and control stand as well-developed disciplines with solid theoretical and methodological foundations. While the development of each of these fields is of great importance, many future problems will require the combined efforts of researchers in all of the disciplines. Among these challenges are the analysis, design, and optimization of large and complex systems, the effective utilization of the capabilities provided by recent developments in digital technology for the design of high-performance control and signal-processing systems, and the application of systems concepts to a variety of applications such as transportation systems, seismic signal processing, and data communication networks.

This series serves several purposes. It not only includes books at the leading edge of research in each field but also emphasizes theoretical research, analytical techniques, and applications that merit the attention of workers in all disciplines. In this way the series should help acquaint researches in each field with other perspectives and techniques and provide cornerstones for the development of new research areas within each discipline and across the boundaries.

Lennart Ljung and Torsten Söderström's book *Theory and Practice of Recursive Identification* stands as a major addition to the literature on system identification and parameter estimation. This topic is a natural one for this series, as the problem of parameter estimation is of great importance in both the fields of signal processing and control. Furthermore, interest in this subject is on the increase, as the availability of inexpensive but computationally powerful digital processors has made feasible the use of advanced and complex adaptive algorithms in a wide variety of applications in which they had not been used or even considered in the past. Consequently Ljung and Söderström's book is a most timely one.

As the authors point out in their preface, the field of recursive identification is filled with a multitude of approaches, perspectives, and techniques whose interrelationships and relative merits are difficult to sort out. As a consequence it has become a decidedly nontrivial task for a newcomer to the field or a nonspecialist to extract the fundamental concepts of recursive identification or to gain enough intuition about a particular technique to be able to use it effectively in practice. For this reason Ljung and Söderström's book is a welcome contribution, as its primary aim is to present a coherent picture of recursive identification.

In doing this the authors have done an outstanding job of constructing and describing a unified framework which not only exposes the crucial issues in the choice and design of an on-line identification algorithm but also provides the reader with a natural and simple frame of reference for understanding the similarities and differences among the many approaches to recursive identification. Furthermore, thanks to careful organization, the authors have produced a book which should have broad appeal. For graduate students and nonspecialists it provides an excellent introduction to the subject. For those primarily interested in using identification algorithms in practice it provides a thorough treatment of the critical aspects of and tradeoffs involved in algorithm design, as well as "user's summaries" which identify those points in each chapter that are of most importance to the practitioner. For the more theoretically inclined, there is a detailed development of convergence analysis for recursive algorithms. And finally, for all who have an interest in identification, be it peripheral or principal, this book should prove to be a valuable reference for many years.

Alan S. Willsky

Preface

The field of recursive identification has been called a "fiddler's paradise" (Åström and Eykhoff, 1971), and it is still often viewed as a long and confusing list of methods and tricks. Though the description was no doubt accurate at the time Åström and Eykhoff's survey was written, we believe that the time has now come to challenge this opinion by providing a comprehensive yet coherent treatment of the field. This has been our motivation for writing this book.

Coherence and unification in the field of recursive identification is not immediate. One reason is that methods and algorithms have been developed in different areas with different applications in mind. The term "recursive identification" is taken from the control literature. In statistical literature the field is usually called "sequential parameter estimation," and in signal processing the methods are known as "adaptive algorithms."

Within these areas, algorithms have been developed and analyzed over the last 30 years. Recently there has been a noticably increased interest in the field from practitioners and industrial "users." This is due to the construction of more complex systems, where adaptive techniques (adaptive control, adaptive signal processing) may be useful or necessary, and, of course, to the availability of microprocessors for the easy implementation of more advanced algorithms. As a consequence, material on recursive identification should be included in undergraduate and graduate courses. With this in mind, the series editor, Alan Willsky, has encouraged us to make this book accessible to a broad audience. This objective perhaps conflicts with our ambition to give a comprehensive treatment of recursive identification. We have tried to solve this conflict by providing bypasses around the more technical portions of the book (see sections 1.4 and 4.1). We have also included a more "leisurely" introduction to the field in chapter 2.

The manuscript of this book has been tested as a text for a first-year graduate course on identification at Stanford University, and for a course for users at Lawrence Livermore Laboratory. For use as a text, the appendixes should be excluded. Depending on whether the emphasis of the course is theory or practice, further reductions in chapters 4–6 could be considered. In a course oriented to practice, chapter 4 could be read according to the "sufficiency path" described in figure 4.1. In a theory course, chapter 5 could be used for illustration, and the algorithms in chapter 6 could be omitted. We have not included exercises in the material. The natural way of getting familiar with recursive identification is to implement and simulate different algorithms; such programming problems are more valuable than formal paper-and-pencil exercises.

As remarked above, the existing literature in the field is extensive. Any attempt to make the reference list comprehensive would therefore be a formidable task. Instead, we have mostly confined ourselves to what appear to be original references and to "further reading" of more detailed accounts of various problems.

Lennart Ljung
Division of Automatic Control
Department of Electrical Engineering
Linköping University, Linköping, Sweden

Torsten Söderström
Department of Automatic Control and Systems Analysis
Institute of Technology
Uppsala University, Uppsala, Sweden

Acknowledgments

Initial conditions play an important role for most systems. We are happy to acknowledge the excellent initial conditions provided for us by Professor Karl Johan Åström and his group at the Lund Institute of Technology in Sweden. They have had a lasting influence on our research in general and on the development of this book in particular.

The structure of the book has emerged from numerous seminars and some short courses on the subject given by the first author around the world. The comments and reactions by the participants in these seminars and courses have helped us greatly.

A large number of researchers have helped us with useful comments on the manuscript in different stages of completion. In the first place we must mention the editor of the series, Dr. Alan Willsky, who has provided most detailed and useful comments throughout the work. We are also grateful to many of our colleagues and friends for helping us with numerous important and clarifying comments. In particular we would like to mention G. Bierman, T. Bohlin, J. Candy, B. Egardt, A. Nehorai, P. Hagander, M. Morf, G. Olsson, U. V. Reddy, J. Salz, C. Samson, S. Shah, B. Widrow, and B. Wittenmark for important and helpful comments.

Dr. F. Soong has provided figure 5.12, which is reproduced from his thesis.

The completion of the book took seven years after it was first planned. Chances are that it would never have been finished if the first author had not had the privilege to spend a sabbatical year at Stanford University in 1980–1981. He would like to thank Professor Thomas Kailath for making this possible and for providing inspiring working conditions.

The manuscript has gone through several versions. They have been expertly typed by Mrs. Ingegerd Stenlund, Miss Ingrid Ringård, and Mrs. Margareta Hallenberg. Our deadlines have been met only because of these persons' willingness to work overtime. We express our sincere appreciation for their help.

We are also indebted to Mrs. Marianne Anse-Lundberg who skilfully and quickly prepared the illustrations.

Finally, we express our gratitude to Dr. Ivar Gustavsson, who took part in the research behind the book and the planning of it. He was an intended coauthor during most of the work. We regret that his obligations in industry finally prevented him from completing his contribution. However, we are sure that he is performing a more important task: To apply and use the methods in the real world–not merely write about them.

Symbols, Abbreviations, and Notational Conventions

Symbols

AsN   asymptotic normal distribution

d dimension of the parameter vector θ

D_c set of parameter vectors describing the convergence points

$D_{\mathcal{M}}$ set of parameter vectors describing the model set

D_s set of parameter vectors decribing models with stable predictors

$\bar{E}$ $\bar{E}f(t) = \lim_{t \to \infty} \dfrac{1}{t} \sum_{k=1}^{t} Ef(k)$ where E = expectation operator

$e(t)$ white noise (a sequence of independent random variables)

F, G, H matrices for state-space models

$g_{\mathcal{M}}(\ ,\ ,\ )$ predictor function

$K_i(t)$ reflection coefficient (see section 6.4)

$L(t)$ gain in algorithm

$l(\ ,\ ,\ )$ loss function

$\mathcal{M}$ model set, model structure

$\mathcal{M}(\theta)$ model corresponding to the parameter vector θ

n model order

N run length (see chapter 5)

$N(m, P)$ normal (Gaussian) distribution of mean value m and covariance matrix P

$O(x)$ $O(x)/x$ bounded when $x \to 0$

$o(x)$ $o(x)/x \to 0$ when $x \to 0$

$P(t)$ $\bar{R}^{-1}(t)$

p dimension of the output vector $y(t)$

$p(\theta \,|\, z^*)$ posterior probability density function

$\mathbf{R}^n$ Euclidean n-dimensional space

R_1, R_2, R_{12} covariance matrices

$R(t)$	Hessian approximation in Gauss-Newton algorithm
$\bar{R}(t)$	$tR(t)$
R_D	matrix corresponding to $R(t)$ in the associated d.e.
r	dimension of the input vector $u(t)$
$r(t)$	scalar factor for stochastic gradient algorithm
$\mathscr{S}$	true system
$S(q^{-1})$	prefilter of data
T	transpose
$T(q^{-1})$	prefilter of data
tr	trace (of a matrix)
t	time variable (integer-valued)
$u(t)$	input signal (column vector of dimension r)
$V_t(\theta)$	loss function at time t
$w(t)$	white noise (a sequence of independent random variables)
$\mathscr{X}$	experimental condition
$y(t)$	output signal (column vector of dimension p)
$y_F(t)$	filtered output
$\hat{y}(t)$	predictor using running estimate
$\hat{y}(t \mid \theta)$	as above using fixed model parameter θ
$z(t)$	measurements $(y^T(t)\ u^T(t))^T$ at time t
z^N	data set made up of $z(1), \ldots, z(N)$
$\bar{\beta}(N, t)$	forgetting profile defined by (2.115), (2.117)
$\beta(N, t)$	forgetting profile defined by (2.128)
$\gamma(t)$	gain (sequence)
$\delta_{t,s}$	Kronecker's delta
$\varepsilon(t)$	prediction error using running estimate
$\varepsilon(t, \theta)$	as above using fixed model parameter θ
$\bar{\varepsilon}(t)$	residual (posterior prediction error)

$\zeta(t)$	vector of instrumental variables used as gradient approximation in the IV method
$\zeta(t, \theta)$	as above for fixed model parameters θ
$\eta(t)$	general gradient approximation using running estimate (in actual algorithms replaced by one of φ, ψ, or ζ)
$\eta(t, \theta)$	as above for fixed model parameters θ
θ	parameter vector of unknown coefficients (column vector of dimension d)
$\hat{\theta}(t)$	recursive estimate of θ based on data up to time t
θ_D	vector corresponding to $\hat{\theta}(t)$ in the associated d.e.
θ_0	true value of the parameter vector θ
θ^*	limit value of $\hat{\theta}(t)$
$\hat{\theta}_t$	off-line estimate of θ based on data up to time t
Λ_0	covariance matrix ($p \times p$-matrix) of prediction errors
$\hat{\Lambda}(t)$	estimated covariance matrix of prediction errors
λ	forgetting factor
$\xi(t)$	state vector in prediction and gradient calculations using running estimate
$\xi(t, \theta)$	as above for fixed model parameters θ
$\varphi(t)$	vector formed from observed data using a running estimate (the gradient approximation used in PLR)
$\varphi(t, \theta)$	as above for fixed model parameters θ
$\varphi_F(t)$	vector of filtered and lagged data
$\psi(t)$	gradient of the predictions computed using running estimate ($d \times p$-matrix)
$\psi(t, \theta)$	as above for fixed model parameter θ
$\triangleq$	defined as
$:=$	assignment operator
$[x]_D$	projection of x into D

Abbreviations

AR   autoregressive

ARMA   autoregressive moving average

ARMAX   autoregressive moving average with exogenous variables

d.e.   differential equation

EKF   extended Kalman filter

ELS   extended least squares

GLS   generalized least squares

IV   instrumental variable

LS   least squares

MA   moving average

MLE   maximum likelihood estimate

PLR   pseudolinear regression

PRBS   pseudorandom binary sequence

RGLS   recursive generalized least squares

RIV   recursive instrumental variable

RLS   recursive least squares

RML   recursive maximum likelihood

RPE   recursive prediction error

RPEM   recursive prediction error method

SISO   single-input/single output

w.p.l   with probability one

w.r.t.   with respect to

Notational Conventions

$H^{-1}(q^{-1})$ $[H(q^{-1})]^{-1}$

$\varphi^{\mathrm{T}}(t)$ $[\varphi(t)]^{\mathrm{T}}$

$A^{-\mathrm{T}}$ $[A^{-1}]^{\mathrm{T}}$

l_θ first derivative of l with respect to θ

$l_{\theta\theta}$ second derivative of l with respect to θ

V' first derivative of V with respect to its argument

To
Ann-Kristin, Johan, Arvid
and
Marianne, Johanna, Andreas

Theory and Practice of Recursive Identification

1 Introduction

1.1 Systems and Models

By *systems* we mean a wide range of more or less complex objects, whose behavior we are interested in studying, affecting, or controlling. From a long list of examples we could mention:

- *A ship:* We would like to control its direction of motion by manipulating the rudder angle.
- *A paper machine:* We would like to control the quality of the manufactured paper by manipulating valves, etc.
- *A telephone communication channel:* We would like to construct a filter to be used at the receiver in order to obtain good reproduction of the transmitted signal.
- *A time series of data* (e.g., sales, unemployment, or rainfall figures): We would like to predict future values of the data in order to act properly now.

This list of examples illustrates tasks typically related to the study and use of systems: *control*, *signal processing* (*filter design*), and *prediction*. These tasks have been extensively studied in control theory, communication theory, signal processing, and statistics. A wide variety of techniques has been developed for solving problems involving these tasks. These techniques all have in common that some knowledge about the system's properties must be used to design the controller/filter/predictor.

The knowledge of the properties of a system will generally be called a *model*; the model may be given in any one of several different forms. We may, for example, distinguish between:

- *"Mental" or "intuitive" models:* Knowledge of the system's behavior is summarized in nonanalytical form in a person's mind. A driver's model of an automobile's dynamics is typically of this character.
- *Graphic models:* Properties of the system are summarized in a graph or in a table. An example could be a graph of a valve's characteristics or a Bode diagram for the frequency response of a linear system.
- *Mathematical models:* One finds a mathematical relationship, often a differential or difference equation, between certain variables. An example of a mathematical model is Newton's laws of motion, which, taken together, relate applied forces and the motion of an object.

It is necessary to have a model of the system in order to solve problems such as those listed above. For many purposes, the model need not be very sophisticated, and most problems are probably solved using only mental models. Mathematical models are necessary, however, when complex design problems are treated. In this book we shall study certain ways of building these mathematical models.

1.2 How to Obtain a Model of a System

There are basically two approaches to the problem of building a mathematical model of a given system.

(1) One can sometimes look more or less directly into the mechanisms that generate signals and variables inside the system. Based on the physical laws and relationships that (are supposed to) govern the system's behavior, a mathematical model can be constructed. This procedure is usually called *modeling*.

(2) Often, such direct modeling may not be possible. The reason may be that the knowledge of the system's mechanism is incomplete; or, the properties exhibited by the system may change in an unpredictable manner. Furthermore, modeling can be quite time-consuming and may lead to models that are unnecessarily complex. In such cases, signals produced by the system can be measured and be used to construct a model. We will use the term *identification* for such a procedure. Clearly, this will cover many types of methods, from elaborate experiments specifically and carefully designed to yield certain information (e.g., wind tunnel experiments for determining aerodynamic properties), to simple transient-response measurements (e.g., change the rudder angle quickly and observe how long it takes before the ship responds).

Techniques to infer a model from measured data typically contains two steps. First a family of candidate models is decided upon. Then we find the particular member of this family that satisfactorily (in some sense) describes the observed data. In this book we shall mostly concentrate on the second step, which in fact is a *parameter estimation* problem. This is not to say that the first step is easy or obvious; it is, however, quite application-dependent, so that it is difficult to give a general discussion of this step.

From this general description, we may say that identification is a link between the mathematical-model world and the real world. As such it is of considerable conceptual as well as practical interest. Some form of identification technique will be a necessary step in any application of theory to real-world systems.

1.3 Why Recursive Identification?

To give a better feeling for the role identification plays in applications we shall consider some problems from different areas. This will also bring out what *recursive* identification is and why it is of interest.

EXAMPLE 1.1 (Ship Steering) A ship's heading angle and position is controlled using the rudder angle. For a large ship, such as a supertanker, this position control could be a fairly difficult problem. The main reason is that the ship's response to a change in the rudder angle is so slow that it is affected by random components of wind and wave motion. Most ships therefore have an autopilot, i.e., a regulator, that measures relevant variables, and, based on these and on information about the desired heading angle, determines the rudder angle. The design of such a regulator must be related to the dynamic properties of the ship. This can be achieved either by basing its design upon a mathematical model of the ship or by experimentally "tuning" its parameters until it yields the desired behavior.

Now the steering dynamics of a ship depends on a number of things. The ship's shape and size, its loading and trim, as well as the water depth, are important factors. Some of these may vary (loading, water depth) during a journey. Obviously, the wind and wave disturbances that affect the steering may also rapidly change. Therefore the regulator must be constantly retuned to match the current dynamics of the system; in fact, it is desirable that the regulator retune itself. Such a ship-steering regulator is described by Åström (1980a) and by van Amerongen (1981). □

Many control problems exhibit features similar to the foregoing example. Airplanes, missiles, and automobiles have dynamic properties that depend on speed, loading, etc. The dynamic properties of electric-motor drives change with the load. Machinery such as that in paper-making plants is affected by many factors that change in an unpredictable manner. The area of *adaptive control* is concerned with the study and design of controllers and regulators that adjust to varying properties of the controlled object. This is currently a very active research area (see, e.g., Åström et al., 1977, and Landau, 1979).

EXAMPLE 1.2 (Short-Term Prediction of Power Demand)   The demand for electrical power from a power system varies over time. The demand changes in a more or less predictable way with time of day and over the courses of the week, month, or year. There is also, however, a substantial random component in the demand. The efficient production of electricity requires good predictions of the power load a few hours ahead, so that the operation of the different power plants in the system can be effectively coordinated.

Now, prediction of the power demand of course requires some sort of a model of its random component. It seems reasonable to suppose that the mechanism that generates this random contribution to the power load depends on circumstances, e.g., the weather, that themselves may vary with time. Therefore it would be desirable to use a predictor that *adapts* itself to changing properties of the signal to be predicted. Such power-demand predictors have been discussed, e.g., by Gupta and Yamada (1972) and by Holst (1977). □

The foregoing is an example of *adaptive prediction*; it has been found that adaptive prediction can be applied to a wide variety of problems. The predictions themselves may be of interest for different reasons; an example of this will now be given.

EXAMPLE 1.3   (Digital Transmission of Speech)   Consider the transmission of speech over a communication channel. This is now more often done digitally, which means that the analog speech signal is quantized to a number of bits, which are transmitted. The transmission line has limited capacity, and it is important to use it as efficiently as possible. If one predicts the "next sampled value" of the signal both at the transmitter and at the receiver, one need transmit only the difference between the actual and the predicted value (the "prediction error"). Since the prediction error is typically much smaller than the signal itself, it requires fewer bits when transmitted; hence the line is more efficiently used. This technique is known as *predictive coding* in communication theory. Now the prediction of the next value very much depends on the character of the transmitted signal. In the case of speech, this character significantly varies with the different sounds (phonemes) being pronounced. Efficient use of the predictive encoding procedure therefore requires that the predictor be adaptive. Adaptive prediction of speech signals is discussed, e.g., in Atal and Schroeder (1970). □

EXAMPLE 1.4 (Channel Equalization)   In a communication network the communication channels distort the transmitted signal. Each channel can be seen as a linear filter with a certain impulse response that in practice differs from the ideal delta function response. If the distortion is serious, the signal must be restored at the receiver. This is accomplished by passing it through a filter whose impulse response resembles the inverse of that of the channel. Such a filter is known as a channel equalizer. If the properties of the communication channel are known, this is a fairly straightforward problem. However, in a network the line between the transmitter and receiver can be quite arbitrary, and then it is desirable that the equalizer can adapt itself to the actual properties of the chosen channel. Such *adaptive equalizers* are discussed, e.g., by Lucky (1965) and Godard (1974). □

The adaptive equalizer treated in the foregoing example belongs to the wide class of algorithms commonly known as *adaptive signal processing* or *adaptive filtering*.

EXAMPLE 1.5 (Monitoring and Failure Detection)   Many systems must be constantly monitored to detect possible failures, or to decide when a repair or replacement must be made. Such monitoring can sometimes be done by manual interference. However, in complex highly automated systems with stringent safety requirements, the monitoring itself must be mechanized. This means that measured signals from the systems must be processed to infer the current (dynamic) properties of the system; based on this data, it is then decided whether the system has undergone critical or undesired changes. The procedure must of course be applied on-line so that any decision is not unnecessarily delayed. □

We have now provided examples of typical systems, control, and signal-processing problems. A feature common to these problems is that a mathematical model is required at some point. We will now elaborate on the construction of models from measured signals—i.e., we will look at *identification* in more detail.

Identification could mean that a batch of data is collected from the system, and that subsequently, as a separate procedure, this batch of data is used to construct a model. Such a procedure is usually called *off-line identification* or *batch identification*.

In our examples, however, the model was needed in order to support decisions that had to be taken "on-line," i.e., during the operation of the system. It is thus necessary to infer the model at the same time as the data is collected. The model is then "updated" at each time instant some new data becomes available. The updating is performed by a recursive algorithm (a formal definition of such an algorithm will be given in section 1.4). We shall use the term *recursive identification* for such a procedure. Synonymous terms are on-line identification, real-time identification, adaptive algorithm, and sequential estimation.

Methods of recursive identification is the topic of this book. A major reason for the interest in such methods is, of course, that they are a key instrument in adaptive control, adaptive filtering, adaptive prediction, and adaptive signal-processing problems. We shall return to such applications in chapter 7.

In addition to on-line decision, we also have the following two reasons for using recursive identification.

(1) *Data compression.* With the processing of data being made on-line, old data can be discarded. The final result is then a model of the system rather than a big batch of data. Since many recursive identification algorithms provide an estimate of the accuracy of the current model, a rational decision of when to stop data acquisition can be made on-line.

(2) *Application to off-line identification.* Methods for off-line identification may process the measured data in different ways. Often, several passes are made through the data to iteratively improve the estimated models. Depending on the complexity of the model, the number of necessary iterations may range from less than ten to a couple of hundred. An alternative to this iterative batch processing is to let the data be processed by a recursive identification algorithm. Then, one would normally go through the data a couple of times to improve the accuracy of the recursive estimates. This has proved to be an efficient alternative to conventional off-line procedures (Young, 1976). Section 7.2 contains a discussion of this application.

There are two disadvantages to recursive identification in contrast to off-line identification. One is that the decision of what model structure to use has to be made a priori, before starting the recursive identification procedure. In the off-line situation different types of models can be tried out. The second disadvantage is that, with a few exceptions, recursive methods do not give as good accuracy of the models as off-line methods. For long data records, the difference need not be significant, though, as we shall see.

1.4 A Recursive Identification Algorithm

In this section we shall define "recursive identification algorithm." The raw material for information processing is of course the recorded data. Let $z(t)$ denote the piece of data received at time t. This is in general a vector, composed of several different measurements. Assuming that the data acquisition takes place in discrete time, as is normally the case, we have at time t received a sequence of measurements $z(1)$, $z(2)$, ..., $z(t)$. Here, for convenience, we enumerate the sampling instants using integers. This does not necessarily imply that the sampling has to be uniform. Let us use a superscript to denote the whole data record:

$$z^t = \{z(t), z(t-1), \ldots, z(1)\}. \tag{1.1}$$

The objective of identification is to infer a model of the system from the record z^t. Normally, the model is parametrized in terms of a parameter vector θ, so the objective really is to determine this vector. We shall discuss in detail models and model parametrization in section 3.2. Let it suffice here to give a simple example.

EXAMPLE 1.6   Consider an electric motor. The voltage applied to it at time t is denoted by $u(t)$ and the angular velocity of its axis at time t is $y(t)$. We assume the relationship between these two variables to be of the form

$$y(t) + ay(t-1) = bu(t-1),$$

where a and b are unknown constants. The data vector is

$$z(t) = \begin{pmatrix} y(t) \\ u(t) \end{pmatrix},$$

and the model parameter vector is

$$\theta = \begin{pmatrix} a \\ b \end{pmatrix}. \quad \square$$

The *identification problem* can thus be phrased as the determination of a mapping from the data z^t to the model parameters θ:

$$z^t \rightarrow \hat{\theta}(t; z^t). \tag{1.2}$$

Here the value $\hat{\theta}(t; z^t)$ is the *estimate* of θ based on the information contained in z^t. The subject of the identification literature is, in fact, to suggest functions (1.2) and to investigate their properties.

In off-line or batch identification, data up to some N is first collected; then

$$\hat{\theta}_N = \hat{\theta}(N; z^N)$$

is computed. In the off-line case, there is, at least conceptually, no constraint as to the character of this mapping.

In on-line or recursive identification, the estimate $\hat{\theta}(t)$ is required for each t. In principle, $\hat{\theta}(t)$ could still be a general function of previous data as in (1.2). However, in practice it is important that memory space and computation time do not increase with t. This introduces restrictions upon how $\hat{\theta}(t)$ may be formed. Basically, we will need to condense the observed data into an auxiliary "memory" quantity $S(t)$ of given and fixed dimensions. This auxiliary vector (or matrix) will then be updated according to an algorithm of the structure

$$\hat{\theta}(t) = F(\hat{\theta}(t-1), S(t), z(t)), \tag{1.3a}$$

$$S(t) = H(S(t-1), \hat{\theta}(t-1), z(t)). \tag{1.3b}$$

Here $F(\cdot, \cdot, \cdot)$ and $H(\cdot, \cdot, \cdot)$ are given functions. We see that the estimate $\hat{\theta}(t)$ is formed from current data, the previous estimate, and the auxiliary variable $S(t)$. The only thing that needs to be stored at time t, consequently, is the information $\{\hat{\theta}(t), S(t)\}$. This quantity is updated with a fixed algorithm, with a number of operations that does not depend on time t. The information contained in the data record z^t has been condensed into $\{\hat{\theta}(t), S(t)\}$, which may or may not be done without loss of relevant information. The data record itself is discarded.

The problem of recursive identification therefore reduces to:

- Choice of a suitable model parametrization.
- Choice of the functions F and H in (1.3).

This is what this book is about.

1.5 Outline of the Book and a Reader's Guide

The statement of the problem of recursive identification at the end of the preceding section may seem simple. However, the literature on recursive identification is extensive, diverse, and sometimes confusing. Many people regard the area to be a "bag of tricks" rather than a theory, and it was called "a fiddler's paradise" in the survey by Åström and Eykhoff (1971). There are several reasons for this situation. One is that the same method is known under different names depending on the model structure it is applied to. Another reason is that slightly different algorithms may result, depending on the approach used in the derivation. Also, different approximations and tricks for performance improvement lead to a myriad of algorithms that in fact are closely related.

Presenting recursive identification as a catalog of existing methods would give a long and confusing list. In this book we take another approach. We try to create a general framework, within which most known methods can be recognized as special cases or seen as arising from particular choices of some "design variables" (including model parametrization). Our point of view can be expressed by saying,

There is only one recursive identification method. It contains some design variables to be chosen by the user.

This statement may seem overly simplified and dogmatic, and we shall not pursue it ad absurdum. The phrasing is intentionally provoking in order to stress our point. We are aware of the fact that there may exist other methods of the general type (1.3) that do not fit into the framework of this book. We do believe, however, that our point of view provides a basis for orienting oneself in the area.

The route we take to explain and exploit our approach is:

I. Develop the general framework. Define "the" recursive identification method. Display its design variables (chapter 3).

II. In order to guide the user's choice of design variables, analyse the properties of the estimates produced by the general algorithm (chapter 4).

III. Discuss the choice of design variables (chapter 5).

IV. Discuss the implementation of algorithms (chapter 6).

V. Discuss various applications of recursive identification (chapter 7).

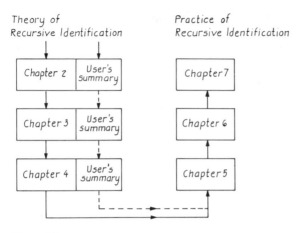

Figure 1.1
The dotted line shows the shortcut to the practice of recursive identification.

A little bit beside this main stream, we give in chapter 2 a survey of typical approaches that have been taken to recursive identification. The objective of chapter 2 is threefold: (1) To provide a less-formal introduction to the world of recursive identification for newcomers, (2) to provide an interface to the literature for those already familiar with the subject, and (3) to show that the character of the resulting algorithm is quite independent of the method of derivation, thus providing motivation for the general framework. This also means that part of the material in chapter 2 will be repeated in later chapters.

Two different themes can be distinguished in the book. Chapters 3 and 4 deal with what we might call the *theory* of recursive identification. Chapters 5–7 address problems associated with the *practice* of recursive identification. No doubt, some readers are mainly interested in the use of the developed methods. They do not care about how things are derived or about the subtleties of analysis, but would like to know the bottom line of the theory. For them, we have provided summary sections in chapters 2–4 as a means of direct access to the practice of recursive identification. See the road map in figure 1.1.

1.6 The Point of Departure

In order to read the book, certain prerequisites are needed. We assume the reader to be familiar with general concepts for stochastic dynamical systems and the mathematical tools used in their analysis. Many of the more specific results will be developed or described in the course of the book. More technical concepts (like stochastic convergence) will be used only in chapter 4, and are not crucial for the understanding of the book. We provide in appendixes 1.A–1.C and in section 3.3 some of the general background and an explanation of the technical concepts that readers of this book should be acquainted with. These are:

Probability theory: Random variables and random vectors, means, covariance matrices, some stochastic convergence concepts (appendix 1.A).

Statistics: Parameter estimation, the maximum likelihood method, the Cramér-Rao bound, the information matrix (appendix 1.B).

Models for stochastic, dynamical systems (including stochastic signals): Black box models, ARMA models, State-space models, the Kalman filter (appendix 1.C).

Off-line identification: Prediction error and maximum likelihood methods, some convergence and consistency results (section 3.3).

2 Approaches to Recursive Identification

2.1 Introduction

One reason for the existence of so many recursive identification algorithms is that several different approaches to the subject can be taken. In this chapter we shall give an overview of four typical frameworks within which recursive identification methods can be developed. The purposes of the present chapter are:

• To give an overview and introduction to some useful concepts and approaches, and to illustrate different interpretations of existing schemes. It thus serves as an informal background to the more formal development of chapter 3.

• To relate this book to the extensive literature on the subject. This is done basically in the annotated bibliography of section 2.8, where it is described how the ideas displayed in this chapter have been used to derive a variety of algorithms.

• To provide some motivation for the general framework used in the remainder of the book.

The purpose of this chapter is *not* to give details. All the algorithms mentioned here will be subsumed in the general framework, and facts and aspects on their implementation, use, and asymptotic properties will be given in later chapters. The approaches to recursive identification that will be reviewed are:

1. Modification of off-line identification methods.
2. Recursive identification as nonlinear filtering (Bayesian approach).
3. Stochastic approximation.
4. Model reference techniques and pseudolinear regressions.

These four approaches cover the derivations of most suggested methods. They will be discussed in the following four sections, respectively. We mentioned in the introduction that an important reason for using recursive identification is that the properties of the system may vary in time. In section 2.6 we shall discuss how each of the aforementioned four approaches can cope with such a situation, and how each can lead to estimation algorithms that track time-varying properties.

A major decision in identification is how to parametrize the properties of the system or signal using a model of suitable structure. In fact, specific "named" methods in the literature are usually associated with a particular model parametrization. In this section we shall, by means of examples, exhibit three different models (see also appendix 1.C).

EXAMPLE 2.1 (A Linear Difference Equation) Consider a dynamical system with input signal $\{u(t)\}$ and output signal $\{y(t)\}$. Suppose that these signals are sampled in discrete time $t = 1, 2, 3, \ldots$ and that the sampled values can be related through the linear difference equation

$$y(t) + a_1 y(t-1) + \cdots + a_n y(t-n) = b_1 u(t-1) + \cdots$$
$$+ b_m u(t-m) + v(t), \tag{2.1}$$

where $v(t)$ is some disturbance of unspecified character. We shall use operator notation for conveniently writing difference equations. Thus let q^{-1} be the backward shift (or delay) operator:

$$q^{-1} y(t) = y(t-1). \tag{2.2}$$

Then (2.1) can be rewritten as

$$A(q^{-1}) y(t) = B(q^{-1}) u(t) + v(t), \tag{2.3}$$

where $A(q^{-1})$ and $B(q^{-1})$ are polynomials in the delay operator:

$$A(q^{-1}) = 1 + a_1 q^{-1} + \cdots + a_n q^{-n},$$
$$B(q^{-1}) = b_1 q^{-1} + b_2 q^{-2} + \cdots + b_m q^{-m}.$$

The *model* (2.1) or (2.3) describes the dynamic relationship between the input and the output signals. It is expressed in terms of the parameter vector

$$\theta^T = (a_1 \ldots a_n \quad b_1 \ldots b_m).$$

We shall frequently express the relation (2.1) or (2.3) in terms of the parameter vector. Introduce the vector of lagged input-output data,

$$\varphi^T(t) = (-y(t-1) \ldots -y(t-n) \ u(t-1) \ldots u(t-m)). \tag{2.4}$$

Then (2.1) can be rewritten as

$$y(t) = \theta^T \varphi(t) + v(t). \tag{2.5}$$

This model describes the observed variable $y(t)$ as an unknown linear combination of the components of the observed vector $\varphi(t)$ plus noise.

Such a model is called a *linear regression* in statistics and is a very common type of model. The components of $\varphi(t)$ are then called *regression variables* or *regressors*.

If the character of the disturbance term $v(t)$ is not specified, we can think of

$$\hat{y}(t \mid \theta) \triangleq \theta^{\mathrm{T}} \varphi(t) \tag{2.6}$$

as a natural guess or "prediction" of what $y(t)$ is going to be, having observed previous values of $y(k)$, $u(k)$, $k = t - 1$, $t - 2$, ... This guess depends, of course, also on the model parameters θ. The expression (2.6) becomes a prediction in the exact statistical sense, if $\{v(t)\}$ in (2.5) is a sequence of independent random variables with zero mean values. We shall use the term "white noise" for such a sequence.

If no input is present in (2.1) ($m = 0$) and $\{v(t)\}$ is considered to be white noise, then (2.1) becomes a model of the signal $\{y(t)\}$:

$$y(t) + a_1 y(t - 1) + \cdots + a_n y(t - n) = v(t). \tag{2.7}$$

Such signal is commonly known as an *autoregressive process of order n*, or an $\mathrm{AR}(n)$ process.

An important feature of the set of models discussed in this example, is that the "prediction" $\hat{y}(t|\theta)$ in (2.6) is linear in the parameter vector θ. This makes the estimation of θ simple.

Since the disturbance term $v(t)$ in the model (2.5) corresponds to an "equation error" in the difference equation (2.1), methods to estimate θ in (2.5) are often known as *equation error methods*. □

EXAMPLE 2.2. (An ARMAX Model)   We could add flexibility to the model (2.3) by also modeling the disturbance term $v(t)$. Suppose that this can be described as a moving average (MA) of a white noise sequence $\{e(t)\}$:

$$v(t) = C(q^{-1})e(t),$$

$$C(q^{-1}) = 1 + c_1 q^{-1} + \cdots + c_r q^{-r}.$$

Then the resulting model is

$$A(q^{-1})y(t) = B(q^{-1})u(t) + C(q^{-1})e(t). \tag{2.8}$$

This is known as an ARMAX model. The reason for this term is that the model is a combination of an autoregressive (AR) part $A(q^{-1})y(t)$, a moving average (MA) part $C(q^{-1})e(t)$, and a control part $B(q^{-1})u(t)$. The control signal is in the econometric literature known as the eXogeneous variable, hence the X.

The dynamics of the model (2.8) are expressed in terms of the parameter vector

$$\theta^T = (a_1 \ \ldots \ a_n \ b_1 \ \ldots \ b_m \ c_1 \ \ldots \ c_r).$$

Since the model (2.8) also provides us with a statistical description of the disturbances, we can compute a properly defined prediction of the output $y(t)$. In example 2.6 we shall do that in a special case, and the general case will be treated in example 3.2 in the next chapter.

When no input is present, the use of (2.8) means that we are describing the output signal $\{y(t)\}$ as an ARMA process. This is a very common type of model for stochastic signals. □

EXAMPLE 2.3. (A State-Space Model) It is common practice in some applications to give linear stochastic dynamical systems in a state-space description

$$x(t + 1) = Fx(t) + Gu(t) + w(t),$$

$$y(t) = Hx(t) + e(t),$$

(2.9)

where $\{w(t)\}$ and $\{e(t)\}$ are sequences of independent random vectors with certain covariance matrices,

$$Ew(t)w^T(t) = R_1, \quad Ee(t)e^T(t) = R_2, \quad Ew(t)e^T(t) = R_{12}.$$

(2.10)

When the matrices F, G, H, R_1, R_2, and R_{12} are all known, predictions and state estimates for the system (2.9) can be computed using the Kalman filter. See appendix 1.C for details. When these matrices are not fully known, (2.9) describes a model with some parameters θ to be determined. This means that the matrices are functions of this parameter vector: $F(\theta)$, $G(\theta)$, etc. Such a model parametrization is especially useful when we have knowledge about basic mechanisms in the system, but certain coefficients are unknown. □

These three examples illustrate the three most widely used models. Other ones will be given in chapter 3. Notice that the models are used to describe a stochastic dynamical *system* with an input $\{u(t)\}$ and an output $\{y(t)\}$, as well as to describe the properties of a stochastic *signal* $\{y(t)\}$, where no input is present. In remainder of this book, rather than repeatedly saying "system or signal", we shall for convenience use the term "system" to cover both cases. We could think of a "system" generating the signal in the latter case.

Remark on notation   Throughout the book we use θ to denote the parameters of a model. The goal of the identification procedure is to determine the "best" values of these parameters. Such estimated values will be denoted by $\hat{\theta}$. Occasionally, when analyzing the properties of different methods, we shall use θ_0 to denote the "true value" of the model parameters, thus assuming that the observed data actually have been generated according to the mechanism of some particular model. This value θ_0 is, of course, available only to the analyst and not to the user. In fact, the user will treat any candidate model $\mathcal{M}(\theta)$ as a correct description of the system when deriving such things as predictions, etc., according to $\mathcal{M}(\theta)$.

2.2 Recursive Algorithms Derived from Off-Line Identification Algorithms

In Chapter 1, we defined off-line identification as a general mapping from measured data. Recursive methods, starting from the data record (1.1), pose further constraints on how the estimates may be computed. An obvious approach to recursive identification is to take any off-line method and modify it, so that it meets the constraints of (1.3). In this section examples of this method of deriving recursive identification algorithms will be given.

There is an extensive literature on off-line identification methods. See, e.g., the survey by Åström and Eykhoff (1971) or the books by Eykhoff (1974), Kashyap and Rao (1976), or Goodwin and Payne (1977). Some formal aspects of off-line methods will be reviewed in section 3.3. Considering the great number of different off-line methods, as well as variants of the same method, one might expect also a rich variety of derived recursive algorithms. We shall in this section limit ourselves to three representative examples; the chapter bibliography will list the sources of many more. The first two examples lead to methods where the off-line estimates can be exactly calculated in a recursive fashion. In the third example approximations have to be introduced.

2.2.1 The Least Squares Method

In this section we consider the difference equation model (2.5) (the linear regression) that we introduced in example 2.1:

$$y(t) = \theta^T \varphi(t) + v(t).\tag{2.11}$$

The parameter vector θ is to be estimated from measurements of $y(t)$, $\varphi(t)$; $t = 1, 2, \ldots N$. A common and natural way is to choose this estimate by minimizing what is left unexplained by the model, viz., the "equation error" $v(t)$. That is, we write down a criterion function

$$V_N(\theta) = \frac{1}{N}\sum_1^N \alpha_t [y(t) - \theta^T \varphi(t)]^2;\tag{2.12}$$

then we minimize this with respect to θ. Here $\{\alpha_t\}$ is a sequence of positive numbers. The inclusion of the coefficients α_t in the criterion (2.12) allows us to give different weights to different observations. In applications, most often α_t is chosen identically equal to 1. As will be explained in the next section, an optimal choice of α_t should be related to the variance of the noise term $v(t)$.

We remarked in example 2.1 that $\hat{y}(t \mid \theta) = \theta^T \varphi(t)$ can be seen as a natural "guess' or "prediction" of $y(t)$, based upon the parameter vector θ. Thus the criterion (2.12) can be seen as an attempt to choose a model that produces the best predictions of the output signal. The criterion $V_N(\theta)$ is quadratic in θ. Therefore it can be minimized analytically, which gives

$$\hat{\theta}(N) = \left[\sum_{t=1}^N \alpha_t \varphi(t)\varphi^T(t)\right]^{-1} \sum_{t=1}^N \alpha_t \varphi(t) y(t),\tag{2.13}$$

provided the inverse exists. This is the celebrated least squares estimate. For our current purposes it is important to note that the expression (2.13) can be rewritten in a recursive fashion. To prove this, we proceed as follows. Denote

$$\bar{R}(t) = \sum_{k=1}^t \alpha_k \varphi(k)\varphi^T(k).$$

Then, from (2.13), we have that

$$\sum_{k=1}^{t-1} \alpha_k \varphi(k) y(k) = \bar{R}(t-1)\hat{\theta}(t-1).$$

(handwritten margin notes:) $\alpha_t = \lambda^{N-t}$, $\lambda = [0.1]$; $\theta(t) = \bar{R}(t)^{-1}$; $2.13 \to \theta(t) = \bar{R}(t)^{-1} \sum_{t=1}^N \alpha_t \varphi(t) y(t)$ $= \bar{R}(t)^{-1}\left[\sum_{k=1}^{t-1} \alpha_k \varphi(k) y(k) + \alpha_t \varphi(t) y(t)\right]$ this is (2.14)

From the definition of $\bar{R}(t)$ it follows that

$$\bar{R}(t-1) = \bar{R}(t) - \alpha_t \varphi(t) \varphi^T(t).$$

Hence

$$\hat{\theta}(t) = \bar{R}^{-1}(t) \left[\sum_{k=1}^{t-1} \alpha_k \varphi(k) y(k) + \alpha_t \varphi(t) y(t) \right]$$

$$= \bar{R}^{-1}(t) [\bar{R}(t-1)\hat{\theta}(t-1) + \alpha_t \varphi(t) y(t)]$$

$$= \bar{R}^{-1}(t) \{\bar{R}(t)\hat{\theta}(t-1) + \alpha_t \varphi(t) [-\varphi^T(t)\hat{\theta}(t-1) + y(t)]\}$$

$$= \hat{\theta}(t-1) + \bar{R}^{-1}(t)\varphi(t)\alpha_t \lfloor y(t) - \hat{\theta}^T(t-1)\varphi(t)], \qquad (2.14a)$$

and

$$\bar{R}(t) = \bar{R}(t-1) + \alpha_t \varphi(t) \varphi^T(t). \qquad (2.14b)$$

Sometimes we may prefer to work with

$$R(t) \triangleq \frac{1}{t}\bar{R}(t).$$

From (2.14b) we easily find that

$$R(t) = \frac{1}{t}[\bar{R}(t-1) + \alpha_t \varphi(t) \varphi^T(t)] = \frac{t-1}{t} R(t-1) + \frac{1}{t}\alpha_t \varphi(t) \varphi^T(t)$$

$$\qquad (2.14')$$

$$= R(t-1) + \frac{1}{t}[\alpha_t \varphi(t) \varphi^T(t) - R(t-1)].$$

The foregoing expressions can be summarized by

$$\hat{\theta}(t) = \hat{\theta}(t-1) + \frac{1}{t} R^{-1}(t)\varphi(t)\alpha_t [y(t) - \theta^T(t-1)\varphi(t)], \qquad (2.15a)$$

$$R(t) = R(t-1) + \frac{1}{t}[\alpha_t \varphi(t) \varphi^T(t) - R(t-1)]. \qquad (2.15b)$$

Equations (2.15) have the form of that we demanded of a recursive algorithm in (1.3). At time t only $\hat{\theta}(t)$, $R(t)$, $y(t)$, and $\varphi(t)$ have to be kept in the memory. Comparing with (1.3), $S(t)$ corresponds to $R(t)$, $\varphi(t)$. All other previous data can be thrown away.

An Equivalent Form   The algorithm (2.15) is not, however, well suited for computation as it stands, since a matrix has to be inverted in each time step. It is more natural to introduce

$$P(t) = \bar{R}^{-1}(t) = \frac{1}{t}R^{-1}(t)$$

and update $P(t)$ directly, instead of using (2.14b). This is accomplished by the so-called matrix inversion lemma, which we now state.

LEMMA 2.1.   Let A, B, C, and D be matrices of compatible dimensions, so that the product BCD and the sum $A + BCD$ exist. Then

$$[A + BCD]^{-1} = A^{-1} - A^{-1}B[DA^{-1}B + C^{-1}]^{-1}DA^{-1} \tag{2.16}$$

Proof Multiply the right-hand side of (2.16) by $A + BCD$ from the right. This gives

$$I + A^{-1}BCD - A^{-1}B[DA^{-1}B + C^{-1}]^{-1}D$$

$$-A^{-1}B[DA^{-1}B + C^{-1}]^{-1}DA^{-1}BCD$$

$$= I + A^{-1}B[DA^{-1}B + C^{-1}]^{-1}\{[DA^{-1}B + C^{-1}]CD - D$$

$$- DA^{-1}BCD\} = I + A^{-1}B[DA^{-1}B + C^{-1}]^{-1}\{0\} = I,$$

which proves (2.16). ∎

Applying (2.16) to (2.14) with

$$A = P(t - 1), \quad B = \varphi(t), \quad C = \alpha_t, \quad D = \varphi^T(t)$$

gives

$$P(t) = [P^{-1}(t - 1) + \varphi(t)\alpha_t\varphi^T(t)]^{-1}$$

$$= P(t - 1) - P(t - 1)\varphi(t)\left[\varphi^T(t)P(t - 1)\varphi(t) + \frac{1}{\alpha_t}\right]^{-1}\varphi^T(t)P(t - 1)$$

$$= P(t - 1) - \frac{P(t - 1)\varphi(t)\varphi^T(t)P(t - 1)}{1/\alpha_t + \varphi^T(t)P(t - 1)\varphi(t)}. \tag{2.17}$$

The advantages of (2.17) over (2.15b) are obvious. The inversion of a square matrix of size dim θ is replaced by inversion of a scalar. From (2.17) we also find that

$$\alpha_t P(t)\varphi(t) = \alpha_t P(t-1)\varphi(t) - \frac{\alpha_t P(t-1)\varphi(t)\varphi^{\mathrm{T}}(t)P(t-1)\varphi(t)}{1/\alpha_t + \varphi^{\mathrm{T}}(t)P(t-1)\varphi(t)}$$

$$= \frac{P(t-1)\varphi(t)}{1/\alpha_t + \varphi^{\mathrm{T}}(t)P(t-1)\varphi(t)}. \tag{2.18}$$

Thus the least squares estimate $\hat{\theta}(t)$ defined by (2.13) can be recursively calculated by means of

$$\hat{\theta}(t) = \hat{\theta}(t-1) + L(t)[y(t) - \hat{\theta}^{\mathrm{T}}(t-1)\varphi(t)], \tag{2.19a}$$

$$L(t) = \frac{P(t-1)\varphi(t)}{1/\alpha_t + \varphi^{\mathrm{T}}(t)P(t-1)\varphi(t)}, \tag{2.19b}$$

$$P(t) = P(t-1) - \frac{P(t-1)\varphi(t)\varphi^{\mathrm{T}}(t)P(t-1)}{1/\alpha_t + \varphi^{\mathrm{T}}(t)P(t-1)\varphi(t)}. \tag{2.19c}$$

These formulas are known as the *recursive least squares* (RLS) *algorithm*. This is one of the most widely used recursive identification methods. It is robust and easily implemented. We shall later on in the book discuss its properties in detail. Let us here only point out two aspects that must be considered in any application of the algorithm.

Initial Conditions   Any recursive algorithm requires some initial value to be started up. In (2.19) we need $\hat{\theta}(0)$ and $P(0)$ and in (2.15) $\hat{\theta}(0)$ and $R(0)$. Since we derived (2.19) from (2.13) under the assumption that $\bar{R}(t)$ is invertible, an exact relationship between these two expressions can hold only if (2.19) is initialized at a time t_0 when $\bar{R}(t_0)$ is invertible. Typically, $R(t)$ becomes invertible at time $t_0 = \dim \varphi(t) = \dim \theta$. Thus, strictly speaking, the proper initial values for (2.19) are obtained if we start the recursion at time t_0, for which

$$P(t_0) = \left[\sum_{k=1}^{t_0} \alpha_k \varphi(k)\varphi^{\mathrm{T}}(k) \right]^{-1},$$

$$\hat{\theta}(t_0) = P(t_0) \sum_{k=1}^{t_0} \alpha_k \varphi(k)y(k). \tag{2.20}$$

It is more common, though, to start the recursion at $t = 0$ with some invertible matrix $P(0)$ and a vector $\hat{\theta}(0)$. The estimates resulting from (2.19) are then

$$\hat{\theta}(t) = \left[P^{-1}(0) + \sum_{k=1}^{t} \alpha_k \varphi(k) \varphi^{\mathrm{T}}(k) \right]^{-1} \left[P^{-1}(0) \hat{\theta}(0) + \sum_{k=1}^{t} \alpha_k \varphi(k) y(k) \right].$$

$$(2.21)$$

This can be seen by verifying that (2.21) obeys the recursion (2.19) with these initial conditions.

By comparing (2.21) to (2.13), we see that the relative importance of the initial values decays with time, as the magnitudes of the sums increase. Also, as $P^{-1}(0) \to 0$ the recursive estimate approaches the off-line estimate. Therefore, a common choice of initial values is to take $P(0) = C \cdot I$ and $\hat{\theta}(0) = 0$, where C is some large constant.

Asymptotic Properties   To investigate how the estimate (2.13) behaves when N becomes large, we assume that the data actually have been generated by

$$y(t) = \theta_0^{\mathrm{T}} \varphi(t) + v(t). \tag{2.22}$$

Inserting this expression for $y(t)$ into (2.13), and noting that $\theta_0^{\mathrm{T}} \varphi(t) = \varphi^{\mathrm{T}}(t) \theta_0$, gives

$$\hat{\theta}(N) = \left[\sum_{t=1}^{N} \alpha_t \varphi(t) \varphi^{\mathrm{T}}(t) \right]^{-1} \left\{ \sum_{t=1}^{N} \alpha_t [\varphi(t) \varphi^{\mathrm{T}}(t) \theta_0 + \varphi(t) v(t)] \right\}$$

$$= \theta_0 + \left[\frac{1}{N} \sum_{t=1}^{N} \alpha_t \varphi(t) \varphi^{\mathrm{T}}(t) \right]^{-1} \frac{1}{N} \sum_{t=1}^{N} \alpha_t \varphi(t) v(t).$$

$$(2.23)$$

Desired properties of $\hat{\theta}(N)$ would be (1) that it is close to θ_0, and (2) that it converges to θ_0 as N approaches infinity. We see that if the "disturbance" $v(t)$ in (2.22) is small compared to $\varphi(t)$, then $\hat{\theta}(N)$ will be close to θ_0. The sum $(1/N) \sum_{t=1}^{N} \alpha_t \varphi(t) v(t)$ will, under weak conditions, converge to its expected value as N approaches infinity, according to the law of large numbers. This expected value depends on the correlation between the disturbance term $v(t)$ and the data vector $\varphi(t)$. It is zero only if $v(t)$ and $\varphi(t)$ are uncorrelated. This is true in the following two typical cases:

• $\{v(t)\}$ is a sequence of independent random variables with zero mean values (white noise). Then $v(t)$ does not depend on what has happened up to time $t - 1$, and hence $Ev(t)\varphi(t) = 0$.

• The input sequence $\{u(t)\}$ is independent of the zero-mean noise sequence $\{v(t)\}$ and $n = 0$ in (2.1), (2.4). Then $\varphi(t)$ contains only u-terms, and hence $Ev(t)\varphi(t) = 0$.

When $n > 0$ so that $\varphi(t)$ contains $y(k)$, $k = t - 1, \ldots, t - n$, and $\{v(t)\}$ is not white noise, then (usually) $Ev(t)\varphi(t) \neq 0$. This follows since $\varphi(t)$ contains $y(t - 1)$, while $y(t - 1)$ contains the term $v(t - 1)$ that is correlated with $v(t)$. This means that we may expect $\hat{\theta}(N)$ to tend to θ_0 as N approaches infinity, usually only in the two cases listed above.

2.2.2 The Instrumental Variable Method

In the foregoing example it was possible to obtain the same estimates in a recursive fashion as for the corresponding off-line method, except for possible effects of initial conditions. This is not the case in general. It is, however, true for the instrumental variable method, which is a modification of the least squares method designed to overcome the convergence problems indicated at the end of section 2.2.1. Consider again the model of example 2.1:

$$y(t) = \theta^T \varphi(t) + v(t),$$

and assume as in (2.22) that the data indeed have been generated by this model or mechanism for a particular value $\theta = \theta_0$.

A disadvantage with the least squares estimate is that in general $\varphi(t)$ and $v(t)$ will be found to be correlated, and then $\hat{\theta}(N)$ will not converge to θ_0. In such a case, we might replace $\varphi(t)$ in (2.13) by a vector $\zeta(t)$, such that

$$\zeta(t) \text{ and } v(t) \text{ are uncorrelated.} \tag{2.24}$$

That is, we try the estimate (here we take $\alpha_t \equiv 1$)

$$\hat{\theta}(N) = \left[\sum_{t=1}^{N} \zeta(t)\varphi^T(t) \right]^{-1} \sum_{t=1}^{N} \zeta(t)y(t). \tag{2.25}$$

By inserting the expression (2.22) for $y(t)$ into (2.25) we obtain

$$\hat{\theta}(N) = \left[\sum_{t=1}^{N} \zeta(t)\varphi^T(t) \right]^{-1} \sum_{t=1}^{N} [\zeta(t)\varphi^T(t)\theta_0 + \zeta(t)v(t)]$$

$$= \theta_0 + \left[\frac{1}{N} \sum_{t=1}^{N} \zeta(t)\varphi^T(t) \right]^{-1} \frac{1}{N} \sum_{t=1}^{N} \zeta(t)v(t). \tag{2.26}$$

We see that $\hat{\theta}(N)$ is likely to tend to θ_0 as N tends to infinity under the following three conditions:

- (2.24) holds,
- $v(t)$ has zero mean, $\hspace{5cm}$ (2.27)
- the matrix $\lim\limits_{N \to \infty} \dfrac{1}{N} \sum\limits_{t=1}^{N} \zeta(t)\varphi^{\mathrm{T}}(t)$ is invertible.

The estimate (2.25) is known as the *instrumental variable (IV) estimate*. The vectors $\zeta(t)$ are referred to as the *instrumental variables*.

It is obvious that the estimate (2.25) can be rewritten in a recursive fashion, just as the least squares estimate in (2.19). We then find that

$$\hat{\theta}(t) = \hat{\theta}(t-1) + L(t)[y(t) - \hat{\theta}^{\mathrm{T}}(t-1)\varphi(t)], \qquad (2.28a)$$

$$L(t) = \frac{P(t-1)\zeta(t)}{1 + \varphi^{\mathrm{T}}(t)P(t-1)\zeta(t)} = P(t)\zeta(t), \qquad (2.28b)$$

$$P(t) = P(t-1) - \frac{P(t-1)\zeta(t)\varphi^{\mathrm{T}}(t)P(t-1)}{1 + \varphi^{\mathrm{T}}(t)P(t-1)\zeta(t)}. \qquad (2.28c)$$

We have not yet discussed the choice of the instrumental variables $\zeta(t)$. Loosely speaking, they should be sufficiently correlated with $\varphi(t)$ to ensure (2.27), but uncorrelated with the system noise terms. A common choice is

$$\zeta^{\mathrm{T}}(t) = (-y_M(t-1) \ \ldots \ -y_M(t-n) \ u(t-1) \ \ldots \ u(t-m)), \qquad (2.29)$$

where $y_M(t)$ is the output of a deterministic system driven by the actual input $u(t)$:

$$y_M(t) + \bar{a}_1 y_M(t-1) + \cdots + \bar{a}_n y_M(t-n)$$
$$= \bar{b}_1 u(t-1) + \cdots + \bar{b}_m u(t-m). \qquad (2.30)$$

For the recursive algorithm (2.28) an often-used approach is to let $\bar{a}_i$ and $\bar{b}_i$ be time-dependent. Then the current estimates $\hat{a}_i(t), \hat{b}_i(t)$ obtained from (2.28) can be used at time t in (2.30). That is, we can write

$$y_M(t) = \hat{\theta}^{\mathrm{T}}(t)\zeta(t). \qquad (2.31)$$

This approach was suggested by Mayne (1967), Wong and Polak (1967), and Young (1965, 1968). The recursive instrumental variable method has been used in many applications, and there exist several variants of the method. We shall return to it a number of times in this book (sections 3.6.3, 4.6, and 5.10).

2.2.3 A Recursive Prediction Error Method

We remarked in section 2.2.1 that the least squares criterion (2.12) can be interpreted as minimization of the error between the "predicted" and the measured values of the output. This idea is a guiding principle in off-line identification, and a similar interpretation can be given to several other methods. The family of methods can be called *prediction error identification methods*. See, e.g., Ljung (1978c) or Åström (1980b) for a general discussion of these methods.

For general models, the criterion will not be quadratic in θ as in (2.12). Then the minimization problem becomes more difficult, and no exact recursive method can be derived. In the following example we shall consider a simple model where this is the case. This will bring out several ideas on how to construct recursive algorithms, that we will utilize later on.

Consider the ARMAX model of example 2.2. We shall discuss how to estimate its parameters a_i, b_i and c_i. In order to simplify the notation, we study only a first-order example:

$$y(t) + ay(t-1) = bu(t-1) + e(t) + ce(t-1). \tag{2.32}$$

Here $\{e(t)\}$ is, as usual, taken to be white noise, i.e., a sequence of independent random variables with zero mean. The parameters to be estimated are collected into the vector

$$\theta = \begin{pmatrix} a \\ b \\ c \end{pmatrix}.$$

Let us first discuss how to determine the prediction of $y(t)$ based on observations of $y(s)$, $u(s)$, $0 \le s \le t-1$, and based on the assumption that the data is indeed produced by (2.32). Denote this prediction by $\hat{y}(t \mid \theta)$. We shall here make an elementary explicit derivation of $\hat{y}(t \mid \theta)$ for this simple special case, while a general case will be treated in example 3.2. From (2.32) we have

$$y(t) = -ay(t-1) + bu(t-1) + ce(t-1) + e(t).$$

Here $y(t - 1)$ and $u(t - 1)$ are known at time t. A good approximation of $e(t - 1)$ can be *computed* from $y(s)$, $u(s)$, $0 \le s \le t - 1$, using (2.32). Denote this approximation by $\hat{e}(t - 1)$. Finally, the value of $e(t)$ cannot be predicted from previous data since it is independent of everything that happened up to time $t - 1$. Hence, the natural prediction of $y(t)$ is

$$\hat{y}(t \mid \theta) = -ay(t - 1) + bu(t - 1) + c\hat{e}(t - 1), \tag{2.33}$$

where $\hat{e}(t - 1)$ is computed recursively from

$$\hat{e}(s) = y(s) + ay(s - 1) - bu(s - 1) - c\hat{e}(s - 1), \tag{2.34}$$

$$s = 1, \ldots, t - 1.$$

The recursion (2.34) requires initial conditions $y(0)$ and $\hat{e}(0)$; but the effect of these decays as $|c|^{t-1}$, so if $|c| < 1$ they can be chosen arbitrarily without affecting the prediction too much. The equations (2.33) and (2.34) can be compressed into one by taking

$$\begin{aligned}
\hat{y}(t \mid \theta) + c\hat{y}(t - 1 \mid \theta) &= -ay(t - 1) - acy(t - 2) \\
&\quad + bu(t - 1) + bcu(t - 2) \\
&\quad + c[\hat{e}(t - 1) + c\hat{e}(t - 2)] \\
&= -ay(t - 1) - acy(t - 2) \\
&\quad + bu(t - 1) + bcu(t - 2) \\
&\quad + cy(t - 1) + acy(t - 2) - bcu(t - 2).
\end{aligned} \tag{2.35}$$

Hence we have

$$\hat{y}(t \mid \theta) + c\hat{y}(t - 1 \mid \theta) = (c - a)y(t - 1) + bu(t - 1), \tag{2.36}$$

where the initial conditions can be taken to be zero. Now, for any given θ, we can compute the prediction $\hat{y}(t \mid \theta)$ from (2.36) using data up to time $t - 1$. We can also evaluate the *prediction error*

$$\varepsilon(t, \theta) \triangleq y(t) - \hat{y}(t \mid \theta) \tag{2.37}$$

according to the model parameters θ. A reasonable criterion of how well the model θ performs is the sum of squared prediction errors

$$V_N(\theta) = \frac{1}{2} \sum_{t=1}^{N} \varepsilon^2(t, \theta). \tag{2.38}$$

In fact, this function is the negative log likelihood function if the variables $\{e(t)\}$ are Gaussian, as we shall prove in section 3.3. The criterion (2.38) for the ARMAX model (2.8) was first suggested and used by Åström and Bohlin (1965).

The off-line estimate $\hat{\theta}_N$ is obtained by the minimization of $V_N(\theta)$. In the present case $\hat{y}(t \mid \theta)$ is a nonlinear function of θ [see (2.36)], and therefore the function $V_N(\theta)$ cannot be minimized analytically. Instead, some numerical minimization routine is used to determine $\hat{\theta}_N$. The numerical minimization of (2.38) typically requires several iterative passes through the data record from $t = 1$ to $t = N$. Thus it cannot be used in a recursive algorithm, which requires a memory vector of fixed size. This means that we cannot expect to be able to determine the sequence of off-line estimates $\hat{\theta}_N$ by recursive methods. Instead we have to be content with methods that determine approximations to the $\hat{\theta}_N$ in a recursive fashion.

We shall here derive a recursive algorithm for the estimation of θ. The derivation follows Åström (1972) and Söderström (1973b). They called it *Recursive Maximum Likelihood* (RML). (The relation between the prediction error concept and the maximum likelihood method is explained in example 3.7.) This method has been applied to both real and simulated data with success. Variants of the algorithm and its derivation will be surveyed in section 2.8.

Let $\hat{\theta}(t - 1)$ be our estimate at time $t - 1$. We wish to obtain a $\hat{\theta}(t)$ that (approximately) minimizes $V_t(\theta)$. By means of a Taylor expansion of $V_t(\theta)$ around $\hat{\theta}(t - 1)$ we obtain

$$V_t(\theta) = V_t(\hat{\theta}(t - 1)) + V_t'(\hat{\theta}(t - 1))[\theta - \hat{\theta}(t - 1)]$$
$$+ \tfrac{1}{2}[\theta - \hat{\theta}(t - 1)]^{\mathrm{T}} V_t''(\hat{\theta}(t - 1))[\theta - \hat{\theta}(t - 1)]$$
$$+ \mathrm{o}(|\theta - \hat{\theta}(t - 1)|^2),$$

where the prime denotes differentiation with respect to θ, and $\mathrm{o}(x)$ denotes a function such that $\mathrm{o}(x)/|x| \to 0$ as $|x| \to 0$. Minimization of this expression with respect to θ gives

$$\hat{\theta}(t) = \hat{\theta}(t - 1) - [V_t''(\hat{\theta}(t - 1))]^{-1} V_t'[\hat{\theta}(t - 1)]^{\mathrm{T}}$$
$$+ \mathrm{o}(|\hat{\theta}(t) - \hat{\theta}(t - 1)|).$$

(2.39)

If we denote the *negative* derivative of $\varepsilon(t, \theta)$ with respect to θ by

$$\psi(t, \theta) \triangleq \left[-\frac{d}{d\theta} \varepsilon(t, \theta) \right]^{\mathrm{T}} \quad \text{(a column vector)},$$

we have

$$[V_t'(\theta)]^{\mathrm{T}} = -\sum_{k=1}^{t} \psi(k, \theta)\varepsilon(k, \theta) = [V_{t-1}'(\theta)]^{\mathrm{T}} - \psi(t, \theta)\varepsilon(t, \theta), \qquad (2.40)$$

and, by differentiating once more,

$$V_t''(\theta) = V_{t-1}''(\theta) + \psi(t, \theta)\psi^{\mathrm{T}}(t, \theta) + \varepsilon''(t, \theta)\varepsilon(t, \theta), \qquad (2.41)$$

where $\varepsilon''(t, \theta)$ is the second-derivative matrix of $\varepsilon(t, \theta)$ with respect to θ.

In order to evaluate (2.39) a number of approximations have to be introduced.

• First we assume that the next estimate $\hat{\theta}(t)$ is to be found in a small neighborhood of $\hat{\theta}(t-1)$. This should be a reasonable approximation if t is large. That assumption leads to the following two approximations:

Neglect $o(|\hat{\theta}(t) - \hat{\theta}(t-1)|)$ in (2.39),   (2.42)

and take

$$V_t''(\hat{\theta}(t)) = V_t''(\hat{\theta}(t-1)). \qquad (2.43)$$

• Then we assume that $\hat{\theta}(t-1)$ is indeed the optimal estimate at time $t-1$, so that

$$V_{t-1}'(\hat{\theta}(t-1)) = 0. \qquad (2.44)$$

• Finally we set

$$\varepsilon''(t, \hat{\theta}(t-1))\varepsilon(t, \hat{\theta}(t-1)) = 0. \qquad (2.45)$$

The rationale for the approximation (2.45) is as follows. Close to the true value θ_0, $\{\varepsilon(t, \theta)\}$ will be *almost* white noise, so that we may approximately consider $\varepsilon(t, \theta)$ to be of zero mean and independent of what happened up to time $t-1$. In particular, it would then be independent of

$$\varepsilon''(t, \theta) \quad \left(= -\frac{d^2}{d\theta^2} \hat{y}(t \mid \theta) \right).$$

The expected value of the left-hand side of (2.45) then is indeed close to zero, so that the last term of (2.41) makes an order of magnitude less contribution to V_t'' than the second term.

With the assumptions (2.45) and (2.43) inserted into (2.41) we can approximately evaluate the second-derivative matrix. Let this approximation be denoted by $\bar{R}(t)$. Then we have

$$\bar{R}(t) = \bar{R}(t-1) + \psi(t, \hat{\theta}(t-1))\psi^{\mathrm{T}}(t, \hat{\theta}(t-1)). \tag{2.46}$$

With the assumption (2.44) inserted into (2.40) we have

$$[V_t'(\hat{\theta}(t-1))]^{\mathrm{T}} = [V_{t-1}'(\hat{\theta}(t-1))]^{\mathrm{T}} - \psi(t, \hat{\theta}(t-1))\varepsilon(t, \hat{\theta}(t-1))$$

$$= -\psi(t, \hat{\theta}(t-1))\varepsilon(t, \hat{\theta}(t-1)).$$

Using this expression for $V_t'(\hat{\theta}(t-1))$ and the approximation $\bar{R}(t)$ for $V_t''(\hat{\theta}(t-1))$ in (2.39) together with the assumption (2.42), we have

$$\hat{\theta}(t) = \hat{\theta}(t-1) + \bar{R}^{-1}(t)\psi(t, \hat{\theta}(t-1))\varepsilon(t, \hat{\theta}(t-1)). \tag{2.47}$$

Notice that for the least squares problems of section 2.2.1 all approximations (2.42)–(2.45) are in fact exact ($\varepsilon'' = 0$ since ε is linear in θ) and the resulting algorithm (2.46), (2.47) coincides with the recursive least squares algorithm (2.14), (2.15).

Now, even if it appears that we can give (2.46), (2.47) in a recursive form, we have not yet discussed how to determine $\varepsilon(t, \hat{\theta}(t-1))$ and $\psi(t, \hat{\theta}(t-1))$. To do this we have to deal with the particular model (2.32).

From (2.37) we have that

$$\psi^{\mathrm{T}}(t, \theta) = -\frac{d}{d\theta}\varepsilon(t, \theta) = \frac{d}{d\theta}\hat{y}(t \mid \theta). \tag{2.48}$$

From (2.36) we find that

$$\frac{d}{da}\hat{y}(t \mid \theta) + c\frac{d}{da}\hat{y}(t-1 \mid \theta) = -y(t-1), \tag{2.49a}$$

$$\frac{d}{db}\hat{y}(t \mid \theta) + c\frac{d}{db}\hat{y}(t-1 \mid \theta) = u(t-1), \tag{2.49b}$$

$$\frac{d}{dc}\hat{y}(t \mid \theta) + \hat{y}(t-1 \mid \theta) + c\frac{d}{dc}\hat{y}(t-1 \mid \theta) = y(t-1). \tag{2.49c}$$

Notice that (2.49c) also can be written

$$\frac{d}{dc}\hat{y}(t \mid \theta) + c\frac{d}{dc}\hat{y}(t-1 \mid \theta) = \varepsilon(t-1, \theta). \tag{2.49c'}$$

We can summarize (2.49) by writing

$$\psi(t, \theta) + c\psi(t - 1, \theta) = \begin{pmatrix} -y(t - 1) \\ u(t - 1) \\ \varepsilon(t - 1, \theta) \end{pmatrix}. \tag{2.50}$$

Thus we can compute $\varepsilon(t, \hat{\theta}(t - 1))$ and $\psi(t, \hat{\theta}(t - 1))$ by solving (2.36), (2.37), and (2.50), respectively. These equations are recursive filters with u and y as inputs and ε and ψ as outputs. The filter coefficients are determined by $\hat{\theta}(t - 1)$. Except when $c = 0$, the filters have infinite impulse responses. This means that the solution must be initialized at time $t = 0$ and that the whole data record $y(s)$, $u(s)$, $0 \leq s \leq t - 1$, is used to determine $\varepsilon(t, \hat{\theta}(t - 1))$ and $\psi(t, \hat{\theta}(t - 1))$. Hence these variables cannot be computed "recursively" in the sense of (1.3) using only $\hat{\theta}(t - 1)$ and a fixed-size memory vector. We must therefore look for approximations that can be computed recursively.

A natural approximation of (2.37) and (2.50) is to make only one time iteration with the current estimates and use previous values of ε and ψ as initial values. This means that $\varepsilon(t, \hat{\theta}(t - 1))$ is approximated by $\varepsilon(t)$, calculated according to

$$\varepsilon(t) = y(t) - \hat{y}(t),$$

$$\hat{y}(t) = -\hat{c}(t - 1)\hat{y}(t - 1)$$
$$+ [\hat{c}(t - 1) - \hat{a}(t - 1)]y(t - 1) + \hat{b}(t - 1)u(t - 1).$$

Here

$$\hat{\theta}(t - 1) = \begin{pmatrix} \hat{a}(t - 1) \\ \hat{b}(t - 1) \\ \hat{c}(t - 1) \end{pmatrix}.$$

If we introduce the vector

$$\varphi(t) = \begin{pmatrix} -y(t - 1) \\ u(t - 1) \\ \varepsilon(t - 1) \end{pmatrix},$$

we can rewrite the foregoing expression for $\varepsilon(t)$ as

$$\varepsilon(t) = y(t) - \hat{\theta}^{\mathsf{T}}(t - 1)\varphi(t). \tag{2.51}$$

Similarly, we see from (2.50) that a natural approximation of $\psi(t, \hat\theta(t-1))$ is $\psi(t)$, computed by means of

$$\psi(t) = -\hat{c}(t-1)\psi(t-1) + \varphi(t). \tag{2.52}$$

Notice that $\varepsilon(t)$ and $\psi(t)$ are computed recursively in (2.51)–(2.52). At time t we need only know $\hat\theta(t-1)$, $\psi(t-1)$, and $\varphi(t)$. The equation (2.52) is a typical approximation of the gradient, which will be used extensively in our algorithms.

Using $\varepsilon(t)$ and $\psi(t)$ in (2.46), (2.47) now gives the recursive algorithm

$$\hat\theta(t) = \hat\theta(t-1) + \frac{1}{t}R^{-1}(t)\psi(t)\varepsilon(t), \tag{2.53a}$$

$$R(t) = R(t-1) + \frac{1}{t}[\psi(t)\psi^T(t) - R(t-1)], \tag{2.53b}$$

$\varepsilon(t)$ and $\psi(t)$ calculated by means of (2.51)–(2.52). $\tag{2.53c}$

Here we used $R(t) = \bar{R}(t)/t$ rather than $\bar{R}(t)$ in the algorithm in order to stress the formal relationship with (2.15).

Just as in section 2.2.1 we can of course introduce

$$P(t) = [\bar{R}(t)]^{-1}, \tag{2.54a}$$

and update $P(t)$ using lemma 2.1:

$$P(t) = P(t-1) - \frac{P(t-1)\psi(t)\psi^T(t)P(t-1)}{1 + \psi^T(t)P(t-1)\psi(t)}. \tag{2.54b}$$

It should be clear from the formulas defining $\varepsilon(t)$ and $\psi(t)$, how the algorithm for an ARMAX model (2.8) of arbitrary order is constructed.

Let us pause here to bring up a minor issue, that however proves to be of some importance later on (see section 5.11). In the calculation of the prediction error $\varepsilon(t)$ in (2.51) we used the estimate $\hat\theta(t-1)$, based on measurements up to and including time $t-1$. This is necessary, since we use $\varepsilon(t)$ to update $\hat\theta$ in (2.53a). However, we also use $\varepsilon(t)$ in $\varphi(t+1)$ to compute $\psi(t+1)$ in (2.52). This variable is needed only after we have updated $\hat\theta(t)$. At that time we can improve on the estimate $\varepsilon(t)$, by taking the new information in $y(t)$ into account. This means that we compute

$$\bar\varepsilon(t) = y(t) - \hat\theta^T(t)\bar\varphi(t), \tag{2.55a}$$

where now

$$\bar{\varphi}(t) = \begin{pmatrix} -y(t-1) \\ u(t-1) \\ \bar{\varepsilon}(t-1) \end{pmatrix}. \tag{2.55b}$$

We shall call $\bar{\varepsilon}(t)$ the *residual* at time t, and $\varepsilon(t)$, defined by

$$\varepsilon(t) = y(t) - \hat{\theta}^{\mathsf{T}}(t-1)\bar{\varphi}(t), \tag{2.55c}$$

the *prediction error* at time t. (Notice that the definition of $\varepsilon(t)$ in (2.55c) actually differs from (2.51) in that $\bar{\varphi}$ is used instead of φ). Other terms that are sometimes used are *a posteriori* and *a priori* prediction errors, respectively.

Remark on Notation   In order not to get too-complicated notation we have actually used the symbol ε for two different quantities. The variable $\varepsilon(t)$ in (2.51) is computed using the previous $\varepsilon(t-1)$. It thus differs from $\varepsilon(t)$ in (2.55c) which uses $\bar{\varepsilon}(t-1)$.

With the residual $\bar{\varepsilon}(t)$ it is natural to compute $\psi(t)$ as

$$\psi(t) = -\hat{c}(t-1)\psi(t-1) + \bar{\varphi}(t). \tag{2.55d}$$

With these improved estimates of $\psi(t)$, we have the modified algorithm of (2.53) and (2.55). It turns out that the slight difference in going from (2.51) and (2.52) to (2.55) actually makes a noticeable improvement in the algorithm's behavior (see section 5.11).

To summarize this section, we have derived the algorithm (2.53) for recursive estimation of the parameters of the ARMAX model (2.32). Let us stress three aspects.

First, the resulting algorithm has a striking formal similarity with the recursive least squares algorithm.

Second, most of the derivation, from (2.38) to (2.47) has been carried out in terms of general properties and relationships between $\varepsilon(t, \theta)$ and $\psi(t, \theta)$. The properties of the actual ARMAX model (2.32) affected only the way these quantities (and their approximations) were computed in (2.49)–(2.52). We may thus expect that the basic ideas of the derivation can be applied to a variety of other models.

Third, the approach was based on local expansion of the criterion function. If one gets close to the true minimizing element $\hat{\theta}_t$, the approximations involved are all very reasonable. We may thus expect good asymptotic properties once we get in the vicinity of the true value. From the given derivation, however, it is impossible to tell anything about the transient behavior of the algorithm.

2.2.4 Summary

In this section we have considered three common off-line identification methods, viz., the least squares method, the instrumental variable method, and the maximum likelihood or prediction error method applied to an ARMAX model. We have seen that the first two can be written exactly in a recursive fashion, while the third can be put into recursive form only at the price of certain approximations. The third example also indicates that the same or similar ideas should be applicable to other off-line methods based on criterion minimization.

2.3 Recursive Identification as Nonlinear Filtering (A Bayesian Approach)

In the Bayesian approach to the parameter estimation problem, the parameter itself is thought of as a random variable. Based on observations of other random variables that are correlated with the parameter, we may infer information about its value. The Kalman filter, for example, is developed in such a Bayesian framework. The unobserved state vector is assumed to be correlated with the output of the system, so that, based on observations of the output, the value of the state vector can be estimated. The Bayesian approach to parameter estimation is therefore related to linear and nonlinear state estimation ("filtering") problems.

Suppose that the true system dynamics can be described in terms of a parameter vector θ. With a Bayesian view of the problem, we thus consider θ to be a random vector with a certain prior distribution. The observations y^t and u^t will obviously be correlated with this θ. [The superscript t refers to the data record up to time t, as in (1.1).] At time t we then ask for the posterior probability density function for θ, i.e., we wish to determine $p(\theta \mid y^t, u^t)$.

An estimate $\hat{\theta}(t)$ can be determined from the posterior distribution in several ways. A common choice is the mean of the posterior distribution, i.e., the conditional expectation:

$$\hat{\theta}(t) = E(\theta \mid y^t, u^t). \tag{2.56}$$

Another choice is to take $\hat{\theta}(t)$ as the value for which the posterior distribution attains its maximum: "the most likely value." This is known as the maximum a posteriori (MAP) estimate and coincides with (2.56) for many symmetric distributions, such as the Gaussian one. The estimate $\hat{\theta}(t)$ defined in (2.56) is also the value that minimizes the parameter error variance $E|\theta - \hat{\theta}(t)|^2$ under very general conditions.

Our problem is consequently to determine how (2.56) or the density function itself evolves with t. This is usually a very difficult problem, and only approximate solutions can be found. If, however, θ is linearly related to the data, such as in example 2.1, and if the noise is Gaussian, an exact solution can be found.

2.3.1 Linear Regression

LEMMA 2.2   Suppose that the data is generated according to

$$y(t) = \varphi^T(t)\theta + e(t) \tag{2.57}$$

where the vector $\varphi(t)$ is a function of y^{t-1}, u^{t-1} and where $\{e(t)\}$ is a sequence of independent Gaussian variables with $Ee(t) = 0$ and $Ee^2(t) = r_2(t)$. Suppose also that the prior distribution of θ is Gaussian with mean θ_0 and covariance matrix P_0. Then the posterior distribution $p(\theta \mid y^t, u^t)$ is also Gaussian with mean $\hat{\theta}(t)$ and covariance matrix $P(t)$, where $\hat{\theta}(t)$ and $P(t)$ are determined according to

$$\hat{\theta}(t) = \hat{\theta}(t-1) + L(t)[y(t) - \hat{\theta}^T(t-1)\varphi(t)],$$

$$L(t) = \frac{1}{r_2(t)}P(t)\varphi(t) = \frac{P(t-1)\varphi(t)}{r_2(t) + \varphi^T(t)P(t-1)\varphi(t)}, \tag{2.58}$$

$$P(t) = P(t-1) - \frac{P(t-1)\varphi(t)\varphi^T(t)P(t-1)}{r_2(t) + \varphi^T(t)P(t-1)\varphi(t)},$$

$$\hat{\theta}(0) = \theta_0, \quad P(0) = P_0.$$

Proof Assume that u^t is a given deterministic sequence. We may then neglect it in the calculations to come (it will be included in the way $\varphi(t)$ is formed from y^{t-1}). The proof will be based on Bayes's rule in the form

$$P(A \mid B, C) = \frac{P(B \mid A, C)P(A \mid C)}{P(B \mid C)}. \tag{2.59}$$

Here $P(A \mid B, C)$ is the probability of the event A, conditioned on B and C. Applying this formula to the posterior density gives

$$p(\theta \mid y^t) = p(\theta \mid y(t), y^{t-1}) = \frac{p(y(t) \mid \theta, y^{t-1})p(\theta \mid y^{t-1})}{p(y(t) \mid y^{t-1})}. \qquad (2.60)$$

We shall prove the desired result by induction.

Step 1 :

$$p(\theta \mid y^0) = \frac{(2\pi)^{-\dim\theta/2}}{\sqrt{\det P(0)}} \exp\left\{-\frac{1}{2}[\theta - \hat\theta(0)]^T P^{-1}(0)[\theta - \hat\theta(0)]\right\}$$

by the assumptions of the lemma.

Step 2: Assume that

$$p(\theta \mid y^{t-1})$$

$$= \frac{(2\pi)^{-\dim\theta/2}}{\sqrt{\det P(t-1)}} \exp\left\{-\frac{1}{2}[\theta - \hat\theta(t-1)]^T P^{-1}(t-1)[\theta - \hat\theta(t-1)]\right\}.$$

We shall now calculate $p(\theta \mid y^t)$ using (2.60). From (2.57) we have

$$e(t) = y(t) - \theta^T \varphi(t).$$

Under the assumption that $\{e(t)\}$ is a sequence of independent random variables with zero means and variances $r_2(t)$, we have

$$p(y(t) \mid \theta, y^{t-1}) = \frac{1}{\sqrt{2\pi r_2(t)}} \exp\left\{-\frac{1}{2r_2(t)}[y(t) - \theta^T\varphi(t)]^2\right\}.$$

Hence, from (2.60),

$$p(\theta \mid y^t) = \text{Norm} \cdot \exp\left\{-\frac{1}{2r_2(t)}[y(t) - \theta^T\varphi(t)]^2\right.$$

$$\left. -\frac{1}{2}[\theta - \hat\theta(t-1)]^T P^{-1}(t-1)[\theta - \hat\theta(t-1)]\right\},$$

where we have not explicitly written out the θ-independent normalization factor.

We shall now try to write the exponent as a quadratic form:

$$-2\log p(\theta \mid y^t) = \text{const} + \frac{1}{r_2(t)}y^2(t)$$

$$-\frac{1}{r_2(t)}y(t)\varphi^T(t)\theta - \frac{1}{r_2(t)}\theta^T\varphi(t)y(t)$$

$$+\frac{1}{r_2(t)}\theta^T\varphi(t)\varphi^T(t)\theta + \theta^TP^{-1}(t-1)\theta$$

$$-\theta^TP^{-1}(t-1)\hat{\theta}(t-1) - \hat{\theta}^T(t-1)P^{-1}(t-1)\theta$$

$$+\hat{\theta}^T(t-1)P^{-1}(t-1)\hat{\theta}(t-1).$$

Define

$$\bar{P}^{-1} = P^{-1}(t-1) + \frac{1}{r_2(t)}\varphi(t)\varphi^T(t); \tag{2.61}$$

then we can rewrite the preceding expression as

$$\text{const} + \frac{1}{r_2(t)}y^2(t) + \hat{\theta}(t-1)^TP^{-1}(t-1)\hat{\theta}(t-1)$$

$$+ \theta^T\bar{P}^{-1}\theta - \theta^T\left[\frac{1}{r_2(t)}\varphi(t)y(t) + P^{-1}(t-1)\hat{\theta}(t-1)\right]$$

$$- \left[\frac{1}{r_2(t)}\varphi(t)y(t) + P^{-1}(t-1)\hat{\theta}(t-1)\right]^T\theta$$

$$= \text{const}' + \left[\theta - \bar{P}\frac{1}{r_2(t)}\varphi(t)y(t) - \bar{P}P^{-1}(t-1)\hat{\theta}(t-1)\right]^T$$

$$\times \bar{P}^{-1}\left[\theta - \bar{P}\frac{1}{r_2(t)}\varphi(t)y(t) - \bar{P}P^{-1}(t-1)\hat{\theta}(t-1)\right],$$

where const' is a new, θ-independent normalization constant. Since

$$\bar{P}\cdot P^{-1}(t-1) = I - \frac{\bar{P}\varphi(t)\varphi^T(t)}{r_2(t)},$$

we may write the expression within parentheses as $\theta - \bar{\theta}$, where

$$\bar{\theta} = \hat{\theta}(t-1) + \frac{1}{r_2(t)}\bar{P}\varphi(t)[y(t) - \hat{\theta}^T(t-1)\varphi(t)]. \tag{2.62}$$

This means that the posterior density at time t, $p(\theta \mid y^t)$, is Gaussian with mean $\bar{\theta}$, given by (2.62), and covariance matrix $\bar{P}$, given by (2.61).

Applying lemma 2.1 to (2.61) gives

$$\bar{P} = P(t-1) - \frac{P(t-1)\varphi(t)\varphi^{\mathrm{T}}(t)P(t-1)}{r_2(t) + \varphi^{\mathrm{T}}(t)P(t-1)\varphi(t)}. \tag{2.63}$$

Moreover, by (2.18) we also have

$$\frac{1}{r_2(t)}\bar{P}\varphi(t) = \frac{P(t-1)\varphi(t)}{r_2(t) + \varphi^{\mathrm{T}}(t)P(t-1)\varphi(t)}.$$

With this we have completed the induction and shown that the posterior density at time t is the one stated in the lemma. ∎

Notice that the algorithm (2.58) that gives us the conditional expectation of the parameter vector under the normality assumption, coincides with the recursive least squares algorithm (2.19) with $r_2(t) = 1/\alpha_t$. This shows that the initial values $\hat{\theta}(0)$ and $P(0)$ in (2.19) can be interpreted as prior knowledge about the parameter vector θ. It also shows that the optimal weighting factor α_t in the least squares criterion (2.12) is the inverse of the variance of the noise term (at least when the noise is white and Gaussian).

2.3.2 Kalman Filter Interpretation

It is interesting to note that the model (2.57) can be seen as a linear state-space model

$$\theta(t+1) = \theta(t), \tag{2.64a}$$

$$y(t) = \varphi^{\mathrm{T}}(t)\theta(t) + e(t). \tag{2.64b}$$

Compare this to the conventional state-space form (2.9):

$$x(t+1) = F(t)x(t) + w(t), \tag{2.65a}$$

$$y(t) = H(t)x(t) + e(t), \tag{2.65b}$$

where $\{w(t)\}$ and $\{e(t)\}$ are mutually independent sequences of independent random vectors with zero mean values and covariance matrices $R_1(t)$ and $r_2(t)$, respectively. With $F(t) = I$, $R_1(t) = 0$, and $H(t) = \varphi^{\mathrm{T}}(t)$ in (2.65), we clearly obtain (2.64). The determination of

$$\hat{x}(t+1) = \mathrm{E}(x(t+1) \mid y^t)$$

for (2.65) is solved by using the well-known Kalman filter (see appendix 1.C)

$$\hat{x}(t+1) = F(t)\hat{x}(t) + K(t)[y(t) - H(t)\hat{x}(t)], \tag{2.66a}$$

$$K(t) = \frac{F(t)P(t)H^{\mathsf{T}}(t)}{r_2(t) + H(t)P(t)H^{\mathsf{T}}(t)}, \tag{2.66b}$$

$$P(t+1) = F(t)P(t)F^{\mathsf{T}}(t) + R_1(t)$$

$$- F(t)P(t)H^{\mathsf{T}}(t)[r_2(t) + H(t)P(t)H^{\mathsf{T}}(t)]^{-1}H(t)P(t)F^{\mathsf{T}}(t). \tag{2.66c}$$

With $F(t) = I$, $R_1(t) = 0$, and $H(t) = \varphi^{\mathsf{T}}(t)$ in (2.66), we obtain the algorithm (2.58).

The discussion so far has taught us two things. First, that the Kalman filter is still valid, even though the H-matrix is realization-dependent. Second, we have obtained an interpretation of the initial conditions $\hat{\theta}(0)$ and $P(0)$ of the recursive least squares algorithm, for the case in which we choose to initialize the algorithm at $t = 0$. The conditions correspond to prior knowledge of the value of the parameter vector. It is reasonable to similarly interpret the more general RML method of section 2.2.3.

2.3.3 A General State-Space Model

Up to this point we have considered only the special case where θ and the data are linearly related. Let us now study a more general situation. Let the model be as in example 2.3:

$$x(t+1) = F(\theta)x(t) + G(\theta)u(t) + w(t),$$
$$\tag{2.67}$$
$$y(t) = H(\theta)x(t) + e(t),$$

where $\{w(t)\}$ and $\{e(t)\}$ are sequences of independent random vectors with zero mean values and covariance matrices $R_1(\theta)$ and $R_2(\theta)$, respectively. Suppose that $x(0)$ has mean value $x_0(\theta)$ and covariance matrix $\Pi_0(\theta)$. The parameter vector θ is unknown; suppose that its prior distribution has mean value θ_0 and covariance matrix P_0. To determine a recursive estimator for θ, we define an extended state vector

$$X(t) = \begin{pmatrix} x(t) \\ \theta \end{pmatrix},$$

and consider the state estimation problem for this extended state. The state obeys the equations

$$X(t + 1) = \bar{F}(X(t), u(t)) + \bar{w}(t),$$

$$y(t) = \bar{H}(X(t)) + e(t),$$

(2.68)

where

$$\bar{F}(X(t), u(t)) = \begin{pmatrix} F(\theta)x(t) + G(\theta)u(t) \\ \theta \end{pmatrix},$$

(2.69a)

$$\bar{w}(t) = \begin{pmatrix} w(t) \\ 0 \end{pmatrix},$$

(2.69b)

$$H(X(t)) = H(\theta)x(t).$$

(2.69c)

The problem in (2.68) is to determine $E(X(t + 1) \mid y^t)$. This in turn, will give the estimate $E(\theta \mid y^t)$ in addition to $\hat{x}(t + 1) = E(x(t + 1) \mid y^t)$. The problem of recursively identifying θ has thus been formulated as a state estimation problem ("a filtering problem"). The state equations (2.68) are however, nonlinear in the state, since X enters nonlinearly in $\bar{F}$ and $\bar{H}$ as seen in (2.69a, c). We are thus faced with a nonlinear filtering problem.

Nonlinear filtering is a subject that has been studied over a long period and many approaches have been suggested and tested (see e.g., Jazwinski, 1970). For a nonlinear system it is not possible, except in a few isolated cases, to get an exact solution with recursive methods; various approximations have to be introduced. With the connection we made here to identification, any nonlinear filtering method applied to (2.68), directly gives rise to a corresponding recursive identification algorithm for estimating θ. Among these, the extended Kalman filter is no doubt the best-known and most widely used example.

EXAMPLE 2.4. (Extended Kalman Filter)   The idea behind the extended Kalman filter (EKF) is to extend linear Kalman filter theory to nonlinear systems. This is achieved by linearization of the system around the current state estimate and application of the Kalman filter to the resulting (time-varying) linear system. The details of the approach are given, e.g., in Theorem 8.1 of Jazwinski (1970). When applied to (2.68)–(2.69), the resulting algorithm is fairly complex. It is given in appendix 2.A. Here, to illustrate the idea and the character of the resulting algorithm, we shall apply the EKF to the simple model, given by

$$x(t + 1) = ax(t) + w(t),$$

$$y(t) = x(t) + e(t).$$

(2.70)

Here $x(t)$ is a one-dimensional state and a is an unknown constant. The sequences $\{w(t)\}$ and $\{e(t)\}$ consist of independent random variables with zero means and variances r_1 and r_2, respectively, which we assume to be known in this example. The problem is to estimate the parameter a from measurements of the output $\{y(t)\}$.

By including a in the state vector we obtain

$$X(t) = \begin{pmatrix} x(t) \\ a \end{pmatrix},$$

$$\left.\begin{aligned} X(t+1) &= \begin{pmatrix} ax(t) \\ a \end{pmatrix} + \begin{pmatrix} w(t) \\ 0 \end{pmatrix} \\ y(t) &- x(t) + e(t) \end{aligned}\right\}$$ (2.71)

Let

$$\hat{X}(t) = \begin{pmatrix} \hat{x}(t) \\ \hat{a}(t-1) \end{pmatrix}$$

be the state estimate based on y^{t-1}. (We use the argument $t-1$ for $\hat{a}$ in order to be consistent with the convention that $\hat{\theta}(t)$ is based on z^t.) In the EKF the estimate is updated as

$$\hat{X}(t+1) = \begin{pmatrix} \hat{a}(t-1)\hat{x}(t) \\ \hat{a}(t-1) \end{pmatrix} + \bar{K}(t)[y(t) - \hat{x}(t)],$$ (2.72)

where the Kalman gain $\bar{K}(t)$ is determined from the linear time-varying system that is obtained by linearization of (2.71) around the current estimate. This linear system has the state transition matrix

$$F(\hat{X}(t)) = \frac{\partial}{\partial X}\begin{pmatrix} ax \\ a \end{pmatrix}\Bigg|_{X=\hat{X}(t)} = \begin{pmatrix} \hat{a}(t-1) & \hat{x}(t) \\ 0 & 1 \end{pmatrix},$$

and the state-to-output matrix

$$H(\hat{X}(t)) = \frac{\partial}{\partial X}x\Bigg|_{X=\hat{X}(t)} = (1 \quad 0).$$

Hence $\bar{K}(t)$ is determined from [see (2.66b, c)]

$$\bar{P}(t+1) = F(\hat{X}(t))\bar{P}(t)F^{\mathrm{T}}(\hat{X}(t)) + \begin{pmatrix} r_1 & 0 \\ 0 & 0 \end{pmatrix}$$

$$- F(\hat{X}(t))\bar{P}(t)H^{\mathrm{T}}(\hat{X}(t))[r_2 + H(\hat{X}(t))\bar{P}(t)H^{\mathrm{T}}(\hat{X}(t))]^{-1}$$
$$\times H(\hat{X}(t))\bar{P}(t)F^{\mathrm{T}}(\hat{X}(t)),$$

$$\bar{K}(t) = F(\hat{X}(t))\bar{P}(t)H^{\mathrm{T}}(\hat{X}(t))[r_2 + H(\hat{X}(t))\bar{P}(t)H^{\mathrm{T}}(\hat{X}(t))]^{-1}.$$

Introducing the block structure

$$\bar{P}(t) = \begin{pmatrix} p_1(t) & p_2(t) \\ p_2(t) & p_3(t) \end{pmatrix}, \quad \bar{K}(t) = \begin{pmatrix} k_1(t) \\ k_2(t) \end{pmatrix},$$

and sorting the foregoing equations now gives

$$k_1(t) = \frac{\hat{a}(t-1)p_1(t) + \hat{x}(t)p_2(t)}{p_1(t) + r_2}, \tag{2.73a}$$

$$k_2(t) = \frac{p_2(t)}{p_1(t) + r_2}, \tag{2.73b}$$

$$p_2(t+1) = [\hat{a}(t) - k_1(t)]p_2(t) + \hat{x}(t)p_3(t), \tag{2.73c}$$

$$p_3(t+1) = p_3(t) - \frac{[p_2(t)]^2}{p_1(t) + r_2}, \tag{2.73d}$$

$$p_1(t+1) = \hat{a}^2(t)p_1(t) + 2\hat{a}(t)p_2(t)\hat{x}(t) + \hat{x}^2(t)p_3(t)$$
$$- k_1^2(t)[p_1(t) + r_2] + r_1. \tag{2.73e}$$

The updating equation (2.72) can now be written

$$\hat{x}(t+1) = \hat{a}(t-1)\hat{x}(t) + k_1(t)[y(t) - \hat{x}(t)], \tag{2.74a}$$

$$\hat{a}(t) = \hat{a}(t-1) + k_2(t)[y(t) - \hat{x}(t)]. \tag{2.74b}$$

Equations (2.73)–(2.74) define the recursive identification algorithm arising from the EKF. They look more complex than those we have encountered previously, but let us point out the following similarities:

• The variable $\hat{x}(t)$ computed from (2.74a) is a prediction of the output $y(t)$, based on previous observations and the sequence of previous estimates. The calculation is of the same character as in (2.51).

• The updating of the parameter estimate (2.74b) is based on the current prediction error $y(t) - \hat{x}(t)$, just as in (2.53a).

• The gain $k_2(t)$ of the updating equation is formed by filtering $y(t)$ through a time-varying, estimate-dependent filter. The output $y(t)$ is filtered to give $\hat{x}(t)$ in (2.74a); $\hat{x}(t)$ is in turn filtered in (2.73c) to give $p_2(t)$, which defines $k_2(t)$. These calculations are much the same as those determining the gain $[\bar{R}(t)]^{-1}\psi(t)$ in (2.53a) [eqs. (2.51)–(2.52)].

• The variables $p_3(t)$ and $p_2(t)$ and hence the gain $k_2(t)$ will tend to zero as $1/t$ when t approaches infinity. This is seen by examination of (2.73c, d). Hence (2.74b) updates a with a gain that decays as $1/t$ just as the gain does in (2.53).

The intriguing resemblance between the EKF algorithm and other recursive algorithms is further clarified in a general context in appendix 3.C.

Remark   The estimate $\hat{x}(t)$ used above is the predicted state estimate based on $y(s)$, $s \leq t - 1$. To stress this, we could use the notation $\hat{x}(t \mid t - 1)$. From (2.73c) we see that $p_2(t)$ then will be constructed from $y(s)$, using measurements only for $s \leq t - 2$. The variable $p_2(t)$ is then used in (2.73b) at a time when $y(t - 1)$ clearly is available; we have thus unnecessarily delayed this piece of information. It seems more reasonable to update the state estimate:

$$\hat{x}(t \mid t) = \hat{x}(t \mid t - 1) + \tilde{k}_1(t)[y(t) - \hat{x}(t \mid t - 1)], \tag{2.74a'}$$

and to use this filtered estimate in

$$p_2(t + 1) = [\hat{a}(t) - k_1(t)]p_2(t) + \hat{x}(t \mid t)p_3(t). \tag{2.73c'}$$

In (2.74a'), the gain $\tilde{k}_1(t)$ is given by

$$\tilde{k}_1(t) = \frac{p_1(t)}{p_1(t) + r_2} \tag{2.73a'}$$

[see also (1.C.17)–(1.C.18)].   □

2.3.4   Summary

We have seen that a Bayesian formulation of the parameter estimation problem leads in a natural way to a nonlinear filtering problem. Various approximative solutions to the filtering problem therefore give corresponding recursive identification algorithms. The extended Kalman filter is a well-known special case.

For a particular structure, where the parameters are linear in data and the noise is white and Gaussian, an exact solution to the Bayesian approach can be given. It coincides with the least squares algorithm of section 2.2.1.

2.4 The Stochastic Approximation Approach

The concept of "stochastic approximation" has been developed in statistics for certain sequential parameter estimation problems. Before giving a formal account of the idea let us consider a simple case.

2.4.1   A Simple Identification Problem

To illustrate how the idea of stochastic approximation relates to recursive identification, let us study the simple difference equation model (2.5):

$$y(t) = \theta^T \varphi(t) + v(t), \tag{2.75}$$

where $y(t)$ and $\varphi(t)$ are measured quantities and θ is to be determined. The variable $v(t)$ is the equation error (see example 2.1) and it is natural to select θ so that the variance of $v(t)$ is minimized, i.e., to seek

$$\min_{\theta} V(\theta), \tag{2.76}$$

where

$$V(\theta) = \tfrac{1}{2}\mathrm{E}[y(t) - \varphi^T(t)\theta]^2. \tag{2.77}$$

The function $V(\theta)$ is quadratic in θ, so therefore (2.76) can be found by solving

$$\left[-\frac{d}{d\theta} V(\theta) \right]^T = \mathrm{E}\varphi(t)[y(t) - \varphi^T(t)\theta] = 0. \tag{2.78}$$

Now the problem (2.76)–(2.78) cannot be solved exactly by the user, since the probability distribution of $(y(t), \varphi(t))$ is not known and therefore the expectations in (2.77) and (2.78) cannot be evaluated. One way around this would be to replace expectations with sample means [i.e., $\mathrm{E}f(t)$ could be approximated by $(1/N)\sum_1^N f(t)$], bringing us to the least squares method of section 2.2.1. The problem (2.76)–(2.78) is, however, of a form to which general stochastic approximation procedures apply. Let us therefore turn to an account of stochastic approximation itself.

2.4.2   Stochastic Approximation

The typical problem for stochastic approximation can be posed as follows. Let $\{e(t)\}$ be a sequence of random variables each of the same distribution, indexed by a discrete time variable t. A function $Q(x, e(t))$ of $e(t)$ and x is given, and a solution to the equation

$$EQ(x, e(t)) = f(x) = 0 \tag{2.79}$$

is sought, where E denotes expectation over $e(t)$. The distribution of $\{e(t)\}$ is not known to the user. The exact form of the function $Q(x, e)$ may also be unknown. The values of $Q(x, e)$ are, however, observed or can be constructed for any chosen x. That is to say that at time t, the user chooses a value x and obtains a value of $Q(x, e(t))$. Comparing (2.79) to (2.78), we see that (2.79) corresponds to

$x = \theta,$

$$e(t) = \begin{pmatrix} y(t) \\ \varphi(t) \end{pmatrix}, \tag{2.80}$$

$$Q(x, e(t)) = \varphi(t)[y(t) - \varphi^\mathsf{T}(t)\theta].$$

In this case $e(t)$ is observed and $Q(x, e)$ is a known function of x and e, but the distribution of $e(t)$ is unknown.

Returning to the general equation (2.79), the problem is to determine a sequence of values $x(t)$, $t = 1, 2, \ldots$, observe the corresponding $Q(x(t), e(t))$, and subsequently infer the solution of (2.79). Conceptually, it is clear that such a thing can be done. A trivial way would be to fix x, make a large number of observations $Q(x, e(t))$ for this x, thereby getting a good estimate of $f(x)$, and repeat the procedure for a number of new x-values until a solution of (2.79) is found. It is, however, more efficient to change the value of x for each observation, in order not to spend a lot of effort to determine $f(x)$ accurately for an x-value far away from a solution to (2.79). Robbins and Monro (1951) suggested the following scheme to solve (2.79) recursively as time evolves:

$$\hat{x}(t) = \hat{x}(t-1) + \gamma(t)Q(\hat{x}(t-1), e(t)), \tag{2.81}$$

where $\{\gamma(t)\}$ is a sequence of positive scalars tending to zero.

The convergence properties of (2.81) have been studied by Robbins and Monro (1951), Blum (1954), and Dvoretzky (1956). They showed that $\hat{x}(t)$ indeed will converge to the solution of (2.79) under certain assumptions. (A typical assumption in these early studies is that $\{e(t)\}$ is a sequence of independent random vectors, which is not the case in our application (2.80))

EXAMPLE 2.5 (Estimation of Mean Value)   As a trivial example of the Robbins-Monro scheme we attempt to solve

$E[e(t) - x] = 0.$

If $Ee(t) = m$, the solution clearly is $x^* = m$. The Robbins-Monro scheme gives

$$\hat{x}(t) = \hat{x}(t - 1) + \gamma(t)[e(t) - \hat{x}(t - 1)].$$

With $\gamma(t) = 1/t$, we find that

$$\hat{x}(t) = \frac{1}{t} \sum_{k=1}^{t} e(k),$$

i.e., $\hat{x}(t)$ is simply the sample mean. □

2.4.3 Stochastic Approximation Applied to Linear Regressions

Let us now return to the case (2.75).

EXAMPLE 2.6. (A Stochastic Approximation Algorithm for a Linear Regression Model) If we apply the Robbins-Monro scheme (2.81) to (2.78) we obtain the algorithm

$$\hat{\theta}(t) = \hat{\theta}(t - 1) + \gamma(t)\varphi(t)[y(t) - \varphi^{\mathrm{T}}(t)\hat{\theta}(t - 1)]. \tag{2.82}$$

This recursive algorithm for estimating the parameter θ in the model (2.75) has been suggested and tested by Saridis and Stein (1968), Sakrison (1967), and Tsypkin (1973). The latter reference gives a very comprehensive treatment of the role of stochastic approximation algorithms in estimation and control. In the control literature algorithms of the type (2.82) are usually known as "stochastic approximation methods." We here prefer to call them "stochastic gradient methods," for reasons that will be explained later, since they represent only one way of applying the idea of stochastic approximation. The algorithm (2.82) has also been widely used in adaptive signal processing. There it is well known as the "LMS algorithm," and was derived and first applied in that context by Widrow and Hoff (1960).

The sequence $\{\gamma(t)\}$ in (2.82) is the "gain sequence." We shall discuss aspects of how to choose this in example 2.11 and, in more detail, in section 5.6. Let us here only remark that some common choices in applications are

$$\gamma(t) = \gamma_0 \quad \text{(constant)}, \tag{2.83a}$$

$$\gamma(t) = \gamma_0/|\varphi(t)|^2 \quad \text{(normalized)}, \tag{2.83b}$$

or

$$\gamma(t) = \left[\sum_{k=1}^{t} |\varphi(k)|^2 \right]^{-1} \qquad \text{(normalized and decreasing)}. \qquad (2.83c)$$

The normalized choices have the advantage of giving an algorithm that is invariant under scaling of the signals $y(t)$ and $\varphi(t)$. Notice that with the choice (2.83c), the algorithm (2.82) can be written

$$\hat{\theta}(t) = \hat{\theta}(t-1) + [r(t)]^{-1} \varphi(t) [y(t) - \varphi^T(t)\hat{\theta}(t-1)], \qquad (2.84a)$$

$$r(t) = r(t-1) + |\varphi(t)|^2. \ \Box \qquad (2.84b)$$

2.4.4 The Robbins-Monro Scheme as a Stochastic Gradient Method

Now, recall that the original problem (2.76)–(2.77) in fact was a minimization problem, and that we approached it by solving for stationary points in (2.78). In terms of the general formulation (2.79) we could think of a minimization problem

$$\min_{x} V(x), \qquad (2.85a)$$

$$V(x) = EH(x, e(t)). \qquad (2.85b)$$

Let

$$-\frac{d}{dx}V(x) = f^T(x),$$

and suppose that the gradient

$$-\frac{\partial}{\partial x}H(x, e(t)) = Q^T(x, e(t))$$

can be obtained for any chosen x. Then (2.85) can be solved by solving the equation

$$0 = \left[-\frac{d}{dx}V(x) \right]^T = f(x) = EQ(x, e(t)), \qquad (2.86)$$

where we have allowed interchange of expectation and differentiation.

We are now back to the formulation (2.79), and the Robbins-Monro scheme (2.81) can thus be regarded as an algorithm to minimize $V(x)$ in (2.85). In this algorithm an adjustment of x is made in a direction that is the negative gradient of the observed criterion function $H(x, e(t))$. "On the average" the adjustments are consequently made in the negative gradient direction of the function $V(x)$. A suitable name for the corresponding algorithm therefore is "stochastic gradient method."

2.4.5 Gradient and Newton Directions

The stochastic gradient method can be seen as a stochastic analog of the method of steepest descent for the numerical minimization of a deterministic function. Let us for future reference quote some fundamentals of the theory of numerical minimization (see, e.g., Luenberger, 1973, for a general treatment). The method of steepest descent is given by

$$x^{(t+1)} = x^{(t)} - \gamma^{(t)} \left[\frac{d}{dx} V(x) \right]^T \Bigg|_{x=x^{(t)}}, \tag{2.87}$$

where $\gamma^{(t)}$ is a positive scalar chosen in a suitable way and $x^{(t)}$ denotes the tth iterate. It is well known that this method is fairly inefficient, in particular when the iterates are getting close to the minimum. So-called Newton methods or quasi-Newton methods give a distinctly better result. In these variants the search direction is modified from the negative gradient direction

$$\left[-\frac{d}{dx} V(x) \right]^T$$

to

$$-\left[\frac{d^2}{dx^2} V(x) \right]^{-1} \left[\frac{d}{dx} V(x) \right]^T. \tag{2.88}$$

Here $d^2 V(x)/dx^2$ is the Hessian, i.e., the second derivative matrix of V. We may call (2.88) the "Newton direction" for convenience. The iteration

$$x^{(t+1)} = x^{(t)} - \left[\frac{d^2}{dx^2} V(x) \right]^{-1} \left[\frac{d}{dx} V(x) \right]^T \Bigg|_{x=x^{(t)}} \tag{2.89}$$

will give convergence in one step to the minimum of $V(x)$, if this function is quadratic in x. Therefore, close to the minimum, where a second-order approximation of $V(x)$ well describes the function, the scheme (2.89) will be very efficient. Far away from the minimum, (2.89) may be inefficient or even diverge. Therefore, the Hessian is usually replaced by a guaranteed positive-definite approximation in order to secure a search direction that points "downhill."

2.4.6   A Stochastic Newton Method

Since the search direction (2.88) in general gives a clear improvement over the negative gradient for deterministic problems, it is reasonable to believe that a similar improvement would be obtained in the stochastic approximation case. Stochastic approximation minimization algorithms have been discussed in a general framework by Kushner and Clark (1978). Here we shall make only the following remarks. Suppose, for the minimization problem (2.85), that for each x an approximation of the Hessian of $V(x)$ can be constructed from previous observations. Denote this approximate Hessian by $\bar{V}''(x, e^t)$, where e^t indicates that the approximation may depend on all previous noise values $e^t = e(t), e(t-1), \ldots, e(1)$. Then a natural variant of the stochastic gradient scheme (2.81) is

$$\hat{x}(t) = \hat{x}(t-1) + \gamma(t)[\bar{V}''(\hat{x}(t-1), e^t)]^{-1}Q(\hat{x}(t-1), e(t)). \qquad (2.90)$$

This scheme could be called a "stochastic Newton algorithm."

Certain theorems regarding the convergence properties of (2.90) can be posed. We shall, however, postpone the formal discussion of asymptotic properties of the schemes until chapter 4.

Let us now apply this algorithm to the problem (2.76)–(2.77).

EXAMPLE 2.6 (CONTINUED)   To derive a Newton algorithm for (2.76)–(2.77) we first find that for the quadratic criterion (2.77) we have

$$\frac{d^2}{d\theta^2}V(\theta) = E\varphi(t)\varphi^T(t). \qquad (2.91)$$

The Hessian consequently is independent of θ. It can, according to (2.91), be determined as the solution R of the equation

$$E[\varphi(t)\varphi^T(t) - R] = 0. \qquad (2.92)$$

Applying the Robbins-Monro procedure to solve (2.92) gives

$$R(t) = R(t-1) + \gamma(t)[\varphi(t)\varphi^T(t) - R(t-1)]. \qquad (2.93a)$$

Our estimate of $d^2V(\theta)/d\theta^2$ at time t is consequently $R(t)$. With this estimate we obtain the stochastic Newton algorithm

$$\hat{\theta}(t) = \hat{\theta}(t-1) + \gamma(t)R^{-1}(t)\varphi(t)[y(t) - \varphi^T(t)\hat{\theta}(t-1)]. \qquad (2.93b)$$

The algorithm (2.93) is closely related to the recursive least squares algorithm of section 2.2.1. In fact, with $\gamma(t) = 1/t$ in (2.93), it coincides with (2.15) for $\alpha_t = 1$. We have consequently obtained yet another interpretation of the RLS scheme: It can be seen as a stochastic Newton method to minimize the criterion (2.77). □

With the same philosophy as in the example above, a more general problem can be approached. Consider the prediction error criterion of section 2.2.3, eq. (2.38). It is reasonable to minimize the criterion

$$V(\theta) = \tfrac{1}{2}E\varepsilon^2(t, \theta). \tag{2.94}$$

If, as before, $\psi(t, \theta)$ denotes the negative gradient of $\varepsilon(t, \theta)$ with respect to θ, we obtain

$$-\left[\frac{d}{d\theta} V(\theta)\right]^{\mathrm{T}} = E\psi(t, \theta)\varepsilon(t, \theta).$$

We can now minimize (2.94) recursively in the same manner as in example 2.6. Stochastic gradient and stochastic Newton methods can be derived, and it turns out that the algorithm (2.53) can be interpreted as a stochastic Newton algorithm for recursive minimization of (2.94). We shall, however, address the problem of minimizing (2.94) using stochastic approximation ideas in a much more general framework in section 3.4, and the details of the arguments will be deferred until then.

2.4.7 Summary

In this section, we have applied ideas from stochastic approximation to recursive minimization of certain criterion functions. We have rederived the recursive least squares method in this framework as a "stochastic Newton method," by which we mean that the parameter adjustments are made in the Newton direction for the criterion. We have also indicated that the approach has a potential for application to more general problems, a fact that we will make use of in section 3.4.

2.5 Pseudolinear Regressions and Model Reference Techniques

In this section we shall describe two additional approaches to the recursive identification problem. They are motivated by different ideas, but the resulting algorithms show some common features, which makes it suitable to discuss them in the same section.

2.5.1 Pseudolinear Regression

Consider the linear regression model

$$y(t) = \theta^T \varphi(t) + v(t) \tag{2.95}$$

that we used in example 2.1. Such models have been widely studied in statistics, and the least squares method of section 2.2.1 is a natural and efficient way of estimating the parameter vector θ. Considering the usefulness of this method, it is tempting to try to cast other models, that are not true linear regressions, in the form of (2.95). This can often be achieved by including in the regression vector $\varphi(t)$ unobservable variables, whose values, however, can be estimated from data. The combined procedure of estimating θ and reconstructing the unobserved φ-components, we shall call a *Pseudolinear Regression* (PLR) method. This term was introduced by Solo (1978). The following example will clarify the details of this approach.

EXAMPLE 2.7. (Extended Least Squares (ELS) Method) Consider the ARMAX model described in example 2.2:

$$y(t) + a_1 y(t - 1) + \cdots + a_n y(t - n)$$

$$= b_1 u(t - 1) + \cdots + b_m u(t - m) + e(t) + c_1 e(t - 1) + \cdots \tag{2.96}$$

$$+ c_r e(t - r).$$

Let us introduce the vectors

$$\varphi_0(t) = (-y(t - 1) \ \ldots \ -y(t - n) \ u(t - 1) \ \ldots \ u(t - m)$$

$$e(t - 1) \ \ldots \ e(t - r))^T$$

and

$$\theta = (a_1 \ \ldots \ a_n \ b_1 \ \ldots \ b_m \ c_1 \ \ldots \ c_r)^T.$$

With this notation, (2.96) can be rewritten as

$$y(t) = \theta^T \varphi_0(t) + e(t). \tag{2.97}$$

This model looks just like the linear regression (2.95), and we can try to apply the recursive least squares algorithm (2.15) to it for estimating θ:

$$\hat{\theta}(t) = \hat{\theta}(t-1) + \frac{1}{t} R^{-1}(t)\varphi_0(t)[y(t) - \hat{\theta}^{\mathrm{T}}(t-1)\varphi_0(t)],$$

$$\tag{2.98}$$

$$R(t) = R(t-1) + \frac{1}{t}[\varphi_0(t)\varphi_0^{\mathrm{T}}(t) - R(t-1)].$$

The problem is, of course, that the variables $e(i)$ entering the φ_0-vector are not measurable, and hence (2.98) cannot be implemented as it stands. We have to replace the components $e(i)$ with some estimate of them. From (2.96) we have

$$e(t) = y(t) + a_1 y(t-1) + \cdots + a_n y(t-n) - b_1 u(t-1) - \cdots$$

$$- b_m u(t-m) - c_1 e(t-1) - \cdots - c_r e(t-r).$$

If we have a sequence of estimates

$$\hat{\theta}(t) = (\hat{a}_1(t) \ \ldots \ \hat{a}_n(t) \ \hat{b}_1(t) \ \ldots \ \hat{b}_m(t) \ \hat{c}_1(t) \ \ldots \ \hat{c}_r(t))^{\mathrm{T}}$$

available, it seems natural to estimate $e(t)$ by $\bar{\varepsilon}(t)$, computed according to

$$\bar{\varepsilon}(t) = y(t) + \hat{a}_1(t)y(t-1) + \cdots + \hat{a}_n(t)y(t-n)$$

$$- \hat{b}_1(t)u(t-1) - \cdots - \hat{b}_m(t)u(t-m)$$

$$\tag{2.99}$$

$$- \hat{c}_1(t)\bar{\varepsilon}(t-1) - \cdots - \hat{c}_r(t)\bar{\varepsilon}(t-r).$$

With

$$\varphi(t) = (-y(t-1) \ \ldots \ -y(t-n) \ u(t-1) \ \ldots \ u(t-m)$$

$$\bar{\varepsilon}(t-1) \ \ldots \ \bar{\varepsilon}(t-r))^{\mathrm{T}},$$

$$\tag{2.100}$$

the equation (2.99) can be written

$$\bar{\varepsilon}(t) = y(t) - \hat{\theta}^{\mathrm{T}}(t)\varphi(t). \tag{2.101}$$

An obvious algorithm for estimating θ is now obtained from (2.98) by replacing $\varphi_0(t)$ by $\varphi(t)$, computed according to (2.100), (2.101). This gives the extended least squares (ELS) algorithm:

$$\varepsilon(t) = y(t) - \hat{\theta}^{\mathrm{T}}(t-1)\varphi(t), \tag{2.102a}$$

$$\hat{\theta}(t) = \hat{\theta}(t-1) + \frac{1}{t}[R(t)]^{-1}\varphi(t)\varepsilon(t), \tag{2.102b}$$

$$R(t) = R(t-1) + \frac{1}{t}\left[\varphi(t)\varphi^{\mathrm{T}}(t) - R(t-1)\right]. \tag{2.102c}$$

An advantage of this algorithm is that it is computationally equivalent to the usual recursive least squares algorithm. The same program can be used, as soon as it is complemented with the recursion (2.101). With $P(t) = \frac{1}{t}R^{-1}(t)$ we know from section 2.2.1 that (2.102) can also be written as

$$\hat{\theta}(t) = \hat{\theta}(t-1) + P(t)\varphi(t)\left[y(t) - \hat{\theta}^{\mathrm{T}}(t-1)\varphi(t)\right],$$
$$P(t) = P(t-1) - \frac{P(t-1)\varphi(t)\varphi^{\mathrm{T}}(t)P(t-1)}{1 + \varphi^{\mathrm{T}}(t)P(t-1)\varphi(t)}. \tag{2.103}$$

It is instructive to compare the algorithm (2.102) with the recursive maximum likelihood method (2.53) that we derived for an ARMAX model in section 2.2.3. We notice that the only difference is that the vector $\varphi(t)$ in (2.102b, c) is replaced by $\psi(t)$ in (2.53a, b). This vector $\psi(t)$ is obtained from $\varphi(t)$ by filtering it through the current estimate of the C-polynomial in (2.52). In chapter 4 we shall investigate what this difference between (2.53) and (2.102) means for the convergence properties of the respective algorithms. □

The extended least squares method is perhaps the best known example of a pseudolinear regression method in system identification. The same idea, however, can be applied to models for many other stochastic dynamical systems. We shall give a more general treatment of this approach in section 3.7.3.

2.5.2 Model Reference Techniques

"Model reference" is a concept that has been extensively used in adaptive control. The idea is to compare the actual output of the plant with that of a reference model (the latter defining the "ideal" output), and make adjustments in the regulator until the plant output coincides with the model output. A similar approach can be taken to the recursive identification problem. For identification, however, the recorded output from the system is compared to that of an adjustable model, and the model parameters are updated until the difference cannot be further improved. The procedure is schematically described in figure 2.1; the details of the approach are illustrated by the following example.

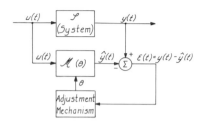

Figure 2.1
The model reference idea.

EXAMPLE 2.8 (A Model Reference Output Error Method)   This method
was described by Landau (1976). It uses the model

$$y(t) = \frac{B(q^{-1})}{F(q^{-1})} u(t) + v(t), \tag{2.104}$$

where

$$B(q^{-1}) = b_1 q^{-1} + \cdots + b_m q^{-m},$$

$$F(q^{-1}) = 1 + f_1 q^{-1} + \cdots + f_n q^{-n},$$

and q^{-1} is, as before, the delay operator. The term $v(t)$ represents some
measurement disturbance of unspecified character. The coefficients b_i
and f_i are to be determined.

Denote the undisturbed output by y_M; this is given by

$$y_M(t) = \frac{B(q^{-1})}{F(q^{-1})} u(t)$$

or

$$y_M(t) = -f_1 y_M(t-1) - \cdots - f_n y_M(t-n)$$
$$+ b_1 u(t-1) + \cdots + b_m u(t-m). \tag{2.105}$$

Based on this expression, and on current estimates of the parameters
$\hat{b}_i(t), \hat{f}_i(t)$, we may calculate the "model output"* at time t (corresponding
to $\hat{y}$ in figure 2.1) as follows:

---

*In model reference adaptive control, the "model output" $\hat{y}(t)$ corresponds to the actual
output of the system, controlled by a regulator that tunes its parameters $\hat{\theta}$. The output
$y(t)$ in figure 2.1 then corresponds to the output of the "ideal" reference model of the
closed-loop behavior.

$$\hat{y}_M(t) + \hat{f}_1(t)\hat{y}_M(t-1) + \cdots + \hat{f}_n(t)\hat{y}_M(t-n)$$
$$= \hat{b}_1(t)u(t-1) + \cdots + \hat{b}_m(t)u(t-m). \tag{2.106}$$

Notice that $\hat{y}_M(t)$ is based only upon the input and estimate sequences. It does not explicitly use $\{y(t)\}$, just as figure 2.1 suggests. With

$$\hat{\theta}(t) = (\hat{f}_1(t) \ \cdots \ \hat{f}_n(t) \ \hat{b}_1(t) \ \cdots \ \hat{b}_m(t))^{\mathrm{T}}$$

and

$$\varphi(t) = (-\hat{y}_M(t-1) \ \cdots \ -\hat{y}_M(t-n) \ u(t-1) \ \cdots \ u(t-m))^{\mathrm{T}},$$

eq. (2.106) can be written

$$\hat{y}_M(t) = \hat{\theta}^{\mathrm{T}}(t)\varphi(t). \tag{2.107}$$

The sequence of estimates $\{\hat{\theta}(t)\}$ will now be defined recursively. At time t, before the estimate $\hat{\theta}(t)$ is available, the model output will be

$$\hat{y}(t) = \hat{\theta}^{\mathrm{T}}(t-1)\varphi(t), \tag{2.108}$$

and a natural updating formula is

$$\hat{\theta}(t) = \hat{\theta}(t-1) + P(t)\varphi(t)[y(t) - \hat{y}(t)], \tag{2.109a}$$

where, in analogy with the RLS algorithm, the gain matrix $P(t)$ can be taken as

$$P(t) = R^{-1}(t) = \left[\sum_{k=1}^{t} \varphi(k)\varphi^{\mathrm{T}}(k) \right]^{-1}, \tag{2.109b}$$

so that

$$R(t) = R(t-1) + \varphi(t)\varphi^{\mathrm{T}}(t). \tag{2.109c}$$

The resulting algorithm (2.109), with $\hat{y}(t)$, $\hat{\theta}(t)$, and $\varphi(t)$ defined according to (2.107)–(2.108), is the one suggested by Landau (1976).

We may note that the model as well as the method very much resemble the recursive least squares method of section 2.2.1. The basic difference is that when going from (2.105) to the prediction (2.106) the previous predictions $\{\hat{y}_M(t-i)\}$ are used rather than the observed outputs. This is, as we remarked above, in accordance with figure 2.1, and motivates the name "output error method" or "model reference method" for the algorithm (2.109). As a consequence, the φ-vector does not directly depend on observed y-variables and hence not on any output noise $v(t)$ added to the measurements. (It does however, indirectly depend on y via $\hat{\theta}$, but this dependence decreases with time.) This also means, as we shall see in chapter 4, that the convergence properties of (2.109) will be less sensitive to the properties of $\{v(t)\}$ than the usual recursive least squares algorithm (2.19). A conceptual relationship with the instrumental variable method is pointed out in example 3.11. □

The model reference algorithm (2.109) in this example can also be interpreted as a pseudolinear regression. If, in (2.105), the sequence $\{y_M(t)\}$ had been a known sequence, the right-hand side would be a linear regression model. Then (2.109) is a pseudolinear regression for this model, with the estimate $\hat{y}_M(t)$ replacing $y_M(t)$ in $\varphi(t)$, in exactly the same way as the extended least squares method (2.102) results when $\bar{\varepsilon}(t)$ replaces $e(t)$ in $\varphi(t)$ [see (2.96) and (2.100)]. Therefore, in the general treatment of pseudolinear regression in section 3.7.3, the algorithm of example 2.8 will appear as a special case.

2.5.3 Summary

In this section we have seen some additional ways of deriving recursive identification methods. The algorithms that are obtained, are similar to the recursive least squares method. The pseudolinear regressions represent a direct extension of the least squares technique, while the model reference approach uses certain analogies with recursive least squares when choosing the updating mechanism. The methods will be treated in a more general framework in section 3.7.3.

2.6 Tracking Time-Varying Systems

In all of the discussion so far, we have assumed that the parameter vector to be estimated is constant, i.e., that the system is time-invariant. As we pointed out in chapter 1, an important reason for using recursive identification in practice is, however, that the dynamics may in fact be changing with time, and tracking will be required. In this section we shall discuss how tracking can be incorporated into recursive identification. Descriptions of time-varying dynamics can be achieved, with different mixes of formality and intuition, in the four approaches that we have discussed. We shall illustrate the ideas by means of examples.

2.6.1 The Bayesian Approach

No doubt, the Bayesian approach of section 2.3 is best capable of giving a formal treatment of a time-varying true parameter vector θ. The reason is that in this approach, the true parameter vector θ is considered to be a random variable. By allowing this vector to be time-varying and by describing how it changes with time, a formal approach to the tracking problem can be taken. We illustrate the idea by an example.

EXAMPLE 2.9 (Linear Regression Model with Stochastically Varying Dynamics)   Consider the model of lemma 2.2 and assume that the true value of the parameter vector θ varies according to

$$\theta(t+1) = \theta(t) + w(t), \tag{2.110}$$

where $\{w(t)\}$ is a sequence of independent Gaussian random vectors such that $w(t)$ has zero mean, and covariance matrix $R_1(t)$. We assume $R_1(t)$ to be known. We then have the following overall description of the system:

$$\begin{aligned}
\theta(t+1) &= \theta(t) + w(t), \\
y(t) &= \varphi^{\mathrm{T}}(t)\theta(t) + e(t), \\
Ew(t)w^{\mathrm{T}}(s) &= R_1(t)\delta_{ts}, \quad Ee(t)e(s) = R_2(t)\delta_{ts}, \\
Ew(t)e(s) &= 0.
\end{aligned} \tag{2.111}$$

Compare with (2.64)! Applying the Kalman filter (2.66) to (2.111) gives the estimates

$$\hat{\theta}(t) = \hat{\theta}(t-1) + L(t)\left[y(t) - \varphi^{\mathrm{T}}(t)\hat{\theta}(t-1)\right], \tag{2.112a}$$

$$L(t) = \frac{P(t-1)\varphi(t)}{R_2(t) + \varphi^{\mathrm{T}}(t)P(t-1)\varphi(t)}, \tag{2.112b}$$

$$P(t) = P(t-1) + R_1(t) - \frac{P(t-1)\varphi(t)\varphi^{\mathrm{T}}(t)P(t-1)}{R_2(t) + \varphi^{\mathrm{T}}(t)P(t-1)\varphi(t)}. \tag{2.112c}$$

It is interesting to note that $\hat{\theta}(t)$ given by (2.112) is still in fact the conditional mean of $\theta(t)$ given the observations; i.e., we have

$$\hat{\theta}(t) = E(\theta(t) \mid y^t). \tag{2.113}$$

Moreover, the posterior distribution of $\theta(t+1)$ given y^t is Gaussian with mean $\hat{\theta}(t)$ and covariance matrix $P(t)$. This follows by going through the original derivation of the Kalman filter (Kalman and Bucy, 1961), as pointed out by Bohlin (1970) and Åström and Wittenmark (1971). Notice, however, that most derivations in textbooks require $\{\varphi(t)\}$ in (2.111) to be a given deterministic sequence. This is not the case in our application, since this vector contains previous values of the output.

Remark A straightforward application of the Kalman filter to (2.111) in fact gives the estimate

$$\hat{\theta}(t + 1 \mid t) = E(\theta(t + 1) \mid y^t).$$

However, in view of (2.110), we have

$$E(\theta(t + 1) \mid y^t) = E(\theta(t) \mid y^t) = \hat{\theta}(t).$$

In the updating formula (2.112) for $\hat{\theta}(t)$, the time indices in P and y have thus been shifted compared to (2.66) in order to give an expression consistent with (2.19). □

In the foregoing example an exact solution of the problem could be obtained. The effect of the varying dynamics on the algorithm is only the additional R_1-term in (2.112c). This term prevents $P(t)$ from tending to zero, and consequently it keeps up the gain vector $L(t)$ in (2.112). This is very natural from an intuitive point of view: When the system is time-varying, the algorithm must be more "alert." The price for being persistently alert ($L(t)$ not tending to zero) is, of course, that the estimates are always sensitive to the random disturbances in the measurements. The estimates will not converge to their true values; the covariance matrix does not tend to zero. There is, as always, a compromise between alertness and noise sensitivity. The optimal compromise in (2.112) is reached via the covariances R_2 and R_1, which can therefore be viewed as design parameters controlling this tradeoff.

2.6.2 The Off-Line Identification Approach

It is not easy to describe formally time-varying dynamics for the off-line-inspired algorithms of section 2.2. The reason is that the off-line philosophy in itself assumes constant systems. However, one may attack the problem in a heuristic way by discounting old measurements. We illustrate the idea with an example.

EXAMPLE 2.10 (Recursive Least Squares Method with Discounted Measurements)   In the least squares criterion (2.12), the sum

$$V_N(\theta) = \frac{1}{N} \sum_1^N \alpha_t [y(t) - \theta^T \varphi(t)]^2 \tag{2.114}$$

is minimized. If we believe that the system is time-varying, the criterion (2.114) gives an estimate of the *average* behavior during the period $1 \leq t \leq N$. To obtain an estimate that is representative for the current (i.e., at time N) properties of the system, it is natural to consider another criterion in which older values are discounted:

$$\tilde{V}_N(\theta) = \sum_1^N \bar{\beta}(N, t)[y(t) - \theta^T \varphi(t)]^2, \tag{2.115}$$

where $\bar{\beta}(N, t)$ typically is increasing in t for given N. Since $\tilde{V}_N(\theta)$ still is quadratic in θ, the minimizing value can easily be computed:

$$\hat{\theta}(N) = \left[\sum_{t=1}^N \bar{\beta}(N, t)\varphi(t)\varphi^T(t) \right]^{-1} \left[\sum_{t=1}^N \bar{\beta}(N, t)\varphi(t)y(t) \right]. \tag{2.116}$$

This expression is the off-line estimate.

The sequence of off-line estimates can be computed recursively only if a certain structure for $\bar{\beta}(t, k)$ is introduced. We thus assume that

$$\bar{\beta}(t, k) = \lambda(t)\bar{\beta}(t - 1, k), \quad 1 \leq k \leq t - 1. \tag{2.117a}$$

This can also be written as

$$\bar{\beta}(t, k) = \left[\prod_{j=k+1}^t \lambda(j) \right] \alpha_k, \quad \text{where} \quad \bar{\beta}(k, k) = \alpha_k. \tag{2.117b}$$

Typically $\lambda(k) \leq 1$. If $\lambda(k) = \lambda$ all k, we obviously get

$$\bar{\beta}(t, k) = \lambda^{t-k} \cdot \alpha_k, \tag{2.118}$$

which gives an exponential forgetting profile in the criterion (2.115). In such a case we refer to λ as the *forgetting factor*.

Introduce

$$\bar{R}(t) = \sum_{k=1}^t \bar{\beta}(t, k)\varphi(k)\varphi^T(k). \tag{2.119}$$

Then

$$\bar{R}(t) = \lambda(t)\bar{R}(t-1) + \alpha_t\varphi(t)\varphi^T(t). \tag{2.120}$$

We now have, as in (2.14a),

$$
\begin{aligned}
\hat{\theta}(t) &= \bar{R}^{-1}(t)\left[\sum_{k=1}^{t-1} \bar{\beta}(t,k)\varphi(k)y(k) + \alpha_t\varphi(t)y(t)\right] \\
&= \bar{R}^{-1}(t)\left[\lambda(t)\sum_{k=1}^{t-1} \bar{\beta}(t-1,k)\varphi(k)y(k) + \alpha_t\varphi(t)y(t)\right] \\
&= \bar{R}^{-1}(t)\left[\lambda(t)\bar{R}(t-1)\hat{\theta}(t-1) + \alpha_t\varphi(t)y(t)\right] \\
&= \bar{R}^{-1}(t)\{\bar{R}(t)\hat{\theta}(t-1) + \alpha_t\varphi(t)[y(t) - \varphi^T(t)\hat{\theta}(t-1)]\} \\
&= \hat{\theta}(t-1) + \bar{R}^{-1}(t)\varphi(t)\alpha_t[y(t) - \varphi^T(t)\hat{\theta}(t-1)].
\end{aligned}
\tag{2.121}
$$

This is exactly the same updating formula as (2.14a). Introduce

$$P(t) = \bar{R}^{-1}(t),$$

and apply lemma 2.1 to (2.120). This gives

$$
P(t) = \frac{1}{\lambda(t)}P(t-1)
$$

$$
- \frac{1}{\lambda(t)}P(t-1)\varphi(t)\left[\varphi^T(t)\frac{1}{\lambda(t)}P(t-1)\varphi(t) + \frac{1}{\alpha_t}\right]^{-1}
$$

$$
\times \varphi^T(t)P(t-1)\frac{1}{\lambda(t)} = \frac{1}{\lambda(t)}\left[P(t-1) - \frac{P(t-1)\varphi(t)\varphi^T(t)P(t-1)}{[\lambda(t)/\alpha_t] + \varphi^T(t)P(t-1)\varphi(t)}\right].
\tag{2.122}
$$

We thus have the algorithm

$$\hat{\theta}(t) = \hat{\theta}(t-1) + L(t)[y(t) - \varphi^T(t)\hat{\theta}(t-1)], \tag{2.123a}$$

$$L(t) = \frac{P(t-1)\varphi(t)}{[\lambda(t)/\alpha_t] + \varphi^T(t)P(t-1)\varphi(t)} = \alpha_t P(t)\varphi(t), \tag{2.123b}$$

$$P(t) = \frac{1}{\lambda(t)}\left[P(t-1) - \frac{P(t-1)\varphi(t)\varphi^T(t)P(t-1)}{[\lambda(t)/\alpha_t] + \varphi^T(t)P(t-1)\varphi(t)}\right]. \tag{2.123c}$$

Consequently, a discounted least squares criterion can be recursively minimized by (2.123) if the forgetting profile is subject to (2.117). Obviously, the usual RLS algorithm (2.19) is a special case of (2.123) with $\lambda(t) \equiv 1$. □

The effect of the forgetting factor $\lambda(t)$ in (2.123c) clearly is that $P(t)$, and hence the gain $L(t)$, is kept larger. If $\lambda(t) \leq 1 - \delta$, $\delta > 0$, then $P(t)$ will not tend to zero and the algorithm will always be alert to track changing dynamics. The effect of $\lambda(t)$ in (2.123c) is therefore similar to that of $R_1(t)$ in (2.112c). The algorithm (2.112) differs from (2.123), though, since the effects of $R_1(t)$ and $\lambda(t)$ are not identical in each. Nonetheless, no major differences in the behavior of the corresponding estimates usually occur. We shall discuss this issue in more detail in section 5.6.

2.6.3 The Stochastic Approximation Approach

Let us now turn to the question of how time-varying dynamics can be handled in the stochastic approximation approach of section 2.4. Again, we do that by studying how the simple regression model is treated.

EXAMPLE 2.11 (Choice of Gain Sequence in Stochastic Approximation Algorithms)   Consider the stochastic Newton algorithm (2.93) that we derived for linear regression models:

$$\hat{\theta}(t) = \hat{\theta}(t-1) + \gamma(t)R^{-1}(t)\varphi(t)[y(t) - \varphi^T(t)\hat{\theta}(t-1)], \qquad (2.124a)$$

$$R(t) = R(t-1) + \gamma(t)[\varphi(t)\varphi^T(t) - R(t-1)]. \qquad (2.124b)$$

The theory of stochastic approximation usually leaves the choice of the gain sequence $\{\gamma(t)\}$ open, as long as it satisfies

$$\gamma(t) \geq 0, \quad \sum_1^\infty \gamma(t) = \infty, \quad \sum_1^\infty \gamma^2(t) < \infty. \qquad (2.125)$$

The second of these conditions implies that we can reach any target, and the third one assumes the gain to go to zero so that the effect of the noisy measurements is eliminated asymptotically.

On the other hand, if the system dynamics is time-varying, and we want to track its variations, we must relax the last condition of (2.125). Most often, one chooses to let $\gamma(t)$ tend to a small positive number γ_0 in such a case, [cf. (2.83a, b)]. The value of γ_0 will be chosen as a tradeoff between tracking capability (γ_0 large) and noise insensitivity (γ_0 small). This is in accordance with the real-time aspects that we have obtained for the other approaches. In fact, the algorithm (2.124) can be exactly converted into the forgetting factor algorithm (2.123). To see this, introduce

$$\bar{R}(t) = \frac{1}{\gamma(t)} R(t).$$

Then

$$\bar{R}(t) = \frac{1}{\gamma(t)} \{R(t-1) + \gamma(t)[\varphi(t)\varphi^{\mathrm{T}}(t) - R(t-1)]\}$$

$$= \frac{\gamma(t-1)}{\gamma(t)} \cdot \frac{1}{\gamma(t-1)} R(t-1) + \varphi(t)\varphi^{\mathrm{T}}(t) - R(t-1)$$

$$= \left[\frac{\gamma(t-1)}{\gamma(t)} - \gamma(t-1)\right] \bar{R}(t-1) + \varphi(t)\varphi^{\mathrm{T}}(t).$$

Now (2.124) can be written

$$\hat{\theta}(t) = \hat{\theta}(t-1) + \bar{R}^{-1}(t)\varphi(t)[y(t) - \varphi^{\mathrm{T}}(t)\hat{\theta}(t-1)], \tag{2.126a}$$

$$\bar{R}(t) = \frac{\gamma(t-1)}{\gamma(t)}[1 - \gamma(t)]\bar{R}(t-1) + \varphi(t)\varphi^{\mathrm{T}}(t). \tag{2.126b}$$

This algorithm coincides with (2.120), (2.121) and hence with (2.123) if we take

$$\lambda(t) = \frac{\gamma(t-1)}{\gamma(t)}[1 - \gamma(t)] \quad \text{and} \quad \alpha_t = 1. \tag{2.127}$$

Consequently, a general sequence $\{\gamma(t)\}$ in (2.124) corresponds to minimizing the discounted criterion (2.115) with the forgetting profile [see (2.117)]

$$\bar{\beta}(N, t) = \prod_{k=t+1}^{N} \frac{\gamma(k-1)}{\gamma(k)}[1 - \gamma(k)]$$

$$= \frac{\gamma(t)}{\gamma(N)} \prod_{k=t+1}^{N} [1 - \gamma(k)]. \tag{2.128a}$$

We shall often have occasion to work with normalized weights given by

$$\beta(N, t) = \gamma(N)\bar{\beta}(N, t). \tag{2.128b}$$

It is easy to verify, e.g., by mathematical induction, that these weights satisfy

$$\sum_{t=1}^{N} \beta(N, t) + \delta(N) = 1, \tag{2.129a}$$

where

$$\delta(N) = \prod_{t=1}^{N} [1 - \gamma(t)]. \tag{2.129b}$$

Obviously it does not make any difference to the minimization problem (2.115) whether β or $\bar{\beta}$ is used. From (2.127) we see in particular that a constant gain γ_0 corresponds to an exponential forgetting factor $\lambda_0 = (1 - \gamma_0)$.

2.6.4 The Model Reference Approach

We have now in three examples treated three of the approaches to recursive identification. The literature on model reference systems has basically been addressing the problem of noise free measurements (and often in continuous time). For this case it is natural to choose the gain to be constant, since that will give tracking abilities in a real-time situation. When measurement noise is brought into the picture in these approaches (see, e.g., Dugard, Landau, and Silviera, 1980), the choice of gain matrix is usually made in analogy with any of the approaches in examples 2.9–2.11.

2.6.5 Summary

In this section, we have discussed how to track time-varying dynamics with real-time recursive identification. The bottom line of any approach to this case is to prevent the gain vector from tending to zero. The size of the gain at which the algorithm settles will be a compromize between tracking ability and noise sensitivity. While these aspects can be handled in an ad hoc way by any approach, only the Bayesian (nonlinear filtering) approach offers a formal solution of the problem. The optimal compromise is then reached, as in the Kalman filter, by solving for the smallest error variance for the estimates.

2.7 User's Summary

The problem of recursive identification can be approached in a number of different ways. The resulting algorithms, however, exhibit very similar features. Another way of expressing this is to say that a given recursive identification algorithm can be interpreted in different ways. We shall here illustrate this in terms of perhaps the best known and most widely used algorithm: the recursive least squares (RLS) method.

The problem is to estimate the parameter vector θ in a linear relation

$$y(t) = \theta^T \varphi(t) + v(t),$$

where $\{y(t)\}$ and $\{\varphi(t)\}$ are measurable. This structure contains, among several others, the following common difference equation model of a dynamical system:

$$y(t) + a_1 y(t-1) + \cdots + a_n y(t-n)$$

$$= b_1 u(t-1) + \cdots + b_m u(t-m) + v(t).$$

Take $\theta^T = (a_1 \ldots a_n \ b_1 \ldots b_m)$ and $\varphi^T(t) = (-y(t-1) \ldots -y(t-n) \ u(t-1) \ldots u(t-m))$. The following RLS algorithm is typical for the recursive estimation of θ:

$$\hat{\theta}(t) = \hat{\theta}(t-1) + P(t)\varphi(t)[y(t) - \hat{\theta}^T(t-1)\varphi(t)],$$

$$P(t) = P(t-1) - \frac{P(t-1)\varphi(t)\varphi^T(t)P(t-1)}{1 + \varphi^T(t)P(t-1)\varphi(t)}.$$

We shall give three independent interpretations of this algorithm:
1. Solving for $\hat{\theta}(t)$ gives

$$\hat{\theta}(t) = \left[P^{-1}(0) + \sum_{k=1}^{t} \varphi(k)\varphi^T(k) \right]^{-1} \left[P^{-1}(0)\hat{\theta}(0) + \sum_{k=1}^{t} \varphi(k)y(k) \right].$$

With $P^{-1}(0) = 0$, this is exactly the θ-value that minimizes the least squares criterion

$$V_t(\theta) = \frac{1}{t} \sum_{k=1}^{t} [y(k) - \theta^T\varphi(k)]^2.$$

2. The estimate $\hat{\theta}(t)$ is the Kalman filter state estimate for the state-space model

$$\theta(t+1) = \theta(t),$$

$$y(t) = \varphi^T(t)\theta(t) + v(t).$$

3. The RLS algorithm is a recursive minimization of the criterion

$$\bar{V}(\theta) = E\tfrac{1}{2}[y(t) - \theta^T\varphi(t)]^2.$$

The factor $\varphi(t)[y(t) - \hat{\theta}^T(t-1)\varphi(t)]$ is then an estimate of the gradient of the criterion, while

$$t \cdot P(t) = \left[\frac{1}{t} P^{-1}(0) + \frac{1}{t} \sum_{1}^{t} \varphi(k) \varphi^{T}(k) \right]^{-1}$$

is the inverse of an estimate of the second derivative (the Hessian) of the criterion. The updating of $\hat{\theta}(t)$ is thus taken in the estimated "Newton" direction with a step size that decays as $1/t$.

Based on each of these three interpretations recursive identification methods can also be derived for other model structures. In the next chapter we shall pursue the third approach in order to develop a general framework for recursive identification.

2.8 Bibliography

General Surveys The survey paper by Åström and Eykhoff (1971) contains an overview of approaches to recursive identification and also contains many references. In Saridis (1974), Isermann et al. (1974), Söderström et al. (1974a, 1978), and Dugard and Landau (1980a) comparative studies of different methods are made.

The Least Squares Method The origins of the least squares method can be traced back to Gauss (1809). It has of course been widely applied to many problems. The recursive algorithm (2.19) to calculate the least squares estimate has apparently been found independently by several authors. The original reference seems to be Plackett (1950). An early and thourough treatment of the least squares method applied to dynamic system identification is Åström (1968). Some basic results from this report are also given in Åström and Eykhoff (1971). There exists many papers dealing with different aspects of the least squares method. Let us mention Ljung (1976b) and Moore (1978) for consistency properties, and Peterka (1975) for implementation aspects. The books by Hsia (1977), Mendel (1973), Unbehauen et al. (1974), and Isermann (1974) also contain comprehensive treatments of the recursive least squares method.

The Instrumental Variable (IV) Method Instrumental variables represent a general tool in statistics for "correlating out" interesting features. See, e.g., Kendall and Stuart (1961) for a discussion. The idea has been applied to system-identification problems in several ways, depending on how the variables are chosen. The instrumental variables defined by (2.29)–(2.31) were suggested by Mayne (1967), Wong and Polak (1967), and Young

(1968) (see also Young, 1965). An alternative with a constant filter in (2.30) has been discussed by Finigan and Rowe (1974).

Another viewpoint is to take all the instrumental variables to be delayed inputs. This was used by Wouters (1972). Also, the correlation method developed by Isermann and Baur (1974) is an IV method with this interpretation of instrumental variables. Other instrumental variables have been used by Banon and Aguilar-Martin (1972), Gentil, Sandraz and Foulard (1973) (inputs and delayed outputs), Stoica and Söderström (1979) (delayed inputs and delayed outputs), and Bauer and Unbehauen (1978) (closed-loop applications). A comparative survey with convergence analysis of these different choices is given by Söderström and Stoica (1981). Recursive instrumental variable methods have been extensively used by P. C. Young (1970). Young has also developed "refined" variants, in which the variables are prefiltered through filters that also are estimated (Young, 1976).

The Recursive Maximum Likelihood Method   The method described in section 2.2.3 was derived by Söderström (1973b) based on an idea by Åström. A similar method was independently derived by Fuhrt (1973). Similar algorithms for another model structure have been derived by Hastings-James and Sage (1969) and by Gertler and Bányász (1974). Both these papers consider a model where the noise term is an autoregression rather than a moving average. The resulting algorithm is also called a "recursive generalized least squares algorithm."

The Bayesian Approach and Nonlinear Filtering (Section 2.3)   The Bayesian interpretation of the recursive least squares algorithm in lemma 2.2 has been pointed out and used by many authors. In an early paper, Ho (1963) showed the close relationships between RLS, the Kalman filter, and the stochastic approximation approach. Bohlin (1970) has given a further treatment of the Bayesian aspects of RLS.

Åström and Wittenmark (1971) stressed that the Kalman filter formula indeed gives the conditional mean also in the least squares identification case in connection with adaptive control applications. Peterka (1979, 1981) has pursued the Bayesian framework in identification applications. A Bayesian convergence analysis of the recursive least squares method was given by Sternby (1977).

The use of the extended Kalman filter for recursive identification seems to have been first suggested by Kopp and Orford (1963) and Cox

(1964). It has subsequently been widely used in many different applications. See, e.g., Sage and Wakefield (1972), Nelson and Stear (1976), and Ljung (1979a) for aspects of this method.

Adaptive Filtering (Section 2.3)   A topic that is related to the nonlinear filtering approach to recursive identification is adaptive filtering. The problem addressed is to find the state estimate for a system like (2.67) with known dynamics (matrices F, G, and H) but unknown noise characteristics (matrices R_1 and R_2). The latter information is then extracted from the data in a recursive fashion. See, e.g., Jazwinski (1969), Mehra (1970), Belanger (1974), and Ohap and Stubberud (1976) for a treatment of this problem.

The Stochastic Approximation Approach (Section 2.4)   The statistical background for stochastic approximation was developed, e.g., in Robbins and Monro (1951), Blum (1954), Albert and Gardner (1967), Dvoretzky (1956), and Fabian (1960, 1968). Tsypkin has in a number of papers and books (e.g., Tsypkin, 1971, 1973) suggested the application of these techniques to a variety of estimation, learning, and control problems. The potential function approach described in Aizerman et al. (1970) and used for various learning problems in system theory is also basically a stochastic approximation scheme.

Stochastic approximation methods have also been derived by Sakrison (1967) and Saridis and Stein (1968). All these algorithms are basically stochastic gradient methods for linear regression models like (2.75). The algorithm is thus essentially given by (2.84). The widely used "LMS" algorithm developed by Widrow and Hoff (1960), and applied to a number of problems in adaptive signal processing is also of the same character. It has been recognized, e.g., in Sakrison (1967) and Polyak and Tsypkin (1979), that using the Newton search direction will improve the properties of the algorithm in this application. In Ljung (1981) the stochastic approximation approach is used to derive recursive identification algorithms for problems other than linear regression models. Methods that are based on random search techniques rather than on the gradient have been developed, e.g., by Kiefer and Wolfowitz (1952). See Saridis (1977) for applications of such techniques to recursive identification.

Pseudolinear Regression (PLR) (Section 2.5)   The extended least squares algorithm was independently derived by Panuska (1968) and Young

(1968). Being a natural extension of the least squares procedure, it has been widely used and rediscovered by many people, e.g., Kashyap (1974), Talmon and van den Boom (1973). The idea of a pseudolinear regression can be applied to structures other than ARMAX models. Talmon and van den Boom (1973) have derived a PLR method for a model where the noise term is described as an ARMA process, rather than as an MA process.

Model Reference Approach (Section 2.5)   Most of the literature on model reference systems concerns adaptive control. Landau (1974, 1979) has pointed out its use for the identification problem and also (Landau, 1976) suggested the procedure quoted in example 2.8.

3 Models and Methods: A General Framework

3.1 Introduction

Chapter 2 contained some excerpts from the long list of existing recursive identification methods. We saw two reasons for this variety of methods: The different approaches that can be taken and the different model sets that can be used.

On the other hand, we also noted that the structures of the resulting algorithms are very similar. This basic structure is

$$\hat{\theta}(t) = \hat{\theta}(t-1) + \gamma(t)P(t)\eta(t)[y(t) - \hat{y}(t)]. \tag{3.1}$$

Here $\hat{\theta}(t)$ is the parameter estimate at time t, $\{\gamma(t)\}$ is a sequence of scalars tending to zero or to some small value, and $P(t)$ is a matrix computed from observations up to time t. The variable $y(t)$ is the system output at time t, and $\hat{y}(t)$ is a prediction of this output based on measurements up to time $t-1$. Finally, $\eta(t)$ is a vector (or, if dim $y > 1$, a matrix) constructed from previous observations, and typically related to the gradient of $\hat{y}(t)$ with respect to θ.

Based on this conclusion from chapter 2, it seems natural to seek a unifying framework within which most of the existing methods can be derived. This is our objective in the present chapter. A general framework will allow the user to see the relationships between different methods, and will guide him in his choice of algorithm.

The development of a unified approach to recursive identification consists of three phases:

1. Define the Framework The diversity of model descriptions for stochastic systems is a major reason for the length of the list of methods. It is therefore necessary to work with a general model description that includes all the common models. In section 3.2 we define this general model description and illustrate how it contains the particular models we used in chapter 2.

It is useful to utilize knowledge and experience from off-line identification when deriving recursive methods. Therefore, in section 3.3, we give a brief account of some major aspects of off-line identification.

2. Derive the Algorithm The approach we choose is to minimize the prediction error variance recursively, using ideas from stochastic approximation (section 2.4). In section 3.4 we derive an algorithm for a quadratic criterion. In section 3.5 a general criterion is treated.

The choice of approach is somewhat arbitrary. Modifying an off-line criterion, as in section 2.2.3, would lead to the same general algorithm, and the nonlinear filtering approach of section 2.3 gives virtually the same result (the differences are explained in appendixes 3.B and 3.C).

3. Apply the Algorithm The general algorithm will then be applied to particular model sets:

• A linear regression model (section 3.6)

• A general single-input/single-output model (section 3.7)

• A state-space model (section 3.8).

For each of these models we first apply the algorithm directly, and then consider those alternatives and variants that are possible and/or natural. In this way, we will be able to cover all the methods reviewed in chapter 2 or mentioned in its bibliography.

The general algorithm can consequently be viewed as *one* method, containing options and choices to be made by the user. This viewpoint is stressed in the User's Summary (section 3.9). The rest of the book is then devoted to discussions, analyses, and suggestions regarding these options and choices.

3.2 Systems and Models

In chapter 2 we worked with the data generated by the system and with models of the system in a somewhat informal way. We will now approach the problem of recursive identification in a slightly more formal fashion. We shall begin by giving more specific meanings to the terms *system, data*, and *model*.

3.2.1 The System and the Data

The *system* is the physical object that generates the observed *output signal*. The output at time t is, as before, denoted by $y(t)$, and is in general a p-dimensional column vector. The data acquisition is assumed to be carried out by sampling, so t is a discrete time index.

Remark There is nothing really that prevents the dimension p from depending on t. That would be the case, e.g., when different measurement

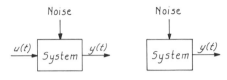

Figure 3.1
Left: A system driven by a known input. *Right:* A system generating the signal *y*.

signals are sampled with different frequencies, which is not uncommon in practice. To keep the notation simple, however, we will not carry a time index on p.

Many systems also have a measurable *input signal*. We denote the input signal at time t by $u(t)$; this is in general an r-dimensional column vector. Where there is an input signal, it can be chosen by the user, in general with the objective of achieving a certain desired behavior of the output signal. This is often accomplished by letting the input be (partly) a feedback from the output. In some identification applications the user may choose the input primarily to obtain a good identification result. We may call such a fortunate situation an "identification experiment" for the system. In most applications of recursive identification, however, the system operates in a "production mode," in which case the identification aspects of the input are fortuitous.

Some systems are not driven by any measurable input. In such a case the output is sometimes called a *time series* or just a *signal*. The identification problem then is to describe the properties of this signal. As we remarked in section 2.1, we prefer, however, to phrase this as describing the "system" that generates the signal. Figure 3.1 illustrates this point of view. In figure 3.1a the system is driven by a measurable input. In figure 3.1b we consider the system to be driven by a signal that is not measurable, but may have known stochastic properties.

In practical applications it may be a substantial part of the identification problem to decide which measurements are worthwhile to collect and regard as outputs, and which signals are to be considered as inputs. It should also be pointed out that the signals $\{u(t)\}$ and $\{y(t)\}$ may very well be obtained from nonlinear transformations of primary measurements. These issues are, however, very problem-dependent, and in the present context we simply regard the sequences

$$\{y(t)\} = y(0), y(1), \ldots, y(t), \ldots$$
$$\{u(t)\} = u(0), u(1), \ldots, u(t), \ldots \tag{3.2}$$

as given data that are to be fed into the appropriate recursive algorithms. For simplicity, we assume in (3.2) that the sampling interval is one time unit. We shall, as before, use the following symbols:

$$z(t) = \begin{pmatrix} y(t) \\ u(t) \end{pmatrix},$$

$$y^t = (y(0) \ \ldots \ y(t)),$$

$$u^t = (u(0) \ \ldots \ u(t)),$$

$$z^t = (z(0) \ \ldots \ z(t)).$$

At present, we do not introduce any restrictions on the data sequences (3.2). We shall in section 4.3 use a fairly weak condition on the probabilistic properties of (3.2) in order to make the analysis possible.

Let us here, however, define an important concept for input sequences u^t. It is fairly obvious that some condition on the input sequence must be introduced in order to secure a reasonable identification result. As a trivial illustration, we may think of an input that is identically zero. Clearly, such an input will not be able to yield full information about the input-output properties of the system. The properties of the input that are required, are, in loose terms, related to the fact that the input should excite all modes of the system. Such an input will be called *persistently exciting*. Other terms that are used are "general enough" or "rich enough." The formal definition is as follows:

Let $\{u(t)\}$ be such that the limits

$$\lim_{N \to \infty} \frac{1}{N} \sum_{t=1}^{N} u(t) u^T(t-j) \triangleq r(j)$$

exist for all $0 \le j \le n$. Form the $n \times n$ block matrix R_n, whose i, k block entry is $r(i - k)$. The sequence $\{u(t)\}$ is then said to be *persistently exciting of order n*, if R_n is nonsingular.

It is perhaps easiest to get an intuitive understanding of this concept in the frequency domain. A signal that is persistently exciting of order n has a spectral density that is nonzero at at least n points (Ljung, 1971).

The requirement of persistent excitation therefore means that the input must have sufficiently rich frequency content.

3.2.2   Models

What is a model of a dynamic and possibly stochastic system? Several different ways of describing input-output relationships have been treated in the literature, and in chapter 2 we used a few of the most common ones (see examples 2.1–2.3). Our objective here is to catch enough of their basic features to allow for the general treatment we are about to develop.

In essence, a model of a dynamical system is a rule which allows us to make some sort of inference based on observations of previous data, about future outputs of the system. A model can thus be said to be a link between the observed past and the unknown future. This idea can be formalized as follows. At time $t - 1$, when the input-output data z^{t-1} has been observed, we may form a "prediction" of the output at time t:

$$g_{\mathcal{M}}(\theta; t, z^{t-1}). \tag{3.3}$$

Here $g_{\mathcal{M}}(\theta; \cdot, \cdot)$ is, for fixed θ, a deterministic function from $\mathbf{R} \times \mathbf{R}^{t(r+p)}$ to $\mathbf{R}^p$. It is parametrized by a finite-dimensional parameter vector θ, belonging to a subset of $\mathbf{R}^d$. We put the word "prediction" within quotes to stress that the function (3.3) need not be derived from formal probabilistic arguments. It merely reflects the user's way of "guessing" the next output, whether it may be a good way or not.

To emphasize the prediction interpretation we shall often use the simpler notation

$$\hat{y}(t \mid \theta) = g_{\mathcal{M}}(\theta; t, z^{t-1}). \tag{3.4}$$

The function $g_{\mathcal{M}}(\theta; t, z^{t-1})$ consequently defines a particular model of the system. This model will be denoted by $\mathcal{M}(\theta)$. The set of models considered will be obtained as θ ranges over a subset $D_{\mathcal{M}}$ of $\mathbf{R}^d$:

$$\mathcal{M} = \{\mathcal{M}(\theta) \mid \theta \in D_{\mathcal{M}}\}. \tag{3.5}$$

We call $\mathcal{M}$ the *model set*.

Remark   In some cases it is meaningful to include in the model assumptions about the error of the prediction (3.4): $y(t) - \hat{y}(t \mid \theta)$. This could be its variance or its distribution. In example 3.7 we shall exhibit a model that includes such assumptions.

The model (3.3) is a predictor given as an explicit function of past data. Most model descriptions are not explicitly given in such a form; see, e.g., those treated in appendix 1.C. Conventional models typically include unmeasurable stochastic disturbance signals, affecting the output. This may often be convenient and may correspond to a certain intuition of how the output signals is generated. When the model is used for control or prediction, though, the unmeasurable signals must be eliminated, and thus a predictor function such as (3.3) must be employed at some point. To illustrate this, let us consider the three model sets we used in chapter 2.

EXAMPLE 3.1 (A Linear Difference Equation)   Consider the model (2.1), (2.3), or (2.5) described in example 2.1:

$$A(q^{-1})y(t) = B(q^{-1})u(t) + v(t), \tag{3.6}$$

or, in other notation,

$$y(t) = \theta^T \varphi(t) + v(t).$$

We noted in that example that, lacking information about the character of the disturbance sequence $\{v(t)\}$, a natural prediction of the output is

$$\hat{y}(t \mid \theta) = \theta^T \varphi(t)$$

$$\begin{aligned} &= -a_1 y(t-1) - a_2 y(t-2) - \cdots - a_n y(t-n) \\ &\quad + b_1 u(t-1) + \cdots + b_m u(t-m) \\ &= g_{\mathscr{M}}(\theta; t, z^{t-1}). \end{aligned} \tag{3.7}$$

In this case the predictor function $g_{\mathscr{M}}(\theta; t, \cdot)$ is a simple linear function of a finite and fixed number of old data $z(k)$. The parameter vector

$$\theta^T = (a_1 \ \ldots \ a_n \ b_1 \ \ldots \ b_m)$$

belongs to $\mathbf{R}^{n+m}$, and the model set that is commonly used for the model (3.6) corresponds to $D_{\mathscr{M}} = \mathbf{R}^{n+m}$. Notice that if we assume $\{v(t)\}$ to be a sequence of independent random variables with zero means, then the function (3.7) is the conditional mean:

$$g_{\mathscr{M}}(\theta; t, z^{t-1}) = E(y(t) \mid z^{t-1}, \theta). \ \square$$

EXAMPLE 3.2 (An ARMAX Model)   Consider now the model (2.8) discussed in example 2.2:

$$A(q^{-1})y(t) = B(q^{-1})u(t) + C(q^{-1})e(t), \qquad (3.8)$$

where, as before $\{e(t)\}$ is a sequence of independent random variables and $A(q^{-1})$, $B(q^{-1})$, and $C(q^{-1})$ are polynomials in the delay operator q^{-1}:

$$A(q^{-1}) = 1 + a_1 q^{-1} + \cdots + a_n q^{-n},$$

$$B(q^{-1}) = b_1 q^{-1} + \cdots + b_m q^{-m},$$

$$C(q^{-1}) = 1 + c_1 q^{-1} + \cdots + c_r q^{-r}.$$

In section 2.2.3 we derived the prediction of the output based on a first-order model (3.8) [see (2.35)]. Here we shall derive the prediction in the general case (Åström, 1970). We introduce the parameter vector

$$\theta^{\mathrm{T}} = (a_1 \ \ldots \ a_n \ b_1 \ \ldots \ b_m \ c_1 \ \ldots \ c_r).$$

From (3.8) we find

$$\frac{A(q^{-1})}{C(q^{-1})}y(t) = \frac{B(q^{-1})}{C(q^{-1})}u(t) + e(t)$$

or

$$y(t) = \left[1 - \frac{A(q^{-1})}{C(q^{-1})}\right]y(t) + \frac{B(q^{-1})}{C(q^{-1})}u(t) + e(t). \qquad (3.9)$$

Since the polynomials $A(q^{-1})$ and $C(q^{-1})$ both start with a 1 we have

$$1 - \frac{A(q^{-1})}{C(q^{-1})} = \sum_{k=1}^{\infty} h_k(\theta)q^{-k}$$

and

$$\frac{B(q^{-1})}{C(q^{-1})} = \sum_{k=1}^{\infty} g_k(\theta)q^{-k}$$

for some θ-dependent sequences $\{h_k(\theta)\}$ and $\{g_k(\theta)\}$. (Recall that the θs are the coefficients of the A-, B-, and C-polynomials.) These sequences will tend to zero exponentially if all the zeros of the polynomial

$$C^*(z) = z^r + c_1 z^{r-1} + \cdots + c_r \qquad (3.10)$$

are inside the unit circle. (The rate of exponential decay of $g_k(\theta)$ and $h_k(\theta)$ is given by the largest absolute value of the zeros of (3.10)). Therefore it is important to restrict θ to the set

$D_s = \{\theta \mid \text{All zeros of } C^*(z) \text{ are inside the unit circle}\}$ (3.11)

to make the above expressions meaningful. We shall therefore in the following assume that $D_{\mathcal{M}} \subset D_s$.

Returning to (3.9), we see that the right-hand side is known at time $t - 1$, except for the term $e(t)$ that is independent of everything that has happened up to time $t - 1$. Therefore, the natural predictor is

$$g_{\mathcal{M}}(\theta; t, z^{t-1}) = \left[1 - \frac{A(q^{-1})}{C(q^{-1})}\right] y(t) + \frac{B(q^{-1})}{C(q^{-1})} u(t)$$

(3.12)

$$= \sum_{k=1}^{\infty} h_k(\theta) y(t - k) + \sum_{k=1}^{\infty} g_k(\theta) u(t - k).$$

Notice that the predictor (3.12) assumes that all data from $t = -\infty$ is known. We generally assume that the data collection is initialized at time 0. This problem is usually resolved in practice by summing only up to $k = t$ in (3.12). Since h_k and g_k decay exponentially, this approximation is reasonable, except possibly for small t.

In practice the explicit expression given by the right-hand side of (3.12) is a bit awkward to work with. Instead, the calculations are organized in the following way. We have

$$\hat{y}(t \mid \theta) = \left[1 - \frac{A(q^{-1})}{C(q^{-1})}\right] y(t) + \frac{B(q^{-1})}{C(q^{-1})} u(t),$$

which gives

$$C(q^{-1})\hat{y}(t \mid \theta) = [C(q^{-1}) - A(q^{-1})] y(t) + B(q^{-1}) u(t).$$ (3.13)

This is a convenient finite difference equation for calculating $\hat{y}(t \mid \theta)$. Often the initial conditions $\hat{y}(t \mid \theta)$, $t = -r + 1, \ldots, 0$ are taken as zero, corresponding to an assumption that $y(t) = u(t) = 0$ for $t < 0$.

Notice that the recursion (3.13) is exponentially stable when $\theta \in D_s$. Constraining θ to this set is no restriction in practice, since, according to the spectral factorization theorem, we can always mirror the zeros of $C(q^{-1})$ inside the unit circle without affecting the second-order properties of the disturbance term (see appendix 1.C). □

EXAMPLE 3.3 (A State-Space Model) The prediction of the output for the state-space model (2.9)

$$x(t + 1) = F(\theta)x(t) + G(\theta)u(t) + w(t),$$
$$y(t) = H(\theta)x(t) + e(t),$$

(3.14)

is obtained by use of the Kalman filter:

$$\hat{x}_\theta(t + 1) = F(\theta)\hat{x}_\theta(t) + G(\theta)u(t) + K_t(\theta)[y(t) - H(\theta)\hat{x}_\theta(t)]$$
$$\hat{y}(t \mid \theta) = H(\theta)\hat{x}_\theta(t),$$

(3.15)

where the matrix $K_t(\theta)$ is computed via a Riccati equation as in (2.66). Notice that in (3.15) $\hat{y}(t \mid \theta)$ is formed by passing $\{y(t)\}$ and $\{u(t)\}$ through a (time-varying) linear filter. Solving this gives

$$\hat{y}(t \mid \theta) = g_{\mathcal{M}}(\theta; t, z^{t-1})$$
$$= \sum_{i=1}^{t-1} [h_i(t, \theta)y(t - i) + g_i(t, \theta)u(t - i)] + r(t, \theta)\hat{x}_\theta(0),$$

(3.16)

where h_i, g_i, and r are determined from the matrices F, G, H, and K in a straightforward manner. This shows how the state-space model (3.14) is contained in the general model description (3.3).

Notice that what matters in the end is the representation (3.15) or (3.16). The assumptions connected with the model (3.14) regarding independent noise sequences, covariance, possible normality, initial estimate information, etc., serve only as a vehicle to arrive at (3.16). These assumptions play a role only to the extent they influence this expression. $\square$

Very general model sets can be described by (3.3). We shall, however, throughout this book confine our explicit treatment to *linear predictor models*. By this is meant that the prediction (3.4) can be obtained by filtering the input-output data through a finite-dimensional linear filter:

$$\varphi(t + 1, \theta) = \mathcal{F}(\theta)\varphi(t, \theta) + \mathcal{G}(\theta)z(t),$$
$$\hat{y}(t \mid \theta) = \mathcal{H}(\theta)\varphi(t, \theta),$$

(3.17)

where $\mathcal{F}$, $\mathcal{G}$, and $\mathcal{H}$ are matrix functions of θ.

Linear models are of course the most common ones in practice. Extensions to nonlinear models are, however, also possible:

• Perhaps the most common nonlinearity in practice consists of non-linear transformations of the primary signals before they enter the linear system (as in the Hammerstein model). Obviously, such nonlinearities are contained in the framework (3.17), simply by interpreting $\{z(t)\}$ as the transformed and possibly constructed data sequence.

• The algorithms we are about to develop can with obvious interpretation also be applied to nonlinear predictor dynamics

$$\varphi(t + 1, \theta) = f(\theta; t, \varphi(t, \theta), z(t)),$$

$$\hat{y}(t \mid \theta) = h(\theta; t, \varphi(t, \theta)).$$

However, in this book we will not address this situation explicitly.

The predictor filter (3.17) has been given in a state-space form. The underlying user model, however, need not be thought of as or parametrized in terms of a state-space model. We have used the notation φ rather than x for the state vector to stress this fact. The representation (3.17) will, however, prove useful for a formal description of the algorithm as well as for the analysis.

Remark on Time-Varying Predictors The predictor (3.17) is a time-invariant filter. This no doubt corresponds to the most common case in practice. However, sometimes predictors are used that are time-varying filters. The Kalman filter (3.15) for the state-space model (3.14) is the by far most important example of this. However, in order not to make the notation and discussion too complex, we defer the treatment of time-varying (Kalman) predictors to section 3.8.3.

Remark on More General Predictors We have discussed the model of the system in terms of the one-step ahead predictor, i.e., a prediction of $y(t)$, based on y^{t-1} and u^{t-1}. This shows up in the general model (3.17) as a delay between $z(t)$ and $\hat{y}(t \mid \theta)$. We could, however, have considered a more general delay structure. Let

$$\hat{y}(t \mid \theta; t - k_1, t - k_2) \tag{3.18a}$$

denote a prediction of $y(t)$ based on

$$y^{t-k_1} \text{ and } u^{t-k_2}, \text{ where } k_1 \geq 1 \text{ and } k_2 \text{ is an arbitrary integer.} \tag{3.18b}$$

When $k_1 = k_2 = k > 0$, this is a k-step-ahead prediction. When $k_1 = 1$ and $k_2 = 0$ it is a one-step-ahead prediction, where a direct feedthrough of $u(t)$ is allowed. (This would, e.g., result from the ARMAX model (3.8) if a term $b_0 u(t)$ was present in the right-hand side.) When the prediction is based only on u^{t-k_2} and $k_2 < 0$, then $\hat{y}(t)$ is actually a "smoothed estimate," which depends on future inputs. This means that the decision about the "prediction" $\hat{y}(t)$ has to be delayed until time $t + k_2$. For several signal processing applications, such a delay is quite acceptable, as we shall see in section 7.5.

Now, the general delay structure (3.18) can easily be incorporated into the model (3.17). The only difference is that $z(t)$ has to be interpreted as

$$z(t) = \begin{pmatrix} y(t - k_1 + 1) \\ u(t - k_2 + 1) \end{pmatrix}. \tag{3.19}$$

The derivation and analysis of our general algorithms will still be valid under (3.19). For a given model, such as the ARMAX one, the actual values of the matrices in the model (3.17) with (3.19) will of course depend on k_1 and k_2. In section 7.4 we give some such expressions for the k-step-ahead predictor for ARMA models.

Obviously, a crucial aspect of the model (3.17) in an identification context is how the prediction $\hat{y}(t \mid \theta)$ depends on θ. To formalize this, we shall assume that the model is differentiable with respect to θ, and as in chapter 2 we shall denote the derivative by

$$\left[\frac{d}{d\theta} \hat{y}(t \mid \theta) \right]^{\mathrm{T}} = \psi(t, \theta) \qquad \text{(a } d \times p\text{-matrix)}. \tag{3.20}$$

For the linear model (3.17), $\psi(t, \theta)$ can be formed from z^t by a finite-dimensional filter by introducing

$$\zeta(t, \theta) = (\varphi^{(1)}(t, \theta) \ \ldots \ \varphi^{(d)}(t, \theta)) \qquad \text{(an } n \times d\text{-matrix)}.$$

where the superscript (i) denotes differentiation with respect to θ_i, the ith component of θ. Since $\mathscr{F}(\theta)$ is a matrix and θ is a vector, the derivative of $\mathscr{F}(\theta)$ with respect to θ will be a quantity with three indices (a tensor). To avoid such complex notation, we introduce the following matrices:

$$\frac{\partial}{\partial\theta}[\mathscr{F}(\theta)\varphi + \mathscr{G}(\theta)z] = M(\theta, \varphi, z) \qquad \text{(an } n \times d\text{-matrix)}, \tag{3.21a}$$

$$= \frac{\partial \mathscr{F}}{\partial \theta} \varphi + \frac{\partial \mathscr{G}}{\partial \theta} z$$

$$\frac{\partial}{\partial\theta}[\mathscr{H}(\theta)\varphi] = D(\theta, \varphi) \qquad \text{(a } p \times d\text{-matrix)}. \tag{3.21b}$$

In other words, (3.21a) means that $M(\theta, \varphi, z)$ is a matrix whose ith column is

$$\left[\frac{\partial}{\partial \theta_i}\mathscr{F}(\theta)\right]\varphi + \left[\frac{\partial}{\partial \theta_i}\mathscr{G}(\theta)\right]z,$$

and analogously for $D(\theta, \varphi)$. Then

$$\frac{\partial g}{\partial \theta_i}\varphi + \mathscr{F}\frac{\partial \varphi}{\partial \theta_i} + \frac{\partial g}{\partial \theta_i}z \quad \textit{differentiate}$$

$$\textit{obtained by} \quad 3.17$$

$$\zeta(t + 1, \theta) = \mathscr{F}(\theta)\zeta(t, \theta) + M(\theta, \varphi(t, \theta), z(t)), \tag{3.22a}$$

$$\psi^{\mathrm{T}}(t, \theta) = \mathscr{H}(\theta)\zeta(t, \theta) + D(\theta, \varphi(t, \theta)). \tag{3.22b}$$

For the theoretical considerations to come, it is more convenient to collect (3.17) and (3.22) into one filter with $z(t)$ as input and $\hat{y}(t \mid \theta)$ and $\psi(t, \theta)$ as output. In order to cast the equations into the conventional form, with a state vector and an output vector, we introduce the state vector

$$\xi(t, \theta) = \begin{pmatrix} \varphi(t, \theta) \\ \mathrm{col}\,\zeta(t, \theta) \end{pmatrix}.$$

The notation $\mathrm{col}\,A$ here means a column vector constructed from the matrix A by stacking its columns under each other. The construction is made for purely formal reasons. Similarily, we introduce the output vector

$$\begin{pmatrix} \hat{y}(t \mid \theta) \\ \mathrm{col}\,\psi(t, \theta) \end{pmatrix}.$$

We can now rewrite (3.17) and (3.22) for some matrices $A(\theta)$, $B(\theta)$, and $C(\theta)$ as

$$\xi(t + 1, \theta) = A(\theta)\xi(t, \theta) + B(\theta)z(t),$$

$$\begin{pmatrix} \hat{y}(t \mid \theta) \\ \mathrm{col}\,\psi(t, \theta) \end{pmatrix} = C(\theta)\xi(t, \theta). \tag{3.23}$$

Comparing (3.23) with (3.22) and (3.17), we see that the $(d + 1)n \times (d + 1)n$-matrix $A(\theta)$ will contain the matrix $\mathscr{F}(\theta)$ in each of its $d + 1$ block-diagonal entries, and has all zeros above the block diagonal. Hence $A(\theta)$ has the same eigenvalues as $\mathscr{F}(\theta)$ (but with higher multiplicities). The stability properties of $A(\theta)$ therefore coincide with those of $\mathscr{F}(\theta)$. Since stability of the predictor will be important in our subsequent discussion, let us introduce the set

$$D_s = \{\theta \mid \mathscr{F}(\theta) \text{ has all eigenvalues strictly inside the unit circle}\}. \tag{3.24}$$

The matrices $A(\theta)$, $B(\theta)$, and $C(\theta)$, and often also $\mathscr{F}(\theta)$, $\mathscr{G}(\theta)$, and $\mathscr{H}(\theta)$, are quite sparse and (3.23) is in most cases inefficient for actual computation of $\hat{y}(t\mid\theta)$ and $\psi(t,\theta)$. The representation (3.23) will nonetheless play an important role when developing the algorithms. It will be used to give a formal description of how to propagate predictions and their gradients in the recursive algorithms to be derived later in this chapter. Therefore it may be useful to illustrate how the models that were discussed previously fit into this framework.

EXAMPLE 3.4 (A Difference Equation Model) For the difference equation (3.6) we may take

$$\varphi^{\mathrm{T}}(t,\theta) = (-y(t-1) \; \ldots \; -y(t-n) \; u(t-1) \; \ldots \; u(t-m))$$

(in fact independent of θ). The matrices in (3.17) are then

$$\mathscr{F}(\theta) = \left(\begin{array}{ccccc|cccc}
0 & 0 & \cdots & 0 & 0 & & & & \\
1 & 0 & \cdots & & 0 & & & & \\
0 & 1 & & & 0 & & 0 & & \\
& \vdots & & & \vdots & & & & \\
0 & & & 1 & 0 & & & & \\
\hline
& & & & & 0 & 0 & \cdots & 0 \\
& & 0 & & & 1 & 0 & \cdots & 0 \\
& & & & & & \vdots & & \\
& & & & & 0 & & & 1 \quad 0
\end{array}\right) \begin{array}{l} \\ \leftarrow \text{row } n+1, \end{array}$$

$$\mathscr{G}(\theta) = \left(\begin{array}{cc}
-1 & 0 \\
0 & 0 \\
\vdots & \vdots \\
0 & 0 \\
0 & 1 \\
0 & 0 \\
\vdots & \vdots \\
0 & 0
\end{array}\right) \leftarrow \text{row } n+1,$$

$$z(t) = \begin{pmatrix} y(t) \\ u(t) \end{pmatrix},$$

$$\mathscr{H}(\theta) = \theta^{\mathrm{T}},$$

and (3.17) will give the prediction. For the gradient, we find that

$$\varphi^{(i)}(t, \theta) = 0, \quad \psi(t, \theta) = \varphi(t, \theta).$$

The matrix $\mathscr{F}(\theta)$ is lower triangular with zeros along its diagonal. Its eigenvalues are therefore all zero, and the set D_s, defined by (3.24), is equal to $\mathbf{R}^d$. □

EXAMPLE 3.5 (An ARMAX Model) Consider the ARMAX model of example 3.2. Let for simplicity $n = m = r = 2$. We introduce the prediction error

$$\varepsilon(t, \theta) = y(t) - \hat{y}(t \mid \theta).$$

Then from (3.13) we have

$$C(q^{-1})\varepsilon(t, \theta) = A(q^{-1})y(t) - B(q^{-1})u(t),$$

which also can be written

$$\varepsilon(t, \theta) = y(t) - \theta^{\mathrm{T}}\varphi(t, \theta) \tag{3.25}$$

with θ^{T} defined as in example 3.2 and with

$$\varphi^{\mathrm{T}}(t, \theta) = (-y(t - 1) \ -y(t - 2) \ u(t - 1) \ u(t - 2)$$
$$\varepsilon(t - 1, \theta) \ \varepsilon(t - 2, \theta)). \tag{3.26}$$

This vector $\varphi(t, \theta)$ can be chosen as the state vector. It obeys (3.17) with

$$\mathscr{F}(\theta) = \begin{pmatrix} 0 & 0 & 0 & 0 & 0 & 0 \\ 1 & 0 & 0 & 0 & 0 & 0 \\ 0 & 0 & 0 & 0 & 0 & 0 \\ 0 & 0 & 1 & 0 & 0 & 0 \\ -a_1 & -a_2 & -b_1 & -b_2 & -c_1 & -c_2 \\ 0 & 0 & 0 & 0 & 1 & 0 \end{pmatrix}, \quad \mathscr{G}(\theta) = \begin{pmatrix} -1 & 0 \\ 0 & 0 \\ 0 & 1 \\ 0 & 0 \\ 1 & 0 \\ 0 & 0 \end{pmatrix},$$

$$\mathscr{H}(\theta) = \theta^{\mathrm{T}}.$$

Examination of the matrix $\mathscr{F}(\theta)$ shows that its eigenvalues are 0 (multiplicity $n + m = 4$) and the zeros of the polynomial $C^*(z)$, given by (3.10). The set (3.24) therefore coincides with (3.11) in this case. □

3.2.3 Summary

In this section we have discussed the input-output sequence $\{z(t)\}$ $(z^T(t) = (y^T(t) \ u^T(t))$ produced by a system. We have also defined a model $\mathcal{M}(\theta)$ in general terms as a predictor of the next system output, and illustrated some consequences of this point of view. The model set $\mathcal{M}$ is obtained as the parameter θ ranges over a set $D_{\mathcal{M}}$. We have confined ourselves to predictors that can be expressed in the form (3.17). Notice, however, that *no* corresponding assumption has been introduced for the system. The discussion on model sets has here been limited to formal aspects. In sections 5.2 and 5.3 user aspects on the choice of model set will be treated.

3.3 Some Aspects of Off-Line Identification

We defined off-line identification in chapter 1 as the determination of a model of a system using a batch of measured data, where the whole batch is available at all stages of the procedure. Off-line identification is a subject that cannot be covered completely in a section of a book. See the survey paper by Åström and Eykhoff (1971), the books by Eykhoff (1974), Goodwin and Payne (1977), or Kashyap and Rao (1976) for comprehensive treatments. The objective of this section is merely to point out some aspects that have useful implications for recursive identification.

3.3.1 Identification as Criterion Minimization

Given a prediction model as in the previous section,

$$\hat{y}(t \mid \theta) = g_{\mathcal{M}}(\theta; t, z^{t-1}), \tag{3.27}$$

a natural measure of its validity is the *prediction error*,

$$\varepsilon(t, \theta) = y(t) - \hat{y}(t \mid \theta). \tag{3.28}$$

This error can be evaluated using the data up to time t. Since $\varepsilon(t, \theta)$ is a p-dimensional vector, it is useful to introduce a scalar measure

$$l(t, \theta, \varepsilon(t, \theta)) \tag{3.29}$$

of "how large" the prediction error is. Here $l(\cdot, \cdot, \cdot)$ is a function from $\mathbf{R} \times \mathbf{R}^d \times \mathbf{R}^p$ to $\mathbf{R}$.

After having recorded data up to time N a natural criterion of the validity of the model $\mathcal{M}(\theta)$ is

$$V_N(\theta, z^N) = \frac{1}{N} \sum_{t=1}^{N} l(t, \theta, \varepsilon(t, \theta)). \tag{3.30}$$

It should be stressed that in all the expressions (3.27)–(3.30), the vector θ is a constant parameter. The function (3.30) is thus a well-defined scalar-valued function of θ, once z^N has been recorded and is known.

The off-line estimate, denoted by $\hat{\theta}_N$, is then obtained by minimization of $V_N(\theta, z^N)$ over $\theta \in D_{\mathcal{M}}$.

A number of common off-line identification methods correspond to minimization of (3.30) for different choices of criterion functions and model sets. Such methods are called prediction error identification methods in Ljung (1976a, 1978c). Here we shall only briefly discuss two aspects of (3.30), namely, the choice of criterion functions (3.29) and the asymptotic statistical properties of the estimate $\hat{\theta}_N$.

3.3.2 Choice of Criterion Function

Two ways of selecting the criterion function will be considered in the following two examples. Further user aspects of this choice will be discussed in section 5.5.

EXAMPLE 3.6 (A Quadratic Criterion)   A natural way of measuring the "size" of the prediction error ε is to use a quadratic norm:

$$l(t, \theta, \varepsilon) = \tfrac{1}{2}\varepsilon^T \Lambda^{-1} \varepsilon, \tag{3.31}$$

where Λ is a positive definite matrix. A possible disadvantage of this criterion function is that it gives a substantial penalty for large errors. This means that the criterion might be sensitive to occasional large measurement errors in $y(t)$. (See section 5.5 for some possible remedies for this problem.) □

EXAMPLE 3.7 (The Maximum Likelihood Criterion)   The likelihood function for an estimation problem is defined in appendix 1.B.4. To derive the likelihood function for our problem we make the additional model assumption

$\{\varepsilon(t, \theta)\}$ is a sequence of independent random vectors, such
that the probability density function of $\varepsilon(t, \theta)$ is $\bar{f}(t, \theta\ x)$,   (3.32)

i.e., such that $P(\varepsilon(t, \theta) \in B) = \displaystyle\int_B \bar{f}(t, \theta, x)\, dx.$

Remark   The assumption (3.32) implies that the model $\mathcal{M}(\theta)$ would give a correct description of the system. Notice, however, that this is a *model assumption* made by the user in order to compute certain quantities. It does not imply that the actual data really is generated according to (3.32). (See also the comment on the notation θ, θ_0 at the end of section 2.1.)

To find the likelihood function we proceed by writing the output at time t as

$$y(t) = g_{\mathcal{M}}(\theta; t, z^{t-1}) + \varepsilon(t, \theta).$$

Hence, under the assumption (3.32), the conditional probability density function of $y(t)$, given z^{t-1}, is

$$f(\theta; x_t \mid z^{t-1}) = P(y(t) = x_t \mid z^{t-1}, \theta) = \bar{f}(t, \theta, x_t - g_{\mathcal{M}}(\theta; t, z^{t-1})).$$

Using Bayes's rule, the joint probability density function of $y(t)$ and $y(t-1)$ given z^{t-2} can be expressed as

$$
\begin{aligned}
f(\theta; x_t, x_{t-1} \mid z^{t-2}) &= P(y(t) = x_t, y(t-1) = x_{t-1} \mid z^{t-2}, \theta) \\
&= P(y(t) = x_t \mid y(t-1) = x_{t-1}, z^{t-2}, \theta) \cdot P(y(t-1) = x_{t-1} \mid z^{t-2}, \theta) \\
&= P(y(t) = x_t \mid z^{t-1}, \theta) \cdot P(y(t-1) = x_{t-1} \mid z^{t-2}, \theta) \\
&= \bar{f}(t, \theta, x_t - g_{\mathcal{M}}(\theta; t, z^{t-1})) \cdot \bar{f}(t-1, \theta, x_{t-1} - g_{\mathcal{M}}(\theta; t-1, z^{t-2})).
\end{aligned}
$$

Here the variable $y(t-1)$ that is implicit in z^{t-1} in the first factor should be replaced by x_{t-1}, since the expression in the second step is conditioned with respect to the event that $y(t-1) = x_{t-1}$. In these calculations we assume u^t to be a given deterministic sequence. (Compare these to similar calculations in the proof of lemma 2.2.) Iterating the foregoing expression from $t = N$ to $t = 1$ gives the joint probability density function of $y(N)$, $y(N-1)$, ..., $y(1)$: $f(\theta; x_N, x_{N-1}, \ldots, x_1)$. By replacing the dummy variables x_i with the corresponding observations $y(i)$, we obtain the likelihood function [see (1.B.18)]. The logarithm of the likelihood function is thus

$$\log(\theta; y(N), y(N-1), \ldots, y(1))$$

$$= \sum_{t=1}^{N} \log \bar{f}(t, \theta, y(t) - g_{\mathcal{M}}(\theta; t, z^{t-1})) \tag{3.33}$$

$$= \sum_{t=1}^{N} \log \bar{f}(t, \theta, \varepsilon(t, \theta)).$$

We have consequently shown that the negative log likelihood function for our estimation problem, given the assumption (3.32), is given by (3.30) (multiplied by N), with

$$l(t, \theta, \varepsilon) = -\log \bar{f}(t, \theta, \varepsilon). \tag{3.34}$$

For this choice of l, $\hat{\theta}_N$ equals the maximum likelihood estimate (MLE).

Remark on the Relationship of the MLE and MAP   The MLE is closely related to the Bayesian maximum a posteriori estimate (MAP) (section 2.3). Using Bayes's rule (2.59) we find that the posterior probability density function for θ is given by

$$P(\theta \mid y^N) = \frac{P(y^N \mid \theta) \cdot P(\theta)}{P(y^N)},$$

where $P(y^N \mid \theta)$ is the likelihood function and $P(\theta)$ is the prior distribution of θ, while $P(y^N)$ is independent of θ. The θ-value that maximizes $P(\theta \mid y^N)$, i.e., the MAP estimate, thus differs from the value that maximizes $p(y^N \mid \theta)$, i.e., the MLE, only via the prior distribution $P(\theta)$. This has an insignificant influence in the case when little prior knowledge of θ is available or when N is large.

Let us now specialize to Gaussian prediction errors. If we assume that $\varepsilon(t, \theta)$ has a Gaussian distribution with zero mean and covariance matrix $\Lambda_t(\theta)$, then

$$l(t, \theta, \varepsilon) = -\log f(t, \theta, \varepsilon)$$
$$= \frac{p}{2}\log 2\pi + \frac{1}{2}\log \det \Lambda_t(\theta) + \frac{1}{2}\varepsilon^T \Lambda_t^{-1}(\theta)\varepsilon. \tag{3.35}$$

If the covariance matrix Λ_t is supposed to be known (independent of θ) then the first two terms of (3.35) do not affect the minimization, and we have obtained a quadratic criterion as in example 3.6. □

3.3.3 Asymptotic Properties of the Off-Line Estimate

We now turn to the asymptotic statistical properties of the estimate $\hat{\theta}_N$. We shall without proofs quote some results from Ljung (1978c) and Ljung and Caines (1979). Suppose that the limit

$$\bar{V}(\theta) = \lim_{N\to\infty} EV_N(\theta, z^N) \tag{3.36}$$

exists, where E is the expectation operator with respect to z^N. The function $\bar{V}(\theta)$ thus is the expected value of the criterion function corresponding to certain fixed values of the model parameters θ. Then, under weak regularity conditions,

$\hat{\theta}_N$ converges w.p. 1* to a minimum of $\bar{V}(\theta)$
as N tends to infinity.   (3.37)

This means that the estimate $\hat{\theta}_N$ converges to the model that gives the best description of the data, where "best" is measured in terms of the criterion (3.36). Notice that (3.37) is true, whether or not the model set $\mathcal{M}$ is capable of a true description of the data. Moreover, if $\hat{\theta}_N$ converges to θ^*, such that the matrix $d^2\bar{V}(\theta^*)/d\theta^2$ is invertible, then

$$\sqrt{N}(\hat{\theta}_N - \theta^*) \in \text{AsN}(0, P),   \tag{3.38}$$

which means that the random variable $\sqrt{N}(\hat{\theta}_N - \theta^*)$ converges in distribution to the normal distribution with zero mean and covariance matrix P [see (1.A.7)]. Here

$$P = [\bar{V}''(\theta^*)]^{-1}\left\{\lim_{N\to\infty} N\cdot\text{E}[V'_N(\theta^*, z^N)]^T V'_N(\theta^*, z^N)\right\}[\bar{V}''(\theta^*)]^{-1},   \tag{3.39}$$

where $'$ and $''$ denotes differentiation once and twice, respectively, with respect to θ. Here V' is a row vector.

3.3.4   Relation to the Cramér-Rao Bound

Suppose there is a value θ_0 in the model set such that (3.32) with $\theta = \theta_0$ indeed gives a correct description of the true data. Then it can be shown that $\theta^* = \theta_0$ and that the matrix P in (3.39) equals the Cramér-Rao lower bound (see section 1.B.3) provided that l is chosen as in (3.34). This means that the estimate $\hat{\theta}_N$ has asymptotically the smallest possible covariance matrix for an unbiased estimate and we say that the estimate is *asymptotically efficient*.

Let us demonstrate this result in a simple special case. Suppose that there exists a value θ_0 such that the prediction errors $\varepsilon(t, \theta_0)$ are independent and Gaussian with zero mean values and *known* covariance matrices Λ_0. The logarithm of the likelihood function is then given by [see (3.35)]

---

*For a definition of w.p.1 (with probability one) see (1.A.6).

$$W_N(\theta, z^N) = -\frac{pN}{2}\log 2\pi - \frac{N}{2}\log \det \Lambda_0 - \sum_{t=1}^{N}\frac{1}{2}\varepsilon^{\mathrm{T}}(t, \theta)\Lambda_0^{-1}\varepsilon(t, \theta).$$

The Fisher information matrix [see (1.B.15)] is thus

$$M_N = \mathrm{E}[W_N'(\theta_0, z^N)]^{\mathrm{T}}[W_N'(\theta_0, z^N)]$$

$$= \frac{1}{N}\sum_{s,t=1}^{N} \mathrm{E}\psi(t, \theta_0)\Lambda_0^{-1}\varepsilon(t, \theta_0)\varepsilon^{\mathrm{T}}(s, \theta_0)\Lambda_0^{-1}\psi^{\mathrm{T}}(s, \theta_0) \qquad (3.40)$$

$$= \frac{1}{N}\sum_{t=1}^{N} \mathrm{E}\psi(t, \theta_0)\Lambda_0^{-1}\psi^{\mathrm{T}}(t, \theta_0),$$

where

$$\psi(t, \theta) = -\left[\frac{d}{d\theta}\varepsilon(t, \theta)\right]^{\mathrm{T}} \qquad (\text{a } d \times p\text{-matrix}),$$

as in (3.20). In (3.40) the third equality follows from the facts that $\varepsilon(t, \theta_0)$ and $\varepsilon(s, \theta_0)$ are independent if $s \neq t$ and that $\varepsilon(t, \theta_0)$ and $\psi(t, \theta_0)$ are independent. The Cramér-Rao inequality then tells us that

$$\mathrm{cov}\{\sqrt{N}(\hat{\theta}_N - \theta_0)\} \geq \left(\frac{1}{N}M_N\right)^{-1} \qquad (3.41)$$

for any unbiased estimator $\hat{\theta}_N$.

Suppose now that we choose $\hat{\theta}_N$ as the estimate given by minimization of (3.30) with

$$l(t, \theta, \varepsilon) = \tfrac{1}{2}\varepsilon^{\mathrm{T}}\Lambda_0^{-1}\varepsilon.$$

We can then evaluate P in (3.39) under the foregoing assumption about $\varepsilon(t, \theta_0)$. The expression within curly brackets in (3.39) is M_N/N according to (3.40). Moreover,

$$\bar{V}''(\theta_0) = \lim_{N\to\infty}\frac{1}{N}M_N; \qquad (3.42)$$

hence

$$P = \left[\lim_{N\to\infty}\frac{1}{N}M_N\right]^{-1}, \qquad (3.43)$$

which shows that P equals the limit of the Cramér-Rao bound (3.41) as N tends to infinity. The estimate $\hat{\theta}_N$ is therefore asymptotically efficient.

3.3.5 Optimal Choice of Weights in Quadratic Criteria

Suppose now that we have chosen a quadratic criterion (3.31):

$$l(t, \theta, \varepsilon) = \tfrac{1}{2}\varepsilon^{\mathrm{T}}\Lambda^{-1}\varepsilon. \tag{3.44a}$$

We shall study what the optimal choice of the weighting matrix Λ^{-1} is. To do that we make the assumption

$\theta^* = \theta_0$ where $\{\varepsilon(t, \theta_0)\}$ is a sequence of independent random vectors with zero means and covariance matrices Λ_0. $\qquad$ (3.44b)

Notice that we do not assume anything about the distribution of $\varepsilon(t, \theta_0)$. With (3.44) the asymptotic covariance matrix (3.39) can be evaluated. With calculations similar to those in the preceding subsection, we find that

$$P = [\mathrm{E}\psi(t, \theta_0)\Lambda^{-1}\psi^{\mathrm{T}}(t, \theta_0)]^{-1}\mathrm{E}\psi(t, \theta_0)\Lambda^{-1}\Lambda_0\Lambda^{-1}\psi^{\mathrm{T}}(t, \theta_0)$$
$$\times [\mathrm{E}\psi(t, \theta_0)\Lambda^{-1}\psi^{\mathrm{T}}(t, \theta_0)]^{-1}, \tag{3.45}$$

where $\psi(t, \theta)$ is the gradient of the predictor as in (3.20).

If we view (3.45) as a function of Λ, it can be established that the minimal value of P is obtained for $\Lambda = \Lambda_0$ (see, e.g., Caines and Ljung, 1976). Then we also have the "best" estimates $\hat{\theta}_N$, in the sense that they are closest to θ_0 according to (3.45). The best choice of quadratic norm in (3.44a) is therefore $\Lambda = \Lambda_0 =$ the true prediction error covariance matrix. This result is in accordance with what we found in chapter 2: The natural choice of weighting factors α_t in the least squares criterion (2.12) is the inverse of the measurement error variance, as in lemma 2.2.

3.3.6 Summary

In this section, we have shown how a class of off-line identification algorithms can be defined in terms of the prediction errors of the models in the model set. We have seen how, e.g., the maximum likelihood method is contained in this class by (3.33) and (3.34). We have also stated some results about the asymptotic properties of the off-line estimates. One objective of subsequent sections in this chapter will be to develop recursive algorithms that produce estimates with the same asymptotic properties.

3.4 A Recursive Gauss-Newton Algorithm for Quadratic Criteria

In this section we shall derive a recursive algorithm for the estimation of those model parameters that minimize a prediction error criterion such as the one discussed in previous section, for general model sets. The development will lead to the basic general algorithm (3.67) that is the main object of study in this book.

In the spirit of the off-line criterion (3.30), we would like to select θ so that the criterion

$$E l(t, \theta, \varepsilon(t, \theta)) \tag{3.46}$$

(with expectation over z^t for fixed values of the model parameter θ) is minimized. Criteria such as (3.46) can be minimized recursively from observations, using the stochastic approximation approach, as explained in section 2.4. We recall from that section that a criterion (2.85)

$$V(x) = E H(x, e(t))$$

can be recursively minimized by the stochastic Newton method (2.90):

$$\hat{x}(t) = \hat{x}(t-1) + \gamma(t) [\bar{V}''(\hat{x}(t-1), e^t)]^{-1} Q(\hat{x}(t-1), e(t)), \tag{3.47}$$

where $-Q(x, e)$ (a column vector) is the gradient of $H(x, e)$ with respect to x, and $\bar{V}''(x, e^t)$ is some approximation of the second derivative of $V(x)$, based on observations up to time t.

In order to be in formal agreement with the development of section 2.4, we shall perform the derivation under the assumption that the expectation (3.46) does not depend on t. This need not be the case in general, and it is in fact sufficient to assume that the limit

$$\bar{E} l(t, \theta, \varepsilon(t, \theta)) \triangleq \lim_{N \to \infty} \frac{1}{N} \sum_{t=1}^{N} E l(t, \theta, \varepsilon(t, \theta)) \triangleq \bar{V}(\theta) \tag{3.48}$$

exists (see chapter 4).

3.4.1 A General Minimization Algorithm for Quadratic Criteria

In this section we shall apply the foregoing idea to the minimization of (3.46) in the special case when l is quadratic in ε:

$$V(\theta) = E l(t, \theta, \varepsilon(t, \theta)) = E \tfrac{1}{2} \varepsilon^T(t, \theta) \Lambda^{-1} \varepsilon(t, \theta). \tag{3.49}$$

Here we have

$$\left[\frac{d}{d\theta}l(t, \theta, \varepsilon(t, \theta))\right]^{\mathrm{T}} = -\psi(t, \theta)\Lambda^{-1}\varepsilon(t, \theta) \quad \text{(a } d \times 1 \text{ column vector),}$$

(3.50)

since, according to (3.20),

$$\frac{d}{d\theta}\varepsilon(t, \theta) = \frac{d}{d\theta}[y(t) - \hat{y}(t \mid \theta)] = -\psi^{\mathrm{T}}(t, \theta).$$

If we denote the second-derivative approximation $\bar{V}''(x, e^t)$ by $R(t)$, the algorithm (3.47) becomes

$$\hat{\theta}(t) = \hat{\theta}(t - 1) + \gamma(t)R^{-1}(t)\psi(t, \hat{\theta}(t - 1))\Lambda^{-1}\varepsilon(t, \hat{\theta}(t - 1)).$$

(3.51)

The quantities $\psi(t, \hat{\theta}(t - 1))$ and $\varepsilon(t, \hat{\theta}(t - 1))$ can be computed using (3.23):

$$\xi(k + 1, \hat{\theta}(t - 1))$$

$$= A(\hat{\theta}(t - 1))\xi(k, \hat{\theta}(t - 1)) + B(\hat{\theta}(t - 1))z(k),$$

(3.52a)

$$k = 0, 1, \ldots, t - 1, \quad \xi(0, \hat{\theta}(t - 1)) = \xi_0;$$

$$\begin{pmatrix} \hat{y}(t \mid \hat{\theta}(t - 1)) \\ \text{col } \psi(t, \hat{\theta}(t - 1)) \end{pmatrix} = C(\hat{\theta}(t - 1))\xi(t, \hat{\theta}(t - 1)),$$

(3.52b)

$$\varepsilon(t, \hat{\theta}(t - 1)) = y(t) - \hat{y}(t \mid \hat{\theta}(t - 1)).$$

(3.52c)

Here the actual model set will determine the matrices A, B, and C, as explained in section 3.2.

The algorithm (3.51)–(3.52) makes perfect sense. The only problem is that it is not recursive in general. The reason is that in order to compute $\xi(t, \hat{\theta}(t - 1))$ we need the whole data record z^t in (3.52a), unless the matrix $A(\hat{\theta}(t - 1))$ happens to be nilpotent (i.e., $[A(\hat{\theta}(t - 1))]^n = 0$ for some n). We encountered the same problem in section 2.2.3 with the algorithm (2.47), and we can solve it in the same way.

From (3.52a) we have

$$\xi(t, \hat{\theta}(t - 1)) = [A(\hat{\theta}(t - 1))]^t \xi_0$$

$$+ \sum_{k=0}^{t-1} [A(\hat{\theta}(t - 1))]^{t-k-1} B(\hat{\theta}(t - 1))z(k).$$

If $\hat{\theta}(t - 1) \in D_s$ (defined by (3.24)), then the factor $[A(\hat{\theta}(t - 1))]^k$ tends to zero exponentially. Consequently, the sum is dominated by its last terms, $k = t - K, t - K + 1, \ldots, t - 1$ for some value K. Also, since $\gamma(t)$ in (3.51) is a small number for large t, the difference between $\hat{\theta}(t - 1)$ and $\hat{\theta}(t - K)$ will be quite small for large t. These two facts together suggest the following approximation of $\xi(t, \hat{\theta}(t - 1))$:

$$\xi(t, \hat{\theta}(t - 1)) \approx \xi(t) \triangleq \sum_{k=0}^{t-1} \left[\prod_{s=k+1}^{t-1} A(\hat{\theta}(s)) \right] B(\hat{\theta}(k)) z(k).$$

The advantage with this expression is of course that $\xi(t)$ can be computed recursively, without having to store the data vector z^t:

$$\xi(t + 1) = A(\hat{\theta}(t))\xi(t) + B(\hat{\theta}(t))z(t). \tag{3.53a}$$

Based on this approximation, we also get the approximations of (3.52b):

$$\begin{pmatrix} \hat{y}(t \mid \hat{\theta}(t - 1)) \\ \mathrm{col}\,\psi(t, \hat{\theta}(t - 1)) \end{pmatrix} \approx \begin{pmatrix} \hat{y}(t) \\ \mathrm{col}\,\psi(t) \end{pmatrix} \triangleq C(\hat{\theta}(t - 1))\xi(t), \tag{3.53b}$$

and also of (3.52c):

$$\varepsilon(t, \hat{\theta}(t - 1)) \approx \varepsilon(t) \triangleq y(t) - \hat{y}(t). \tag{3.53c}$$

Using $\varepsilon(t)$ and $\psi(t)$ in (3.51) gives

$$\hat{\theta}(t) = \hat{\theta}(t - 1) + \gamma(t)R^{-1}(t)\psi(t)\Lambda^{-1}\varepsilon(t). \tag{3.54}$$

This algorithm (3.53)–(3.54) is now truly recursive, in that we only need to store $\xi(t)$, $\hat{\theta}(t)$, and $z(t)$ at time t.

3.4.2 The Gauss-Newton Search Direction

In a stochastic Newton algorithm, the matrix $R(t)$ in (3.51) should be an approximation of the second derivative (the Hessian) of the criterion. Let us now discuss how to choose $R(t)$.

We find that the Hessian of $V(\theta)$ [defined in (3.49)] is given by

$$\frac{d^2}{d\theta^2} V(\theta) = \mathrm{E}\psi(t, \theta)\Lambda^{-1}\psi^{\mathrm{T}}(t, \theta) + \mathrm{E}\left\{ \left[\frac{d^2}{d\theta^2} \varepsilon^{\mathrm{T}}(t, \theta) \right] \Lambda^{-1}\varepsilon(t, \theta) \right\}. \tag{3.55}$$

Here the second derivative of ε is in fact a tensor; and the correct interpretation of the last term of (3.55) is that it is a matrix whose i, j-component is given by

$$\mathrm{E} \sum_{k,l=1}^{p} \left[\frac{d}{d\theta_i} \frac{d}{d\theta_j} \varepsilon_k(t, \theta) \right] (\Lambda^{-1})_{kl} \varepsilon_l(t, \theta).$$

Suppose that there exists a value $\theta_0 \in D_{\mathcal{M}}$, that gives a correct description of the system, so that $\{\varepsilon(t, \theta_0)\}$ is a sequence of independent random vectors each of zero mean. This implies that $\varepsilon(t, \theta_0)$ is independent of z^{t-1}, and hence of

$$\frac{d}{d\theta_i} \frac{d}{d\theta_j} \varepsilon(t, \theta) = \frac{d}{d\theta_i} \frac{d}{d\theta_j} [y(t) - g_{\mathcal{M}}(\theta; t, z^{t-1})]$$

$$= -\frac{d}{d\theta_i} \frac{d}{d\theta_j} g_{\mathcal{M}}(\theta; t, z^{t-1}). \qquad \longrightarrow 0 \quad \text{at minimum}$$

At the "true" minimum θ_0, the last term of (3.55) consequently is equal to zero ⌊see also (2.45)⌋. A suitable guaranteed positive semidefinite approximation of the Hessian is therefore

$$\frac{d^2}{d\theta^2} V(\theta) \approx \mathrm{E}\psi(t, \theta)\Lambda^{-1}\psi^{\mathrm{T}}(t, \theta). \qquad (3.56)$$

This approximation is good close to the minimum, where a true Hessian is more important than elsewhere (see the discussion in section 2.4.). The search direction that is obtained when using (3.56) as an approximation of the Hessian is often referred to as the *Gauss-Newton direction*.

A natural approximation of the Hessian at $\theta = \hat{\theta}(t-1)$, based on the observation z^t, is then obtained by replacing expectation in (3.56) by the sample mean:

$$R(t) = \frac{1}{t} \sum_{k=1}^{t} \psi(k, \hat{\theta}(t-1))\Lambda^{-1}\psi^{\mathrm{T}}(k, \hat{\theta}(t-1)). \qquad (3.57)$$

This matrix cannot, though, be used in a recursive algorithm, since, as we have seen, $\psi(k, \hat{\theta}(t-1))$ cannot be computed recursively. It has to be replaced by

$$R(t) = \frac{1}{t} \sum_{k=1}^{t} \psi(k)\Lambda^{-1}\psi^{\mathrm{T}}(k), \qquad (3.58)$$

where the $\psi(k)$ are determined as in (3.53a, b). However, since the first terms of the sum (3.58) are computed for parameter estimates far from $\hat{\theta}(t-1)$, (3.58) is usually not a good approximation of (3.57). It is better to use a weighted mean where more weight is put on the last values:

$$R(t) = \sum_{k=1}^{t} \beta(t, k)\psi(k)\Lambda^{-1}\psi^{T}(k) + \delta(t)R_0. \tag{3.59}$$

Here we have also added the possibility that we have some *prior* information about $R(t)$, viz., R_0.

The weighting coefficients $\beta(t, k)$ and $\delta(t)$ should be such that

$$\delta(t) + \sum_{k=1}^{t} \beta(t, k) = 1 \quad \text{for all } t, \tag{3.60}$$

and such that $R(t)$ can be computed recursively. From (2.128)–(2.129) we know one such choice of weighting coefficients. If $\beta(t, k)$ is chosen as in (2.128) it is easy to verify that the expression (3.59) can be written recursively as

$$R(t) = R(t - 1) + \gamma(t)[\psi(t)\Lambda^{-1}\psi^{T}(t) - R(t - 1)],$$
$$R(0) = R_0. \tag{3.61}$$

From (2.128) we see that the choice $\gamma(t) = 1/t$ makes $\beta(t, k) = 1/t$ and $\delta(t) = 0$. When $\gamma(t)$ is chosen larger than $1/t$, then more weight is put to recent measurements in (3.59), i.e., $\{\beta(t, k)\}$ is increasing in k. In the notation we have assumed that the sequence $\{\gamma(t)\}$ in (3.61) is equal to the one in (3.54). This is no doubt the most common situation in practice. The possibility of using different step sizes in (3.54) and (3.61) should however be pointed out.

We have now complemented the recursive algorithm (3.53)–(3.54) with a proper updating of the matrix $R(t)$. Before we summarize the algorithm we shall bring up two further points; namely, the choice of Λ and the assumption that $\hat{\theta}(t) \in D_s$.

3.4.3 Choice of Weighting Matrix

So far, we have considered Λ to be a given constant matrix. When the system has scalar output and hence Λ is a scalar, then a constant Λ acts only as scaling factor and can be taken to be 1. (Λ then corresponds to the number α we used in section 2.2.1). For a multioutput system the choice of Λ will affect the accuracy of the obtained estimates. We have pointed out in the previous section that the optimal choice of Λ in the criterion is the covariance matrix of the true prediction errors

$$\Lambda_0 = E[\varepsilon(t, \theta_0)\varepsilon^{T}(t, \theta_0)], \tag{3.62}$$

where θ_0 is such that (3.44b) holds. This choice gives the smallest covariance matrix of the parameter estimates in the off-line case. Since Λ_0 typically is unknown, a reasonable choice of Λ in the algorithms (3.54) and (3.61) should be to replace it by an estimate $\hat{\Lambda}(t)$, where

$$\hat{\Lambda}(t) = \hat{\Lambda}(t-1) + \gamma(t)[\varepsilon(t)\varepsilon^{\mathrm{T}}(t) - \hat{\Lambda}(t-1)]. \qquad (3.63)$$

Remark It can be argued that the recursion (3.63) slightly over-estimates Λ_0. The reason is that $\varepsilon(t) = \varepsilon(t, \theta_0) + \tilde{\varepsilon}(t)$, where the quantity $\tilde{\varepsilon}(t)$ is independent of $\varepsilon(t, \theta_0)$. The matrix $\tilde{\varepsilon}(t)\tilde{\varepsilon}^{\mathrm{T}}(t)$ therefore gives an undesired positive semidefinite contribution to $\hat{\Lambda}$. It can be shown that

$$E\tilde{\varepsilon}(t)\tilde{\varepsilon}^{\mathrm{T}}(t) \approx E\psi^{\mathrm{T}}(t)\gamma(t)R^{-1}(t)\psi(t) \quad \text{asymptotically.}$$

3.4.4 Projection into the Stability Region

The model set $\mathcal{M}$ is, as we defined it in (3.5), obtained as θ ranges over the set $D_{\mathcal{M}}$. The generation of the prediction is stable only for $\theta \in D_s$, where D_s is given by (3.24). It is natural to restrict the model set to those models that give stable predictors, i.e., to require that

$$D_{\mathcal{M}} \subset D_s. \qquad (3.64)$$

In fact, in the derivation of the algorithm we used an assumption that $\hat{\theta}(t) \in D_s$ to justify the approximate calculation of $\varepsilon(t)$ and $\psi(t)$ in (3.53). It may therefore be essential to prevent $\hat{\theta}(t)$ from getting outside D_s, or, more specifically, to keep $\hat{\theta}(t)$ in $D_{\mathcal{M}}$. This may be accomplished by a projection facility of the type

$$\hat{\theta}(t) = [\hat{\theta}(t-1) + \gamma(t)R^{-1}(t)\psi(t)\varepsilon(t)]_{D_{\mathcal{M}}}, \qquad (3.65)$$

where

$$[x]_{D_{\mathcal{M}}} = \begin{cases} x \text{ if } x \in D_{\mathcal{M}} \\ \text{a value strictly interior to } D_{\mathcal{M}} \text{ if } x \notin D_{\mathcal{M}}. \end{cases} \qquad (3.66)$$

We shall generally assume that such a projection facility is included in the algorithm, even when this is not explicitly stated. The exact way of implementing the projection will be discussed in section 6.6.

3.4.5 Summary of the Algorithm

We can now summarize the general algorithm derived in this section. It consists of equations (3.53), (3.65), (3.61), and (3.63). At time $t-1$

we have stored $\hat{\theta}(t-1), R(t-1), \hat{\Lambda}(t-1)$, and $\xi(t)$. With this information we can compute $\hat{y}(t)$ and $\psi(t)$. At time t we receive the input output data $z(t)$ and make the following calculations:

$$\varepsilon(t) = y(t) - \hat{y}(t), \tag{3.67a}$$

Variance $\quad \hat{\Lambda}(t) = \hat{\Lambda}(t-1) + \gamma(t)[\varepsilon(t)\varepsilon^T(t) - \hat{\Lambda}(t-1)], \tag{3.67b}$

Hessian $\quad R(t) = R(t-1) + \gamma(t)[\psi(t)\hat{\Lambda}^{-1}(t)\psi^T(t) - R(t-1)], \tag{3.67c}$

$$\hat{\theta}(t) = [\hat{\theta}(t-1) + \gamma(t)R^{-1}(t)\psi(t)\hat{\Lambda}^{-1}(t)\varepsilon(t)]_{D_{\mathcal{M}}}, \tag{3.67d}$$

$$\xi(t+1) = A(\hat{\theta}(t))\xi(t) + B(\hat{\theta}(t))z(t), \tag{3.67e}$$

$$\begin{pmatrix} \hat{y}(t+1) \\ \mathrm{col}\,\psi(t+1) \end{pmatrix} = C(\hat{\theta}(t))\xi(t+1). \tag{3.67f}$$

This is the basic algorithm we are going to study in this book. Notice that (3.67a, c, d) has a structure similar to the recursive least squares algorithm (2.15).

3.4.6 An Algebraically Equivalent Rearrangement of the Algorithm

In practice, the algorithm is not implemented in a straightforward way from (3.67) with matrix inversions R^{-1} and $\hat{\Lambda}^{-1}$. We shall discuss such aspects in more detail in chapter 6. Here we shall just point out how the matrix inversion lemma (lemma 2.1) is used to derive an equivalent form of (3.67), corresponding to the version (2.19) of the recursive least squares method. Introduce

$$P(t) = \gamma(t)R^{-1}(t). \tag{3.68}$$

Then from (3.67c) we obtain, as in section 2.6 [see (2.123c) and (2.127)],

$$P(t) = \frac{1}{\lambda(t)}\{P(t-1) - P(t-1)\psi(t)[\psi^T(t)P(t-1)\psi(t)$$

$$+ \lambda(t)\hat{\Lambda}(t)]^{-1}\psi^T(t)P(t-1)\},$$

where

$$\lambda(t) = \gamma(t-1)[1 - \gamma(t)]/\gamma(t). \tag{3.69}$$

Moreover, using this expression for $P(t)$ we can write [see (2.18)]

$$L(t) \triangleq \gamma(t) R^{-1}(t) \psi(t) \hat{\Lambda}^{-1}(t)$$

$$= P(t)\psi(t)\hat{\Lambda}^{-1}(t)$$

$$= P(t-1)\psi(t)[\psi^{T}(t)P(t-1)\psi(t) + \lambda(t)\hat{\Lambda}(t)]^{-1}.$$

Hence (3.67) can also be written

$\varepsilon(t) = y(t) - \hat{y}(t),$	(3.70a)
$\hat{\Lambda}(t) = \hat{\Lambda}(t-1) + \gamma(t)[\varepsilon(t)\varepsilon^{T}(t) - \hat{\Lambda}(t-1)],$	(3.70b)
$S(t) = \psi^{T}(t)P(t-1)\psi(t) + \lambda(t)\hat{\Lambda}(t),$	(3.70c)
$L(t) = P(t-1)\psi(t)S^{-1}(t),$	(3.70d)
$\theta(t) = [\theta(t-1) + L(t)\varepsilon(t)]_{D_{\mathcal{M}}},$	(3.70e)
$P(t) = [P(t-1) - L(t)S(t)L^{T}(t)]/\lambda(t),$	(3.70f)
$\xi(t+1) = A(\hat{\theta}(t))\xi(t) + B(\hat{\theta}(t))z(t),$	(3.70g)
$\begin{pmatrix} \hat{y}(t+1) \\ \mathrm{col}\,\psi(t+1) \end{pmatrix} = C(\hat{\theta}(t))\xi(t+1).$	(3.70h)

This algorithm is equivalent to the stochastic Gauss-Newton algorithm (3.67). In appendix 3.A we discuss another way of computing the Gauss-Newton direction that is asymptotically equivalent to (3.70) and has some importance for comparisons with the extended Kalman filter.

Remark Notice that if $y(t)$ is a scalar in (3.70), then $\hat{\Lambda}(t)$ is a scalar. If it were constant, it would just scale $P(t)$ and $S(t)$ and thus would not affect the gain $L(t)$. Hence, in the single-output case, the equation (3.70b) is often dispensed with and $\hat{\Lambda}(t)$ replaced by unity. However, the time-varying $\hat{\Lambda}(t)$ does have an effect on the algorithm, even in the scalar output case. Then it corresponds to the scalar $1/\alpha_t$ that we used in the least squares criterion (2.12). We may thus say that more weight is being put on the measurement of time t if $\hat{\Lambda}(t)$ is small, than if it is large.

3.4.7 Summary

In this section we have derived the recursive identification algorithm (3.67) and an equivalent version (3.70). The rationale behind this is that the algorithm is a recursive way of minimizing the quadratic criterion

(3.49), using a stochastic Gauss-Newton search method. The structure of the resulting algorithm strongly resembles the recursive least squares algorithm of section 2.2.1.

3.5 A Recursive Prediction Error Identification Algorithm for General Criteria

3.5.1 General Criteria

Consider now the general criterion function (3.46):

$$V(\theta) = El(t, \theta, \varepsilon(t, \theta)). \tag{3.71a}$$

The recursive minimization of this criterion is entirely analogous to the development in section 3.4. We have

$$\left[\frac{d}{d\theta} l(t, \theta, \varepsilon(t, \theta))\right]^{\mathrm{T}} = l_\theta^{\mathrm{T}}(t, \theta, \varepsilon(t, \theta)) - \psi(t, \theta) l_\varepsilon^{\mathrm{T}}(t, \theta, \varepsilon(t, \theta)), \tag{3.71b}$$

where l_θ and l_ε denote the partial derivatives (d- and p-dimensional row vectors) with respect to θ and ε. The approach of section 3.4 now leads to the algorithm

$$\hat{\theta}(t) = \hat{\theta}(t-1) + \gamma(t) R^{-1}(t) \left[- l_\theta^{\mathrm{T}}(t, \hat{\theta}(t-1), \varepsilon(t)) \right.$$
$$\left. + \psi(t) l_\varepsilon^{\mathrm{T}}(t, \hat{\theta}(t-1), \varepsilon(t)) \right], \tag{3.72}$$

where $\varepsilon(t)$ and $\psi(t)$ are given by (3.67a, e, f).

A common special case of (3.71) is when l depends only on ε:

$$V(\theta) = El(\varepsilon(t, \theta)). \tag{3.71'}$$

The first term of (3.71b) is then zero and the corresponding algorithm is then simply

$$\hat{\theta}(t) = \hat{\theta}(t-1) + \gamma(t) R^{-1}(t) \psi(t) l_\varepsilon^{\mathrm{T}}(\varepsilon(t)), \tag{3.72'}$$

the only difference from (3.67d) being that $\hat{\Lambda}^{-1}(t)\varepsilon(t)$ is replaced by $l_\varepsilon^{\mathrm{T}}(\varepsilon(t))$.

3.5.2 General Search Directions

In (3.72) $R(t)$ is a matrix that modifies the gradient search direction to perhaps a more suitable one. As in the previous section, we may choose

$R(t)$ as an approximation of the Hessian of $V(\theta)$, which would lead to an expression

? not yet convinced

$$R(t) = R(t-1) + \gamma(t)[l_{\theta\theta}(t, \hat{\theta}(t-1), \varepsilon(t))$$
$$+ \psi(t)l_{\varepsilon\varepsilon}(t, \hat{\theta}(t-1), \varepsilon(t))\psi^T(t) - R(t-1)] \qquad (3.73)$$

where $l_{\theta\theta}$ and $l_{\varepsilon\varepsilon}$ are the second derivatives w.r.t. θ and ε. Even though this gives a Newton-type updating direction, which is known to be efficient in nonlinear programming problems, we are of course not confined to the choice (3.73). Any updating direction that forms a sharp angle with the gradient will, on the average, move $\hat{\theta}$ "downhill" in terms of the criterion. This means that we may take the matrix $R(t)$ in (3.72) to be any positive definite matrix, and still ensure that the criterion is minimized. Hence the equation (3.73) could be replaced by a general variant

$$R(t) = R(t-1) + \gamma(t)H(R(t-1), \hat{\theta}(t-1), \varepsilon(t), \psi(t)), \qquad (3.73')$$

where H is a function such that the positive definiteness of $R(t)$ is guaranteed.

3.5.3 Stochastic Gradient Algorithms

An often-made choice of $R(t)$ is to let it be a multiple of the identity matrix. This makes the updating direction coincide with the negative gradient of the criterion. The scaling of the identity matrix is often chosen as

$$R(t) = r(t)I,$$
$$r(t) = r(t-1) + \gamma(t)\{\mathrm{tr}\,[l_{\theta\theta}(t, \hat{\theta}(t-1), \varepsilon(t)) \qquad (3.73'')$$
$$+ \psi(t)l_{\varepsilon\varepsilon}(t, \hat{\theta}(t-1), \varepsilon(t))\psi^T(t)] - r(t-1)\}.$$

In the quadratic case, (3.49), this gives the algorithm

$$\hat{\theta}(t) = \hat{\theta}(t-1) + \gamma(t)\frac{1}{r(t)}\psi(t)\hat{\Lambda}^{-1}(t)\varepsilon(t),$$
$$\qquad (3.74)$$
$$r(t) = r(t-1) + \gamma(t)\{\mathrm{tr}[\psi(t)\hat{\Lambda}^{-1}(t)\psi^T(t)] - r(t-1)\}.$$

We shall call such algorithms with the choice (3.73'') *stochastic gradient algorithms*. In the control literature, they are commonly known as "stochastic approximation algorithms," but this name does not indicate the updating direction.

In section 5.7 we shall discuss some user aspects of the choice of $H(\cdot, \cdot, \cdot, \cdot)$ in (3.73′) and in particular some common alternatives to the Gauss-Newton updating direction.

3.5.4 Summary

In this section we have concluded the derivation of a general recursive prediction error algorithm. The algorithm (3.67), (3.70), or (3.72), (3.73) is well-defined. It can be applied to general model sets. The underlying model set shows up only in the matrices $A(\cdot)$, $B(\cdot)$, and $C(\cdot)$, and hence the only part of the algorithm that depends on the model set is (3.67e, f). We shall in the following three sections study some specific algorithms that arise when (3.67) is applied to particular, common model sets.

3.6 Application to Linear Regression Models

So far we have not discussed the structure of the model set (3.4),

$$\hat{y}(t \mid \theta) = g_{\mathscr{M}}(\theta; t, z^{t-1}),$$

other than in terms of the general representation (3.23). Clearly, the complexity of the algorithm (3.67) will greatly depend on the choice of the model set. Also the expressions for the gradient $\psi(t, \theta) = d\hat{y}(t \mid \theta)/d\theta$ and the stability set D_s (defined by (3.24)) are direct consequences of the structure of (3.4). In this and the following two sections we shall consider particular and commonly used model sets. We shall give explicit expressions for the calculations corresponding to (3.67e, f) in the general algorithm and point out certain approximate versions of (3.67) that have been suggested. In this section we shall study perhaps the simplest special case of (3.4): when the prediction is linear in θ. Such models are known as linear regression models in statistics and have been thoroughly studied in that field. The linear difference equation model we discussed in example 2.1 is a special case of this type. When the general algorithm (3.67) is applied to linear regression models, essentially the recursive least squares (RLS) algorithm (see section 2.2.1) will be obtained. The model set will be discussed in somewhat more detail in section 3.6.1, while the application of (3.67) will be illustrated in section 3.6.2. Some useful approximations of the gradient will be treated in section 3.6.3 and the instrumental variable (IV) method of section 2.2.2 will be interpreted as an approximate recursive prediction error method.

3.6.1 The Model Set

A predictor that is linear (or affine) in the unknown parameter vector can be expressed as

$$\hat{y}(t \mid \theta) = g_{\mathscr{M}}(\theta; t, z^{t-1}) = \varphi^{\mathrm{T}}(t)\theta + \mu(t). \tag{3.75}$$

Here $\varphi(t)$ is a $d \times p$-dimensional matrix function of t and z^{t-1}. [Recall that $d = \dim \theta$, $p = \dim y(t)$ and $z^{\mathrm{T}}(t) = (y^{\mathrm{T}}(t)\ u^{\mathrm{T}}(t))$]. Moreover, θ is a $d \times 1$ column vector and $\mu(t)$ is a known $p \times 1$ column vector function of t and z^{t-1}. To satisfy the linear filter expression (3.17) for the predictor, the quantities $\varphi(t)$ and $\mu(t)$ should strictly speaking be linear in z^{t-1}. This restriction is, however, immaterial, since, as we remarked before, we may think of $z(t)$ as an arbitrary (nonlinear) transformation of primary measurement data up to time t.

We sometimes have occasion to study linear regressions of the form

$$\hat{y}(t \mid \theta) = \theta^{\mathrm{T}}\varphi(t) + \mu(t). \tag{3.76}$$

We shall illustrate an important application of (3.76) in example 3.8. The difference compared to (3.75) is that θ is an $n' \times p$-matrix and $\varphi(t)$ is an $n' \times 1$ column vector function of t and z^{t-1} in (3.76). The kth row of (3.76) reads

$$\hat{y}_k(t \mid \theta) = \theta_k^{\mathrm{T}}\varphi(t) + \mu_k(t) = \varphi^{\mathrm{T}}(t)\theta_k + \mu_k(t), \tag{3.77}$$

where $\hat{y}_k$ is the kth component of $\hat{y}$ and θ_k is the kth column of θ. This is a linear regression model with a parameter vector θ_k that is independent of the models corresponding to other rows of (3.76). Notice, though, that all the models (3.77), $k = 1, \ldots, p$, have the same regression vector $\varphi(t)$. We can thus regard (3.76) as a collection of p independent linear regressions with the same regression vector. Each of these regressions are of the form (3.75). This is an interpretation that will be useful when deriving the estimation algorithm for (3.76) in section 3.6.2. The interest in (3.76) lies in the fact that the $n' \cdot p$ parameters of the matrix θ can be estimated quite efficiently. The algorithm complexity is determined by n', rather than by the total number of parameters.

We shall now give some specific examples of (3.75) and (3.76).

EXAMPLE 3.8 (Linear Difference Equations) (see also Examples 2.1 and 3.1) Consider a model with p-dimensional output $y(t)$ and r-dimensional input $u(t)$ described by

$$y(t) + A_1 y(t - 1) + \cdots + A_n y(t - n) = B_1 u(t - 1) + \cdots$$
$$+ B_m u(t - m) + v(t). \tag{3.78}$$

Here A_k are $p \times p$-matrices and B_k are $p \times r$-matrices. Suppose that all elements of these matrices are unknown and to be estimated. The variable $v(t)$ is a p-dimensional disturbance term ("equation error") that is either of unspecified character or supposed to be a sequence of independent random vectors each of zero mean. In either case, as we found in example 3.1, a reasonable predictor is given by

$$\hat{y}(t \mid \theta) = \theta^{\mathrm{T}} \varphi(t), \tag{3.79}$$

with

$$\theta^{\mathrm{T}} = (A_1 \ \ldots \ A_n \ B_1 \ \ldots \ B_m) \qquad \text{(a } p \times (np - mr)\text{-matrix)},$$

$$\varphi^{\mathrm{T}}(t) = (-y^{\mathrm{T}}(t - 1) \ \ldots \ -y^{\mathrm{T}}(t - n) \ u^{\mathrm{T}}(t - 1) \ \ldots \ u^{\mathrm{T}}(t - m)).$$

The multivariable difference equation consequently is an example when a description like (3.76) is possible.

The form (3.75) can of course also be chosen. We illustrate it for $p = 2, n = 2, r = 1, m = 1$:

$$\hat{y}(t \mid \theta) = \begin{bmatrix} -y_1(t-1) & -y_2(t-1) & 0 & 0 & -y_1(t-2) & -y_2(t-2) & 0 & 0 & u_1(t-1) & 0 \\ 0 & 0 & -y_1(t-1) & -y_2(t-1) & 0 & 0 & -y_1(t-2) & -y_2(t-2) & 0 & u_1(t-1) \end{bmatrix} \begin{bmatrix} a_{11}^{(1)} \\ a_{12}^{(1)} \\ a_{21}^{(1)} \\ a_{22}^{(1)} \\ a_{11}^{(2)} \\ a_{12}^{(2)} \\ a_{21}^{(2)} \\ a_{22}^{(2)} \\ b_{11}^{(1)} \\ b_{21}^{(1)} \end{bmatrix}, \tag{3.80}$$

where $A_k = (a_{ij}^{(k)})$, $B_k = (b_{ij}^{(k)})$.

Models of the type (3.78) are frequently used in many different areas. If no input sequence $\{u(t)\}$ is present, we have the familiar autoregressive (AR) representation of a (multivariate) time series $\{y(t)\}$. When $n = 0$, the output is modeled by a finite memory impulse response $\{B_k, k = 1, \ldots, m\}$.

Sometimes in practice, an unknown constant D is added to the right-hand side of (3.78) to match the constants (levels) of the measured signals. (In other words, $v(t)$ may have an unknown mean value.) This is easily incorporated in the predictor (3.79) by including D in the θ-matrix and extending the φ-vector with p entries, each of the value 1. □

EXAMPLE 3.9 (Linear difference Equation with Partly Known Coefficients) Sometimes, when models like (3.78) are used, certain of the matrix elements may be known or fixed to given values. The reason may be that the values of the parameters are known a priori as physical constants; or, we would like to impose a certain structure on the model. A very common case is when it is known that there exists a certain time delay in the system from some of the inputs. Then some of the leading $b_{ij}^{(k)}$ elements are set equal to zero. Such cases can be handled by means of (3.75), if the known or fixed elements in θ are deleted from θ together with the corresponding elements of $\varphi(t)$. These known or fixed elements are then collected into the vector $\mu(t)$, as we now illustrate.

Suppose it is known in (3.80) that

$$a_{21}^{(2)} = 0, \quad a_{21}^{(1)} = \alpha, \quad a_{11}^{(2)} = 1, \quad b_{11}^{(1)} = \beta.$$

The other matrix elements are to be estimated. Then we may write

$$\hat{y}(t|\theta) = \begin{bmatrix} -y_1(t-1) & -y_2(t-1) & 0 & -y_2(t-2) & 0 & 0 \\ 0 & 0 & -y_2(t-1) & 0 & -y_2(t-2) & u_1(t-1) \end{bmatrix} \begin{bmatrix} a_{11}^{(1)} \\ a_{12}^{(1)} \\ a_{22}^{(1)} \\ a_{12}^{(2)} \\ a_{22}^{(2)} \\ b_{21}^{(1)} \end{bmatrix} + \begin{bmatrix} -y_1(t-2) + \beta u(t-1) \\ -\alpha y_1(t-1) \end{bmatrix} \triangleq \varphi^T(t)\theta + \mu(t). \qquad \square \quad (3.81)$$

The Gradient of the Prediction   It is easy to determine the gradient for the model (3.75). It is given by

$$\left[\frac{d}{d\theta}\hat{y}(t\,|\,\theta)\right]^T = \psi(t,\theta) = \varphi(t). \qquad (3.82)$$

The stability region of the predictor $\hat{y}(t\,|\,\theta)$ is easy to determine for the linear difference equations. Since only finitely many past $y(t)$- and $u(t)$-data enter the predictor, it is stable for all θ:

$$D_s = \mathbf{R}^d. \qquad (3.83)$$

The foregoing examples indicate that linear regression models arc capable of describing several model structures that are common in systems, control, and communications applications. We may repeat once more that the "regressor" variables in $\varphi(t)$ may in practice be nonlinear functions of recorded data. Hence, the models are useful also in handling nonlinearities.

The main disadvantage with models such as (3.78) is that they do not allow for modeling of the noise characteristics $\{v(t)\}$. The predictor (3.79) is reasonable only if the noise level is small (good signal-to-noise ratio) or if $\{v(t)\}$ is unpredictable from past data. The latter case essentially corresponds to an assumption of white equation error noise. Often, however, there is no specific, physical reason to assume that the equation error is white, which is much less realistic than assuming, e.g., that the measurement error, when recording $y(t)$, is white.

3.6.2 The Recursive Prediction Error Algorithm

Structure (3.75) Application of the recursive Gauss-Newton algorithm (3.67) or (3.70) to the model (3.75) is immediate. Equations (3.67 c–f) are in this case trivial; the prediction and its gradient φ are directly formed from the data. We obtain

$$\varepsilon(t) = y(t) - \varphi^{\mathrm{T}}(t)\hat{\theta}(t-1) - \mu(t), \tag{3.84a}$$

$$\hat{\Lambda}(t) = \hat{\Lambda}(t-1) + \gamma(t)[\varepsilon(t)\varepsilon^{\mathrm{T}}(t) - \hat{\Lambda}(t-1)], \tag{3.84b}$$

$$R(t) = R(t-1) + \gamma(t)[\varphi(t)\hat{\Lambda}^{-1}(t)\varphi^{\mathrm{T}}(t) - R(t-1)], \tag{3.84c}$$

$$\hat{\theta}(t) = \hat{\theta}(t-1) + \gamma(t)R^{-1}(t)\varphi(t)\hat{\Lambda}^{-1}(t)\varepsilon(t), \tag{3.84d}$$

which of course also can be written in forms corresponding to (3.70). We may also choose to work with a fixed matrix Λ and skip (3.84b).

Structure (3.76) Consider now the structure (3.76),

$$\hat{y}(t \mid \theta) = \theta^{\mathrm{T}}\varphi(t) + \mu(t).$$

Its kth row is given by (3.77), and hence the structure is a linear regression of the type we have just discussed. The kth column θ_k of θ is thus estimated by

$$\varepsilon_k(t) = y_k(t) - \varphi^{\mathrm{T}}(t)\hat{\theta}_k(t-1) - \mu_k(t), \tag{3.85a}$$

$$R(t) = R(t-1) + \gamma(t)[\varphi(t)\varphi^{\mathrm{T}}(t) - R(t-1)], \tag{3.85b}$$

$$\hat{\theta}_k(t) = \hat{\theta}_k(t-1) + \gamma(t)R^{-1}(t)\varphi(t)\varepsilon_k(t). \tag{3.85c}$$

Here we dispensed with the scaling $\hat{\Lambda}$, since ε_k is a scalar. Equation (3.85) gives us p recursions, one for each column of θ. Since the recursions have the R-matrix and the φ-vector in common, they can be collected into

$$\varepsilon(t) = y(t) - \hat{\theta}^{\mathrm{T}}(t-1)\varphi(t) - \mu(t), \tag{3.86a}$$

$$R(t) = R(t-1) + \gamma(t)[\varphi(t)\varphi^{\mathrm{T}}(t) - R(t-1)], \tag{3.86b}$$

$$\hat{\theta}(t) = \hat{\theta}(t-1) + \gamma(t)R^{-1}(t)\varphi(t)\varepsilon^{\mathrm{T}}(t). \tag{3.86c}$$

The dimension of the matrix R is $n' \times n'$. The complexity of the algorithm is thus essentially determined by n' rather than by the total number of parameters $p \cdot n'$.

With this we have concluded the application of the Gauss-Newton, recursive prediction error method to linear regression models. We have seen that for a quadratic criterion we obtain the well-known recursive least squares method. This is of course what we expected, in accordance with section 2.2.1. If we choose a general criterion $l(\varepsilon)$, as in (3.71′), rather than the quadratic criterion, the only difference in the algorithm is that $\hat{\Lambda}^{-1}(t)\varepsilon(t)$ in (3.84d), and $\varepsilon(t)$ in (3.86c) are replaced by $l_\varepsilon^{\mathrm{T}}(\varepsilon(t))$, where l_ε is the $1 \times p$ derivative vector of the scalar function l with respect to ε (see 3.72′). Moreover, stochastic gradient variants (3.74) of (3.84) and (3.86) are obtained by replacing the matrix $R(t)$ by $I \cdot \mathrm{tr}\, R(t)$. These variants are generalizations of the "stochastic gradient" algorithm (2.84), discussed in section 2.4.

3.6.3   An Approximate Gradient: The Instrumental Variable Method

(The discussion of this subsection is an extension of section 2.2.2.) The recursive prediction error method of section 3.6.2 aims at minimizing the criterion

$$\mathrm{E}\tfrac{1}{2}\varepsilon^{\mathrm{T}}(t, \theta)\varepsilon(t, \theta), \tag{3.87a}$$

where, in the case of a linear regression,

$$\varepsilon(t, \theta) = y(t) - \varphi^{\mathrm{T}}(t)\theta.$$

(See (3.75); we take, for simplicity, $\mu(t) = 0$ in this subsection.) The θ-value that minimizes (3.87a) is the solution of

$$\mathrm{E}\varphi(t)[y(t) - \varphi^{\mathrm{T}}(t)\theta] = 0. \tag{3.87b}$$

Suppose now that the actual output is given by

$$y(t) = \varphi^{\mathrm{T}}(t)\theta_0 + v(t) \tag{3.88}$$

for some θ_0 and stochastic disturbance sequence $\{v(t)\}$. We will refer to θ_0 as the *true value* of θ.

Now, we see that θ_0 minimizes the prediction error criterion, i.e., θ_0 is a solution to (3.87b), *only if*

$$E\varphi(t)v(t) = 0,$$

which means that $v(t)$ must be (zero mean and) uncorrelated with $\varphi(t)$. As we noted in section 2.2.1, this will normally be the case only if

$\{v(t)\}$ is white noise

or

$n = 0$ in (3.78) and $\{v(t)\}$ and $\{u(t)\}$ are independent.

Therefore, in order to obtain a sequence of estimates that converges to θ_0 as t approaches infinity, we should instead solve

$$E\zeta(t)[y(t) - \varphi^T(t)\theta] = 0 \tag{3.89}$$

rather than (3.87b). The vector or matrix $\zeta(t)$ should be uncorrelated with $v(t)$, but correlated with the gradient:

$$E\zeta(t)v(t) = 0 \tag{3.90a}$$

$$E\zeta(t)\varphi^T(t) \text{ positive definite or at least nonsingular} \tag{3.90b}$$

in order for (3.89) to have the solution $\theta = \theta_0$ when (3.88) holds.

This gives the recursive estimation algorithm

$$\hat{\theta}(t) = \hat{\theta}(t-1) + \gamma(t)R^{-1}(t)\zeta(t)[y(t) - \varphi^T(t)\hat{\theta}(t-1)]. \tag{3.91}$$

The vector $\zeta(t)$ is called the *instrumental variable* (IV) and the algorithm (3.91) is a *recursive instrumental variable* (RIV) algorithm. We shall discuss various choices of $\zeta(t)$ below, but let us first comment on the matrix $R(t)$ in (3.91).

Choice of $R(t)$ A couple of different choices of the matrix $R(t)$ are possible. One choice is to make $\hat{\theta}(t)$ equal to the off-line IV estimate:

$$\hat{\theta}(t) = \left[\sum_{k=1}^{t} \beta(t,k)\zeta(k)\varphi^T(k)\right]^{-1} \sum_{k=1}^{t} \beta(t,k)\zeta(k)y(k),$$

where $\beta(t,k)$ are the weighting coefficient corresponding to the sequence $\{\gamma(t)\}$, as in (2.128). This is achieved, as we saw in section 2.2.2, by taking

$$R(t) = R(t-1) + \gamma(t)[\zeta(t)\varphi^T(t) - R(t-1)]. \tag{3.92a}$$

Another choice is to go all the way and replace the gradient $\varphi(t)$ by $\zeta(t)$ in both positions above:

$$R(t) = R(t-1) + \gamma(t)[\zeta(t)\zeta^{\mathrm{T}}(t) - R(t-1)]. \tag{3.92b}$$

This variant will be referred to as the *symmetric RIV algorithm*. Other possibilities are "stochastic approximation" variants where $R(t)$ is given by a properly scaled identity matrix, analogously to (3.74).

Choice of Instrumental Variables   Many ways of choosing the instrumental variables $\zeta(t)$ have been proposed in the literature. Some typical choices for single-input/single-output systems will be reviewed here. Some more alternatives can be found in the survey paper by Söderström and Stoica (1981). Generally speaking, the elements of $\zeta(t)$ are usually chosen as delayed and possibly filtered values of the input and output.

Instrumental Variables Computed from Inputs by Constant Filters   One quite common choice of instrumental variables is to take

$$\zeta(t) = (-x(t-1) \ \ldots \ -x(t-n) \ u(t-1) \ \ldots \ u(t-m))^{\mathrm{T}} \tag{3.93a}$$

[See, e.g., Finigan and Rowe (1974)], where $x(t)$ is obtained from the input by filtering according to

$$\bar{A}(q^{-1})x(t) = \bar{B}(q^{-1})u(t), \tag{3.93b}$$

with

$$\bar{A}(q^{-1}) = 1 + \bar{a}_1 q^{-1} + \ldots + \bar{a}_n q^{-n},$$
$$\bar{B}(q^{-1}) = \bar{b}_1 q^{-1} + \ldots + \bar{b}_m q^{-m}. \tag{3.93c}$$

It has been argued that if the true system is subject to

$$A_0(q^{-1})y(t) = B_0(q^{-1})u(t) + v(t),$$

with $\{v(t)\}$ being a disturbance that is uncorrelated with the input, then a desirable choice would be given by

$$\bar{A}(q^{-1}) = A_0(q^{-1}), \quad \bar{B}(q^{-1}) = B_0(q^{-1}).$$

The rationale for such a choice will be explained in section 4.6.3.

Instrumental Variables Computed from Inputs by Adaptive Filters   When the true system parameters are not known, the foregoing choice cannot be realized. Instead, an adaptive way of forming the instrumental variables can be used; e.g., the last available estimates are used to compute the instrumental variables. This means that $\zeta(t)$ and $x(t)$ are recursively defined by (3.93a), together with

$$x(t) = \zeta^{\mathrm{T}}(t)\hat{\theta}(t). \tag{3.94}$$

This choice has been proposed by Wong and Polak (1967) and Young (1970).

Instrumental Variables Consisting of Delayed Inputs Another way of choosing the polynomials $\bar{A}(q^{-1})$ and $\bar{B}(q^{-1})$ in (3.93b) is to take $\bar{A}(q^{-1}) = 1$, $\bar{B}(q^{-1}) = q^{-m}$. After reordering the elements in $\zeta(t)$, this means that

$$\zeta(t) = (u(t-1) \ \ldots \ u(t-n-m))^{\mathrm{T}}, \tag{3.95a}$$

i.e., the vector $\zeta(t)$ consists only of delayed inputs. This choice is closely related to correlation analysis, which can be seen as follows. With the weighting coefficients $\beta(t, k) = 1$, and with (3.95a), the off-line IV estimate can be written as

$$
\begin{pmatrix}
-\hat{R}_{yu}^t(0) & \cdots & -\hat{R}_{yu}^t(1-n) & \hat{R}_u^t(0) & \cdots & \hat{R}_u^t(1-m) \\
\vdots & & \vdots & \vdots & & \vdots \\
-\hat{R}_{yu}^t(n+m-1) & \cdots & -\hat{R}_{yu}^t(m) & \hat{R}_u^t(n+m-1) & \cdots & \hat{R}_u^t(n)
\end{pmatrix}
\begin{pmatrix}
\hat{a}_1(t) \\
\vdots \\
\hat{a}_n(t) \\
\hat{b}_1(t) \\
\vdots \\
\hat{b}_m(t)
\end{pmatrix}
$$

$$
=
\begin{pmatrix}
\hat{R}_{yu}^t(1) \\
\vdots \\
\hat{R}_{yu}^t(n+m)
\end{pmatrix}, \tag{3.95b}
$$

with

$$\hat{R}_{yu}^t(\tau) = \frac{1}{t}\sum_{k=1}^{t} y(k)u(k-\tau), \quad \hat{R}_u^t(\tau) = \frac{1}{t}\sum_{k=1}^{t} u(k)u(k-\tau) \tag{3.95c}$$

being estimated covariance functions. A variant of this correlation method has been proposed by Isermann and Baur (1974). In their treatment the vector $\zeta(t)$ is extended and may contain more than $n + m$ elements. The system (3.95b) will then be overdetermined, and can be solved in a least squares sense.

It is easy to update the covariance estimates recursively. We have, e.g.,

$$\hat{R}_{yu}^{t+1}(\tau) = \hat{R}_{yu}^{t}(\tau) + \frac{1}{t+1}[y(t+1)u(t+1-\tau) - \hat{R}_{yu}^{t}(\tau)]. \tag{3.95d}$$

In the recursive method of Isermann and Baur the covariance elements are updated at every sampling interval, while the linear system (3.95b) is solved more infrequently.

Refined Instrumental Variables A refined IV method has been proposed by Young (1976); in his method, prefiltering of the data is included. Then the algorithm will be [see (3.91), (3.92)]

$$\hat{\theta}(t) = \hat{\theta}(t-1) + \gamma(t)R^{-1}(t)\zeta(t)[y_F(t) - \varphi_F^{T}(t)\hat{\theta}(t-1)], \tag{3.96a}$$

$$R(t) = R(t-1) + \gamma(t)[\zeta(t)\varphi_F^{T}(t) - R(t-1)], \tag{3.96b}$$

$$y_F(t) = T(q^{-1})y(t), \qquad \varphi_F(t) = T(q^{-1})\varphi(t), \tag{3.96c}$$

where $T(q^{-1})$ is the prefilter.

We have here not discussed if the above choices of $\zeta(t)$, equations (3.93)–(3.95), actually satisfy (3.90). That will be done in some detail in section 4.6.

The multioutput case can be arranged according to (3.76), (3.86) also for the RIV algorithm. Moreover, variants where the prediction error $\varepsilon(t) = y(t) - \varphi^{T}(t)\hat{\theta}(t-1)$ is replaced by a general nonlinear function $l_{\varepsilon}^{T}(\varepsilon(t))$ can be useful in some contexts.

So we may say that we have found it suitable to interpret the RIV algorithm (3.91) as a recursive prediction error algorithm where the gradient $\varphi(t)$ has been replaced by an approximation $\zeta(t)$. It must be stressed, however, that the word "approximation" does not at all have a pejorative meaning in this context. The replacement is done deliberately to make the estimate converge to the true value θ_0 rather than to the minimum of the prediction error criterion (3.87). The actual convergence properties of the algorithm (3.91)–(3.92) will be discussed in section 4.6.

3.6.4 Summary

We have in this section studied predictors $\hat{y}(t \mid \theta)$ that are linear in θ. Such models are known in statistics as linear regression models. For dynamical systems, linear difference equations with an unstructured equation error are described by linear regressions. The recursive prediction error method applied to these models is the recursive least squares (RLS) method (3.84) and variants thereof. If the gradient $\varphi(t)$ is replaced by an instrumental variable vector or matrix $\zeta(t)$ that is uncorrelated with the equation error, we obtain the recursive instrumental variable (RIV) method.

3.7 Application to a General SISO Model

In many contexts it is suitable to model the system as a "black box," i.e., to describe its input-output behavior without going into the mechanisms of how it operates. In this section we shall study several common black box models and how to estimate their parameters. Examples are discussed in section 3.7.1, where we also give a unified description of such models. The general recursive Gauss-Newton prediction error algorithm (3.67) is then applied to these models in section 3.7.2. In section 3.7.3 we describe a common way of approximating the gradient that leads to so called pseudolinear regressions. Some useful formulas relating certain variables in pseudolinear regression are given in appendix 3.D.

3.7.1 The Model Set

We mentioned in connection with the difference equation model (3.78) that the main limitation with linear regression models is that the disturbance terms $\{v(t)\}$ cannot be modeled. Several different ways to include a model of the noise have been suggested. We shall in this section give a treatment of such models. Mainly for notational reasons, we shall confine ourselves to the single-input/single-output (SISO) case, i.e., $p = r = 1$. The results can be generalized to the multivariable case, in much the same way as in example 3.8 [see e.g., Gauthier and Landau (1978)].

In the section we shall make extensive use of operator polynomial notation, which we used earlier in examples 2.1 and 3.1 and 2.2 and 3.2. We thus have the delay operator q^{-1} and the polynomials

$$A(q^{-1}) = 1 + a_1 q^{-1} + \cdots + a_{n_a} q^{-n_a}, \tag{3.97a}$$

$$B(q^{-1}) = b_1 q^{-1} + \cdots + b_{n_b} q^{-n_b}, \tag{3.97b}$$

$$C(q^{-1}) = 1 + c_1 q^{-1} + \cdots + c_{n_c} q^{-n_c}, \tag{3.97c}$$

$$D(q^{-1}) = 1 + d_1 q^{-1} + \cdots + d_{n_d} q^{-n_d}, \tag{3.97d}$$

$$F(q^{-1}) = 1 + f_1 q^{-1} + \cdots + f_{n_f} q^{-n_f}. \tag{3.97e}$$

Some Different Models The simple linear regression model can be written in the foregoing notation as in example 2.1:

$$A(q^{-1})y(t) = B(q^{-1})u(t) + v(t). \tag{3.98}$$

Perhaps the most common noise model is to describe $\{v(t)\}$ as a moving average of a white noise sequence $\{e(t)\}$:

$$A(q^{-1})y(t) = B(q^{-1})u(t) + C(q^{-1})e(t). \tag{3.99}$$

This yields the familiar ARMAX model already discussed in examples 2.1 and 3.1. As an alternative, we could choose to model $v(t)$ as an auto regression, which gives the model

$$A(q^{-1})y(t) = B(q^{-1})u(t) + \frac{1}{D(q^{-1})}e(t). \tag{3.100}$$

Such a model was used by Clarke (1967) in the so-called generalized least squares (GLS) method. In econometric literature the model (3.100) is also known as a dynamic adjustment (DA) model.

A further alternative is of course to describe $\{v(t)\}$ in (3.98) as an ARMA process:

$$A(q^{-1})y(t) = B(q^{-1})u(t) + \frac{C(q^{-1})}{D(q^{-1})}e(t). \tag{3.101}$$

This model has been used by Talmon and van den Boom (1973).

In (3.98) the disturbance $v(t)$ enters as an equation error. From a physical point of view it is often more natural to work with disturbances that are "measurement errors" or "output errors":

$$y(t) = \frac{B(q^{-1})}{F(q^{-1})}u(t) + v(t). \tag{3.102}$$

The difference between this model and (3.98) is also illustrated in figure 3.2. Identification methods that use the model (3.102) are often known as output error methods or model reference identification methods (see also figure 2.1 in section 2.5).

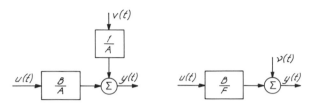

Figure 3.2
Left: Equation error model (3.98). *Right:* Output error model (3.102).

The measurement noise $v(t)$ in (3.102) can of course be further modeled. If we choose to represent it as an ARMA process we obtain the model

$$y(t) = \frac{B(q^{-1})}{F(q^{-1})}u(t) + \frac{C(q^{-1})}{D(q^{-1})}e(t). \qquad (3.103)$$

In many ways this model is a natural one from a physical point of view: the dynamics from u to y is modeled separately from the measurement noise. It has been used, e.g., by Bohlin (1971) for identification problems and by Young and Jakeman (1979) for a refined instrumental variable recursive identification algorithm. It also corresponds to the type of models that are used by Box and Jenkins (1970) for modeling stochastic dynamical systems. We shall refer to (3.103) as the *Box-Jenkins model.*

A General Model Description We have now listed six different models for identification of SISO stochastic systems. They are all commonly used in different identification procedures. They are black boxes, which means that no internal structure has been imposed on them. In order to give a general treatment of black box SISO models, we shall consider the model set

$$A(q^{-1})y(t) = \frac{B(q^{-1})}{F(q^{-1})}u(t) + \frac{C(q^{-1})}{D(q^{-1})}e(t), \qquad (3.104)$$

where $\{e(t)\}$ is a sequence of independent random variables each of zero mean. This model obviously contains all the foregoing ones as special cases. As it stands, (3.104) is too general. In most applications one or several of the polynomials A, F, C, and D would be fixed to unity. For our purposes, however, it is convenient to develop the recursive prediction error algorithm for the general case (3.104). Algorithms of practical interest will then be obtained as simple special cases.

An Expression for the Prediction   We shall now proceed to derive the one-step-ahead prediction and its gradient for the model (3.104). The calculations will show some resemblance to those of example 3.2.

From (3.104) we have

$$\frac{A(q^{-1})D(q^{-1})}{C(q^{-1})}y(t) = \frac{B(q^{-1})D(q^{-1})}{F(q^{-1})C(q^{-1})}u(t) + e(t),$$

which also can be written as

$$y(t) = \left[1 - \frac{A(q^{-1})D(q^{-1})}{C(q^{-1})}\right]y(t) + \frac{D(q^{-1})B(q^{-1})}{C(q^{-1})F(q^{-1})}u(t) + e(t).$$

Notice that since the polynomials A, C, D, and F all start with a 1, the first term on the right-hand side contains $y(s)$ only up to (and including) time $t - 1$. Similarily, since $B(q^{-1})$ starts with $b_1 q^{-1}$, the second term contains only $u(s)$ for $s \leq t - 1$. Since $\{e(t)\}$ is supposed to be white noise, the term $e(t)$ is not predictable from data up to time $t - 1$. Hence the best one-step-ahead predictor for (3.104) is given by

$$\hat{y}(t \mid \theta) = \left[1 - \frac{D(q^{-1})A(q^{-1})}{C(q^{-1})}\right]y(t) + \frac{D(q^{-1})B(q^{-1})}{C(q^{-1})F(q^{-1})}u(t), \qquad (3.105)$$

where we have used the parameter vector θ to denote the unknown parameters:

$$\theta^{\mathrm{T}} = (a_1 \ \ldots \ a_{n_a} \ b_1 \ \ldots \ b_{n_b} \ f_1 \ \ldots \ f_{n_f} \ c_1 \ \ldots \ c_{n_c} \ d_1 \ \ldots \ d_{n_d}). \quad (3.106)$$

The expression (3.105) assumes that all previous data $z(k)$, $k \leq t - 1$, are available. In practice, unobserved data $z(k)$, $k \leq 0$, are often replaced by zeros. The predictor (3.105) is then only approximate, but when the polynomials

$$C^*(z) = z^{n_c} + c_1 z^{n_c - 1} + \cdots + c_{n_c},$$

$$F^*(z) = z^{n_f} + f_1 z^{n_f - 1} + \cdots + f_{n_f}$$

have all zeros inside the unit circle, the difference is negligible, except in the transient phase. The stability set for the predictor (3.105) is clearly given by

$$D_s = \{\theta \mid C^*(z) \cdot F^*(z) \text{ has all zeros inside the unit circle}\}. \qquad (3.107)$$

We will encounter the product $C(q^{-1})F(q^{-1})$ several times in this section, so let us introduce the following notation for it:

$$G(q^{-1}) = C(q^{-1})F(q^{-1}). \tag{3.108}$$

For practical calculations, the predictor (3.105) is normally written as a recursion

$$G(q^{-1})\hat{y}(t \mid \theta)$$
$$= F(q^{-1})[C(q^{-1}) - D(q^{-1})A(q^{-1})]y(t) + D(q^{-1})B(q^{-1})u(t). \tag{3.109}$$

From (3.104) and (3.105) we find that the prediction error

$$\varepsilon(t, \theta) = y(t) - \hat{y}(t \mid \theta)$$

can be written

$$\varepsilon(t, \theta) = \frac{D(q^{-1})}{C(q^{-1})} \left[A(q^{-1})y(t) - \frac{B(q^{-1})}{F(q^{-1})} u(t) \right]. \tag{3.110}$$

It is convenient to introduce the auxiliary variables

$$w(t, \theta) = \frac{B(q^{-1})}{F(q^{-1})} u(t) \tag{3.111}$$

and

$$v(t, \theta) = A(q^{-1})y(t) - w(t, \theta). \tag{3.112}$$

Then

$$\varepsilon(t, \theta) = y(t) - \hat{y}(t \mid \theta) = \frac{D(q^{-1})}{C(q^{-1})} v(t, \theta). \tag{3.113}$$

It is straightforward to put the linear filter (3.105) in the form (3.17). The state vector would then be

$$\varphi(t, \theta) = (-y(t - 1) \ \ldots \ -y(t - n_a) \ u(t - 1) \ \ldots \ u(t - n_b)$$
$$- w(t - 1, \theta) \ \ldots \ -w(t - n_f, \theta) \ \varepsilon(t - 1, \theta) \ \ldots \ \varepsilon(t - n_c, \theta)$$
$$- v(t - 1, \theta) \ \ldots \ -v(t - n_d, \theta))^{\mathrm{T}}. \tag{3.114}$$

The state equation that $\varphi(t, \theta)$ obeys follows from the definitions (3.111)–(3.113). There is, however, no point in writing it down explicitly, since it in practice is more efficient to generate $\varphi(t, \theta)$ directly from (3.111)–(3.113).

With (3.105) and (3.114) we can give a convenient expression for the prediction. To find this, we proceed as follows. From (3.111) and (3.113) we obtain

$$_{(115)} \; w(t, \theta) = b_1 u(t-1) + \cdots + b_{n_b} u(t - n_b) - f_1 w(t - 1, \theta) - \cdots$$
$$- f_{n_f} w(t - n_f, \theta) \tag{3.115}$$

and

$$_{(113)} \; \varepsilon(t, \theta) = v(t, \theta) + d_1 v(t - 1, \theta) + \cdots + d_{n_d} v(t - n_d, \theta)$$
$$- c_1 \varepsilon(t - 1, \theta) - \cdots - c_{n_c} \varepsilon(t - n_c, \theta). \tag{3.116}$$

Now inserting

$$v(t, \theta) = y(t) + a_1 y(t - 1) + \cdots + a_{n_a} y(t - n_a) - w(t, 0) \;_{112-}$$

into (3.116) and substituting $w(t, \theta)$ with the expression (3.115), we find that

$$\varepsilon(t, \theta) = y(t) - \theta^T \varphi(t, \theta). \tag{3.117}$$

Hence $\quad_{/106}$

$$\hat{y}(t \mid \theta) = \theta^T \varphi(t, \theta). \tag{3.118}$$

An Expression for the Gradient   In order to apply the recursive prediction error method we need also the gradient $\psi(t, \theta)$ of the prediction. For the derivative w.r.t. a_i we find by differentiating (3.109)

$$G(q^{-1}) \frac{\partial}{\partial a_i} \hat{y}(t \mid \theta) = -F(q^{-1}) D(q^{-1}) y(t - i)$$

or

$$\frac{\partial}{\partial a_i} \hat{y}(t \mid \theta) = -\frac{F(q^{-1}) D(q^{-1})}{F(q^{-1}) C(q^{-1})} y(t - i) = -\frac{D(q^{-1})}{C(q^{-1})} y(t - i). \tag{3.119a}$$

Similarily, we find for the other derivatives

$$\frac{\partial}{\partial b_i} \hat{y}(t \mid \theta) = \frac{D(q^{-1})}{G(q^{-1})} u(t - i), \tag{3.119b}$$

$$\frac{\partial}{\partial f_i} \hat{y}(t \mid \theta) = -\frac{D(q^{-1})}{G(q^{-1})} w(t - i, \theta), \tag{3.119c}$$

note that
$$G = CF \text{ in } 109$$

$$\frac{\partial}{\partial c_i} \hat{y}(t \mid \theta) = \frac{1}{C(q^{-1})} \varepsilon(t - i, \theta), \tag{3.119d}$$

$$\frac{\partial}{\partial d_i} \hat{y}(t \mid \theta) = -\frac{1}{C(q^{-1})} v(t - i, \theta). \tag{3.119e}$$

The gradient vector $\psi(t, \theta)$ is thus obtained analogously to $\varphi(t, \theta)$ by letting the components pass through filters that depend on θ, given by (3.119).

We have now derived a set of recursions, (3.109) and (3.119), that allows us to calculate $\hat{y}(t \mid \theta)$ and $\psi(t, \theta)$ from z^t for any given parameter vector θ. These recursions can, of course, also be summarized in a state-space form (3.22). In the present context it is however more convenient to work with (3.109) and (3.119).

Some Special Cases   Before turning to the recursive estimation algorithm we shall comment on some special cases of (3.104). When

$$A(q^{-1}) = 1, \quad C(q^{-1}) = 1, \quad D(q^{-1}) = 1,$$

i.e., when the model is given by (3.102), we find that

$$\varphi^{\mathrm{T}}(t, \theta) = (u(t - 1) \ \ldots \ u(t - n_b) \ -w(t - 1, \theta) \ \ldots \ -w(t - n_f, \theta)).$$

This vector is formed entirely from the input sequence u^t, in view of (3.111). The prediction (3.109) is given by

$$\hat{y}(t \mid \theta) = \frac{B(q^{-1})}{F(q^{-1})} u(t), \tag{3.120}$$

and is simply the output the model would produce for the given input. The prediction error is then the difference between measured output and model output, which explains the term "output error methods" for algorithms that use the model (3.102). Moreover, for this model we find from (3.119) that

$$\psi(t, \theta) = \frac{1}{F(q^{-1})} \varphi(t, \theta). \tag{3.121}$$

(Notice that equations (3.119a, d, e) do not apply, since the parameters a_i, c_i, and d_i do not belong to θ.)

If we apply (3.119) to the ARMAX case (3.99) $(F(q^{-1}) = 1, D(q^{-1}) = 1)$ we find

$$\psi(t, \theta) = \frac{1}{C(q^{-1})} \varphi(t, \theta), \tag{3.122}$$

just as in example 3.5.

3.7.2 The Recursive Prediction Error Method

Having established how to compute the prediction $\hat{y}(t \mid \theta)$ and its gradient $\psi(t, \theta)$ for the general model, it is now straightforward to apply the general recursive prediction error algorithm (3.67) to the model (3.104). In the recursions that define $\hat{y}(t \mid \theta)$ and $\psi(t, \theta)$, corresponding to (3.67e, f) we simply use the latest available estimate of θ. Let us denote the components of $\hat{\theta}(t)$ by $\hat{a}_i(t)$, $\hat{b}_i(t)$, etc., and let $\hat{g}_i(t)$ be the coefficients of the polynomial $\hat{C}_t(q^{-1}) \cdot \hat{F}_t(q^{-1})$. Since the output is scalar we dispense with the matrix scaling $\Lambda(t)$, as commented upon previously. We now obtain the following explicit form of (3.67):

$\varepsilon(t) = y(t) - \hat{y}(t),$ *predition error* *if* $n_a \cdots n_f = 0$ *the series is* *gone* $\tag{3.123a}$

$$R(t) = R(t-1) + \gamma(t)[\psi(t)\psi^{\mathrm{T}}(t) - R(t-1)], \tag{3.123b}$$

$$\hat{\theta}(t) = \hat{\theta}(t-1) + \gamma(t)R^{-1}(t)\psi(t)\varepsilon(t), \tag{3.123c}$$

$$w(t) = \hat{b}_1(t)u(t-1) + \cdots + \hat{b}_{n_b}(t)u(t-n_b)$$
$$- \hat{f}_1(t)w(t-1) - \cdots - \hat{f}_{n_f}(t)w(t-n_f), \tag{3.124a}$$

$$v(t) = y(t) + \hat{a}_1(t)y(t-1) + \cdots + \hat{a}_{n_a}(t)y(t-n_a) - w(t), \tag{3.124b}$$

$\bar{\varepsilon}(t) = v(t) + \hat{d}_1(t)v(t-1) + \cdots + \hat{d}_{n_d}(t)v(t-n_d)$

$\varepsilon(t)$ *model error* $- \hat{c}_1(t)\bar{\varepsilon}(t-1) - \cdots - \hat{c}_{n_c}(t)\bar{\varepsilon}(t-n_c), \tag{3.124c}$

$$\varphi^{\mathrm{T}}(t+1) = (-y(t) \ \ldots \ -y(t-n_a+1) \ \underline{u(t) \ \ldots \ u(t-n_b+1)}_{\text{gone if } n_b = 0}$$
$$-w(t) \ \ldots \ -w(t-n_f+1) \ \bar{\varepsilon}(t) \ \ldots \ \bar{\varepsilon}(t-n_c+1)$$
$$-v(t) \ \ldots \ -v(t-n_d+1)), \tag{3.124d}$$

$$\hat{y}(t+1) = \hat{\theta}^{\mathrm{T}}(t)\varphi(t+1), \tag{3.124e}$$

$$\tilde{y}(t) = y(t) + \hat{d}_1(t)y(t-1) + \cdots + \hat{d}_{n_d}(t)y(t-n_d)$$
$$- \hat{c}_1(t)\tilde{y}(t-1) - \cdots - \hat{c}_{n_c}(t)\tilde{y}(t-n_c), \tag{3.125a}$$

$$\tilde{u}(t) = u(t) + \hat{d}_1(t)u(t-1) + \cdots + \hat{d}_{n_d}(t)u(t-n_d)$$
$$- \hat{g}_1(t)\tilde{u}(t-1) - \cdots - \hat{g}_{n_g}(t)\tilde{u}(t-n_g), \tag{3.125b}$$

$$\tilde{w}(t) = w(t) + \hat{d}_1(t)w(t-1) + \cdots + \hat{d}_{n_d}(t)w(t-n_d)$$
$$- \hat{g}_1(t)\tilde{w}(t-1) - \cdots - \hat{g}_{n_g}(t)\tilde{w}(t-n_g), \tag{3.125c}$$

$$\tilde{\varepsilon}(t) = \bar{\varepsilon}(t) - \hat{c}_1(t)\tilde{\varepsilon}(t-1) - \cdots - \hat{c}_{n_c}(t)\tilde{\varepsilon}(t-n_c), \tag{3.125d}$$

$$\tilde{v}(t) = v(t) - \hat{c}_1(t)\tilde{v}(t-1) - \cdots - \hat{c}_{n_c}(t)\tilde{v}(t-n_c), \tag{3.125e}$$

$$\psi^{\mathrm{T}}(t+1) = (-\tilde{y}(t) \ \ldots \ -\tilde{y}(t-n_a+1) \ \tilde{u}(t) \ \ldots \ \tilde{u}(t-n_b+1)$$
$$-\tilde{w}(t) \ \ldots \ -\tilde{w}(t-n_f+1) \ \tilde{\varepsilon}(t) \ \ldots \ \tilde{\varepsilon}(t-n_c+1)$$
$$-\tilde{v}(t) \ \ldots \ -\tilde{v}(t-n_d+1)). \tag{3.125f}$$

In this algorithm, (3.124) that calculates $\hat{y}(t+1)$ and (3.125) that calculates $\psi(t+1)$ correspond to (3.67e, f). We could of course, also have written (3.123b, c) in the form (3.70). Minimization of a general criterion and/or using the gradient-updating direction gives obvious variants of (3.123) corresponding to (3.72)–(3.73″).

Notice that in the calculation of φ and ψ the latest possible estimates of $\hat{\theta}$ are used. This is the reason for making a distinction between the "prediction error" $\varepsilon(t)$ (calculated using $\hat{\theta}(t-1)$) that is required for updating θ and the "residual" $\bar{\varepsilon}(t)$ (calculated using $\hat{\theta}(t)$). We commented upon this distinction also in section 2.2.3. It will turn out that the extra effort involved in making this distinction (compared to using $\varepsilon(t)$ in (3.124d), (3.125) instead of $\bar{\varepsilon}(t)$) is worthwhile in practice; see section 5.11.

The algorithm (3.123)–(3.125) for the general model (3.104) is well known in some special cases:

1. For $F(q^{-1}) = C(q^{-1}) = D(q^{-1}) = 1$ the algorithm is, of course, the recursive least squares algorithm of section 2.2.1.

2. For $F(q^{-1}) = D(q^{-1}) = 1$ (an ARMAX model), the algorithm is the RML (recursive maximum likelihood) algorithm derived in section 2.2.3.

3. For $A(q^{-1}) = C(q^{-1}) = D(q^{-1}) = 1$ it corresponds to an algorithm suggested by White (1975).

4. For $F(q^{-1}) = C(q^{-1}) = 1$ it is a recursive maximum likelihood method derived by Gertler and Bányász (1974).

5. Consider again the model $F(q^{-1}) = C(q^{-1}) = 1$. The matrix $R(t)$ defined by (3.123b) is formed from a φ-vector which reads

$$\varphi^T(t) = (\varphi_1^T(t) \ \ \varphi_2^T(t)),$$

where

$$\varphi_1^T(t) = (-y(t-1) \ \ldots \ -y(t-n_a) \ u(t-1) \ \ldots \ u(t-n_b)),$$

$$\varphi_2^T(t) = (-v(t-1) \ \ldots \ -v(t-n_d)).$$

Suppose that the off-diagonal blocks of $R(t)$ (corresponding to the cross terms $\varphi_1 \times \varphi_2$) are forced to zero, so that $R(t)$ consists of two diagonal blocks each updated, analogously to (3.123b), with $\varphi_1(t)$ and $\varphi_2(t)$ respectively. The resulting algorithm is then the recursive generalized least squares algorithm derived by Hastings-James and Sage (1969). This method was called RGLS in Söderström et al. (1974a, 1978).

6. If $A(q^{-1}) = 1$ and the R-matrix in (3.123) is taken as block diagonal (with blocks corresponding to B, F and C, D respectively, analogously to case (5)), the refined instrumental variable method of Young and Jakeman (1979) is obtained (see appendix 4.E for details).

Some Polynomial Known   The general model (3.104) contains several special cases. The ones we discussed in the beginning of section 3.7.1, i.e., (3.98)–(3.103), correspond to the assumption that certain of the polynomials in (3.104) are known to be unity. We could of course also assume that they have other values. The derivation of the identification algorithm is still given by the general recipe: Compute the prediction and its gradient and use approximations of them to update the estimates. The procedure is illustrated in the following example.

EXAMPLE 3.10 (An ARMAX Model with a Known C-polynomial)   Suppose that the model is given by

$$A(q^{-1})y(t) = B(q^{-1})u(t) + C_*(q^{-1})e(t), \qquad \varepsilon(t) - \hat{y}(t|\theta) - y(t) \tag{3.126}$$

where the polynomial C_* is known and A and B are to be determined. The one-step-ahead predictor is given by (3.13):

$$C_*(q^{-1})\hat{y}(t \mid \theta) = [C_*(q^{-1}) - A(q^{-1})]y(t) + B(q^{-1})u(t),$$

$$\theta = (a_1 \ a_2 \ \ldots \ a_{n_a} \ b_1 \ b_2 \ \ldots \ b_{n_b})^T.$$

This expression can also be written as

$$\hat{y}(t \mid \theta) = y(t) - y_*(t) + \theta^\mathrm{T} \varphi_*(t), \tag{3.127a}$$

where

$$y_*(t) = \frac{1}{C_*(q^{-1})} y(t), \qquad u_*(t) = \frac{1}{C_*(q^{-1})} u(t), \tag{3.127b}$$

$$\varphi_*(t) = (-y_*(t-1) \ \ldots \ -y_*(t-n_a) \ u_*(t-1) \ \ldots \ u_*(t-n_b))^\mathrm{T}. \tag{3.127c}$$

From (3.127) it is immediately seen that

$$\left[\frac{d}{d\theta} \hat{y}(t \mid \theta) \right]^\mathrm{T} = \varphi_*(t).$$

The recursive prediction error algorithm is then, for (3.126),

$$\hat{\theta}(t) = \hat{\theta}(t-1) + \gamma(t) R^{-1}(t) \varphi_*(t) \varepsilon(t), \tag{3.128a}$$

$$R(t) = R(t-1) + \gamma(t) [\varphi_*(t) \varphi_*^\mathrm{T}(t) - R(t-1)], \tag{3.128b}$$

$$\varepsilon(t) = y(t) - \hat{y}(t \mid \hat{\theta}(t-1)) = y_*(t) - \hat{\theta}^\mathrm{T}(t-1) \varphi_*(t), \tag{3.128c}$$

where the last equality follows from (3.127a). The algorithm (3.128) is the same as we would have obtained by first filtering all quantities in the model (3.126) through $1/C_*$ (thus obtaining a model with white equation error) and then using the recursive least squares method on the filtered variables. This latter procedure is a fairly natural ad hoc approach when the equation error in models such as (3.98) has known covariance properties. The method has been called generalized least squares (Mendel, 1973), but in recursive identification this term is usually reserved for another procedure (see Clarke, 1967). We have here derived this filtered ad hoc solution in a formal way as a recursive prediction error algorithm. This illustrates the power of the general prediction error approach.

In most cases it is not very natural to assume that the noise characteristics are known. One special case is however of interest. We may know that the only noise source in the system is white output measurement noise. This means that $C(q^{-1})$ is known to be equal to $A(q^{-1})$. The value of A is, however, not known. Based on (3.127)–(3.128) an approach to this problem is obvious: Replace C_* in (3.127b, c) by the current estimate of the A-polynomial and use (3.127)–(3.128) to estimate A and B. This procedure is a recursive variant of the Steiglitz-McBride algorithm (Steiglitz and McBride, 1965). □

3.7.3   An Approximate Gradient: Pseudolinear Regression

Consider again the model set (3.104). We noted that the prediction can be written, as in (3.118),

$$\hat{y}(t \mid \theta) = \theta^{\mathrm{T}}\varphi(t, \theta).$$

If, when calculating the gradient of the prediction, the implicit θ-dependence in $\varphi(t, \theta)$ is neglected, we would obtain an approximate expression

$$\left[\frac{d}{d\theta}\hat{y}(t \mid \theta)\right]^{\mathrm{T}} \approx \varphi(t, \theta). \tag{3.129}$$

Comparing this to (3.119), we see that the quality of the approximation (3.129) will depend on how close the polynomials C, D, and F are to unity.

   If we use the approximation (3.129) in the gradient calculations instead of $\psi(t, \theta)$, we get the following recursive algorithm

$$\varepsilon(t) = y(t) - \hat{\theta}^{\mathrm{T}}(t - 1)\varphi(t), \tag{3.130a}$$

$$R(t) = R(t - 1) + \gamma(t)[\varphi(t)\varphi^{\mathrm{T}}(t) - R(t - 1)], \tag{3.130b}$$

$$\hat{\theta}(t) = \hat{\theta}(t - 1) + \gamma(t)R^{-1}(t)\varphi(t)\varepsilon(t), \tag{3.130c}$$

where $\varphi(t)$ is given by (3.124). This algorithm is a pseudolinear regression (PLR) for the model (3.118), according to the discussion in section 2.5.1. It is simpler than the recursive prediction error algorithm (3.123)–(3.125), in that the filters (3.125) are not required.

   The interpretation of this last algorithm as an approximate prediction error method is of conceptual interest, since it shows the relationship between the approaches. But it should perhaps be pointed out that these algorithms are not necessarily inferior to the true prediction error algorithm. The interpretation in terms of "simple approximations" of the gradient refers to our general framework. The algorithm (3.130) has been derived in a number of special cases in other frameworks and often performs well in simulations and applications. See section 5.9 for a further discussion of this point.

   In some suggested variants of (3.130), the data vectors are filtered through fixed filters chosen by the user:

$$\varepsilon(t) = y(t) - \hat{\theta}^{\mathrm{T}}(t-1)\varphi(t), \tag{3.131a}$$

$$\varepsilon_F(t) = T(q^{-1})\varepsilon(t), \tag{3.131b}$$

$$\varphi_F(t) = S(q^{-1})\varphi(t), \tag{3.131c}$$

$$R(t) = R(t-1) + \gamma(t)[\varphi_F(t)\varphi_F^{\mathrm{T}}(t) - R(t-1)], \tag{3.131d}$$

$$\hat{\theta}(t) = \hat{\theta}(t-1) + \gamma(t)R^{-1}(t)\varphi_F(t)\varepsilon_F(t), \tag{3.131e}$$

where $\varphi(t)$ is given by (3.124).

The algorithm (3.130) or (3.131) is well known in a number of special cases:

1. The extended least squares method (see example 2.7) is obtained as (3.130) for $F(q^{-1}) = D(q^{-1}) = 1$, and for the filters $T(q^{-1}) = S(q^{-1}) = 1$.
2. For the special case $F(q^{-1}) = 1$, $T(q^{-1}) = S(q^{-1}) = 1$ the algorithm (3.130) was called "the extended matrix method" by Talmon and van den Boom (1973).
3. The model-reference-based method suggested by Landau (1976) (see example 2.8) is obtained as (3.131) in the special case $A(q^{-1}) = C(q^{-1}) = D(q^{-1}) = 1$. Then $T(q^{-1})$ is taken as a polynomial and $S(q^{-1}) = 1$. This algorithm has also been called HARF (Hyperstable Adaptive Recursive Filter) by Johnson (1979), and a stochastic gradient variant is called SHARF (Simple HARF) (Johnson et al., 1981). The stochastic gradient version has also been suggested by Feintuch (1976).
4. For $F(q^{-1}) = C(q^{-1}) = 1$ (and filters $S(q^{-1}) = T(q^{-1}) = 1$) the algorithm is that suggested by Bethoux (1974).
5. For $A(q^{-1}) = C(q^{-1}) = 1$ the modified method of Landau (1978) is obtained.

A natural extension of the algorithm (3.131) would be to let the filters S and T be time-varying and dependent upon the current estimate. In fact, the recursive prediction error algorithm (3.123)–(3.125) is obtained for such properly chosen S-filters. It is of interest to note that the instrumental variable method described in section 3.6.3 can be interpreted within the framework of (3.131).

EXAMPLE 3.11 (An Interpretation of the Instrumental Variable Method) Consider the model corresponding to the special case

$$A(q^{-1}) = C(q^{-1}) = D(q^{-1}) = 1.$$

Choose in the algorithm (3.131) the filters $S(q^{-1}) = 1$ and $T(q^{-1}) = \hat{F}_{t-1}(q^{-1})$ (the current estimate of the F-polynomial). Then

$$\varphi^{T}(t) = (u(t-1) \ \ldots \ u(t-n_b) \ -w(t-1) \ \ldots \ -w(t-n_f)),$$

where

$$w(t) = \frac{\hat{B}_{t-1}(q^{-1})}{\hat{F}_{t-1}(q^{-1})} u(t).$$

Here $\hat{B}_{t-1}(q^{-1})$ denotes the current estimate of the B-polynomial. Moreover,

$$\varepsilon(t) = y(t) - w(t)$$

and

$$\varepsilon_F(t) = \hat{F}_{t-1}(q^{-1})y(t) - \hat{F}_{t-1}(q^{-1})w(t)$$
$$= \hat{F}_{t-1}(q^{-1})y(t) - \hat{B}_{t-1}(q^{-1})u(t).$$

Comparing this algorithm with the instrumental variable choice (3.94), we find that $\varphi(t)$ given as above is identical to $\zeta(t)$ when the instrumental variables are taken as inputs and "model outputs." Moreover, $\varepsilon_F(t)$ is equal to the equation error:

$$\varepsilon_F(t) = y(t) - \hat{\theta}^{T}(t-1)\bar{\varphi}(t)$$

with

$$\bar{\varphi}^{T}(t) - (-y(t-1) \ \ldots \ -y(t-n_f) \ u(t-1) \ \ldots \ -u(t-n_b)).$$

Hence the algorithm (3.131) with the above choices of $S(q^{-1})$ and $T(q^{-1})$ coincides with the symmetric recursive instrumental variable method (3.91), (3.92b), (3.94). We have consequently obtained yet another interpretation of the RIV method: as an output error, pseudolinear regression method with a particular prediction error filter. $\square$

3.7.4 Summary

We have in this section discussed a generalized SISO model, that contains most commonly used ones:

$$A(q^{-1})y(t) = \frac{B(q^{-1})}{F(q^{-1})} u(t) + \frac{C(q^{-1})}{D(q^{-1})} e(t).$$

To this model we have applied the general prediction error method, yielding the algorithm (3.123)–(3.125). We have also described a natural approximation of the gradient of the predictor, which leads to a pseudo-linear regression (3.131). These algorithms are given by equations (3.104), (3.110)–(3.118), (3.123)–(3.125), and (3.131). For future quick reference we summarize them below in somewhat symbolic form.

Model: $Ay = \dfrac{B}{F}u + \dfrac{C}{D}e.$

Let $\quad \bar{w} = \dfrac{B}{F}u \quad \bar{v} = Ay - \bar{w} \quad \bar{\varepsilon} = \dfrac{D}{C}\bar{v}$

$$\theta^{\mathrm{T}} = (a \ b \ f \ c \ d)$$

$$\varphi^{\mathrm{T}} = (-y \ u \ -\bar{w} \ \bar{\varepsilon} \ -\bar{v}) \qquad\qquad (3.132)$$

$$\psi^{\mathrm{T}} = \left(-\dfrac{D}{C}y \ \dfrac{D}{CF}u \ -\dfrac{D}{CF}\bar{w} \ \dfrac{1}{C}\bar{\varepsilon} \ -\dfrac{1}{C}\bar{v}\right).$$

Then $\quad \hat{y} = \theta^{\mathrm{T}}\varphi \quad \varepsilon = y - \hat{y}.$

Algorithm: RPEM   PLR

$$\hat{\theta} := \hat{\theta} + \gamma R^{-1}\psi\varepsilon \qquad\qquad \hat{\theta} := \hat{\theta} + \gamma R^{-1}\varphi\varepsilon$$

$$R := R + \gamma(\psi \ \psi^{\mathrm{T}} - R) \qquad R := R + \gamma(\varphi \ \varphi^{\mathrm{T}} - R)$$

Within the two families of methods, some of the most commonly used algorithms can be recognized. We summarize this in table 3.1.

3.8 Application to State-Space Models

In this section we shall discuss the estimation of parameters in state-space models. The model set and corresponding predictors are described in section 3.8.1. Application of the recursive prediction error algorithm to a so-called innovations model is described in section 3.8.2. Algorithms for the general state-space model are discussed in section 3.8.3 and 3.8.4, with several of the technical details deferred to appendixes 3.B and 3.C.

3.8.1   The Model Set

We have in examples 2.3 and 3.3 described the general stochastic state-space model that is widely used in systems theory and applications:

Table 3.1
Special cases of the general model (3.104) and the corresponding RPEM and PLR algorithms

Special Case	RPEM	PLR
$A = F = C = D = 1$	RLS	RLS
$F = C = D = 1$	RLS	RLS
$F = D = 1$	RML (RML2) (Söderström, 1973b; Fuhrt, 1973)	ELS (RML1, AML) (Young, 1968; Panuska, 1968)
$F = C = 1$	RGLS, RML (Hastings-James and Sage, 1969; Gertler and Bányász, 1974)	(Bethoux, 1976)
$A = C = D = 1$	(White, 1975)	Model reference identification; HARF, SHARF (Landau, 1976; Johnson, 1979; Feintuch, 1976)
$A = 1$	Refined IV (Young, 1976); Young and Jakeman, 1979)	
$F = 1$		Extended matrix method (Talmon and van den Boom, 1973)

Note: See list of symbols for all acronyms.

$$x(t + 1) = F(\theta)x(t) + G(\theta)u(t) + w(t),$$
$$y(t) = H(\theta)x(t) + e(t), \tag{3.133a}$$

where $\{w(t)\}$ and $\{e(t)\}$ are sequences of independent random vectors, each vector being of zero mean; the covariance matrices are

$$Ew(t)w^T(t) = R_1(\theta),$$
$$Ee(t)e^T(t) = R_2(\theta), \tag{3.133b}$$
$$Ew(t)e^T(t) = R_{12}(\theta).$$

The initial value $x(0)$ has the properties

$$Ex(0) = x_0(\theta),$$
$$E[x(0) - x_0(\theta)][x(0) - x_0(\theta)]^T = \Pi(\theta). \tag{3.133c}$$

The unknown parameter vector θ may enter the matrices F, G, H, R_1, R_2, and R_{12} in an arbitrary way. We shall only assume that the matrix entries are differentiable with respect to θ. The parametrization of the matrices may be based on different philosophies. The model (3.133) can have been obtained as a result of sampling a continuous-time stochastic state-space model. Then the matrix elements of (3.133) are well defined, but rather complicated, functions of the parameters of the original continuous-time model. (The formulas for this are given in Åström, 1970.) The model (3.133) could also be a canonical representation of an input-output model. Then the parameter vector θ consists of the parameters of the original input-output model. In (1.C.22) we give an example of a state-space representation of a single-input/single-output ARMAX model.

For a given fixed value of θ, the predictor corresponding to the model (3.133) is

$$\hat{x}(t + 1, \theta) = [F(\theta) - K_\theta(t)H(\theta)]\hat{x}(t, \theta) + G(\theta)u(t) + K_\theta(t)y(t)$$

$$\hat{y}(t \mid \theta) = H(\theta)\hat{x}(t, \theta); \quad \hat{x}(0, \theta) = x_0(\theta) \tag{3.134}$$

(see appendix 1.C), where $K_\theta(t)$ is the (time-varying) Kalman gain, determined from $F(\theta)$, $G(\theta)$, $H(\theta)$, $R_1(\theta)$, $R_2(\theta)$, $R_{12}(\theta)$, and $\Pi(\theta)$ via the Riccati equation in the well-known way:

$$P_\theta(t + 1) = F(\theta)P_\theta(t)F^T(\theta) + R_1(\theta) - K_\theta(t)S_\theta(t)K_\theta^T(t), \tag{3.135a}$$

$$S_\theta(t) = H(\theta)P_\theta(t)H^T(\theta) + R_2(\theta), \tag{3.135b}$$

$$K_\theta(t) = [F(\theta)P_\theta(t)H^T(\theta) + R_{12}(\theta)]S_\theta^{-1}(t), \tag{3.135c}$$

$$P_\theta(0) = \Pi(\theta). \tag{3.135d}$$

Here $S_\theta(t)$ is the covariance matrix of the prediction error sequence $\varepsilon(t, \theta) = y(t) - \hat{y}(t \mid \theta)$ (provided that θ gives a correct description of the system). Under weak conditions, the Kalman gain $K_\theta(t)$ will converge to a limit

$$K_\theta(t) \to \bar{K}_\theta \quad \text{as } t \to \infty$$

that can be determined from the equations

$$\bar{P}_\theta = F(\theta)\bar{P}_\theta F^T(\theta) + R_1(\theta) - \bar{K}_\theta \bar{S}_\theta \bar{K}_\theta^T,$$

$$\bar{S}_\theta = H(\theta)\bar{P}_\theta H^T(\theta) + R_2(\theta), \tag{3.136}$$

$$\bar{K}_\theta = [F(\theta)\bar{P}_\theta H^T(\theta) + R_{12}(\theta)]\bar{S}_\theta^{-1}.$$

The time-varying predictor (3.134) then approaches the time-invariant one:

$$\hat{x}(t + 1, \theta) = [F(\theta) - \bar{K}_\theta H(\theta)]\hat{x}(t, \theta) + G(\theta)u(t) + \bar{K}_\theta y(t),$$

$$\hat{y}(t \mid \theta) = H(\theta)\hat{x}(t, \theta). \tag{3.137}$$

In most applications of the Kalman filter actually the time-invariant limit (3.137) is implemented instead of (3.134).

When using the predictor (3.137) we should, by solving (3.136), determine $\bar{K}_\theta$ from $F(\theta)$, $G(\theta)$, $H(\theta)$, $R_1(\theta)$, $R_2(\theta)$, and $R_{12}(\theta)$. The only way that the noise covariances R_i affect the predictor is via $\bar{K}_\theta$ in (3.137). If these matrices do not contain a lot of a priori structure it seems more reasonable to parametrize $\bar{K}_\theta$ directly and explicitly rather than via $R_1(\theta)$, $R_2(\theta)$, and $R_{12}(\theta)$. That would give $K(\theta)$ as a gain matrix whose entries are direct functions of θ, and a model

$$\hat{x}(t + 1, \theta) = [F(\theta) - K(\theta)H(\theta)]\hat{x}(t, \theta) + G(\theta)u(t) + K(\theta)y(t),$$

$$\hat{y}(t \mid \theta) = H(\theta)\hat{x}(t, \theta). \tag{3.138}$$

This model can equivalently be written

$$\hat{x}(t + 1, \theta) = F(\theta)\hat{x}(t, \theta) + G(\theta)u(t) + K(\theta)v(t),$$

$$y(t) = H(\theta)\hat{x}(t, \theta) + v(t), \tag{3.139}$$

where $v(t)$ corresponds to the prediction error or *innovation* $y(t) - \hat{y}(t \mid \theta)$. For this reason (3.139) is usually known as a state-space *innovations model*, or a model in *innovations form*. It can also be regarded as a special case of the general model (3.133) corresponding to special choices of the R_i-matrices in (3.133b).

Notice that while the filters (3.137) and (3.138) are equivalent for a given θ, the important difference for recursive identification is that the innovations representation gives a Kalman gain matrix that is explicitly parametrized, while the general model (3.133) requires indirect calculation of this gain. We shall discuss user aspects of the choice between (3.133) and (3.139) in section 5.2.

3.8.2   The Recursive Prediction Error Method Applied to the Innovations Model

The predictor model (3.138) is explicitly given in the form (3.17), which was used when deriving the general algorithm. Application of the recursive prediction error algorithm (3.67) to (3.138) is therefore straightforward.

The matrix $\mathscr{F}(\theta)$ in (3.17) corresponds to $F(\theta) - K(\theta)H(\theta)$, the matrix $\mathscr{G}(\theta)$ to $(G(\theta)\ K(\theta))$, and $\mathscr{H}(\theta)$ to $H(\theta)$. We have only to derive an expression for the gradient

$$\psi(t, \theta) = \left[\frac{d}{d\theta}\hat{y}(t \mid \theta)\right]^{\mathrm{T}}.$$

Computation of ψ   We obtain from (3.138)

$$\psi^{\mathrm{T}}(t, \theta) = \frac{d}{d\theta}\left[H(\theta)\hat{x}(t, \theta)\right].$$

To handle this expression we introduce (recall that $n = \dim x$, $d = \dim \theta$, $p = \dim y$, $r = \dim u$)

$$W(t, \theta) = \frac{d}{d\theta}\hat{x}(t, \theta) \quad \text{(an } n \times d\text{-matrix)} \tag{3.140a}$$

and

$$D(\hat{\theta}, x) = \frac{\partial}{\partial\theta}[H(\theta)x]|_{\theta=\hat{\theta}} \quad \text{(a } p \times d\text{-matrix)}. \tag{3.140b}$$

Then

$$\psi^{\mathrm{T}}(t, \theta) = H(\theta)W(t, \theta) + D(\theta, \hat{x}(t, \theta)). \tag{3.141}$$

Introduce also

$$\varepsilon(t, \theta) = y(t) - H(\theta)\hat{x}(t, \theta), \tag{3.142a}$$

which gives

$$\frac{d}{d\theta}\varepsilon(t, \theta) = -\psi^{\mathrm{T}}(t, \theta). \tag{3.142b}$$

We now must find an expression for $W(t, \theta)$. To do this we differentiate (3.138):

$$W(t + 1, \theta) = \frac{d}{d\theta}[F(\theta)\hat{x}(t, \theta) - K(\theta)H(\theta)\hat{x}(t, \theta) + G(\theta)u(t) + K(\theta)y(t)]$$

$$= \frac{d}{d\theta}[F(\theta)\hat{x}(t, \theta) + G(\theta)u(t) + K(\theta)\varepsilon(t, \theta)].$$

Let us introduce the matrix

$$\overline{M}(\hat{\theta}, x, u, \varepsilon) \triangleq \frac{\partial}{\partial \theta}[F(\theta)x + G(\theta)u + K(\theta)\varepsilon]|_{\theta=\hat{\theta}} \quad \text{(an } n \times d\text{-matrix)},$$

$$(3.143)$$

This gives, using (3.141), (3.142b),

$$W(t + 1, \theta) = [F(\theta) - K(\theta)H(\theta)]W(t, \theta)$$
$$+ \overline{M}(\theta, \hat{x}(t, \theta), u(t), \varepsilon(t, \theta)) - K(\theta)D(\theta, \hat{x}(t, \theta)).$$

$$(3.144)$$

The equations (3.138), (3.141)–(3.144) now form a set for computing $\hat{y}(t \mid \theta)$ and $\psi(t, \theta)$ from y^t, u^t, corresponding to (3.22) in the general case.

The Identification Algorithm The recursive prediction error algorithm (3.67) applied to the innovations model (3.138) thus gives the following method:

$$\varepsilon(t) = y(t) - \hat{y}(t), \tag{3.145a}$$

$$\hat{\Lambda}(t) = \hat{\Lambda}(t - 1) + \gamma(t)[\varepsilon(t)\varepsilon^{\mathrm{T}}(t) - \hat{\Lambda}(t - 1)], \tag{3.145b}$$

$$R(t) = R(t - 1) + \gamma(t)[\psi(t)\hat{\Lambda}^{-1}(t)\psi^{\mathrm{T}}(t) - R(t - 1)], \tag{3.145c}$$

$$\hat{\theta}(t) = \hat{\theta}(t - 1) + \gamma(t)R^{-1}(t)\psi(t)\hat{\Lambda}^{-1}(t)\varepsilon(t), \tag{3.145d}$$

$$\hat{x}(t + 1) = F_t\hat{x}(t) + G_t u(t) + K_t\varepsilon(t), \tag{3.145e}$$

$$\hat{y}(t + 1) = H_t\hat{x}(t + 1), \tag{3.145f}$$

$$W(t + 1) = [F_t - K_t H_t]W(t) + \overline{M}_t - K_t D_t, \tag{3.145g}$$

$$\psi(t + 1) = W^{\mathrm{T}}(t + 1)H_t^{\mathrm{T}} + D^{\mathrm{T}}(\hat{\theta}(t), \hat{x}(t + 1)). \tag{3.145h}$$

Here

$$F_t = F(\hat{\theta}(t)), \quad G_t = G(\hat{\theta}(t)),$$

$$H_t = H(\hat{\theta}(t)), \quad K_t = K(\hat{\theta}(t)),$$

$$\overline{M}_t = \overline{M}(\hat{\theta}(t), \hat{x}(t), u(t), \varepsilon(t)),$$

$$D_t = D(\hat{\theta}(t), \hat{x}(t)).$$

In this case it is obvious that the stability region for the predictor is given by

$$D_s = \{\theta \mid F(\theta) - K(\theta)H(\theta) \text{ has all eigenvalues strictly inside the unit circle}\}. \tag{3.146}$$

Hence the right-hand side of (3.145d) should be projected into D_s as in (3.67d).

The philosophy behind the ordering of equations in (3.145) can be expressed as follows: "Always use the latest available estimate for θ and do not update a quantity before it is needed."

EXAMPLE 3.12 (An ARMA Model)   The process

$$y(t) + ay(t - 1) = e(t) + ce(t - 1) \tag{3.147}$$

can be realized in state-space form as

$$x(t + 1) = -ax(t) + (c - a)e(t),$$
$$y(t) = x(t) + e(t) \tag{3.148}$$

[see (1.C.21)]. Here we have

$$\theta = \begin{pmatrix} a \\ c \end{pmatrix}, \quad F(\theta) = -a, \quad \bar{K}(\theta) = c - a, \quad G(\theta) = 0,$$

$$H(\theta) = 1, \quad \bar{M}(\theta, x, \varepsilon) = (-x - \varepsilon \quad \varepsilon), \quad D(\theta, x) = (0 \ 0),$$

which gives the algorithm

$$\hat{x}(t) = -\hat{a}_{t-1}\hat{x}(t - 1) + (\hat{c}_{t-1} - \hat{a}_{t-1})\varepsilon(t - 1),$$

$$W(t) = -\hat{c}_{t-1}W(t - 1) + (-\hat{x}(t - 1) - \varepsilon(t - 1) \quad - \varepsilon(t - 1)),$$

$$\psi(t) = W^{\mathrm{T}}(t),$$

$$\varepsilon(t) = y(t) - \hat{x}(t),$$

$$R(t) = R(t - 1) + \gamma(t)[\psi(t)\psi^{\mathrm{T}}(t) - R(t - 1)],$$

$$\begin{pmatrix} \hat{a}_t \\ \hat{c}_t \end{pmatrix} = \begin{pmatrix} \hat{a}_{t-1} \\ \hat{c}_{t-1} \end{pmatrix} + \gamma(t)R^{-1}(t)\psi(t)\varepsilon(t).$$

Since

$$-\hat{x}(t - 1) - \varepsilon(t - 1) = -y(t - 1),$$

we find that

$$\psi(t) + \hat{c}_{t-1}\psi(t - 1) = \begin{pmatrix} -y(t - 1) \\ \varepsilon(t - 1) \end{pmatrix}.$$

The algorithm is therefore identical to the recursive maximum likelihood algorithm of section 2.2.3, and hence to the general algorithm (3.123)–(3.125) of section 3.7.2, applied to the model (3.147). □

3.8.3 The Recursive Prediction Error Method Applied to the General State-Space Model

Let us now consider the general model (3.133). In predictor form it is given by (3.134)–(3.135). There is a slight complication with this predictor, in that the dynamics $[F(\theta) - K_\theta(t)H(\theta)]$ are not a direct function of θ as in (3.17). Instead, the dynamics are determined via an intermediate θ-dependent equation (3.135). However, the gradient of the prediction

$$\psi(t, \theta) = \left[\frac{d}{d\theta}\hat{y}(t \mid \theta)\right]^{\mathrm{T}}.$$

can still be computed as in (3.140)–(3.143):

$$W(t + 1, \theta) = [F(\theta) - K_\theta(t)H(\theta)]W(t, \theta) + M(\theta, \hat{x}(t, \theta), u(t))$$
$$+ \left[\frac{\partial}{\partial\theta}K_\theta(t)\right]\varepsilon(t, \theta) - K_\theta(t)D(\theta, \hat{x}(t, \theta)), \tag{3.149a}$$

$$\psi^{\mathrm{T}}(t, \theta) = H(\theta)W(t, \theta) + D(\theta, \hat{x}(t, \theta)), \tag{3.149b}$$

where $D(\theta, x)$ is given by (3.140b) and where

$$M(\theta, x, u) = \frac{\partial}{\partial\theta}[F(\theta)x + G(\theta)u]. \tag{3.150}$$

The components of the gradient of $K_\theta(t)$ are obtained by differentiating (3.135). This gives a straightforward but lengthy set of equations.

Consider now the recursive identification situation, where ψ and ε are to be calculated using a sequence of estimates $\{\hat{\theta}(t)\}$. Let K_t denote the Kalman gain that is obtained from (3.135) by replacing θ in each time step by the latest available estimate $\hat{\theta}(t)$. Also let $\mathcal{K}_t^{(i)}$ denote the variables obtained by differentiating $K_\theta(t)$ w.r.t. θ_i (ith component) and, in the equations resulting from (3.135), replacing θ by the latest available estimates $\hat{\theta}(t)$. (See equation (3.B.3) for explicit expressions.) A reasonable approximation of the gradient

$$\frac{\partial}{\partial\theta}\hat{y}(t \mid \theta)\big|_{\theta=\hat{\theta}(t)}$$

derived following the philosophy of this chapter will now be given by using (3.145e–h), with $\overline{M}_t$ replaced by

$$M_t^* = M(\hat{\theta}(t), \, \hat{x}(t), \, u(t)) + \mathscr{K}_t \varepsilon(t). \tag{3.151}$$

Here, the interpretation of the last term is that its ith column is $\mathscr{K}_t^{(i)} \varepsilon(t)$. With this modification of $\overline{M}_t$, the resulting algorithm is essentially given by (3.145). A further difference is that the Riccati equation (3.135) provides us with $S(t)$, which is an estimate of the covariance matrix of the innovations. It is natural to replace $\hat{\Lambda}(t)$ calculated according to (3.145b) by $S(t)$ in (3.145c, d). The explicit form of the algorithm is given in appendix 3.B.

3.8.4   An Approximate Gradient: The Extended Kalman Filter

We derived in example 2.4 and appendix 2.A another algorithm for recursive identification of the parameters in the general state-space model (3.133): The extended Kalman filter (EKF). It is of interest to compare the EKF method with the recursive prediction error algorithm of section 3.8.3. The details of this comparison are given in appendix 3.C. The bottom line of the analysis is that the EKF is virtually identical to the recursive prediction error algorithm, except for the important fact that the term $\mathscr{K}_t \varepsilon(t)$ of the expression (3.151) is missing in the EKF. The EKF can therefore be seen as an recursive prediction error algorithm, where the coupling between the parameters θ and the gain $K_\theta(t)$ of the Kalman filter has been neglected. It is true that such an omission is tempting, since the calculation of $\mathscr{K}_t$ may be very complex for the model (3.133). However, leaving the term out may influence the convergence properties of the algorithm. This is illustrated in appendix 4.G.

3.9   User's Summary

The aim of this chapter has been to provide a framework for a systematic discussion of recursive identification methods. Three aims have been met:

• We have derived a general recursive identification method that can be applied to any set of (linear) models.

• The general method has been and will be used as a framework for a unified discussion of many special recursive methods.

- Specific algorithms are obtained by making certain choices within the general method. Consequently, the discussion of various different methods can be carried out in terms of these "user choices."

The model set is defined in general terms as a one-step-ahead predictor $\hat{y}(t \mid \theta)$ that depends on the model parameter vector θ. We have considered the case where this prediction is formed using a linear finite-dimensional filter acting on the observed input-output data $\{z(t)\}$:

$$\varphi(t + 1, \theta) = \mathscr{F}(\theta)\varphi(t, \theta) + \mathscr{G}(\theta)z(t),$$
$$\hat{y}(t \mid \theta) = \mathscr{H}(\theta)\varphi(t, \theta). \tag{3.152}$$

Consequently $\hat{y}(t \mid \theta)$ is the one-step-ahead prediction of $y(t)$ that is obtained by processing the observed data z^{t-1} through the constant model (3.152). We have considered three particular examples of model sets:

- Linear regression models:

$$\hat{y}(t \mid \theta) = \varphi^{T}(t)\theta, \tag{3.153}$$

where $\varphi(t)$ is a function of z^{t-1}.

- A general single-input/single-output model:

$$A(q^{-1})y(t) = \frac{B(q^{-1})}{F(q^{-1})}u(t) + \frac{C(q^{-1})}{D(q^{-1})}e(t). \tag{3.154}$$

- State-space models:

$$x(t + 1) = F(\theta)x(t) + G(\theta)u(t) + w(t),$$
$$y(t) = H(\theta)x(t) + e(t). \tag{3.155}$$

For the general model (3.152) we have shown how the gradient of $\hat{y}(t \mid \theta)$ w.r.t. θ, denoted by $\psi(t, \theta)$, can be computed by means of

$$\xi(t + 1, \theta) = A(\theta)\xi(t, \theta) + B(\theta)z(t),$$
$$\begin{pmatrix} \hat{y}(t \mid \theta) \\ \mathrm{col}\,\psi(t, \theta) \end{pmatrix} = C(\theta)\xi(t, \theta). \tag{3.156}$$

Here "col" means that the columns of the $d \times p$-matrix ψ have been stacked on top of each other.

For this general model (3.152) we derived the following algorithm [see (3.67)]:

$$\varepsilon(t) = y(t) - \hat{y}(t), \tag{3.157a}$$

$$\hat{\Lambda}(t) = \hat{\Lambda}(t-1) + \gamma(t)[\varepsilon(t)\varepsilon^T(t) - \hat{\Lambda}(t-1)], \tag{3.157b}$$

$$\hat{\theta}(t) = [\hat{\theta}(t-1) + \gamma(t)R^{-1}(t)\psi(t)\hat{\Lambda}^{-1}(t)\varepsilon(t)]_{D_{\mathscr{M}}}, \tag{3.157c}$$

$$\xi(t+1) = A(\hat{\theta}(t))\xi(t) + B(\hat{\theta}(t))z(t), \tag{3.157d}$$

$$\begin{pmatrix} \hat{y}(t+1) \\ \operatorname{col}\psi(t+1) \end{pmatrix} = C(\hat{\theta}(t))\xi(t+1). \tag{3.157e}$$

Here $\{\gamma(t)\}$ is a sequence of positive scalars. The bracket in the right-hand side of (3.157c) means that the argument should be projected into that region of $\mathbf{R}^d$ where the linear predictor model (3.152) is stable. $R(t)$ is a positive definite matrix that modifies the search direction. We have discussed the following two particular choices:

- Gauss-Newton direction:

$$R(t) = R(t-1) + \gamma(t)[\psi(t)\hat{\Lambda}^{-1}(t)\psi^T(t) - R(t-1)]. \tag{3.158a}$$

- Gradient direction:

$$R(t) = r(t) \cdot I \qquad (r(t) \text{ a scalar}),$$

$$r(t) = r(t-1) + \gamma(t)[\operatorname{tr}\psi(t)\hat{\Lambda}^{-1}(t)\psi^T(t) - r(t-1)]. \tag{3.158b}$$

The use of a model of the general form (3.152) together with the representation of the prediction $\hat{y}(t \mid \theta)$ and its gradient $\psi(t, \theta)$ in (3.156), and the subsequent calculation of the approximations $\hat{y}$ and ψ in (3.157d, e) may seem complicated. The idea behind (3.156) and (3.157d, e) is, however, simple:

Derive an expression for how the prediction $\hat{y}(t \mid \theta)$ depends on the model parameters. Then derive an expression for the gradient $\psi(t, \theta)$ of $\hat{y}(t \mid \theta)$ with respect to θ. These expressions will be filters that depend on θ and that have observed data as inputs. Then $\hat{y}(t)$ and $\psi(t)$ are determined from these expressions by replacing past $\hat{y}(t - k \mid \theta)$ and $\psi(t - k, \theta)$ with $\hat{y}(t - k)$ and $\psi(t - k)$ and by replacing θ with its most recent estimate.

The algorithm (3.157) aims at minimizing the quadratic criterion

$$\mathrm{E}\tfrac{1}{2}\varepsilon^T(t, \theta)\Lambda_0^{-1}\varepsilon(t, \theta),$$

where Λ_0 is the covariance matrix of the prediction errors. If we instead aim at minimizing the general criterion

$$El(\varepsilon(t, \theta)), \tag{3.159}$$

the only difference is that $\hat{\Lambda}^{-1}(t)\varepsilon(t)$ in (3.157c) must be replaced by $l_\varepsilon^T(\varepsilon(t))$, so that that equation becomes

$$\hat{\theta}(t) = [\hat{\theta}(t-1) + \gamma(t)R^{-1}(t)\psi(t)l_\varepsilon^T(\varepsilon(t))]_{D_\mathcal{M}}. \tag{3.157c'}$$

Also, corresponding normalizations in (3.158) should be used [see (3.73)].

In this chapter we have given explicit expressions for (3.157d, e) when the method is applied to the model sets (3.153)–(3.155). Within the general algorithm (3.157) we have recognized a number of well-known methods, as shown in sections 3.6–3.8. These are either special cases of (3.157), or can be seen as certain approximations of it. The algorithm (3.157) should thus be interpreted as a family of methods, with a number of choices that have to be made by the user. Much of the discussion to follow in this book will deal with these user choices. Let us therefore list them here.

1. *Choice of model set:* The first choice the user has to make is that of the model set. In our terminology, this means that the matrices $\mathcal{F}(\theta)$, $\mathcal{G}(\theta)$, and $\mathcal{H}(\theta)$ in (3.152) have to be chosen. In particular, for the model set (3.154) one has to decide which polynomials should be chosen as unity.

2. *Choice of input signal:* The character of the input signal $\{u(t)\}$ may affect the resulting estimates $\hat{\theta}(t)$ quite substantially. Depending on the application, the input sequence may be at the user's disposal, and then it should be chosen properly.

3. *Choice of criterion function $l(t, \theta, \varepsilon)$:* The criterion function $l(t, \theta, \varepsilon)$ by which the "size" of the prediction error is measured will affect the algorithm and hence the properties of the estimates.

4. *Choice of gain sequence $\{\gamma(t)\}$:* The step size or gain sequence $\{\gamma(t)\}$ (or, equivalently, the forgetting factors $\{\lambda(t)\}$, related to $\{\gamma(t)\}$ by (3.69)) strongly influences the behavior of the algorithm. In an algorithm designed to track time-varying parameters $\gamma(t)$ should not tend to zero, as discussed in section 2.6. However, even for time-invariant parameters, the choice of $\{\gamma(t)\}$ greatly affects the convergence rate.

5. *Choice of search direction:* The algorithm (3.157) leaves the choice of the matrix $R(t)$ free (as long as it is positive definite). The matrix defines

the updating direction. We have in particular mentioned the gradient and the Gauss-Newton directions. Other choices of R can also be used.

6. *Choice of initial conditions:* To initialize (3.157), the values $\hat{\theta}(0)$, $R(0)$, and $\xi(0)$ are required, and they have to be chosen by the user.

Depending on the particular model set used, further options and choices may be possible:

7. *Options of using "approximate" gradients of the predictor:*
7a: For the linear regression model (3.153) we have the option to use an instrumental variable approximation of the gradient of the predictor (section 3.6.3). That is, $\zeta(t)$ replaces $\psi(t)$ in the algorithm, and the form of $\zeta(t)$ has also to be chosen.
7b: For the single-input/single-output model (3.154) we have the option to use a pseudolinear approximation, $\varphi(t)$ of the gradient ($\varphi(t)$ replaces $\psi(t)$ in the algorithm; see section 3.7.3). Then we also have the further option of using filters as in (3.131).
7c: For the state-space model (3.155) we have the option of deleting the gradient of the Kalman gain in the expression for the gradient of the predictor (section 3.8.4 and appendix 3.C). This gives (essentially) the extended Kalman filter.

8. Finally, we might add the choice between using prediction errors $\varepsilon(t)$ or residuals $\bar{\varepsilon}(t)$ in connection with the single-input/single-output model (see (3.124)–(3.125) and the discussion in section 2.2.3).

Out of these choices and options, the first three ones are common for any identification problem, recursive or not. The options 7 have, as we shall see, important implications for the convergence properties of the algorithm. Choices 4 and 5 are very important for the transient behavior of the algorithm, but with reasonable choices they do not affect the convergence properties. Choice 6 has only a transient effect. Choice 8 influences the transient behavior and may also have importance for overall stability properties (see lemma 4.2).

The algorithm (3.157) with its eight choices and options in fact covers most of the recursive parameter algorithms discussed in the literature and used in practice. Specific algorithms, with certain names attached to themselves, correspond to specific combinations of choices 1, 3, 4, 5, 7 and 8. We have given a few examples of this in sections 3.6–3.8. Therefore, the question of which algorithm to use in a given situation can be answered in terms of choices 1–8. In chapter 5 we will give a comprehensive discussion of various aspects of each of these choices. In order to give such guidance, however, we must first develop analytic results concerning the asymptotic properties of the algorithms under discussion.

3.10 Bibliography

Section 3.2   The concept of persistent excitation was introduced by Åström and Bohlin (1965). A discussion of the concept can be found, e.g., in Ljung (1971). The general model description (3.4) was used in Ljung (1976a, 1978c).

Section 3.3   A general discussion of prediction error methods can be found in Åström (1980b), Caines (1978) and Ljung (1978c). The latter reference contains a proof of (3.37). The result (3.38) is shown in Ljung and Caines (1979). The maximum likelihood method was first used in system identification (for an ARMAX model) by Åström and Bohlin (1965). The calculations leading to (3.45) and the optimization of this expression are given in Caines and Ljung (1976).

Sections 3.4, 3.5   The derivation of the general recursive prediction error method follows Ljung (1978d, 1981). Moore and Weiss (1979) have discussed a similar algorithm for the quadratic case using arguments such as those in section 2.2.3.

Section 3.6   Recursive algorithms for linear regression models are discussed in detail in Mendel (1973) and Strejc (1980). Some instrumental variable variants for the multivariable case are discussed and analysed in Stoica and Söderström (1982a). The choice (3.95a) of instrumental variables was suggested by Wouters (1972).

Section 3.7   The framework in this section for treating several black box models follows Ljung (1977a, 1978e, 1979b). The unified features of these algorithms have also been stressed, by Solo (1978, 1980) and Dugard and Landau (1980a). The table on p. 447 in the latter refers to different algorithms that relate to our classification as follows: Alg. 1: PLR, $A = C = D = 1$; Alg. 2: PLR, $A = C = 1$; Alg. 4: PLR, $F = D = 1$; Alg. 5: RPEM, $F = D - 1$; Alg. 6: PLR, $F = C = 1$. Alg. 3 is a reparametrized version of Alg. 4, estimating A, B, and G in a model $Ay = Bu + (A + G)e$.

Section 3.8   The off-line estimation of parameters in a continuous-time state-space model is discussed in Åström and Källström (1973) and Mehra and Tyler (1973). The same technique can also be applied to recursive identification, and is in fact somewhat simpler, since the system does not have to be sampled. The continuous-time model can be propagated by numerical integration using the current estimates between sampling instants. State-space models based on canonical parametrization is extensively treated in the adaptive observer literature. See Dugard, Landau, and Silviera (1980) for a recent survey. As seen from example 3.12, such methods are closely related to the algorithms of section 3.7. The effect of the auxiliary signals that are injected into the observers is to make the observer output be a prediction based on the latest parameter estimates. Recursive identification for canonical, multivariable state-space models is treated in Moore and Ledwich (1980) and Ljung (1978f). Our discussion of the innovations form algorithm and the extended Kalman filter is based on Ljung (1979a).

4 Analysis

4.1 Introduction

In the previous chapter we have discussed a framework for developing and describing recursive identification algorithms. We have also studied a number of specific identification schemes within this framework. We now turn to the logical questions: How do these algorithms perform, i.e., what are the properties of the estimates $\{\hat{\theta}(t)\}$? How do the choices and options that we listed in section 3.9 affect these properties?

Such questions can be answered in several ways. One way is to apply the algorithms to known data sequences and to evaluate the obtained estimates $\{\hat{\theta}(t)\}$. This is known as *simulation*. Simulation is a very useful tool for investigating recursive identification algorithms. In chapter 5 we shall evaluate the algorithms and the effect of various choices mainly by simulation. However, a serious limitation of simulation is that it may not be conclusive. It is difficult to tell whether a simulation result has universal implications, or merely reflects properties of the chosen data sequence. To obtain results of more general validity we must use *analysis*. In the present context, analysis means that we make certain assumptions about the data set $\{z(t)\}$ and try to calculate what the resulting properties of the estimate sequence $\{\hat{\theta}(t)\}$ are. Since the mappings from z^t to $\hat{\theta}(t)$ that we have discussed are nonlinear and time-varying, their analysis is quite difficult except in certain special cases. In general we can analytically describe only the *asymptotic properties* of $\hat{\theta}(t)$, i.e., the properties as t approaches infinity. Such results will be given in this chapter under the assumption that the gain sequence $\{\gamma(t)\}$ tends to zero. There are thus also limitations to the analytic treatment. Let us comment upon two of these limitations here.

(1) Although analysis can provide us with information about asymptotic properties such as the value to which $\{\hat{\theta}(t)\}$ will converge, the asymptotic rate of convergence, and the asymptotic covariance matrix, it usually does not provide any hints about how large t has to be for the results to be applicable. It may be $t \approx 100$ or $t \approx 10^6$, which clearly makes a big difference to the user. Therefore, to get some insight into the practical convergence rate, transient behavior, and finite-sample properties, the analysis must be complemented by simulation studies.

(2) The assumption that the gain sequence tends to zero is another limitation of the analysis. The properties of the system or signal may be varying with time. Then, as we remarked in section 2.6, we must have a

gain that does not tend to zero in order to track the variations. In example 2.9 we gave explicit and simple analytical results in a special case of this kind. Apart from that, few results have been given for the problem where the true system is time-varying. When the system is slowly time-varying the gain should tend to a small value $\gamma_0 > 0$. The analysis we provide in this chapter for the case $\gamma(t) \to 0$ will thus have implications also for how the algorithms behave in such a case when γ_0 is small enough.

There is of course a certain amount of mathematical satisfaction associated with analytical results, pushing assumptions to a minimum, etc. The value of analysis for the user, however, is related to its implications for the choice and understanding of algorithms. In our case the bottom line of analysis is fairly easy to express, while the analysis itself is rather complex and technical. We shall therefore, in the next section, provide a preview of the results that are obtained later on in this chapter. In this preview the user-oriented implications of the analysis will be pointed out.

The present chapter is structured as follows. After the preview of results in section 4.2, we discuss tools for convergence analysis in section 4.3. Sections 4.3.1 and 4.3.2 will give the reader sufficient insight into the character of the tools to be able to follow the rest of the chapter. Recursive prediction error methods are studied in section 4.4. Pseudolinear regressions are studied in section 4.5, and instrumental variable techniques are analysed in section 4.6. A user's summary of the chapter is given in section 4.7.

Finally, we recognize that the analysis is technically difficult. Portions of this chapter are no doubt more difficult to read than the rest of the book. At the same time, it is necessary to understand the basic analytical results in order to follow the discussion in the remaining chapters. We have tried to organize this chapter so that the reader can acquire such an understanding, without having to get involved in the technicalities of how things are proved. This is illustrated in figure 4.1.

4.2 Asymptotic Properties of Recursive Identification Methods: A Preview

The general family of recursive algorithms was defined by (3.157). Here the gradient of the prediction with respect to the parameters played a crucial role. This gradient was denoted by $\psi(t, \theta)$, and the approximation

A Necessary Path
Through Chapter 4

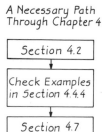

A Sufficient Path
Through Chapter 4

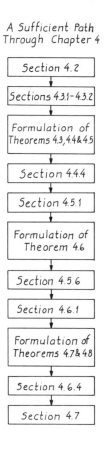

Figure 4.1
A necessary and a sufficient path through chapter 4. "Necessary" and "sufficient" mean
"necessary/sufficient in order to understand the rest of the book."

proposed in the recursive prediction error algorithm was denoted by
$\psi(t)$. In chapter 3 we also discussed some other approximation of ψ; see
choice 7 in section 3.9. In the pseudolinear regression approach, $\psi(t)$
was replaced by a vector $\varphi(t)$ [see (3.130)] and in the recursive instrumental variable approach it was replaced by $\zeta(t)$ [see (3.91)]. It turns out
that the convergence properties of the algorithm will critically depend
on which of these approximations is used for $\psi(t)$. Therefore we shall
discuss the convergence properties separately for each of these choices.

    In this section we will not state the underlying assumption under which
the results hold. They are, however, quite weak, and should not be con-

sidered as restrictive when it comes to practical use of the algorithms—
with the important exception that adaptive control applications are not
included. This means that the results of the present chapter assume that
the generation of the input sequence is not based on the current estimates.
Other types of feedback are, however, allowed. Applications to adaptive
control are treated separately in section 7.3.

4.2.1 Recursive Prediction Error Methods

The general recursive prediction error method (3.72) was designed to
minimize the criterion

$$\bar{V}(\theta) = \bar{E}l(t, \theta, \varepsilon(t, \theta)), \tag{4.1}$$

where $\varepsilon(t, \theta)$ is the prediction error associated with the parameter value θ:

$$\varepsilon(t, \theta) = y(t) - \hat{y}(t \mid \theta), \tag{4.2}$$

and $\bar{E}$ is defined as follows:

$$\bar{E}f(t) = \lim_{N \to \infty} \frac{1}{N} \sum_{t=1}^{N} Ef(t). \tag{4.3}$$

This is done by recursively adjusting the estimates in what is believed to
be a descent direction (the negative gradient direction modified by multiplication with a strictly positive definite matrix R^{-1}). The convergence
result is easy to express. The algorithm does exactly what it is required
to: *The estimate $\hat{\theta}(t)$ will converge w.p. 1 to a local minimum of $\bar{V}(\theta)$ as t
approaches infinity.* (This will be shown in theorems 4.3 and 4.4 in section
4.4.2.) The convergence properties of the recursive algorithm consequently
coincide with those of the off-line method, described in section 3.3.

There are two issues to be discussed in connection with this result. One
regards the word "local." Since the algorithm is based on local search,
we cannot expect global convergence to a global minimum, if $\bar{V}(\theta)$ has
several local minima. The number of local minima will depend on the
character of the model set used. Some results about existence and nonexistence of undesired local minima for certain model sets will be reviewed
in section 5.3.

The other issue is that the quoted convergence result holds whether or
not the true system belongs to the model set. This means that even if the
true system is more complex then our models, the identification procedure

will pick the best approximation of the system. "Best" here refers to the prediction performance as measured by the criterion (4.1). Therefore the algorithm is not sensitive to particular assumptions related to the true system (which are difficult to verify): it does the best it possibly can. This is a valuable robustness property.

Notice, however, that what is the best model approximation of a given system in general depends on the input signal used. An approximation that is good for a certain sinusoidal input may be bad for a white-noise input. The recursive algorithm converges to that approximation that is best under the input signal used during the experiment. If the true system is a member of the model set, though, the "best approximation" is equal to the true system. Then the recursive identification algorithm will converge to the true system description, regardless of the input, as long as it is general enough (see section 3.2.1). Such issues are discussed in section 4.4.4.

The asymptotic distribution of the estimate $\hat{\theta}(t)$ can also be determined. Suppose we use a Gauss-Newton algorithm (3.72)–(3.73) with a gain sequence such that $t \cdot \gamma(t) \to 1$ as $t \to \infty$. If θ^* denotes the value to which $\hat{\theta}(t)$ converges we then have

$$\sqrt{t}[\hat{\theta}(t) - \theta^*] \in \text{As N}(0, P), \tag{4.4}$$

which means that the left-hand side converges in distribution to the normal distribution with zero mean and covariance matrix P. This is proved in theorem 4.5 in section 4.4.3. In practical terms this implies that probabilistic questions related to $\hat{\theta}(t)$ (such as, "What is the probability that $\hat{\theta}(t)$ differs from θ^* by more than 10%?") can be answered, for sufficiently large t, using tables for the normal distribution. Such results are of course useful when giving confidence intervals for the estimates.

The matrix P is the same as in the corresponding result (3.39) for the off-line case. This means that *the asymptotic properties of the recursive algorithm are the same as those of the corresponding off-line method.* This may seem surprising, since information is continually being thrown away in the recursive algorithm. The heuristic explanation is that when $\hat{\theta}(t)$ comes close to θ^*, a quadratic Taylor expansion in θ of the criterion becomes very accurate. The recursive Gauss-Newton algorithm can then be regarded as a recursive least squares (RLS) algorithm for this quadratic criterion, and RLS does not destroy information, as we saw in section 2.2.1. It should be stressed, though, that this result is an asymptotic one. It does not imply that the recursive estimate is as good as the off-line one after only a finite number of data has been processed.

In the case where $\theta^* = \theta_0$ gives white prediction errors (i.e., $\{\varepsilon(t, \theta_0)\}$ is a sequence of independent random vectors each of zero mean and covariance matrix Λ_0), and the algorithm (3.67) is used, we furthermore have

$$R^{-1}(t) \to P = [\bar{E}\psi(t, \theta_0)\Lambda_0^{-1}\psi^T(t, \theta_0)]^{-1} \quad \text{w.p.1 as } t \to \infty, \tag{4.5}$$

where $R(t)$ is the matrix in the recursive Gauss-Newton algorithm and P is the asymptotic covariance matrix. This means that RPEM provides its own estimates of the covariance matrix for $\hat{\theta}(t)$.

4.2.2 Pseudolinear Regressions

Consider now the case that the gradient vector $\psi(t)$ in the algorithm (3.157) is replaced by the pseudolinear regression (PLR) approximation $\varphi(t)$ as described in section 3.7.3. The vector $\psi(t)$ can be obtained from $\varphi(t)$ by filtering its entries through certain filters associated with the estimate $\hat{\theta}(t)$. The exact form of these filters depends on the particular model set that is used. For example, if we use an ARMAX model,

$$A(q^{-1})y(t) = B(q^{-1})u(t) + C(q^{-1})e(t),$$

the relationship is

$$\psi(t) = \frac{1}{\hat{C}_t(q^{-1})} \varphi(t) \tag{4.6}$$

[see (3.122)]. Here $\hat{C}_t(q^{-1})$ is the current estimate of the C-polynomial. Similarily, for the output error model

$$y(t) = \frac{B(q^{-1})}{F(q^{-1})} u(t) + e(t)$$

the relation is

$$\psi(t) = \frac{1}{\hat{F}_t(q^{-1})} \varphi(t). \tag{4.7}$$

We might say that the PLR approximation neglects the filtering that should be done in (4.6) and (4.7).

One might then suspect that the success of the PLR method will depend on how crude an approximation this is. In fact, if, in the ARMAX case, the true system is given by

$$A_0(q^{-1})y(t) = B_0(q^{-1})u(t) + C_0(q^{-1})e(t), \tag{4.8}$$

then a sufficient condition for convergence of $\hat{\theta}(t)$ to the true parameters θ_0 is that

$$|C_0(e^{i\omega}) - 1| < 1 \quad \forall \omega. \tag{4.9}$$

provided the input is general enough. This condition can obviously be interpreted as a measure of how good an approximation it is to replace $\psi(t)$ by $\varphi(t)$ close to $\hat{\theta}(t) = \theta_0$. Notice that (4.9) can also be written as

$$\text{Re} \left[\frac{1}{C_0(e^{i\omega})} - \frac{1}{2} \right] > 0 \quad \forall \omega, \tag{4.10}$$

which is often expressed as "the filter $\dfrac{1}{C_0(q^{-1})} - \dfrac{1}{2}$ is strictly positive real." The analogous result holds for the output error model. These results are proved in section 4.5.2.

While the aforementioned conditions are sufficient for convergence it is also known that PLR algorithms may not converge. In fact, if

$$\text{Re}\, C_0(e^{i\omega}) > 0 \quad \forall \omega \tag{4.11}$$

does not hold, we can always find an A-polynomial, a B-polynomial and a (well-behaved) input signal such that the probability that $\hat{\theta}(t)$ converges to the desired value θ_0 is zero. This means that as a condition on the C-polynomial alone, (4.11) is necessary to assure convergence. These results are proved in section 4.5.4.

Notice that these convergence results apply only if the true system indeed belongs to the model set. If the true system is more complicated than the models in the model set, there are no general results on to what model the algorithm converges.

No explicit expression for the asymptotic covariance matrix for the estimates obtained by PLR is known in general. Notice in particular that it is *not true* for a PLR algorithm that $R(t)$ converges to the asymptotic covariance matrix (in contrast to the case (4.5) for RPEM).

4.2.3 Instrumental Variable Methods

We introduced in section 3.6.3 the instrumental variable (IV) method as a way to estimate the parameter vector θ_0 in a relationship

$$y(t) = \theta_0^T \varphi(t) + v(t). \tag{4.12}$$

Here $\{v(t)\}$ is a sequence of zero-mean disturbances of unspecified character, not necessarily white noise. Convergence and asymptotic properties of the nonsymmetric method (3.96), i.e., of

$$\hat{\theta}(t) = \hat{\theta}(t-1) + \gamma(t) R^{-1}(t) \zeta(t) [y_F(t) - \hat{\theta}^T(t-1) \varphi_F(t)],$$

$$R(t) = R(t-1) + \gamma(t)[\zeta(t) \varphi_F^T(t) - R(t-1)],$$

$$y_F(t) = T(q^{-1}) y(t), \qquad \varphi_F(t) = T(q^{-1}) \varphi(t),$$

$$t \cdot \gamma(t) \to 1 \text{ as } t \to \infty,$$

can be established by examination of the off-line expression

$$\hat{\theta}(t) = \left[\sum_1^t \zeta(k) \varphi_F^T(k)\right]^{-1} \sum_1^t \zeta(k) y_F(k), \tag{4.13}$$

when the filters involved in the generation of $\{\zeta(t)\}$ are time-invariant. The result is that if (4.12) describes the true system, then $\hat{\theta}(t)$ will tend to θ_0 as t approaches infinity if

$$\bar{E}\zeta(t)\varphi^T(t) \text{ is nonsingular} \tag{4.14}$$

and

$$\bar{E}\zeta(t)v(t) = 0. \tag{4.15}$$

These conditions (4.14) and (4.15), in turn, are generically satisfied for all the choices of $\zeta(t)$ and prefilters mentioned in section 3.6.3, based on time-invariant filters, provided that the system operates in an open-loop mode and that the model orders *coincide* with the true system orders. "Generically" should here be interpreted as follows: Conditions (4.14) and (4.15) depend on the properties of the true system (4.12) as well as on the choice of instrumental variables and filters. If, for a given choice of $\{\zeta(t)\}$ and $T(q^{-1})$, the true parameter θ_0 is picked at random, then the probability that (4.14) and (4.15) will not hold is zero. This is discussed in section 4.6.2. The conditions for convergence are consequently quite liberal.

Furthermore, it can be shown that

$$\sqrt{t}[\hat{\theta}(t) - \theta_0] \in \text{AsN}(0, \bar{P})$$

where an explicit expression for $\bar{P}$ can be given in terms of the true system

(4.12), the properties of $\{v(t)\}$, and the chosen instrumental variables and filters (see theorem 4.8). It turns out that the "size" of $\bar{P}$ significantly depends on the choice of $\{\zeta(t)\}$ and $T(q^{-1})$. It is therefore of interest to make optimal choices of these variables to minimize $\bar{P}$. Such choices, however, depend on the properties of the true system (4.12). It turns out that the choice of prefilter $T(q^{-1})$ is related to the properties of $\{v(t)\}$, and the choice of $\{\zeta(t)\}$ is related to the true value θ_0. These properties are, however, unknown a priori. Hence a natural idea is to generate $\{\zeta(t)\}$ and choose $T(q^{-1})$ based on $\hat{\theta}(t)$ and on current estimates of the properties of $\{v(t)\}$, Young (1976). A closer look at these choices shows that the resulting algorithm is in fact identical to the recursive prediction error method for the model (3.154) with $A(q^{-1}) = 1$. This is discussed in appendix 4.E.

4.3 Tools for Convergence Analysis

4.3.1 Introduction

The asymptotic analysis of this chapter will deal both with convergence and with asymptotic distribution. In this section we shall discuss tools for the convergence part of the analysis. A major goal is to have the analysis be applicable to general algorithms. Let us therefore here extract an algorithmic structure that covers the different families of methods previously described.

The basic structure for algorithms related to quadratic criteria is, as we found in section 3.9,

$$\varepsilon(t) = y(t) - \hat{y}(t), \tag{4.16a}$$

$$R(t) = R(t-1) + \gamma(t)[\eta(t)\Lambda^{-1}(t)\eta^{\mathsf{T}}(t) - R(t-1)], \tag{4.16b}$$

$$\hat{\theta}(t) = \hat{\theta}(t-1) + \gamma(t)R^{-1}(t)\eta(t)\Lambda^{-1}(t)\varepsilon(t), \tag{4.16c}$$

$$\xi(t+1) = A(\hat{\theta}(t))\xi(t) + B(\hat{\theta}(t))z(t), \tag{4.16d}$$

$$\begin{pmatrix} \hat{y}(t+1) \\ \operatorname{col} \eta(t+1) \end{pmatrix} = C(\hat{\theta}(t))\xi(t+1). \tag{4.16e}$$

Here, in the parameter updating equation (4.16c), $\eta(t)$ is a vector that is related to the gradient of the prediction $\hat{y}(t)$ with respect to $\hat{\theta}$. We have in chapter 3 discussed three specific choices of $\eta(t)$ (ψ, φ, and ζ; see section 3.9).

When aiming at minimizing general criteria $\bar{E}l(t, \theta, \varepsilon(t, \theta))$, using a general search direction, the equations (4.16b, c) should be replaced by

$$R(t) = R(t - 1) + \gamma(t)H(t, R(t - 1), \hat{\theta}(t - 1), \varepsilon(t), \eta(t)), \qquad (4.16b')$$

$$\hat{\theta}(t) = \hat{\theta}(t - 1) + \gamma(t)R^{-1}(t)h(t, \hat{\theta}(t - 1), \varepsilon(t), \eta(t)). \qquad (4.16c')$$

Here the functions H and h are related to the criterion function l by expressions like (3.72), and (3.73).

Convergence analysis of (4.16) is in general difficult. A major reason for that is the coupling between (4.16c) and (4.16d, e): The quantities $\eta(t)$ and $\varepsilon(t)$ are formed using (implicitly) all previous estimates $\hat{\theta}(k)$, $k = 1, \ldots, t - 1$, which makes the mapping from z^t to $\hat{\theta}(t)$ fairly complex. In the literature the following approaches have been taken to treat this problem.

1. *Associate a deterministic differential equation with (4.16). Its stability properties will be tied to the convergence properties of (4.16).* This approach to convergence analysis was suggested in Ljung (1974, 1977b). It can be applied to the general algorithms (4.16').

2. *Introduce a function that plays the role of a stochastic Lyapunov function for the problem (4.16). Use martingale theory to study the convergence of this function and hence of (4.16).* This approach was apparently first used in Moore and Ledwich (1980) and Solo (1978, 1979). The approach has so far been applied only to pseudolinear regressions.

3. *Examine the expression for a "summed regression" $R(t)\hat{\theta}(t)$.* This approach was apparently suggested by Hannan (1980) and subsequently used by Solo (1978). It has so far been applied only to Gauss-Newton type algorithms for fairly simple special cases.

In this chapter we shall primarily use approach (1), due to its general applicability. In appendix 4.C we will illustrate the second approach by giving an alternative proof for a key lemma in the analysis of pseudolinear regressions. The remainder of this section will be devoted to an outline of the theory behind approach (1). The approach is based on a theorem in Ljung (1977b), which can be regarded as the counterpart of the martingale convergence theorem used in approach (2). That theorem is quoted in section 4.3.3. We will in section 4.3.2 give a heuristic outline of the proof and some intuitive explanations of the results. The required regularity conditions on (4.16′) will be stated and discussed in section 4.3.4, and the basic theorem will be formulated in section 4.3.5.

This section is structured so that the reader who is not concerned with the formalities and technicalities in the convergence proofs may read only section 4.3.2. This should give sufficient background for an understanding of the results and proofs of sections 4.4 and 4.5.

4.3.2   An Associated Differential Equation: A Heuristic Discussion

Consider the algorithm (4.16). It is a recursive stochastic time-varying difference equation. We shall in this section heuristically investigate how it is likely to behave when t becomes large.

For sufficiently large t, the step size $\gamma(t)$ in (4.16c) will be arbitrarily small, due to our assumption that $\gamma(t) \to 0$ as $t \to \infty$. Then the estimates $\{\hat{\theta}(t)\}$ will change more and more slowly. Let us study the consequences of this fact for (4.16d, e). We have from (4.16d)

$$\xi(t) = \sum_{j=0}^{t-1} \left[\prod_{k=j+1}^{t-1} A(\hat{\theta}(k)) \right] B(\hat{\theta}(j))z(j). \tag{4.17}$$

Suppose now that $\hat{\theta}(k)$ belongs to a small neighborhood of a value $\bar{\theta}$ for $t - K \leq k \leq t - 1$, such that $\bar{\theta} \in D_s$. (Remember that D_s, defined by (3.24) is the set of those θ for which $A(\theta)$ has all eigenvalues strictly inside the unit circle.) Then, if the neighborhood is small enough, we can write

$$\prod_{k=t-K}^{t-1} A(\hat{\theta}(k)) \approx A(\bar{\theta})^K, \tag{4.18}$$

which has a norm smaller than $C \cdot \lambda^K$ for some $\lambda < 1$. For large enough K, we may thus approximate (4.17) as

$$\xi(t) \approx \sum_{j=t-K}^{t-1} A(\bar{\theta})^{t-j-1} B(\bar{\theta})z(j). \tag{4.19}$$

Now we can add terms corresponding to $A(\bar{\theta})^{t-j} B(\bar{\theta})z(j)$ for $j < t - K$ to this sum. That will only make an arbitrarily small change, since $A(\bar{\theta})$ is stable. We thus have

$$\xi(t) \approx \xi(t, \bar{\theta}) \triangleq \sum_{j=0}^{t-1} A(\bar{\theta})^{t-j-1} B(\bar{\theta})z(j), \tag{4.20}$$

which can be written recursively as

$$\xi(t+1, \bar{\theta}) = A(\bar{\theta})\xi(t, \bar{\theta}) + B(\bar{\theta})z(t),$$
$$\xi(0, \bar{\theta}) = 0.$$
(4.21)

As a consequence we also have

$$\hat{y}(t) \approx \hat{y}(t \mid \bar{\theta}), \quad \eta(t) \approx \eta(t, \bar{\theta}),$$
$$\varepsilon(t) \approx \varepsilon(t, \bar{\theta}),$$
(4.22a)

where

$$\begin{pmatrix} \hat{y}(t \mid \bar{\theta}) \\ \operatorname{col}\eta(t, \bar{\theta}) \end{pmatrix} = C(\bar{\theta})\xi(t, \bar{\theta}),$$
(4.22b)

$$\varepsilon(t, \bar{\theta}) = y(t) - \hat{y}(t \mid \bar{\theta}).$$
(4.22c)

When $\hat{\theta}(t)$ is close to $\bar{\theta}$ and $R(t)$ is close to $\bar{R}$ and t is large we can consequently use the approximation (4.22a) to conclude that (4.16b, c) approximately behave like

$$\hat{\theta}(t) \approx \hat{\theta}(t-1) + \gamma(t)\bar{R}^{-1}\eta(t, \bar{\theta})\Lambda^{-1}\varepsilon(t, \bar{\theta}),$$
(4.23a)

$$R(t) \approx R(t-1) + \gamma(t)[\eta(t, \bar{\theta})\Lambda^{-1}\eta^{\mathrm{T}}(t, \bar{\theta}) - \bar{R}].$$
(4.23b)

Remark Notice that in going from (4.16b, c) to (4.23) we have simply reversed the chain of arguments that took us from (3.51) to (3.54) in section 3.4

Introduce the expected values

$$f(\bar{\theta}) \triangleq \mathrm{E}\eta(t, \bar{\theta})\Lambda^{-1}\varepsilon(t, \bar{\theta}),$$
(4.24a)

$$G(\bar{\theta}) \triangleq \mathrm{E}\eta(t, \bar{\theta})\Lambda^{-1}\eta^{\mathrm{T}}(t, \bar{\theta}),$$
(4.24b)

mean of $\eta\Lambda^{-1}\varepsilon = f(\theta)$
$\eta^{-1}\eta^{T} = G(\theta)$

where expectation is over z^t. Since t is large, we have neglected the transients in (4.21) and assumed the right-hand side of (4.24) to be time-invariant. We thus have

$$\hat{\theta}(t) \approx \hat{\theta}(t-1) + \gamma(t)\bar{R}^{-1}f(\bar{\theta}) + \gamma(t)v(t),$$
$$R(t) \approx R(t-1) + \gamma(t)[G(\bar{\theta}) - \bar{R}] + \gamma(t)w(t),$$
(4.25)

where $\{v(t)\}$ and $\{w(t)\}$ are zero-mean random variables. Now, let $\Delta\tau$ be a small number and let t, t' be defined by

mean removed $f(\theta)$
$G(\theta)$

$$\sum_{k=t}^{t'} \gamma(k) = \Delta\tau. \tag{4.26}$$

If $\hat{\theta}(t) = \bar{\theta}$ and $R(t) = \bar{R}$, we then have from (4.25)

$$\theta(t') \approx \bar{\theta} + \Delta\tau\bar{R}^{-1}f(\bar{\theta}) + \sum_{k=t}^{t'} \gamma(k)v(k),$$

$$R(t') \approx \bar{R} + \Delta\tau[G(\bar{\theta}) - \bar{R}] + \sum_{k=t}^{t'} \gamma(t)w(t). \tag{4.27}$$

Since $v(k)$ and $w(k)$ have zero means, the contribution from the third terms of the right-hand side will be of an order of magnitude less than those from the second terms. Therefore

$$\hat{\theta}(t') \approx \bar{\theta} + \Delta\tau\bar{R}^{-1}f(\bar{\theta}),$$

$$\bar{R}(t') \approx \bar{R} + \Delta\tau[G(\bar{\theta}) - \bar{R}]. \tag{4.28}$$

With a change of time scale, according to (4.26) such that $t \leftrightarrow \tau$ and $t' \leftrightarrow \tau + \Delta\tau$ we could regard (4.28), for small $\Delta\tau$, as a scheme to solve the differential equation

$$\frac{d}{d\tau}\theta_D(\tau) = R_D^{-1}(\tau)f(\theta_D(\tau)),$$

$$\frac{d}{d\tau}R_D(\tau) = G(\theta_D(\tau)) - R_D(\tau). \tag{4.29}$$

Here we use subscript D to distinguish the solution of (4.29) from the variables in the algorithm (4.16). The chain of arguments suggests that if for some large t_0

$$\hat{\theta}(t_0) = \theta_D(\tau_0), \qquad R(t_0) = R_D(\tau_0), \qquad \sum_{k=1}^{t_0} \gamma(k) = \tau_0,$$

then for $t > t_0$

$$\hat{\theta}(t) \approx \theta_D(\tau), \qquad R(t) \approx R_D(\tau), \qquad \sum_{k=1}^{t} \gamma(k) = \tau. \tag{4.30}$$

These arguments have of course been entirely heuristic. They point, however, to the result (4.30), that asymptotically the algorithm (4.16) can be linked to the differential equation (4.29). The estimates $\hat{\theta}(t)$ should, in some sense, follow the trajectories of the d.e. (4.29) asymptotically.

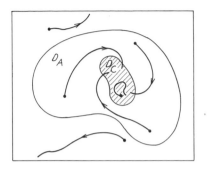

Figure 4.2
An invariant set D_c and its domain of attraction D_A.

Now, the asymptotic behavior of a d.e. is usually expressed in terms of stability. Before giving details of the nature of the link between (4.16) and (4.29), let us therefore pause to give a brief background of stability theory. For details, see Hahn (1967).

A Brief Account of Stability Theory   Consider a d.e. $[\dot{x} = dx(\tau)/d\tau]$

$$\dot{x} = f(x). \tag{4.31}$$

A set D_c is called an *invariant set* for the d.e. (4.31) if any trajectory that starts in D_c remains there:

$$x(0) \in D_c \Rightarrow x(\tau) \in D_c \text{ for all } \tau. \tag{4.32}$$

A *stationary point* x^* of the d.e. (4.31) is a point such that

$$f(x^*) = 0. \tag{4.33}$$

A stationary point is thus an invariant set, since

$$x(0) = x^* \Rightarrow \dot{x} = f(x^*) = 0 \Rightarrow x(\tau) = x^* \text{ for all } \tau.$$

Any invariant set D_c has a *domain of attraction* D_A, which is such that any trajectory that starts in D_A will converge to D_c as $\tau \to \infty$:

$$x(0) \in D_A \Rightarrow x(\tau) \to D_c \text{ as } \tau \to \infty. \tag{4.34}$$

Obviously, $D_A \supset D_c$. These concepts are illustrated in figure 4.2. If D_A contains a neighborhood of D_c ("is strictly larger than D_c") the set D_c is a stable invariant set. If D_A equals the whole set for which the d.e. is defined, we speak of "global asymptotic stability" of D_c.

Now, how can stability be proven? The usual way is to use Lyapunov theory. Let $V(x)$ be a positive function

$$V(x) \geq 0 \text{ for all } x \tag{4.35}$$

such that when evaluated along solutions to (4.31) it is decreasing:

$$\frac{d}{d\tau} V(x(\tau)) = \frac{d}{dx} V(x(\tau)) \cdot \frac{d}{d\tau} x(\tau)$$

$$= V'(x(\tau))f(x(\tau)) \leq 0 \text{ for all } x(\tau) \tag{4.36}$$

and

$$\frac{d}{d\tau} V(x(\tau)) = 0 \Rightarrow x(\tau) \in D_c. \tag{4.37}$$

Such a function is called a *Lyapunov function*. We see that outside D_c the function $V(x(\tau))$ is strictly decreasing as a function of τ. But since V is bounded from below, it cannot continue to decrease indefinitely. Hence $x(\tau)$ must tend to D_c. The condition (4.35)–(4.37) thus guarantees that D_c is a globally asymptotically stable invariant set.

To establish that D_A is a domain of attraction of D_c, we require (4.36) to hold only for $x(\tau) \in D_A$, but introduce the condition

$$C \geq V(x) \geq 0, \; V(x) = C \text{ for } x \in \partial D_A \tag{4.38}$$

(∂D_A = the boundary of D_A) in order to assure that the trajectories do not leave D_A. It can also be shown that if an invariant set D_c has a domain of attraction D_A, then there exists a function $V(x)$ with the above properties.

Connection between the Algorithm and the Differential Equation   We now return to the link (4.30) between the algorithm (4.16) and the d.e. (4.29). With the stability language just reviewed, we could phrase our conjectures about the connections between (4.16) and (4.29) as follows:

(A) Suppose D_c is an invariant set for the d.e. (4.29) and D_A is its domain of attraction. Then, provided we know that $\hat{\theta}(t) \in D_A$ "sufficiently often," the estimate $\hat{\theta}(t)$ will tend to D_c w.p.1 as t approaches infinity.

(B) Only stable, stationary points of (4.29) are possible convergence points for the algorithm (4.16).

(C) The trajectories $\theta_D(\tau)$ of (4.29) are the "asymptotic paths" of the estimates $\hat{\theta}(t)$, generated by (4.16).

Notice that the statement (A) is also a statement of local convergence. If a stationary point θ^* is locally stable, it will have a domain of attraction that contains a neighborhood of θ^*. Thus if the sequence of estimates belongs to this neighborhood sufficiently often, it will converge to θ^*.

All the above conclusions (A)–(C) can be proven to be formally correct. We shall give the formal result corresponding to (A) in the next three subsections.

We summarize this discussion with the following recipe for convergence analysis of (4.16):

1. Compute the prediction errors $\varepsilon(t, \theta)$ and gradient approximations $\eta(t, \theta)$ that would be obtained for a fixed and constant model θ.

2. Evaluate the average updating direction for the algorithm, based on these variables [see (4.24)].

3. Define a differential equation that has this direction as the right hand side:

$$\dot{\theta}_D = R_D^{-1}f(\theta_D),$$
$$\dot{R}_D = G(\theta_D) - R_D. \tag{4.39}$$

4. Study the stability properties of this differential equation.

Steps 1–3 are often quite easy, since $\varepsilon(t)$ and $\eta(t)$ are typically constructed as approximations of $\varepsilon(t, \theta)$ and $\eta(t, \theta)$ when the algorithm is derived. (see section 3.4).

We illustrate the outlined procedure by an example.

EXAMPLE 4.1 Let us apply the foregoing recipe to a simple recursive prediction error algorithm, estimating the parameter c in a first-order moving average model

$$y(t) = e(t) + ce(t - 1). \tag{4.40}$$

This algorithm is a special case of the algorithm developed in section 3.7.2. With $n_a = n_f = n_b = n_d = 0$ and $n_c = 1$ we obtain from (3.123)–(3.125)

$$\varepsilon(t) = y(t) - \hat{y}(t), \tag{4.41a}$$

$$R(t) = R(t - 1) + \gamma(t)[\psi^2(t) - R(t - 1)], \tag{4.41b}$$

$$\hat{c}(t) = \hat{c}(t-1) + \gamma(t)\frac{1}{R(t)}\psi(t)\varepsilon(t), \tag{4.41c}$$

$$\bar{\varepsilon}(t) = y(t) - \hat{c}(t)\bar{\varepsilon}(t-1), \tag{4.41d}$$

$$\hat{y}(t+1) = \hat{c}(t)\bar{\varepsilon}(t), \tag{4.41e}$$

$$\psi(t+1) = \bar{\varepsilon}(t) - \hat{c}(t)\psi(t). \tag{4.41f}$$

Here we write $\hat{c}$ for the estimate $\hat{\theta}$ and note that ψ and R are scalars.

Step 1 Compute the prediction errors $\varepsilon(t, c)$ and gradient approximations $\psi(t, c)$ that would be obtained for a fixed and constant model in (4.41d, f):

$$\varepsilon(t, c) = y(t) - c\varepsilon(t-1, c),$$

i.e.,

$$\varepsilon(t, c) = \frac{1}{1 + cq^{-1}}y(t). \tag{4.42}$$

Similarily, from (4.41f) and (4.42) we would get the gradient approximation

$$\psi(t+1, c) = \varepsilon(t, c) - c\psi(t, c),$$

i.e.,

$$\psi(t, c) = \frac{1}{1 + cq^{-1}}\varepsilon(t-1, c). \tag{4.43}$$

Step 2 Evaluate the average updating direction for the algorithm, based on these variables: The average updating direction for a given c in (4.41c) is the expected value of the quantity $\psi(t)\varepsilon(t)$, evaluated with the variables (4.42) and (4.43):

$$f(c) = E\psi(t, c)\varepsilon(t, c). \tag{4.44}$$

From (4.42) and (4.43) we know that both ψ and ε are formed by filtering $\{y(t)\}$. The variable (4.44) will therefore depend on the covariance properties of this signal. If $\{y(t)\}$ is a stationary stochastic process, then (4.44) can be evaluated using complex integrals, as shown in (1.A.8):

$$f(c) = \int_{-\pi}^{\pi} \frac{e^{i\omega}}{(1 + ce^{i\omega})^2} \cdot \frac{1}{(1 + ce^{-i\omega})}\Phi_y(\omega)d\omega, \tag{4.45}$$

where $\Phi_y(\omega)$ is the spectrum of $\{y(t)\}$. The actual value of $f(c)$ thus depends on the properties of the true signal $y(t)$. If we suppose that this signal can be described by

$$y(t) = e(t) + 0.5e(t - 1), \quad \Rightarrow \quad \phi_y = (1+0.5z^{-1})(1+0.5z)\phi_\varepsilon \tag{4.46}$$

where $\{e(t)\}$ is a sequence of independent random variables each of zero mean value and variance 1, then calculation gives

$$f(c) = \frac{1 + c^2 - 2.5c}{2(1 - c^2)^2} \cdot \int_{-\pi}^{\pi} \frac{z}{(1 + cz)^2} \frac{(1+0.5z^{-1})(1+0.5z)}{1 + cz^{-1}} \, dz \tag{4.47}$$

For the average updating direction in (4.41b), we similarily obtain

$$G(c) = \mathrm{E}[\psi^2(t, c)] \tag{4.48}$$
$$= \frac{1.25(1 + c^2) - 2c}{(1 - c^2)^3},$$

where the second equality follows if (4.46) holds.

Step 3 Define a d.e. that has this direction as its right-hand side. This gives

$$\frac{d}{d\tau}c_D(\tau) = \frac{1}{R_D(\tau)}f(c_D(\tau)), \tag{4.49a}$$

$$\frac{d}{d\tau}R_D(\tau) = G(c_D(\tau)) - R_D(\tau). \tag{4.49b}$$

Step 4 Study the stability properties of this d.e.

Solving the equation numerically. We start by solving the d.e. (4.49) numerically for some different initial conditions of $c_D(0)$ and $R_D(0)$, using the expressions (4.47) and (4.48). These solutions are shown in figure 4.3 as a function of time. In figure 4.4 we have plotted $c_D(\tau)$ against $R_D(\tau)$ in a phase diagram. From these figures we see that the point $c = 0.5$, $R = 4/3$ appears to be a globally asymptotically stable stationary point of the d.e. (4.49), since all trajectories end up at this point.

Examining the right-hand side of the d.e. The function $f(c)$ given by (4.47) is plotted in figure 4.5. We see that f is positive for $c < 0.5$, and negative for $c > 0.5$. Since $R_D(\tau)$ in (4.49) is always positive, this means that, in the interval $|c| < 1$, the solution $c_D(\tau)$ is increasing whenever it is less than 0.5 and decreasing whenever it is greater than 0.5. This clearly

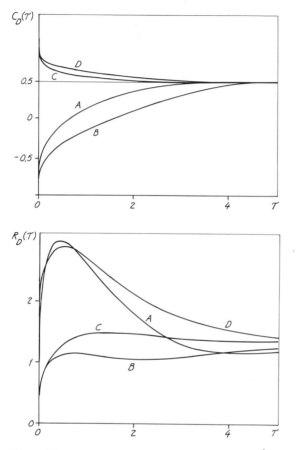

Figure 4.3
The solution of (4.49) as a function of τ for different initial conditions:

A: $c_D(0) = -0.8,\quad R_D(0) = 1$
B: $c_D(0) = -0.8,\quad R_D(0) = 0.2$
C: $c_D(0) = 0.9,\quad R_D(0) = 0.2$
D: $c_D(0) = 0.8,\quad R_D(0) = 2$

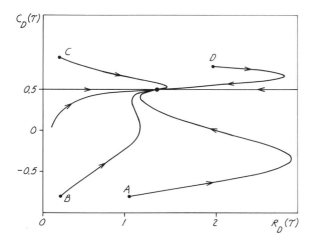

Figure 4.4
The solution of (4.49) where c_D is plotted against R_D for the same inital conditions as in figure 4.3.

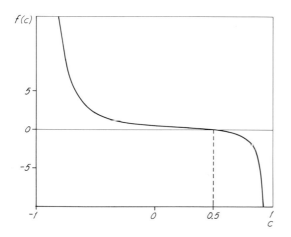

Figure 4.5
The function $f(c)$ given by (4.47).

implies that $c_D(\tau)$ must converge to 0.5, thus confirming what figures 4.3 and 4.4 show.

Using a Lyapunov Function. What we said in the previous paragraphs can be formalized using Lyapunov theory. Try the function

$$W(c) = \tfrac{1}{2}(c - 0.5)^2$$

as a Lyapunov function. We have

$$\frac{d}{d\tau} W(c_D(\tau)) = W'(c_D(\tau))\frac{d}{d\tau}c_D(\tau) = [c_D(\tau) - 0.5] \cdot \frac{1}{R_D(\tau)}f(c_D(\tau))$$

$$= \frac{1}{2R_D(\tau)}[c_D(\tau) - 0.5]^2 \cdot \frac{[c_D(\tau) - 2]}{[1 - c_D^2(\tau)]^2}. \tag{4.50}$$

Hence

$$\frac{d}{d\tau} W(c_D(\tau)) \le 0 \text{ always in } |c_D| < 1$$

and $\tag{4.51}$

$$\frac{d}{d\tau} W(c_D(\tau)) = 0 \Rightarrow c_D(\tau) = 0.5.$$

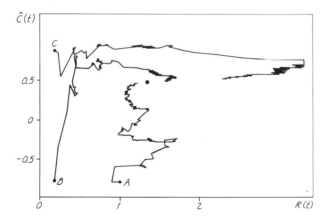

Figure 4.6a
Estimates obtained from the algorithm (4.41) with $\gamma(t) = 1/t$. Here $\hat{c}(t)$ is plotted against $R(t)$. Algorithm initialized at $t = 10$ and run for 1,000 steps.

Compare this to figure 4.4. The initial values A–D are the same in that figure and these figures 4.6a–c.

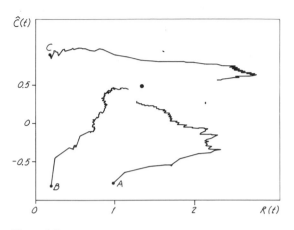

Figure 4.6b
As figure 4.6a, with algorithm initialized at $t = 100$ and run for 1,000 steps.

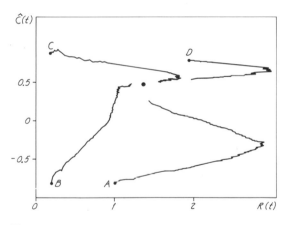

Figure 4.6c
As figure 4.6a, with algorithm initialized at $t = 1,000$ and run for 10,000 steps.

According to the stability theory given earlier in this section, this implies that $c_D(\tau) \to 0.5$ as $\tau \to \infty$. When $c_D(\tau) \to 0.5$ it follows from (4.49b) that $R_D(\tau)$ must tend to $G(0.5) = 4/3$. The solution $(c, R) = (0.5, 4/3)$ is thus asymptotically stable (with domain of attraction $|c| < 1$; $R > 0$).

Comparing the Algorithm with the Trajectories of the d.e. To illustrate the intuitive claim (C) that "the trajectories of (4.49) are the asymptotic paths of the algorithm (4.41)" let us compare the trajectories in figures 4.3 and 4.4 with simulations of the algorithm. In figures 4.6a–c simulations are shown. Comparing with figure 4.4 we see that the estimates from the algorithm stay close to the trajectories of the d.e. when initialized at $t = 1,000$. When initialized at $t = 10$ the estimates quickly move away from the corresponding trajectory. This illustrates the asymptotic nature of the link between (4.41) and (4.49). □

In the foregoing example, we performed some fairly detailed analysis based on an explicit evaluation of the right-hand side of the d.e. Fortunately, it is often not necessary to do such detailed calculations. As we shall see in sections 4.4.2 and 4.5.2, most of the analysis can be carried out on the basis of more general relationships.

4.3.3 An Associated Differential Equation: The Basic Theorem

Based on heuristic arguments in the previous section we described a link between the recursive identification algorithms under discussion and a certain associated differential equation. The analysis suggested that the convergence of the algorithm could be studied in terms of the stability properties of the differential equation, as summarized in claims (A)–(C).

We shall in this section arrive at these claims in a formal manner. These formal results for the analysis of recursive stochastic algorithms were developed by Ljung (1977b, 1978a). Using a different approach, similar results were proved in Kushner and Clark (1978). The applicability of the results to the particular algorithms under discussion in this book is described in section 4.3.5.

In Ljung (1977b) the following algorithm is studied:

$$x(t) = x(t - 1) + \gamma(t)Q(t, x(t - 1), \varphi(t)), \qquad (4.52a)$$

$$\varphi(t) = A(x(t - 1))\varphi(t - 1) + B(x(t - 1))e(t). \qquad (4.52b)$$

The following regularity conditions are introduced (D_R is a subset of the x-space, where the conditions are assumed to hold.):

C1: The function $Q(t, x, \varphi)$ is Lipschitz continuous in x and φ in any neighborhood of $(\bar{x}, \bar{\varphi})$, where $\bar{x} \in D_R$ and $\bar{\varphi}$ is arbitrary:

$$|Q(t, x_1, \varphi_1) - Q(t, x_2, \varphi_2)| \le K(\bar{x}, \bar{\varphi}, \rho, v)[|x_1 - x_2| + |\varphi_1 - \varphi_2|]$$

for $|x_i - \bar{x}| \le \rho$, $|\varphi_i - \bar{\varphi}| \le v$,

where $\rho = \rho(\bar{x}) > 0$, $v = v(\bar{\varphi}) > 0$. The Lipschitz constant K may thus depend on the neighborhood.

C2: The matrix functions $A(x)$ and $B(x)$ are Lipschitz continuous in x for $x \in D_R$.

C3: The sequence $\{e(t)\}$ is such that

$$\sum_{k=1}^{t} \beta(t, k) Q(k, \bar{x}, \varphi(k, \bar{x})) \to f(\bar{x}) \text{ as } t \to \infty \text{ for all } \bar{x} \in D_R.$$

Here $\beta(t, k)$ are the weights corresponding to $\{\gamma(t)\}$ [see (2.128)], and $\varphi(t, \bar{x})$ is defined by

$$\varphi(t, \bar{x}) = A(\bar{x})\varphi(t - 1, \bar{x}) + B(\bar{x})e(t), \quad \varphi(0, \bar{x}) = 0.$$

C4: For all $x \in D_R$ we have, for some $C(\bar{x}, \lambda, c)$,

$$\sum_{k=1}^{t} \beta(t, k)[1 + v(k, \lambda, c)] \cdot K(\bar{x}, \varphi(k, \bar{x}), \rho(\bar{x}), v(k, \lambda, c))$$

$$\to C(\bar{x}, \lambda, c) < \infty \text{ as } t \to \infty.$$

Here λ is the maximum eigenvalue norm of $A(\bar{x})$, $v(t, \lambda, c)$ is defined by

$$v(t, \lambda, c) \triangleq c \sum_{k=1}^{t} \lambda^{t-k} |e(k)|,$$

and K is the Lipschitz constant defined in C1.

C5: $\sum_{1}^{\infty} \gamma(t) = \infty$.

C6: $\gamma(t) \to 0$ as $t \to \infty$.

The interpretation of these conditions is as follows. C1 and C2 assure that the functions are smooth so that a small change in x or φ cannot give a big change in the estimates. Recall that $\Sigma \beta(t, k) = 1$ if $\gamma(1) = 1$ according to (2.129). This means that the variable

$$\sum_{k=1}^{t} \beta(t, k) Q(k, \bar{x}, \varphi(k, \bar{x}))$$

in C3 is a weighted average of the terms $\{Q(k, \bar{x}, \varphi(k, \bar{x}))\}$. These in turn are the average asymptotic updating steps that the algorithm uses when $\{x(k)\}$ stays close enough to a certain value $\bar{x}$. Hence the limit $f(\bar{x})$, that is assumed to exist in C3 represents an average, asymptotic update direction for (4.52a). Comparing with the heuristic discussion in the previous subsection, this value therefore corresponds to the expectation (4.24). C4 is a technical way of expressing that no large deviations from the average behavior must occur. Finally, C5 expresses that the estimates must be able to travel an arbitrary distance, and C6 implies that we may have convergence despite random fluctuations in Q.

C3 and C4 may seem to be complex and difficult to verify. We shall, however, show in section 4.3.5 that they are satisfied for the algorithm (4.16′).

Several different results on convergence and asymptotic paths of (4.52) are proved in Ljung (1977b). Here, we shall concentrate on convergence and quote the following result (corollary 1 in Ljung, 1977b, p.555).

THEOREM 4.1   Consider the algorithm (4.52) subject to the conditions C1–C6. Let $\bar{D}$ be a closed subset of D_R. Assume that there is a constant $C < \infty$ and a subsequence $\{t_k\}$ (that may depend on the particular sequence $\{e(r)\}$) such that

$$x(t_k) \in \bar{D} \quad \text{and} \quad |\varphi(t_k)| < C. \tag{4.53}$$

Assume also that there exists a twice-differentiable function $V(x)$ in D_R, such that, with $f(x)$ given by C3,

$$V'(x)f(x) \leq 0, \quad x \in D_R. \tag{4.54}$$

Then either

$$x(t) \to D_c = \{x \,|\, x \in D_R \text{ and } V'(x)f(x) = 0\} \text{ as } t \to \infty \tag{4.55a}$$

or

$$\{x(t)\} \text{ has a cluster point on the boundary of } D_R. \tag{4.55b}$$

The proof of this theorem is given in Ljung (1977b) and follows the heuristic outline of section 4.3.2. That $\{x(t)\}$ has a cluster point on the boundary of D_R means that there exists a subsequence $\{t_j\}$ such that $x(t_j)$ tends to the boundary of D_R as j approaches infinity.

In this theorem, no explicit reference is made to any d.e. The asymptotic, average updating direction for the algorithm is, according to C3, given by $f(x)$, so the corresponding d.e. is

$$\frac{d}{d\tau}x(\tau) = f(x(\tau)). \tag{4.56}$$

We then recognize in the assumption (4.54) the Lyapunov condition (4.36). Hence (4.54) assures that all trajectories of the d.e. (4.56) that start in D_R will either leave D_R or converge to D_c as time tends to infinity. It is often useful and suggestive to think of (4.54) as such a stability condition on the trajectories of (4.56).

Let us stress that theorem 4.1 holds for any sequence $\{e(t)\}$, subject to C3 and C4. If $\{e(t)\}$ is regarded as a stochastic process, such that C3 and C4 hold w.p.1, then the conclusion of the theorem also holds w.p.1.

To illustrate the verification of C1–C6 and the application of the theorem we give a trivial example.

EXAMPLE 4.2 (Estimation of Mean Value)   Consider the algorithm in example 2.5 for estimating the mean of a random variable:

p43

$$x(t) = x(t - 1) + \frac{1}{t}[e(t) - x(t - 1)].$$

mean   input sequence

We see that this algorithm is obtained from the general one by

$$Q(t, x, \varphi) = \varphi - x, \quad \varphi(t) = e(t) \text{ (i.e. } A(x) = 0, B(x) = 1), \text{ and } \gamma(t) = 1/t.$$

Conditions C1 and C2 are trivially satisfied with $D_R = R$. In C1 we can take $K(x, \varphi, \rho, v) = 1$. In C3 we find that $\varphi(t, x) = e(t)$ and that $\beta(t, k) = 1/t$ for $\gamma(t) = 1/t$, so the condition is that

$$\frac{1}{t} \sum_{k=1}^{t} [e(k) - x] = \frac{1}{t} \sum_{k=1}^{t} e(k) - x$$

should converge. This is the case if $\{e(t)\}$ is such that

$$\frac{1}{t} \sum_{k=1}^{t} e(k) \to m \quad \text{as } t \to \infty. \tag{4.57}$$

Then $f(x) = m - x$. To verify C4 we note that since $A(x) = 0$, its maximum eigenvalue λ is also 0. Hence $v(t, \lambda, c) = 0$, and we had $K(x, \varphi, \rho, v) = 1$. The condition thus reads

$$\sum_{k=1}^{t} \beta(t, k) \to C \quad \text{as } t \to \infty,$$

which is trivially satisfied with $C = 1$. Finally, C5 and C6 are satisfied for $\gamma(t) = 1/t$.

Thus we can take $\bar{D} = D_R = R$ so (4.53) is satisfied. The existence of a bounded subsequence of $\{e(t)\}$ follows from the assumption (4.57).

The associated d.e. now is

$$\dot{x} = m - x,$$

which definitely is globally asymptotically stable with $x = m$ as stationary point. We could, e.g., take $V(x) = (m - x)^2/2$, which gives

$$V'(x)f(x) = -(m - x)^2,$$

which is negative for all x. The theorem thus tells us that $x(t)$ will either tend to infinity or to m as t approaches infinity.

In this trivial case we have of course not accomplished anything by using theorem 4.1, since the assumption (4.57) about $\{e(t)\}$ directly implies that $x(t)$ tends to m. However, the example illustrates the basic steps in applying the theorem. □

In the proof of theorem 4.1 it is shown that the estimates $\{x(t)\}$ follow the trajectories of (4.56) more and more tightly as t increases (see lemma 1 in Ljung, 1977b). Intuitively, we may picture the sequence of estimate as eventually being "caught" by a trajectory and forced to follow it. This applies only in the set D_R, where the regularity conditions C1–C4 are assumed to hold. Outside D_R the d.e. (4.56) is not defined and hence does not describe the behavior of the algorithm. Thus we must assure by (4.53) that the estimates are inside $\bar{D} \subset D_R$ infinitely often, so that they will eventually be captured by a trajectory. This also explains the alternative asymptotic result (4.55b). Even if the stability of the d.e. (4.56) is assured by (4.54), we have not said that its domain of attraction coincides with $\bar{D}$. There may therefore exist trajectories that start in $\bar{D}$ and then leave D_R. The sequence $\{x(t)\}$ could follow such a path and leave $\bar{D}$ and D_R. Once out of D_R we have no control over the estimate sequence, except that by (4.53) we know that it will come back into $\bar{D}$. But that could be anywhere along the aforementioned trajectory, and the story could be repeated. This indicates that if the sequence of estimates leaves $\bar{D}$ an infinite number of times (so that it will have a cluster point on the boundary of $\bar{D}$), then it may have a cluster point also inside $\bar{D}$.

Projection Algorithms   A remedy for the foregoing situation is to force the estimates to remain inside a compact subset $\bar{D}$ of D_R by projecting them according to the rule

$$\bar{x}(t) = x(t - 1) + \gamma(t)Q(t, x(t - 1), \varphi(t)),$$

$$x(t) = \begin{cases} \bar{x}(t) & \text{if } \bar{x}(t) \in \bar{D} \\ x(t - 1) & \text{if } \bar{x}(t) \notin \bar{D} \end{cases}. \tag{4.58}$$

(Several other projection rules are possible.) Then we assure that the regularity conditions C1–C4 are always applicable and that the d.e. (4.56) will thus describe the asymptotic behavior of $\{x(t)\}$ without reservation.

This means that if no trajectory of the d.e. that starts in $\bar{D}$ ever leaves $\bar{D}$, then $\{x(t)\}$ will not have a cluster point on the boundary. However, if some trajectory points out of $\bar{D}$, then $\{x(t)\}$ may follow that. With the projection (4.58), $\{x(t)\}$ will then get stuck at the intersection of the boundary and the trajectory. In that case $\{x(t)\}$ converges to the boundary of $\bar{D}$ (and may consequently not have any cluster point inside $\bar{D}$). This is a stronger result than (4.55b). We summarize it as a corollary.

COROLLARY TO THEOREM 4.1   Consider the algorithm (4.52b), (4.58) under the same assumptions as in theorem 4.1. Then either

$$x(t) \rightarrow D_c \text{ as } t \rightarrow \infty \tag{4.59a}$$

or

$$x(t) \rightarrow \delta\bar{D} \text{ as } t \rightarrow \infty, \tag{4.59b}$$

where $\delta\bar{D}$ is the boundary of $\bar{D}$.

The conclusion (4.59b) may hold only if there is a trajectory of the d.e. that leaves $\bar{D}$ (see theorem 4 in Ljung, 1977b). If the projection is done so that the existence of a bounded subsequence of $\{\varphi(t)\}$ is guaranteed, then of course the assumption (4.53) can be dispensed with in the corollary.

With Theorem 4.1 we have obtained a formal version of the heuristic convergence claim (A) of section 4.3.2. In Ljung (1977b), corresponding formalizations are given also of claims (B) and (C) under conditions that are of the same type, but not identical to C1–C6. Our main concern in this chapter is with convergence, and therefore we shall not spend time on giving these details here, but just quote informal versions. The result corresponding to claim (B) is

RESULT 4.1 (Theorem 2 in Ljung, 1977b)   Suppose that $x(t)$ given by (4.52) converges to x^* with a probability greater than zero. Then

$$f(x^*) = 0 \tag{4.60a}$$

and

$$H(x^*) = \frac{d}{dx}f(x)\bigg|_{x=x^*} \qquad \text{has all its eigenvalues in the left half-plane.} \tag{4.60b}$$

Here $f(x)$ is the function defined in condition C3.

The result corresponding to claim (C) is

RESULT 4.2 (Theorem 3 in Ljung, 1977b)   The probability that $\{x(t)\}$ does not remain in an ε-neighborhood of the corresponding trajectory of (4.56) over the interval $t = t_0$ to $t = N$ is bounded by

$$\frac{K}{\varepsilon^{4p}} \sum_{j=t_0}^{N} \gamma(j)^p.$$

Our next goal is to apply the results quoted in this section to the general recursive identification algorithm (4.16′). This will be achieved by verifying that the regularity conditions C1–C6 are satisfied for (4.16′). We shall first, in section 4.3.4, state a number of assumptions about (4.16′). Then in section 4.3.5 we shall show that these assumptions imply that the conditions of theorem 4.1 are satisfied.

4.3.4   Regularity Conditions for the General Recursive Identification Algorithm

In this section we shall impose a number of conditions on the algorithm (4.16′), which will ensure the applicability of theorem 4.1.

The model set determines the form of the matrices $A(\theta)$, $B(\theta)$, and $C(\theta)$ in (4.16d, e). We shall work with the following conditions on these:

M1: $D_{\mathcal{M}}$ is a compact set of $\mathbf{R}^d$, such that $\theta \in D_{\mathcal{M}} \Rightarrow A(\theta)$ has all eigenvalues strictly inside the unit circle.

M2: The matrices $A(\theta)$, $B(\theta)$, and $C(\theta)$ are continuously differentiable w.r.t. θ for $\theta \in D_{\mathcal{M}}$.

Condition M1 defines $D_{\mathcal{M}}$ as a compact subset of the stability region D_s defined in (3.24). We have earlier (section 3.4) remarked that it may be important to ascertain that the estimates are confined to such a region. Let us also stress that by definition D_s is the stability region for the *predictor*, not for the system dynamics, and that constraining θ to D_s does not impose any serious restriction on the model. Condition M2 is a weak condition, that should not be restrictive in practice.

When we consider the general algorithm (4.16b′, c′), we must also introduce two smoothness conditions for the function h and H. These are as follows.

Cr1: $h(t, \theta, \varepsilon, \eta)$ is differentiable w.r.t. θ, ε, and η, such that, for some $C < \infty$,

$$|h(t, \theta, \varepsilon, \eta)| + |h_\theta(t, \theta, \varepsilon, \eta)| \leq C(1 + |\varepsilon|^2 + |\eta|^2)$$

and

$$|h_\varepsilon(t, \theta, \varepsilon, \eta)| + |h_\eta(t, \theta, \varepsilon, \eta)| \leq C(1 + |\varepsilon| + |\eta|)$$

for $\theta \in D_{\mathcal{M}}$.

Cr2: $H(t, R, \theta, \varepsilon, \eta)$ is differentiable w.r.t. R, θ, ε, and η such that, for some $C < \infty$,

$$|H(t, R, \theta, \varepsilon, \eta)| \leq C(1 + |\varepsilon|^2 + |\eta|^2 + |R|),$$

$$|H_R(t, R, \theta, \varepsilon, \eta)| + |H_\theta(t, R, \theta, \varepsilon, \eta)| \leq C(1 + |\varepsilon|^2 + |\eta|^2),$$

and

$$|H_\varepsilon(t, R, \theta, \varepsilon, \eta)| + |H_\eta(t, R, \theta, \varepsilon, \eta)| \leq C(1 + |\varepsilon| + |\eta|)$$

for $\theta \in D_{\mathcal{M}}$.

Here h_θ denotes the partial derivative w.r.t. θ, etc.

In the algorithm (4.16) we have

$$H(t, R, \theta, \varepsilon, \eta) = \eta \Lambda^{-1} \eta^T - R \tag{4.61a}$$

and

$$h(t, \theta, \varepsilon, \eta) = \eta \Lambda^{-1} \varepsilon, \tag{4.61b}$$

which clearly satisfy Cr1 and Cr2. If the matrix Λ is a function of t, we must require that its inverse is uniformly bounded.

In the general Gauss-Newton algorithm (3.72)–(3.73) for minimization of a criterion $El(t, \theta, \varepsilon(t, \theta))$ we have

$$H(t, R, \theta, \varepsilon, \eta) = \eta l_{\varepsilon\varepsilon}(t, \theta, \varepsilon)\eta^T + l_{\theta\theta}(t, \theta, \varepsilon) - R, \quad {}^{3.73} \tag{4.62a}$$

$$h(t, \theta, \varepsilon, \eta) = \eta l_\varepsilon^T(t, \theta, \varepsilon) - l_\theta^T(t, \theta, \varepsilon). \quad {}^{7.72} \tag{4.62b}$$

In this case conditions Cr1 and Cr2, respectively, are implied by the following conditions on the criterion function $l(t, \theta, \varepsilon)$.

Cr3: The function $l(t, \theta, \varepsilon)$ is twice-continuously differentiable w.r.t. θ and ε and

$$|l_\theta(t, \theta, \varepsilon)| + |l_{\theta\theta}(t, \theta, \varepsilon)| \leq C(1 + |\varepsilon|)^2,$$

$$|l_\varepsilon(t, \theta, \varepsilon)| + |l_{\varepsilon\theta}(t, \theta, \varepsilon)| \leq C(1 + |\varepsilon|),$$

$$|l_{\varepsilon\varepsilon}(t, \theta, \varepsilon)| \le C$$
for $\theta \in D_{\mathcal{M}}$.

Cr4: The function $l(t, \theta, \varepsilon)$ is three times continuously differentiable w.r.t. θ and ε and

$$|l_{\varepsilon\varepsilon}(t, \theta, \varepsilon)| + |l_{\varepsilon\varepsilon\varepsilon}(t, \theta, \varepsilon)| + |l_{\varepsilon\varepsilon\theta}(t, \theta, \varepsilon)| \le C,$$
$$|l_{\theta\theta}(t, \theta, \varepsilon)| + |l_{\theta\theta\theta}(t, \theta, \varepsilon)| \le C(1 + |\varepsilon|)^2,$$
$$|l_{\theta\theta\varepsilon}(t, \theta, \varepsilon)| \le C$$
for $\theta \in D_{\mathcal{M}}$.

In the algorithm (4.16′) the inverse of the matrix $R(t)$ is used. To ensure that no problems arise here we introduce the following condition.

R1: The generation of the matrix $R(t)$ by (4.16b′) is such that $R(t)$ is symmetric and $R(t) \ge \delta I \, \forall t$ for some $\delta > 0$.*

In section 4.4 we shall discuss how this condition is assured by a certain modification of the algorithms derived in chapter 3. We shall in section 6.5 explain why such a modification should in any case be introduced for practical reasons and how it is best implemented.

Regarding the gain sequence $\{\gamma(t)\}$ we shall impose the condition:

G1: $\lim\limits_{t \to \infty} t \cdot \gamma(t) = \mu > 0$.

This condition restricts the choice of $\gamma(t)$ to asymptotically behave like μ/t. We know from section 2.8 that the effect of $\{\gamma(t)\}$ in (4.16′) can be interpreted as a "forgetting" of old data. Only the choice $\gamma(t) = 1/t$ gives equal weight to all observations. Since, for the convergence theory, we suppose the system to be time-invariant, it makes sense to asymptotically give equal weight to the observations. Then G1 is a natural condition. It is, however, not necessary for our analysis—it basically just makes certain expressions nice. Later in this section we shall remark on this explicitly.

Conditions on the Data   We have not yet introduced any conditions on the observed data $\{z(t)\}$. Such conditions will now be discussed.

Let the variables $\varepsilon(t, \theta)$, $\hat{y}(t \mid \theta)$ and $\eta(t, \theta)$ be obtained from z^t by the linear filter

---

* $A \ge B$ means that the matrix $A - B$ is positive semidefinite.

$$\xi(t + 1, \theta) = A(\theta)\xi(t, \theta) + B(\theta)z(t), \qquad \xi(0, \theta) = 0,$$

$$\begin{pmatrix} \hat{y}(t \mid \theta) \\ \mathrm{col}\, \eta(t, \theta) \end{pmatrix} = C(\theta)\xi(t, \theta),$$

(4.63)

$$\varepsilon(t, \theta) = y(t) - \hat{y}(t \mid \theta).$$

Here θ is any value in $D_{\mathscr{M}}$. These variables played an important role in our heuristic analysis in section 4.3.2. They correspond to what the time-varying filter (4.16d, e) in the algorithm would produce for a constant estimate θ. The sequences $\{\varepsilon(t, \theta)\}$ and $\{\eta(t, \theta)\}$ are consequently obtained by filtering the data sequence $\{z(t)\}$. Their properties will depend both on $\{z(t)\}$ and on $\mathscr{M}(\theta)$.

Our first condition on $\{z(t)\}$ will be phrased in terms of convergence properties for these variables.

A1: The data sequence $\{z(t)\}$ is such that the following limits exist for all $\theta \in D_{\mathscr{M}}$.

(a) $\displaystyle \lim_{N \to \infty} \frac{1}{N} \sum_{t=1}^{N} h(t, \theta, \varepsilon(t, \theta), \eta(t, \theta)) \triangleq f(\theta).$

(b) $\displaystyle \lim_{N \to \infty} \frac{1}{N} \sum_{t=1}^{N} H(t, R, \theta, \varepsilon(t, \theta), \eta(t, \theta)) \triangleq F(R, \theta).$

(c) $\displaystyle \lim_{N \to \infty} \sup \frac{1}{N} \sum_{t=1}^{N} [1 + |z(t)|]^3 < \infty.$

This condition A1 (that corresponds to C3 of section 4.3.3) on the data sequence $\{z(t)\}$ is certainly implicit, and one may wonder how it can be verified. We shall return to this question shortly.

We recognize in A1(a) and A1(b) the "average adjustments" that would be made in the algorithm if θ were held constant. Our heuristic derivation of the differential equation in section 4.3.2 was based on this concept. We notice that in the case (4.61) of a Gauss-Newton method for a quadratic criterion, assumptions A1(a, b) are of the simple form

$$\lim_{N \to \infty} \frac{1}{N} \sum_{1}^{N} \eta(t, \theta)\Lambda^{-1}\varepsilon(t, \theta) = f(\theta),$$

(4.64a)

$$\lim_{N \to \infty} \frac{1}{N} \sum_{1}^{N} \eta(t, \theta)\Lambda^{-1}\eta^{\mathrm{T}}(t, \theta) = G(\theta);$$

(4.64b)

hence

$F(R, \theta) = G(\theta) - R.$

Remark   The equal weights in the averages above are a consequence of condition G1. If we used an arbitrary gain sequence $\{\gamma(t)\}$ with corresponding weights $\{\beta(t, k)\}$ according to (2.128), we would have to replace the sums in A1 with accordingly weighted averages.

Condition A1 refers to a particular observation sequence $\{z(t)\}$. It is often useful to regard the output sequence, and sometimes also the input sequence, as a stochastic process, to account for the random disturbances during the data collection. If we do that, the observed sequence $\{z(t)\}$ will be a realization of that stochastic process, and A1 refers to that particular realization. We shall then demand that condition A1 holds w.p.1, i.e., that almost all realizations of the process $\{z(t)\}$ are such that A1 holds. Let us discuss what conditions must be imposed on $\{z(t)\}$ in order to assure this.

Introduce

$$h_t \triangleq h(t, \theta, \varepsilon(t, \theta), \eta(t, \theta)). \tag{4.65}$$

Then A1(a) will hold w.p.1 if the following two conditions are satisfied:

$$\frac{1}{N} \sum_{t=1}^{N} (h_t - Eh_t) \to 0 \text{ w.p.1 as } N \to \infty \tag{4.66a}$$

and

$$\frac{1}{N} \sum_{t=1}^{N} Eh_t \to f \text{ as } N \to \infty. \tag{4.66b}$$

The sum in (4.66b) that consists of expected values is of course non-stochastic. Equation (4.66a) is the arithmetic mean of a sequence of random variables, $\{h_t - Eh_t\}$, where each variable is of zero mean. The property that such a sum converges to zero w.p.1 is known as "the strong law of large numbers" in probability theory. It is a well-known elementary theorem that (4.66a) holds when the random variables are independent (see, e.g., Chung, 1968). In our application, however, the sequences will not be independent. For the strong law of large numbers to hold, though, it is sufficient that variables for time instances far away from each other be almost independent. We have the following result by Cramér and Leadbetter (1967), pages 94–96 (where it is proved for the continuous-time case; but it is easy to see that it is valid with trivial modifications also for the discrete-time case):

Let $\{x(t)\}$ be a sequence of random variables each of zero mean and suppose that

$$|Ex(t)x(s)| \le C\frac{t^p + s^p}{1 + |t - s|^q}, \quad 0 \le 2 \cdot p < q < 1. \tag{4.67}$$

Then

$$\frac{1}{N} \sum_{t=1}^{N} x(t) \to 0 \text{ w.p.1 as } N \to \infty. \tag{4.68}$$

We conclude from this that A1 should hold w.p.1 if we impose two conditions: One corresponding to (4.66b) about the behavior of the expected values (condition A2 below), and one requiring h_t and h_s to be almost independent when $|t - s|$ is large (condition S1 below), so that (4.66a) will hold. We first formulate the two conditions and then prove a lemma showing that they indeed imply A1 w.p.1.
The first condition is

A2: The following limits exist:
$$\bar{E}h(t, \theta, \varepsilon(t, \theta), \eta(t, \theta)) = f(\theta),$$
$$\bar{E}H(t, R, \theta, \varepsilon(t, \theta), \eta(t, \theta)) = F(R, \theta).$$

Here, as before,

$$\bar{E}f(t) \triangleq \lim_{N \to \infty} \frac{1}{N} \sum_{t=1}^{N} Ef(t),$$

with an implied assumption that the limit exists. Expectation is here over the stochastic process $\{z(t)\}$. We may very well choose to let $\{z(t)\}$ contain deterministic parts, such as the input sequence.

To assure that h_t and h_s are asymptotically independent we introduce a condition on $\{z(t)\}$, basically requiring that old values of $z(t_0)$ are almost independent of the present value $z(t)$; $t \gg t_0$. We express this formally as follows.

S1: For each $t, s, t \ge s$, there exists a random vector $z_s^0(t)$ that belongs to the σ-algebra generated by z^t but is independent of z^s (for $s = t$ take $z_s^0(t) = 0$), such that

$E|z(t) - z_s^0(t)|^4 < C \cdot \lambda^{t-s}, \quad C < \infty, \quad \lambda < 1.$

Think of $z_s^0(t)$ as an approximation to $z(t)$ computed from the new information obtained during the time interval $[s, t]$. The condition then requires that this approximation is "very good" if the time s is remote enough. That is to say that what happened before time s has a very small influence on what is going on at time t, $t \gg s$. This appears to be a reasonable condition for most data sequences. We could phrase it as requiring the data generation to be "exponentially stable."

Remark   Conditions like S1 that require a sequence of random variables to be asymptotically independent are called *mixing conditions* in probability theory. Our way of phrasing it, however, is not the conventional one.

We can now give a result assuring that the condition A1 holds w.p.1:

LEMMA 4.1   Suppose that Cr1, Cr2, A2 and S1 hold. Then A1 holds w.p.1.

The proof is given in appendix 4.A.
We have now imposed a series of conditions on the algorithm $(4.16')$: M1, M2, Cr1, Cr2, R1, G1, A2, and S1. The following example demonstrates how the conditions can be verified.

EXAMPLE 4.3 (Example 4.1, continued)   Let us verify that the regularity conditions are satisfied for our simple example (4.41), where $\theta = c$ and $\eta = \psi$. This algorithm corresponds to the general one $(4.16')$ with

$$\xi(t) = \begin{pmatrix} \bar{\varepsilon}(t-1) \\ \psi(t) \end{pmatrix}, \qquad A(c) = \begin{pmatrix} -c & 0 \\ -c & -c \end{pmatrix},$$

$$B(c) = \begin{pmatrix} 1 \\ 1 \end{pmatrix}, \qquad C(c) = \begin{pmatrix} c & 0 \\ 0 & 1 \end{pmatrix},$$

$$z(t) = y(t), \qquad h(t, c, \varepsilon, \psi) = \psi\varepsilon, \qquad H(t, R, c, \varepsilon, \psi) = \psi^2 - R.$$

Condition M1: The eigenvalues of $A(c)$ are in this case both equal to $-c$. The condition thus is that $D_{\mathcal{M}} = \{c | |c| \leq 1 - \delta\}$ for some (small) positive number δ. From figure 4.5 we see that the d.e. will anyway not be defined outside this set.

Condition M2: The differentiability condition is obviously satisfied for the matrices $A(c)$, $B(c)$, and $C(c)$.

Condition Cr1: From $h(t, c, \varepsilon, \psi) = \psi\varepsilon$ we find, with $C = 1$,

$$|h| + |h_c| = |\psi\varepsilon| + |0| \leq C(1 + |\psi|^2 + |\varepsilon|^2)$$

and

$$|h_\psi| + |h_\varepsilon| = |\varepsilon| + |\psi| \leq C(1 + |\varepsilon| + |\psi|),$$

so Cr1 holds.

Condition Cr2: From $H(t, R, c, \varepsilon, \psi) = R - \psi^2$ we find

$$|H| + |H_R| + |H_c| = |R - \psi^2| + |1| + |0| \leq C(1 + |\varepsilon|^2 + |\psi|^2 + |R|)$$

and

$$|H_\varepsilon| + |H_\psi| = |0| + |2\psi| \leq C(1 + |\varepsilon| + |\psi|),$$

so Cr2 holds.

Condition R1: We would modify (4.41b) to

$$R(t) = \max\{\delta, R(t-1) + \gamma(t)[\psi^2(t) - R(t-1)]\}$$

to ensure this condition. See also (4.95) in section 4.4.2.

Condition G1: This is just a condition on the choice of $\gamma(t)$.

Condition A1: This means that the limits

(a) $\lim\limits_{N\to\infty} \dfrac{1}{N} \sum\limits_{t=1}^{N} \psi(t, c)\varepsilon(t, c),$

(b) $\lim\limits_{N\to\infty} \dfrac{1}{N} \sum\limits_{t=1}^{N} [R - \psi^2(t, c)],$

and

(c) $\lim\limits_{N\to\infty} \sup \dfrac{1}{N} \sum\limits_{t=1}^{N} (1 + |y(t)|^3) < \infty$

should exist. From (4.42) and (4.43) we know that ψ and ε are just filtered versions of the signal $\{y(t)\}$. Whether these limits exist is thus entirely dependent on the properties of this signal. The requirement is basically that sample covariances of filtered y-signals should converge. This is a weak condition and can be assured by a variety of assumption on $\{y(t)\}$. Let us show that S1 and A2 hold so that lemma 4.1 can be applied. For the simple case (4.46) in example 4.1,

$$y(t) = e(t) + 0.5e(t-1),$$

we find that $y(t)$ is independent of everything that happened up to time $t - 2$. Hence we can take, in condition S1, $y_s^0(t) = y(t)$ when $s \le t - 2$, so this condition is certainly satisfied if the fourth moment of $y(t)$ is bounded. In (4.46) the process $y(t)$ is stationary. Hence so will the filtered processes $\psi(t, c)$ and $\varepsilon(t, c)$ be (apart from possible, exponentially decaying transients). This means that the limits in A2,

$$\lim_{N \to \infty} \frac{1}{N} \sum_{t=1}^{N} E\psi(t, c)\varepsilon(t, c) \quad \text{and} \quad \lim_{N \to \infty} \frac{1}{N} \sum_{t=1}^{N} E\psi^2(t, c)$$

will exist, and since S1 holds, we find from lemma 4.1 that condition A1 holds w.p.1. □

It is worth pointing out the philosophy underlying conditions M1, M2, Cr1, Cr2, R1 and G1 on one hand and A1, A2 and S1 on the other. The former conditions all deal with functions and variables chosen by the user, and they can always be verified in a strict sense. The latter conditions deal with properties of the true data $\{z(t)\}$, and hence might better be regarded as hypotheses. This means that we can never achieve a formal verification of them. What we do is to make certain hypotheses about the data generation and then check that these assumptions imply the conditions we need. From that point of view S1 and A2 are as good as any other hypothesis about $\{z(t)\}$. S1 says that the data generation should be stable and A2 that it should be asymptotically mean stationary. Both these assumptions appear to be reasonable.

Another often-made assumption about the data generation is that the system can be described by one of the models in the model set (or, perhaps, by a higher order model). For the general model (3.17) that we have used, this would mean that

$$y(t) = \hat{y}(t \mid \theta_0) + e(t)$$

where $\{e(t)\}$ is a sequence of independent random vectors each of zero mean and with covariance matrix Λ_0, and where $\hat{y}(t \mid \theta_0)$ is given by (3.17). Then from (3.17)

$$\varphi(t + 1, \theta_0) = [\mathscr{F}(\theta_0) + \mathscr{G}_1(\theta_0)\mathscr{H}(\theta_0)]\varphi(t, \theta_0) + \mathscr{G}(\theta_0)\begin{pmatrix} e(t) \\ u(t) \end{pmatrix},$$

$$y(t) = \mathscr{H}(\theta_0)\varphi(t, \theta_0) + e(t), \tag{4.69}$$

where $\mathscr{G}_1$ is the first p columns of $\mathscr{G}$.

It is now easy to establish that S1 and A2 will hold for data generated as in (4.69) if

- $\mathcal{F}(\theta_0) + \mathcal{G}_1(\theta_0)\mathcal{H}(\theta_0)$ is asymptotically stable.

- $e(t)$ has bounded fourth moments.

- u^t is a deterministic input sequence, such that the limits

$$\lim_{N \to \infty} \frac{1}{N} \sum_{t=1}^{N} u(t)u^{\mathrm{T}}(t + k)$$

exist for all k.

(Let in S1

$$z_s^0(t) = \begin{pmatrix} y_s^0(t) \\ u(t) \end{pmatrix},$$

where $y_s^0(t)$ is the output of (4.69) if initialized at time s with $\varphi(s, \theta_0) = 0$.)

The only situation where it is not realistic to assume S1 and A2 a priori is when the generation of $\{z(t)\}$ may depend on past estimates, such as in adaptive control applications. Then asymptotic stationarity of $\{z(t)\}$ would be a result, rather than an assumption in the convergence analysis. Such situations can, however, be handled with a modified but similar approach, as explained in section 7.3.

4.3.5 A Theorem for the Convergence Analysis of the General Recursive Identification Algorithm

In this section we shall discuss the application of theorem 4.1 to our algorithm (4.16′):

$$\hat{\theta}(t) = \hat{\theta}(t - 1) + \frac{\mu(t)}{t} R^{-1}(t)h(t, \hat{\theta}(t - 1), \varepsilon(t), \eta(t)), \tag{4.70a}$$

$$R(t) = R(t - 1) + \frac{\mu(t)}{t} H(t, R(t - 1), \hat{\theta}(t - 1), \varepsilon(t), \eta(t)), \tag{4.70b}$$

$$\varepsilon(t) = y(t) - \hat{y}(t), \tag{4.70c}$$

$$\xi(t + 1) = A(\hat{\theta}(t))\xi(t) + B(\hat{\theta}(t))z(t), \tag{4.70d}$$

$$\begin{pmatrix} \hat{y}(t + 1) \\ \mathrm{col}\,\eta(t + 1) \end{pmatrix} = C(\hat{\theta}(t))\xi(t + 1). \tag{4.70e}$$

We shall verify that the regularity conditions C1–C6 in section 4.3.3 are satisfied for (4.70) provided that the conditions introduced in section 4.3.4 hold. In (4.70a, b) we have written $\mu(t)/t = \gamma(t)$, keeping in mind that $\mu(t) \to \mu$ according to condition G1.

We first slightly redefine (4.70c, d, e) to include $y(t)$ in the $\xi(t)$-vector:

$$\xi^*(t) = \begin{pmatrix} \xi(t) \\ z(t) \end{pmatrix}, \qquad A^*(\theta) = \begin{pmatrix} A(\theta) & B(\theta) \\ 0 & 0 \end{pmatrix}, \quad \overset{\text{part of}}{z(t)}$$

$$B^* = \begin{pmatrix} 0 \\ I \end{pmatrix}, \qquad C^*(\theta) = \begin{pmatrix} C_1^*(\theta) \\ C_2^*(\theta) \end{pmatrix} = \begin{pmatrix} -C_1(\theta) & I & 0 \\ C_2(\theta) & 0 & 0 \end{pmatrix}.$$

Then

$$\xi^*(t) = A^*(\hat{\theta}(t-1))\xi^*(t-1) + B^*z(t), \tag{4.70d'}$$

$$\begin{pmatrix} \varepsilon(t) \\ \operatorname{col}\eta(t) \end{pmatrix} = \begin{pmatrix} C_1^*(\hat{\theta}(t-1)) \\ C_2^*(\hat{\theta}(t-1)) \end{pmatrix} \xi^*(t). \tag{4.70e'}$$

The algorithm (4.70) can now be identified with the general structure (4.52) by

$$x = \begin{pmatrix} \theta \\ \operatorname{col} R \end{pmatrix}, \qquad \varphi = \xi^*, \tag{4.71}$$

and

$$Q(t, x, \varphi) = \begin{pmatrix} \mu(t)R^{-1}h(t, \theta, C_1^*(\theta)\xi^*, C_2^*(\theta)\xi^*) \\ \mu(t)\operatorname{col} H(t, R, \theta, C_1^*(\theta)\xi^*, C_2^*(\theta)\xi^*) \end{pmatrix}, \tag{4.72}$$

$$\gamma(t) = \frac{1}{t}, \tag{4.73}$$

$$A(x) = A^*(\theta), \qquad B(x) = B^*, \qquad e(t) = z(t). \tag{4.74}$$

(We will drop the superscript * henceforth.)

Condition C1 requires Lipschitz continuity of $Q(t, x, \varphi)$ with $K(\theta, R, \xi, \rho, v)$ as a uniform Lipschitz constant in a ρ-neighborhood of θ, R and a v-neighborhood of ξ. To verify this, and to find an expression for K, we may simply differentiate (4.72) w.r.t. R, θ, and ξ and evaluate upper bounds for it in these neighborhoods. Straightforward calculations, using Cr1, Cr2, and M2 [and using the facts that $C(\theta)$ and its derivative is uniformly bounded in $D_{\mathscr{M}}$ (according to M2), that $\mu(t) < \bar{\mu}$ (according to

G1), and that $|R^{-1}| < \delta^{-1}$ (according to R1)], give

$$K(\theta, R, \xi, \rho, v) = C(1 + v + |\xi|)^2, \qquad \theta \in D_\mathcal{M}, \tag{4.75}$$

where C is a constant computed from δ, $\bar{\mu}$, and the constants in Cr1 and Cr2. Hence C1 holds with (4.75) as a Lipschitz constant.

Condition C2 follows directly from M2.

To prove C3 we have to show that, in our notation, with $\gamma(t) = 1/t$,

$$\lim_{N \to \infty} \frac{1}{N} \sum_{t=1}^{N} \mu(t) R^{-1} h(t, \theta, \varepsilon(t, \theta), \eta(t, \theta)) \quad \text{exists for } \theta \in D_\mathcal{M},$$

$$\lim_{N \to \infty} \frac{1}{N} \sum_{t=1}^{N} \mu(t) H(t, R, \theta, \varepsilon(t, \theta), \eta(t, \theta)) \quad \text{exists for } \theta \in D_\mathcal{M}.$$

This is almost A1(a, b), except for the factors $\mu(t)$ and R^{-1}. We have, with $f(\theta)$ defined as in A1(a),

$$\lim_{N \to \infty} \left| \frac{1}{N} \sum_1^N \mu(t) R^{-1} h(t, \theta, \varepsilon(t, \theta), \eta(t, \theta)) - \mu R^{-1} f(\theta) \right|$$

$$= \lim_{N \to \infty} \left| \frac{1}{N} \sum_1^N [\mu(t) - \mu] R^{-1} h(t, \theta, \varepsilon(t, \theta), \eta(t, \theta)) \right|$$

$$\leq \lim_{N \to \infty} \left| \frac{1}{N} \sum_1^N \delta^{-2} [\mu(t) - \mu]^3 \right|^{1/3}$$

$$\times \lim_{N \to \infty} \left| \frac{1}{N} \sum_1^N C[1 + |\varepsilon(t, 0)|^2 + |\eta(t, \theta)|^2]^{3/2} \right|^{2/3}.$$

Here the first equality follows from A1(a), and the inequality is Hölder's inequality using R1 and Cr1. Now the first factor on the right-hand side tends to zero according to G1, and the second one is bounded according to A1(c) (since $\varepsilon(t, \theta)$ and $\eta(t, \theta)$ are obtained by exponentially stable filtering of z^t). Hence

$$\lim_{N \to \infty} \frac{1}{N} \sum_1^N \mu(t) R^{-1} h(t, \theta, \varepsilon(t, \theta), \eta(t, \theta)) = \mu R^{-1} f(\theta). \tag{4.76a}$$

Similarly,

$$\lim_{N \to \infty} \frac{1}{N} \sum_1^N \mu(t) H(t, R, \theta, \varepsilon(t, \theta), \eta(t, \theta)) = \mu F(R, \theta) \tag{4.76b}$$

in view of A1(b). Consequently C3 holds, with the limits given by (4.76).

Let us now consider C4. With the expression (4.75) for K, C4 means that

$$\frac{1}{N} \sum_{1}^{N} C[1 + v(t, \lambda, c)] \cdot [1 + v(t, \lambda, c) + |\xi(t, \theta)|]^2$$

where μ β

should be bounded, where

$$v(t, \lambda, c) = c \sum_{k=1}^{t} \lambda^{t-k} |z(k)|.$$

We have

$$\frac{1}{N} C \sum_{t=1}^{N} \left\{ \left[1 + \sum_{s=1}^{t} c\lambda^{t-s} |z(s)| + |\xi(t, \theta)| \right]^2 \cdot \left[1 + \sum_{s=1}^{t} c\lambda^{t-s} |z(s)| \right] \right\}$$

$$\leq C_1 \frac{1}{N} \sum_{t=1}^{N} [1 + |z(t)|]^3 \leq \text{const}, \tag{4.77}$$

where the second inequality follows from A1(c). In the first inequality we used

$$|\xi(t, \theta)| = \left| \sum_{s=1}^{t} A(\theta)^{t-s} B(\theta) z(s) \right|$$

$$\leq \sum_{s=1}^{t} |A(\theta)|^{t-s} |B(\theta)| |z(s)| \leq C_2 \sum_{s=1}^{t} \lambda_1^{t-s} |z(s)|,$$

according to M1 and the definition of $\xi(t, \theta)$. With this expression inserted, the second step in (4.77) follows by calculation. The constant C_1 is formed from C, C_2, λ, and λ_1. Hence C4 holds.

Finally, C5 and C6 are trivially satisfied for $\gamma(t) = 1/t$.

We have now verified all the underlying regularity assumption of theorem 4.1. The right-hand side of the d.e. corresponding to (4.56) will be given by the limit in C3, i.e., by (4.76). The d.e. is consequently

$$\frac{d}{d\tau} \theta_D(\tau) = \mu R_D^{-1}(\tau) f(\theta_D(\tau)), \tag{4.78a}$$

$$\frac{d}{d\tau} R_D(\tau) = \mu F(R_D(\tau), \theta_D(\tau)), \tag{4.78b}$$

where we have written (4.78b) with R as a matrix d.e. Obviously the positive scalar μ only acts like a time scaling and does not affect the stability properties of (4.78).

With the conditions verified, we can now apply Theorem 4.1 and its corollary to the algorithm (4.70).

THEOREM 4.2 Consider the algorithm (4.70) where a projection is used to keep $\hat{\theta}(t)$ in $D_{\mathcal{M}}$, so that there is a bounded subsequence of $\xi(t)$. Assume that M1, M2, Cr1, Cr2, R1, G1, and A1 hold (see section 4.3.4). Assume also that there exists a positive function $V(\theta, R)$, such that

$$\frac{d}{d\tau} V(\theta_D(\tau), R_D(\tau)) \le 0 \quad \text{for } \theta_D \in D_{\mathcal{M}} \tag{4.79}$$

when evaluated along solutions to the d.e.

$$\frac{d}{d\tau} \theta_D(\tau) = R_D^{-1}(\tau) f(\theta_D(\tau)),$$

$$\tag{4.80}$$

$$\frac{d}{d\tau} R_D(\tau) = F(R_D(\tau), \theta_D(\tau)),$$

where f and F are defined by A1. Let

$$D_c = \{\theta, R \Big| \frac{d}{d\tau} V(\theta_D(\tau), R_D(\tau)) = 0\}. \tag{4.81}$$

Then as $t \to \infty$ either $\{\hat{\theta}(t), R(t)\}$ tend to D_c or $\{\hat{\theta}(t)\}$ tends to the boundary of $D_{\mathcal{M}}$.

Remark 1 If the sequence $\{z(t)\}$ is a stochastic process and A1 holds w.p.1, then the conclusion of the theorem holds w.p.1. Notice that this is implied by A2, Cr1, Cr2, and S1 according to lemma 4.1.

Remark 2 Notice that (4.79), (4.81) imply that the trajectories of the d.e. (4.80) that remain in $D_{\mathcal{M}}$ converge to D_c as $\tau \to \infty$. Conversely, as we pointed out in section 4.3.2, if the d.e. (4.80) has an invariant set D_c with a domain of attraction that includes $D_{\mathcal{M}}$, then the existence of a function V with the property (4.79) follows.

With theorem 4.2 we have obtained a formal version of the heuristic claim (A) of section 4.3.2. For the claim (B) we obtain from result 4.1 (section 4.3.3):

RESULT 4.3   Under conditions of the same character as M1, M2, Cr1, Cr2, R1, G1, A2, and S1 the algorithm (4.70) can only converge to values (θ^*, R^*) that are *stable stationary points* of the d.e. (4.80).

This result can be used to improve on the convergence result in theorem 4.2, by excluding from D_c unstable stationary points. It can also be used to prove failure of convergence to the desired limit.

In the case of a quadratic criterion, the algorithm has the form (4.16). Then, according to (4.64) the d.e. (4.80) is

$$\dot{\theta} = R^{-1}f(\theta), \tag{4.82a}$$

$$\dot{R} = G(\theta) - R. \tag{4.82b}$$

The stationary points of (4.82) are

$$\{\theta, R \,|\, f(\theta) = 0 \text{ and } R = G(\theta)\}.$$

Let us linearize (4.82) around a stationary point $\theta = \theta^*$, $R = R^* = G(\theta^*)$. We then obtain

$$\dot{\Delta\theta} = [G(\theta^*)]^{-1} H(\theta^*)\Delta\theta,$$
$$\dot{\Delta R} = -\Delta R + G'(\theta^*)\Delta\theta, \tag{4.83}$$

where $\Delta\theta = \theta - \theta^*$, $\Delta R = R - G(\theta^*)$, and

$$H(\theta^*) = \frac{d}{d\theta}f(\theta)\bigg|_{\theta=\theta^*}, \tag{4.84}$$

and the term $G'(\theta^*)\Delta\theta$ is a formal way to write the second term in the Taylor expansion of $G(\theta^* + \Delta\theta)$. Clearly, the stability properties of the linearized equation (4.83) are entirely determined by the matrix

$$L^* = [G(\theta^*)]^{-1} H(\theta^*). \tag{4.85}$$

Result 4.3 can therefore be rephrased for the algorithm (4.16) as

$\hat{\theta}(t)$ can only converge to a value θ^*, such that $f(\theta^*) = 0$ and such that L^* (given by (4.84) and (4.85)) has all eigenvalues in the left half-plane (including the imaginary axis).

This result will be of some use in section 4.5 when we study pseudolinear regressions.

With this we conclude the discussion of tools for convergence analysis. The heuristic treatment was summarized in conclusions A–C in section 4.3.2. The formal treatment is summarized in theorem 4.2. We shall now use the tools to analyse the general algorithm (4.16′) with the different choices of "gradient vector" $\eta(t)$: $\psi(t)$ (RPEM), $\varphi(t)$ (PLR), and $\zeta(t)$

(RIV), respectively. This will be done in sections 4.4, 4.5, and 4.6, respectively.

4.4 Analysis of Recursive Prediction Error Algorithms

4.4.1 Introduction

In this section we shall analyse the general recursive prediction error algorithm (3.72)–(3.73′):

$$\varepsilon(t) = y(t) - \hat{y}(t), \tag{4.86a}$$

$$R(t) = R(t-1) + \gamma(t)H(t, R(t-1), \hat{\theta}(t-1), \varepsilon(t), \psi(t)), \tag{4.86b}$$

$$\hat{\theta}(t) = \hat{\theta}(t-1) + \gamma(t)R^{-1}(t)\left[-l_{\hat{\theta}}^{\mathrm{T}}(t, \hat{\theta}(t-1), \varepsilon(t))\right.$$
$$\left. + \psi(t)l_{\varepsilon}^{\mathrm{T}}(t, \hat{\theta}(t-1), \varepsilon(t))\right], \tag{4.86c}$$

$$\xi(t+1) = A(\hat{\theta}(t))\xi(t) + B(\hat{\theta}(t))z(t), \tag{4.86d}$$

$$\begin{pmatrix} \hat{y}(t+1) \\ \mathrm{col}\,\psi(t+1) \end{pmatrix} = C(\hat{\theta}(t))\xi(t+1). \tag{4.86e}$$

This algorithm is a special case of (4.16′) or (4.70) with the choices

$$\eta(t) = \psi(t) \tag{4.87a}$$

and

$$h(t, \theta, \varepsilon, \psi) = -l_{\hat{\theta}}^{\mathrm{T}}(t, \theta, \varepsilon) + \psi l_{\varepsilon}^{\mathrm{T}}(t, \theta, \varepsilon). \tag{4.87b}$$

Recall that the RPE algorithm aims at minimizing a criterion

$$\overline{E}l(t, \theta, \varepsilon(t, \theta)).$$

It can thus be regarded as a recursive counterpart of the off-line prediction error method described in section 3.3., which minimizes the criterion (3.30):

$$V_N(\theta, z^N) = \frac{1}{N}\sum_{t=1}^{N} l(t, \theta, \varepsilon(t, \theta)).$$

The result of the analysis is that the asymptotic properties of RPEM in fact coincide with those of the off-line prediction error method. We shall treat the convergence properties in section 4.4.2 and the asymptotic distribution in section 4.4.3. The treatment in these two sections will basically be formal and a discussion of the significance, implications, and use of the analytic results will be deferred to section 4.4.4.

In the following sections we make frequent references to the conditions introduced in section 4.3.4. We therefore provide a quick guide to these conditions:

M1: The model predictor is stable for $\theta \in D_{\mathcal{M}}$.

M2: The model predictor is differentiable with respect to the model parameters θ.

Cr2: The function H in (4.86b) is smooth enough.

Cr3 and Cr4: The criterion function l is smooth enough.

R1: $R(t) \geq \delta I$.

G1: $t \cdot \gamma(t) \to \mu$ as $t \to \infty$.

A2: The data generation is asymptotically mean stationary. ("Data generation" refers to the measurements $\{z(t)\}$.)

S1: The data generation is exponentially stable.

4.4.2   Convergence of Recursive Prediction Error Algorithms

We start with a simple example.

EXAMPLE 4.4 (Example 4.1 continued)   Let us once more consider the simple RPE algorithm (4.41). We established in example 4.3 that the regularity conditions of theorem 4.2 hold, and we showed in example 4.1 that the d.e. is stable. Theorem 4.2 then tells us that $\hat{c}(t)$, as generated by (4.41) will converge to 0.5 with probability one as t approaches infinity. This value is the true one, according to (4.46).

The stability of the d.e. was proven using fairly detailed computations. We shall now investigate whether it could be established more simply. In the example we have, according to (4.42) and (4.46),

$$\varepsilon(t, c) = \frac{1}{1 + cq^{-1}} y(t) = \frac{1 + 0.5q^{-1}}{1 + cq^{-1}} e(t).$$

The variance of the prediction error is given by

$$V(c) = \frac{1}{2} E\varepsilon^2(t, c) = \frac{1}{2} \frac{1.25 - c}{1 - c^2}. \tag{4.88}$$

$Var(ax+b\theta)$
$= a^2 Var X + b^2 Var \theta$
$+ 2ab \, cov(x, \theta)$

Simple calculation reveals that

$$\frac{d}{dc} V(c) = -\frac{1 + c^2 - 2.5c}{2(1 - c^2)^2} = -f(c), \tag{4.89}$$

where the last equality follows from (4.47). The average updating direction $f(c)$ is thus *the negative gradient of the prediction error variance $V(c)$*. This is no surprise, since the RPEM algorithm (4.41) indeed was designed to make adjustments in a negative gradient direction.

But if (4.89) holds we could use $V(c)$ as a Lyapunov function for the d.e. (4.49) rather than $\frac{1}{2}(c - 0.5)^2$ (which we used in example 4.1), since

$$\frac{d}{d\tau} V(c_D(\tau)) = \frac{d}{dc} V(c_D(\tau)) \cdot \frac{d}{d\tau} c_D(\tau) = - \left| f(c_D(\tau)) \right|^2 \cdot \frac{1}{R_D(\tau)},$$

$$\underset{-f(c)}{} \quad \underset{4.49a.}{}$$

yielding the same conclusions as in (4.50)–(4.51). This choice of Lyapunov function has promise of general applicability. □

General Algorithm   The foregoing example indicates that the criterion $\bar{E} l(t, \theta, \varepsilon(t, \theta))$ can be used as a Lyapunov function to prove stability of the d.e. associated with (4.86). This seems reasonable, since the algorithm is designed to update the estimates in a descent direction.

We introduce the following condition.

A3: The limits

(a) $\bar{E} l(t, \theta, \varepsilon(t, \theta)) = \bar{V}(\theta)$,

(b) $E \left[\dfrac{d}{d\theta} l(t, \theta, \varepsilon(t, \theta)) \right]^{\mathrm{T}} = -f(\theta)$,

$$\underset{= l_\theta + l_\varepsilon \frac{d\varepsilon}{d\theta}}{} \quad = l_\theta - l_\varepsilon \psi$$

(c) $\bar{E} H(t, R, \theta, \varepsilon(t, \theta), \psi(t, \theta)) = F(R, \theta)$,

exist for all $\theta \in D_{\mathcal{M}}$. (The symbol $\bar{E}$ was defined by (4.3).) Here H is the function in (4.86b). Notice that A3(b) and A3(c) are just restatements of A2 in view of (4.87b).

We have thus introduced a new condition on the criterion function. This condition will hold when the data sequence is asymptotically mean stationary. Then the filtered sequence $\{\varepsilon(t, \theta)\}$ is also stationary in the sense that the limit in A3(a) will exist.

We note that if A3 holds, then under weak conditions

$$V_N(\theta, z^N) \to \bar{V}(\theta) \quad \text{w.p.1 as } N \to \infty$$

(see Ljung (1978c) for a proof). This means that the criterion function $\bar{V}(\theta)$ will also be the limit of the off-line criterion (3.30). Theorem 4.2 applied to (4.86) now gives the following result.

THEOREM 4.3   Consider the algorithm (4.86). Assume that it includes a projection to keep $\hat{\theta}(t)$ inside $D_{\mathcal{M}}$ and to assure a bounded subsequence of $\{\xi(t)\}$. (The subsequence may depend on the realization of $\{z(t)\}$.) Assume that M1, M2, Cr2, Cr3, A3, S1, R1, and G1 hold. Then $\{\hat{\theta}(t)\}$, w.p.1, converges either to the set

$$D_c = \left\{ \theta \left| \frac{d}{d\theta} \bar{V}(\theta) = 0 \right. \right\}$$

or to the boundary of $D_{\mathcal{M}}$ as t approaches infinity.

Before giving the proof let us stress a number of important points.

• The theorem is formulated for a general search direction. The only condition is that $R(t)$ be positive definite. This means that the theorem applies to the Gauss-Newton scheme (3.67) and the stochastic gradient scheme (3.74) as well as to other descent search directions.

• Nothing is assumed about the true system, other than that the data generation be stable (condition S1) and "asymptotically mean stationary" (condition A3). The system itself may be much more complex than the resulting model, but this model is the "best approximation" of the system in terms of the chosen criterion. Of course, it must be realized that the true system hides behind the symbol $\bar{E}$ since all expectations are with respect to $\{z(t)\}$. The function $\bar{V}(\theta)$ will thus depend on the system and will not be accessible to the user before the data is collected. (If it were, there would be no need for identification).

• The theorem states that the estimate $\hat{\theta}(t)$ converges to a stationary point of the criterion function. In view of result 4.3 (p. 177), this can be strengthened to convergence to a local minimum of the function $\bar{V}(\theta)$.

Proof of Theorem 4.3   Condition A3 ensures that A2 holds, and Cr3 implies Cr1. Hence lemma 4.1 shows that A1 holds w.p.1. All conditions of theorem 4.2 are therefore satisfied, and we need only find a function V, subject to (4.79).

Our candidate for this function clearly is $\bar{V}(\theta)$ defined in A3. To verify (4.79) we must first establish that

$$f^{\mathsf{T}}(\theta) = -\frac{d}{d\theta} \bar{V}(\theta), \tag{4.90}$$

where $f(\theta)$ is defined in A3. By definition,

$$\bar{V}(\theta) = \lim_{N \to \infty} \mathrm{E} V_N(\theta, z^N).$$

Moreover, by differentiating $V_N(\theta, z^N)$ we obtain

$$\frac{d}{d\theta} V_N(\theta, z^N) = \frac{1}{N} \sum_1^N \frac{d}{d\theta} l(t, \theta, \varepsilon(t, \theta)). \tag{4.91}$$

Hence, in order to prove (4.90), we only have to show that the limit operation, the differentiation w.r.t θ, and the E-operator commute when applied to (4.91). Since $\frac{d}{d\theta} l(t, \theta, \varepsilon(t, \theta))$ is dominated by a θ-independent integrable function according to Cr3 and S1, we have

$$\frac{d}{d\theta} \mathrm{E} l(t, \theta, \varepsilon(t, \theta)) = \mathrm{E} \frac{d}{d\theta} l(t, \theta, \varepsilon(t, \theta))$$

and

$$\frac{d}{d\theta} \frac{1}{N} \sum_1^N \mathrm{E} l(t, \theta, \varepsilon(t, \theta)) = \frac{1}{N} \sum_1^N \mathrm{E} \frac{d}{d\theta} l(t, \theta, \varepsilon(t, \theta)).$$

The right-hand side of the above expression converges according to A3. In fact, the convergence is uniform in $\theta \in D_{\mathcal{M}}$, since the second derivative of $\mathrm{E} l(t, \theta, \varepsilon(t, \theta))$ is bounded according to Cr3 and S1. Therefore, the limit operation commutes with differentiation and we have established (4.90).

We now verify (4.79). We have that the d.e. associated with (4.86) is

$$\frac{d}{d\tau} \theta_D(\tau) = R_D^{-1}(\tau) f(\theta_D(\tau)) = -R_D^{-1}(\tau) [\bar{V}'(\theta_D(\tau))]^\mathrm{T},$$

$$\tag{4.92}$$

$$\frac{d}{d\tau} R_D(\tau) = F(R_D(\tau), \theta_D(\tau)).$$

Hence, along trajectories of (4.92),

$$\frac{d}{d\tau} \bar{V}(\theta_D(\tau)) = \bar{V}'(\theta_D(\tau)) \cdot \frac{d}{d\tau} \theta_D(\tau)$$

$$= -\bar{V}'(\theta_D(\tau)) R_D^{-1}(\tau) [\bar{V}'(\theta_D(\tau))]^\mathrm{T} \le 0,$$

where the inequality follows from R1. Moreover, equality is obtained only for $\theta_D(\tau) \in D_c$. We have thus proved that (4.79) holds. Hence theorem 4.2 implies the conclusion of theorem 4.3. ∎

Theorem 4.3 tells us that the estimate will converge to a stationary point (local minimum) of the chosen criterion $\bar{V}(\theta)$, or it will get stuck at the boundary. According to our discussion preceding the corollary in section 4.3.3, the estimates can get stuck at the boundary only if there is a trajectory to (4.92) that points out from $D_\mathcal{M}$, i.e., if $\bar{V}(\theta)$ is decreasing as θ leaves $D_\mathcal{M}$ at some point. Now, at the boundary of the stability region D_s, $\bar{V}(\theta)$ tends to infinity. Hence, if the boundary of $D_\mathcal{M}$ is chosen close enough to the boundary of D_s, no trajectory will point out from $D_\mathcal{M}$, and the algorithm will not converge to the boundary of $D_\mathcal{M}$.

Recall that $\bar{V}(\theta)$ is also the limit of the off-line criterion, and the off-line estimate is usually found by local numerical minimization of $V_N(\theta, z^N)$, as we discussed in section 3.3. The convergence properties of the recursive estimate and the off-line one will thus coincide.

In section 3.3 (example 3.7) we pointed out that $V_N(\theta, z^N)$ is the negative log likelihood function if the function $l(\cdot, \cdot, \cdot)$ is chosen in a certain way. Theorem 4.3 thus implies convergence to a stationary point (local maximum) of the expected value of the likelihood function, i.e., it implies that $\hat{\theta}(t)$ will asymptotically coincide with the maximum likelihood estimate.

The only flaw in the result is that we cannot guarantee convergence to the *global* minimum of the criterion. If other local minima exist, the estimate may converge to one of these. This must be kept in mind when chosing criterion function and model set, which both may affect the possible existence of "undesired" stationary points of $\bar{V}(\theta)$, as we will see in sections 5.2 and 5.3.

Gauss-Newton Algorithm with Constant Weighting Matrix   Consider now the stochastic Gauss-Newton algorithm for a quadratic criterion, (3.54), (3.61):

$$\hat{\theta}(t) = \hat{\theta}(t-1) + \gamma(t)R^{-1}(t)\psi(t)\Lambda^{-1}\varepsilon(t), \tag{4.93a}$$

$$R(t) = R(t-1) + \gamma(t)[\psi(t)\Lambda^{-1}\psi^T(t) - R(t-1)]. \tag{4.93b}$$

For this case obviously Cr2 and Cr3 are satisfied, and A3 takes the form

$$\tfrac{1}{2}\bar{E}\varepsilon^T(t, \theta)\Lambda^{-1}\varepsilon(t, \theta) = \bar{V}(\theta),$$

$$\bar{E}\psi(t, \theta)\Lambda^{-1}\varepsilon(t, \theta) = f(\theta) \quad (= -[\bar{V}'(\theta)]^T), \tag{4.94}$$

$$\bar{E}\psi(t, \theta)\Lambda^{-1}\psi^T(t, \theta) = G(\theta).$$

Now condition R1 (i.e. $R(t) \geq \delta I$) is not automatically satisfied for

(4.93b). The reason is that $G(\theta)$ (defined in (4.94)) may be singular for certain θ, i.e., that there is a certain linear dependence among the entries of $\psi(t, \theta)$. (We shall see in sections 4.4.4 and 5.4 that this will happen if the input is not exciting the system, or if the model contains too many parameters). This means that (4.93b) has to be modified to

$$\bar{R}(t) = R(t - 1) + \gamma(t)[\psi(t)\Lambda^{-1}\psi^{T}(t) - R(t - 1)],$$

$$R(t) = \begin{cases} \bar{R}(t) \text{ if } \bar{R}(t) \geq \delta I \\ R(t) + M_{\delta}(t) \quad \text{otherwise} \end{cases},$$

(4.95)

where $M_{\delta}(t)$ is chosen so that $R(t) \geq \delta I$. A modification of this sort should always be included in any implementation of Gauss-Newton algorithms. It is required to ensure good numerical behavior. Specific ways of implementing (4.95) will be discussed in section 6.5. With the modification (4.95), theorem 4.3 can be applied to (4.93), with the result that $\hat{\theta}(t)$ will converge to a stationary point (local minimum) of

$$\tfrac{1}{2}\bar{E}\varepsilon^{T}(t, \theta)\Lambda^{-1}\varepsilon(t, \theta).$$

Gauss-Newton Algorithm with Estimated Weighting Matrix   In section 3.4 we found it suitable to replace Λ by an estimate of the prediction error covariance matrix [see (3.67)]:

$$\hat{\theta}(t) = \hat{\theta}(t - 1) + \gamma(t)R^{-1}(t)\psi(t)\hat{\Lambda}^{-1}(t)\varepsilon(t), \tag{4.96a}$$

$$R(t) = R(t - 1) + \gamma(t)[\psi(t)\hat{\Lambda}^{-1}(t)\psi^{T}(t) - R(t - 1)], \tag{4.96b}$$

$$\hat{\Lambda}(t) = \hat{\Lambda}(t - 1) + \gamma(t)[\varepsilon(t)\varepsilon^{T}(t) - \hat{\Lambda}(t - 1)]. \tag{4.96c}$$

We will assume that (4.96b) in fact is implemented in its modified form (4.95). A corresponding modification for (4.96c) is normally not required, since in the typical case

$$E\varepsilon(t, \theta)\varepsilon^{T}(t, \theta) \geq \delta I, \quad \delta > 0 \text{ for all } \theta, \tag{4.97}$$

meaning that no linear combinations of the output vector $y(t)$ can be predicted exactly.

For the algorithm (4.96) it is not immediately clear what criterion we are minimizing, since the replacement of Λ by $\hat{\Lambda}(t)$ was made by means of an ad hoc argument. The answer is that the algorithm minimizes the criterion

$$\bar{V}(\theta, \Lambda) = \tfrac{1}{2}\bar{E}\varepsilon^T(t, \theta)\Lambda^{-1}\varepsilon(t, \theta) + \tfrac{1}{2}\log\det\Lambda. \tag{4.98}$$

This is proved in the following theorem. Notice that this criterion is the expected value of the negative log likelihood function if the prediction errors are assumed to be Gaussian with unknown covariance matrix Λ, see (3.35).

THEOREM 4.4   Consider the Gauss-Newton algorithm (4.96), subject to the same assumptions as in theorem 4.3 as well as to (4.97). Then $\{\hat{\theta}(t),$ $\hat{\Lambda}(t)\}$ converges w.p.1, either to the set of stationary points (local minima) of $\bar{V}(\theta, \Lambda)$, defined by (4.98), or to the boundary of D_M as t approaches infinity.

Proof   The algorithm (4.96) is slightly more complex than (4.70), in that it estimates not only $\hat{\theta}(t)$ and $R(t)$ but also the matrix $\hat{\Lambda}(t)$. Therefore also col Λ has to be included in the estimate vector x when forming (4.71). The regularity conditions will still hold with this modified x, as is easy to verify. The d.e. then becomes

$$\dot{\theta}_D = R_D^{-1}f(\theta_D, \Lambda_D),$$

$$\dot{\Lambda}_D = H_D(\theta) - \Lambda_D, \tag{4.99}$$

$$\dot{R}_D = G(\theta_D, \Lambda_D) - R_{D'}$$

where

$$H(\theta) = \bar{E}\varepsilon(t, \theta)\varepsilon^T(t, \theta) \qquad (\geq \delta I),$$

$$G(\theta, \Lambda) = \bar{E}\psi(t, \theta)\Lambda^{-1}\psi^T(t, \theta)$$

and

$$f(\theta, \Lambda) = \bar{E}\psi(t, \theta)\Lambda^{-1}\varepsilon(t, \theta).$$

As in the proof of theorem 4.3 we have

$$f(\theta, \Lambda) = -\left[\frac{\partial}{\partial\theta}\bar{V}(\theta, \Lambda)\right]^T. \tag{4.100}$$

We shall now verify that, along solutions to (4.99),

$$\frac{d}{d\tau}\bar{V}(\theta_D(\tau), \Lambda_D(\tau)) \leq 0$$

for all θ and Λ with equality if and only if (θ, Λ) is a stationary point of $\bar{V}$.

We will use the differentiation formula

$$\frac{d}{d\alpha} \log \det \Lambda(\alpha) = \mathrm{tr}\left[\Lambda^{-1} \cdot \frac{d}{d\alpha}\Lambda(\alpha)\right], \qquad (4.101)$$

which can be readily established. We now have

$$\frac{d}{d\tau}\bar{V}(\theta_D(\tau), \Lambda_D(\tau))$$

$$= \frac{1}{2}\frac{d}{d\tau}\overline{\mathrm{E}}\varepsilon^{\mathrm{T}}(t, \theta_D(\tau))\Lambda_D^{-1}(\tau)\varepsilon(t, \theta_D(\tau)) + \frac{1}{2}\frac{d}{d\tau}\log \det \Lambda_D(\tau)$$

$$= -\frac{1}{2}\left[\frac{d}{d\tau}\theta_D(\tau)\right]^{\mathrm{T}}\overline{\mathrm{E}}\psi(t, \theta_D(\tau))\Lambda_D^{-1}(\tau)\varepsilon(t, \theta_D(\tau))$$

$$\quad -\frac{1}{2}\left[\overline{\mathrm{E}}\varepsilon^{\mathrm{T}}(t, \theta_D(\tau))\Lambda_D^{-1}(\tau)\psi^{\mathrm{T}}(t, \theta_D(\tau))\right]\frac{d}{d\tau}\theta_D(\tau)$$

$$\quad -\frac{1}{2}\overline{\mathrm{E}}\left\{\varepsilon^{\mathrm{T}}(t, \theta_D(\tau))\Lambda_D^{-1}(\tau)\left[\frac{d}{d\tau}\Lambda_D(\tau)\right]\Lambda_D^{-1}(\tau)\varepsilon(t, \theta_D(\tau))\right\}$$

$$\quad +\frac{1}{2}\mathrm{tr}\left[\Lambda_D^{-1}(\tau)\frac{d}{d\tau}\Lambda_D(\tau)\right]$$

$$= -f^{\mathrm{T}}(\theta_D(\tau), \Lambda_D(\tau))R_D^{-1}(\tau)f(\theta_D(\tau), \Lambda_D(\tau))$$

$$\quad -\frac{1}{2}\mathrm{tr}\left\{\overline{\mathrm{E}}[\varepsilon(t, \theta_D(\tau))\varepsilon^{\mathrm{T}}(t, \theta_D(\tau))]\Lambda_D^{-1}(\tau)[H(\theta_D(\tau)) - \Lambda_D(\tau)]\Lambda_D^{-1}(\tau)\right\}$$

$$\quad +\frac{1}{2}\mathrm{tr}\left\{\Lambda_D^{-1}(\tau)[H(\theta_D(\tau)) - \Lambda_D(\tau)]\right\}.$$

Here, in the second equality we used (4.101) and in the third one, (4.99), and (4.100), together with the fact that $A^{\mathrm{T}}B = \mathrm{tr}\,BA^{\mathrm{T}}$ for column vectors A and B. The two last terms can be written (suppressing the argument τ)

$$-\tfrac{1}{2}\mathrm{tr}\left\{H(\theta)\Lambda^{-1}[H(\theta) - \Lambda]\Lambda^{-1} - \Lambda^{-1}[H(\theta) - \Lambda]\right\}$$

$$= -\tfrac{1}{2}\mathrm{tr}\left[H(\theta) - \Lambda\right]\Lambda^{-1}[H(\theta) - \Lambda]\Lambda^{-1}$$

$$= -\tfrac{1}{2}\mathrm{tr}\,\Lambda^{-1/2}[H(\theta) - \Lambda]\Lambda^{-1}[H(\theta) - \Lambda]\Lambda^{-1/2},$$

where $\Lambda^{-1/2}$ is a symmetric square root of Λ^{-1}. This is of the form

$$-\tfrac{1}{2}\operatorname{tr} A^{\mathrm{T}} A = -\tfrac{1}{2}\sum_{i,j}|a_{ij}|^2$$

for the symmetric matrix

$$A = \Lambda^{-1/2}[H(\theta) - \Lambda]\Lambda^{-1/2}.$$

These facts show that

$$\frac{d}{d\tau}\bar{V}(\theta_D(\tau), \Lambda_D(\tau)) \le 0$$

with equality only for θ, Λ such that

$$f(\theta, \Lambda) = -\left[\frac{d}{d\theta}\bar{V}(\theta, \Lambda)\right]^{\mathrm{T}} = 0$$

and

$$H(\theta) - \Lambda = 2\frac{\partial}{\partial\Lambda^{-1}}\bar{V}(\theta, \Lambda) = 0,$$

i.e., for θ, Λ that are stationary points of $\bar{V}(\theta, \Lambda)$. This proves the theorem, according to theorem 4.2. ∎

This concludes the derivation of convergence results for RPEM. In section 4.4.4 we shall discuss the practical implications of these results; but first we will investigate the asymptotic distribution of the estimates.

4.4.3 Asymptotic Distribution of Recursive Prediction Error Estimates

Once the convergence properties of a recursive estimate are known, the next natural question is to determine "how fast" it converges to the limit. This is usually expressed in terms of an asymptotic distribution for the estimate. In section 3.3 we quoted such a result for the off-line prediction error estimate (see (3.38)):

$$\sqrt{N}(\hat{\theta}_N - \theta^*) \in \mathrm{AsN}(0, P). \tag{4.102}$$

This expression means that the random variable $\sqrt{N}(\hat{\theta}_N - \theta^*)$ converges in distribution to a normal distribution with zero mean and covariance matrix P. A general expression for P is given by (3.39). We shall here first evaluate this expression under the following conditions about the limiting prediction errors:

L1: The limit $\theta^* = \theta_0$ is such that

$$\varepsilon(t, \theta_0) = e(t) + r(t),$$

where $\{e(t)\}$ is a sequence of independent random variables each of zero mean and covariance Λ_0, and $\{r(t)\}$ is a sequence of random variables such that for some $\beta > 0$

$$t^{1/2+\beta} r(t) \to 0 \quad \text{w.p.1} \quad \text{as } t \to \infty.$$

L2: The random variables $l_\varepsilon(t, \theta_0, e(t))$, $l_{\varepsilon\theta}(t, \theta_0, e(t))$, and $l_\theta(t, \theta_0, e(t))$ have zero means and the $d \times d$-matrix

$$G(\theta_0) = \overline{E}[l_{\theta\theta}(t, \theta_0, e(t)) + \psi(t, \theta_0)l_{\varepsilon\varepsilon}(t, \theta_0, e(t))\psi^T(t, \theta_0)]$$

is invertible.

Let us comment on these conditions. L1 requires that the limit θ_0 give prediction errors that are asymptotically white. This means that θ_0 gives a description of the system that is asymptotically correct. The term $r(t)$ is included to allow for the fact that there may be transient phenomena in the true system (such as initial conditions) that are not accurately described by θ_0. For all practical purposes we may think of L1 as a condition that θ_0 is the true value of the model parameter. To illustrate the significance of L2, consider the special case $l(t, \theta, \varepsilon) = \frac{1}{2}\varepsilon^T\Lambda^{-1}\varepsilon$. Then the first part of L2 is automatically satisfied if L1 holds. The second part, which amounts to saying that $\psi(t, \theta_0)\Lambda^{-1}\psi^T(t, \theta_0)$ $(= G(\theta_0))$ is invertible, poses certain conditions on the model parametrization and on the input properties. These matters are discussed in some detail in section 4.4.4 (see examples 4.8 and 4.10).

Asymptotic Covariance Matrix for the Off-Line Estimate We shall now evaluate (3.39) under conditions L1, L2, and Cr4 (p. 166). With the loss function

$$V_N(\theta, z^N) = \frac{1}{N} \sum_1^N l(t, \theta, \varepsilon(t, \theta))$$

and with

$$\overline{V}(\theta) = \overline{E}V_N(\theta, z^N),$$

we obtain

$$[V_N'(\theta, z^N)]^{\mathrm{T}} = \frac{1}{N} \sum_1^N l_\theta^{\mathrm{T}}(t, \theta, \varepsilon(t, \theta)) - \psi(t, \theta) l_\varepsilon^{\mathrm{T}}(t, \theta, \varepsilon(t, \theta)),$$

$$\bar{V}''(\theta) = \lim_{N\to\infty} \mathrm{E}\frac{1}{N} \sum_1^N [l_{\theta\theta}(t, \theta, \varepsilon(t, \theta))$$

$$+ \psi(t, \theta) l_{\varepsilon\varepsilon}(t, \theta, \varepsilon(t, \theta)) \psi^{\mathrm{T}}(t, \theta) - l_{\theta\varepsilon}(t, \theta, \varepsilon(t, \theta)) \psi^{\mathrm{T}}(t, \theta)$$

$$- \psi(t, \theta) l_{\varepsilon\theta}(t, \theta, \varepsilon(t, \theta)) - \psi_\theta(t, \theta) l_\varepsilon^{\mathrm{T}}(t, \theta, \varepsilon(t, \theta))].$$

Now when we evaluate $\bar{V}''(\theta)$ at $\theta = \theta_0$ the three last terms will not give any contribution to the limit, in view of L1, L2, and Cr4 (expand in Taylor series around $\varepsilon = e$, use independence of $e(t)$ and $\psi(t, \theta)$, and use Cauchy's inequality for the remainder term). Hence

$$\bar{V}''(\theta_0) = \bar{\mathrm{E}}[l_{\theta\theta}(t, \theta_0, e(t)) + \psi(t, \theta_0) l_{\varepsilon\varepsilon}(t, \theta_0, e(t)) \psi^{\mathrm{T}}(t, \theta_0)].$$

Also

$$\lim_{N\to\infty} N \cdot \mathrm{E}[V_N'(\theta_0, z^N)]^{\mathrm{T}}[V_N'(\theta_0, z^N)]$$

$$= \lim_{N\to\infty} \frac{1}{N} \sum_{t,s=1}^N \mathrm{E}\{[l_\theta^{\mathrm{T}}(t, \theta_0, \varepsilon(t, \theta_0)) - \psi(t, \theta_0) l_\varepsilon^{\mathrm{T}}(t, \theta_0, \varepsilon(t, \theta_0))]$$

$$\times [l_\theta^{\mathrm{T}}(s, \theta_0, \varepsilon(s, \theta_0)) - \psi(s, \theta_0) l_\varepsilon^{\mathrm{T}}(s, \theta_0, \varepsilon(s, \theta_0))]^{\mathrm{T}}\}$$

$$= \bar{\mathrm{E}}\{[l_\theta^{\mathrm{T}}(t, \theta_0, e(t)) - \psi(t, \theta_0) l_\varepsilon^{\mathrm{T}}(t, \theta_0, e(t))]$$

$$\times [l_\theta^{\mathrm{T}}(t, \theta_0, e(t)) - \psi(t, \theta_0) l_\varepsilon^{\mathrm{T}}(t, \theta_0, e(t))]^{\mathrm{T}}\} \triangleq Q(\theta_0). \tag{4.103}$$

Here, the second step follows since the effect of the cross terms $s \neq t$ vanishes asymptotically due to L1, L2, and Cr4.

The asymptotic covariance matrix P in (4.102) for the off-line estimate is consequently given by

$$P = [G(\theta_0)]^{-1} Q(\theta_0) [G(\theta_0)]^{-1} \tag{4.104}$$

under conditions L1, L2, and Cr4.

Specializing to a quadratic criterion

$$l(t, \theta, \varepsilon) = \tfrac{1}{2}\varepsilon^{\mathrm{T}}\Lambda^{-1}\varepsilon$$

gives

$$G(\theta_0) = \bar{\mathrm{E}}\psi(t, \theta_0)\Lambda^{-1}\psi^{\mathrm{T}}(t, \theta_0), \tag{4.105a}$$

$$Q(\theta_0) = \bar{E}\psi(t, \theta_0)\Lambda^{-1}\Lambda_0\Lambda^{-1}\psi^T(t, \theta_0), \tag{4.105b}$$

which, of course, coincides with the expression (3.45).

Asymptotic Distribution for the Gauss-Newton Recursive Prediction Error Algorithm We shall now derive the asymptotic distribution for the *recursive* prediction error algorithm. The rather remarkable result is that, if the Gauss-Newton search direction (3.73) is used and if the gain sequence (asymptotically) is $\gamma(t) = 1/t$, then the result (4.102), (4.104) holds also for the recursively computed estimate.

Before formulating the theorem, we shall discuss a complication in the recursive case. We know from the previous section that the estimate will converge w.p.1 to a local minimum of the criterion function. In the general case we cannot exclude the possibility that several distinct local minima exist. Hence we may converge to different points, depending on the realization, and results such as (4.102) would not be possible. (Strictly speaking, the same applies to the off-line situation, where the numerically obtained estimate $\hat{\theta}_N$ may differ from the globally minimizing value $\hat{\theta}_N$, to which (4.102) refers.) To resolve this problem we shall henceforth consider only a subset of realizations Ω such that $\hat{\theta}(t) \to \theta_0$ as $t \to \infty$ almost everywhere on Ω. All probabilistic quantifiers, such as "E," "w.p.1," etc., will refer to Ω. We express this by the phrase "conditioned on the event $\hat{\theta}(t) \to \theta_0$."

We shall prove the result under the following restriction about the data $\{z(t)\}$.

S2: The sequence of data is subject to

$$|z(t)| \le \alpha_1,$$

where α_1 is a t-independent random variable with finite variance.

The restriction that the data sequence is bounded is of course not unrealistic, but it does exclude, e.g., that $\{z(t)\}$ be a normal process. However, S2 simplifies the proof.

For the asymptotic normality result, we shall also strengthen the condition on the model set.

M3: The matrices $A(\theta)$, $B(\theta)$, and $C(\theta)$ defining the model set in (4.86) are twice continuously differentiable for $\theta \in D_{\mathscr{M}}$.

The main result of this section can now be formulated. This theorem and its proof are inspired by a corresponding result in a special case by Solo (1981).

THEOREM 4.5   Consider the recursive Gauss-Newton prediction error algorithm

$$\hat{\theta}(t) = \hat{\theta}(t-1) + \frac{1}{t}R^{-1}(t)[-l_\theta^T(t, \hat{\theta}(t-1), \varepsilon(t))$$

$$+ \psi(t)l_\theta^T(t, \hat{\theta}(t-1), \varepsilon(t))],$$

$$R(t) = R(t-1) + \frac{1}{t}[l_{\theta\theta}(t, \hat{\theta}(t-1), \varepsilon(t))$$

$$+ \psi(t)l_{\varepsilon\varepsilon}(t, \hat{\theta}(t-1), \varepsilon(t))\psi^T(t) - R(t-1)],$$

where $\{\varepsilon(t)\}$ and $\{\psi(t)\}$ are determined as in (4.86a, d, e). Assume that the limits

$$G(\theta_0) = \bar{E}[l_{\theta\theta}(t, \theta_0, \varepsilon(t, \theta_0)) + \psi(t, \theta_0)l_{\varepsilon\varepsilon}(t, \theta_0, \varepsilon(t, \theta_0))\psi^T(t, \theta_0)],$$

$$Q(\theta_0) = \bar{E}\{[l_\theta^T(t, \theta_0, \varepsilon(t, \theta_0)) - \psi(t, \theta_0)l_\varepsilon^T(t, \theta_0, \varepsilon(t, \theta_0))]$$

$$\times [l_\theta^T(t, \theta_0, \varepsilon(t, \theta_0)) - \psi(t, \theta_0)l_\varepsilon^T(t, \theta_0, \varepsilon(t, \theta_0))]^T\}$$

exist, and that S1, M1, and Cr4 (all three defined in section 4.3.4) as well as L1, L2, S2, and M3 (defined in this section) hold. Then, conditioned on the event that $\hat{\theta}(t) \to \theta_0$, we have

$$\sqrt{t}[\hat{\theta}(t) - \theta_0] \in \text{AsN}(0, P)$$

where

$$P = [G(\theta_0)]^{-1}Q(\theta_0)[G(\theta_0)]^{-1}.$$

Moreover for any $\delta > 0$,

$$t^{+1/2-\delta}|\hat{\theta}(t) - \theta_0| \to 0 \text{ w.p.1 as } t \to \infty.$$

The proof of this theorem is given in appendix 4.B.

The important point is, of course, that the asymptotic covariance matrix for RPEM under the conditions of the theorem is the same as the off-line expression (4.104). We shall elaborate on this fact later in this section.

Gauss-Newton Recursive Prediction Error Algorithm with Estimated Weighting Matrix for Quadratic Criteria   Let us now consider the algorithm (4.96), where the prediction error variance is estimated:

$$\hat{\theta}(t) = \hat{\theta}(t-1) + \frac{1}{t} R^{-1}(t)\psi(t)\hat{\Lambda}^{-1}(t)\varepsilon(t),$$

$$\hat{\Lambda}(t) = \hat{\Lambda}(t-1) + \frac{1}{t}[\varepsilon(t)\varepsilon^{T}(t) - \hat{\Lambda}(t-1)], \tag{4.106}$$

$$R(t) = R(t-1) + \frac{1}{t}[\psi(t)\hat{\Lambda}^{-1}(t)\psi^{T}(t) - R(t-1)].$$

We showed in theorem 4.4 that $\hat{\theta}(t)$ will tend to a minimum of the function (4.98). For its asymptotic distribution we have the following corollary.

COROLLARY 4.5   Consider the algorithm (4.106), subject to the same assumptions as in theorem 4.5. Assume that Λ_0 is invertible. Then, conditioned on the event that $\hat{\theta}(t) \to \theta_0$, we have

$$\sqrt{t}[\hat{\theta}(t) - \theta_0] \in \text{AsN}(0, P)$$

where

$$P = [\bar{E}\psi(t, \theta_0)\Lambda_0^{-1}\psi^{T}(t, \theta_0)]^{-1}. \tag{4.107}$$

The proof is given in appendix 4.B.

Notice that P given by (4.107) is also the limit of $R^{-1}(t)$, according to lemma 4.B.2. Therefore the algorithm (4.106) provides its own estimate of the covariance matrix of $\hat{\theta}(t)$:

$$\text{cov}(\hat{\theta}(t)) \approx \frac{1}{t} R^{-1}(t). \tag{4.108}$$

In the implementation (3.70) of (4.106) we actually work with this estimate

$$P(t) = \frac{1}{t} R^{-1}(t)$$

explicitly in the algorithm. Notice the scaling Λ_0^{-1} in (4.107)! If the output is scalar-valued so that $\Lambda_0 = \bar{E}\varepsilon^2(t, \theta_0)$ is a scalar, then we can write

$$P = [\bar{E}\varepsilon^2(t, \theta_0)][\bar{E}\psi(t, \theta_0)\psi^{T}(t, \theta_0)]^{-1}.$$

When the R-equation in (4.106) is implemented without the $\hat{\Lambda}$-scaling, as may be the case when $p = 1$, it is important to remember this scale factor before using (4.108).

The asymptotic covariance matrices for the recursive estimate $\hat{\theta}(t)$ and for the off-line estimate $\hat{\theta}_t$ coincide under the assumptions of the theorem. This matrix also equals the Cramér-Rao lower bound, as we remarked in section 3.3, if the function $l(\,\cdot\,,\,\cdot\,,\,\cdot\,)$ is chosen as the negative logarithm of the probability density function of the prediction errors. Therefore we cannot hope for a better asymptotic behaviour than what the algorithm (4.106) achieves. In this algorithm the search direction is the Gauss-Newton one, and the gain sequence is $\gamma(t) = 1/t$. We may ask how important these two facts are for the result to hold.

First, the Gauss-Newton direction is absolutely crucial for lemma 4.B.4 to hold. Other (asymptotic) search directions will give a contribution of order $|\tilde{\theta}(k)|$ to the term in question. It is therefore to be expected that such directions will give strictly larger covariance matrices.

Second, suppose we use a general gain sequence $\{\gamma(t)\}$ in the Gauss-Newton scheme (4.106) with corresponding weights

$$\beta(t, k) = \gamma(t) \prod_{s=k+1}^{t} [1 - \gamma(s)], \quad \beta(t, t) = \gamma(t), \quad \gamma(1) = 1 \tag{4.109}$$

[see (2.128)]. Then averages like

$$\frac{1}{t} \sum_{1}^{t} f(k)$$

in the proof of the theorem will be replaced by weighted averages

$$\sum_{1}^{t} \beta(t, k) f(k), \quad \text{where} \quad \sum_{k=1}^{t} \beta(t, k) = 1. \tag{4.110}$$

Pursuing these weights through the proof in appendix 4.B shows that the theorem will hold with the P-matrix replaced by

$$\Gamma \cdot P \quad \text{where} \quad \Gamma = \lim_{t \to \infty} t \cdot \sum_{k=1}^{t} \beta^2(t, k). \tag{4.111}$$

Here, of course, with $\gamma(t) = 1/t$ we have $\beta(t, k) = 1/t$ and $\Gamma = 1$. It is also straightforward to verify, using (4.110) and Schwarz' inequality, that $\Gamma \geq 1$ for all $\gamma(t)$-sequences. Moreover, $\Gamma = 1$ when $t \cdot \gamma(t) \to 1$ as $t \to \infty$.

The fact that we obtain the same asymptotic distribution for the off-line and the recursive estimates may seem surprising. In the recursive algorithm we are constantly throwing away information, and we might have expected a worse asymptotic performance. The intuitive explanation of theorem 4.5 is, however, that the Gauss-Newton algorithm can be seen as a least squares algorithm for a quadratically expanded approximation of the criterion function around the current estimate. As the estimates come close to the minimum this approximation becomes better and better, and asymptotically no information is wasted. Compare the calculations in section 2.2.3, where the approximations (2.42)–(2.45) become exact as $\hat{\theta}(t)$ tends to the minimum.

Notice, however, that at any given finite t the off-line estimate $\hat{\theta}_t$ is usually strictly better than the recursive estimate $\hat{\theta}(t)$. The theory is not able to tell at what t the asymptotic distribution becomes applicable and whether this is significantly later for $\hat{\theta}(t)$ than for $\hat{\theta}_t$. This question must be left to simulation studies (see tables 5.4 and 5.5 in section 5.7). This also means that while it is natural, from theorem 4.5, to use the Gauss-Newton search direction and the step size $\gamma(t) = 1/t$ asymptotically, other choices could be advantageous in transient stages.

4.4.4 A Discussion of the Asymptotic Properties

In the two preceding sections we have proved some quite general results about the asymptotic properties of the estimates obtained by recursive prediction error identification. These results were given in theorems 4.3, 4.4, and 4.5. In this section we shall discuss some practical aspects of these results; in particular, how they relate to the choice of model set and experimental condition.

Let us first point out that we introduced several conditions for the general case studied in this section. The results of the theorems might hold under less restrictive assumptions when applied to particular members in the algorithm family. For example, convergence and asymptotic distribution for the recursive least squares (RLS) method, described in section 2.2.1, can more easily be analyzed using the explicit (off-line) expression for the estimate, (2.13). From this it is clear that projection is not required in RLS, and that the estimates will converge to the only stationary point of the criterion (which is quadratic in θ), viz, to the global minimum.

The general convergence result (theorem 4.3) is easy to understand:

The estimates converge to a local minimum of the chosen criterion function $\bar{V}(\theta) = \bar{E}l(t, \theta, \varepsilon(t, \theta))$. Thus, an important question is whether other local minima (stationary points) than the global one exist. One can think of several ways of testing and manipulating the estimates so that they are not caught in "false" local minima. For example, the correlation of the prediction errors can be monitored and appropriate action can be taken if they are not found to be "white enough." However, such procedures have not been extensively tested so far.

Some results concerning the existence and nonexistence of local minima for particular model sets are summarized in section 5.3. The current knowledge of such results is, however, fairly limited. On the other hand, the problems with "false" local minima should not be exaggerated. When it comes to black box models, which perhaps are the most common ones in on-line applications, the practical experience is good. In most reported cases convergence to the global minimum has been obtained.

When convergence is to the global minimum, theorem 4.3 can be formulated in the following more suggestive way: The estimate converges to the best possible predictor, i.e., to the best possible approximation of the system (in the sense of the chosen criterion). When the system has no input signal $\{u(t)\}$, i.e., when the properties of a time series, or signal, $\{y(t)\}$ are modeled, this is a very strong robustness result. It means that we obtain a very meaningful approximate description of the signal. For a system with an input, the algorithm still makes the best of the situation when the system is more complex than the model set: it chooses the best approximation. This may have some surprising effects on the parameter estimates of the model, which we illustrate in the following simple example.

EXAMPLE 4.5 (Bias in LS Estimates)   Suppose that the system is given by

$$y(t) + a_0 y(t - 1) = b_0 u(t - 1) + e(t) + c_0 e(t - 1), \tag{4.112}$$

where $\{u(t)\}$ and $\{e(t)\}$ are independent sequences of independent random variables each of zero mean and unit variance. Let the model set be given by

$$y(t) + ay(t - 1) = bu(t - 1) + e(t). \tag{4.113}$$

It is easy to verify that values of a and b that give the best predictions when applied to (4.112) are

$$a^* = a_0 - \frac{c_0}{r_0},$$

$$b^* = b_0,$$ (4.114)

where

$$r_0 = Ey^2(t) = \frac{b_0^2 + c_0(c_0 - a_0) - a_0 c_0 + 1}{1 - a_0^2}.$$

[handwritten margin notes:]
var (ax+by)
= a² var x + b² var y
+ 2ab cov(x, y)

To show (4.114) we can reason as follows: The true prediction for (4.112) is given by

$$\hat{y}(t \mid \theta_0) = -a_0 y(t-1) + b_0 u(t-1) + c_0 \hat{e}(t-1),$$

where $\hat{e}(t-1)$ is computed from y^{t-1} and u^{t-1} using (4.112). Our model set, however, only allows a structure

$$\hat{y}(t \mid \theta) = -ay(t-1) + bu(t-1),$$

and cannot accomodate the term $\hat{e}(t-1)$. It has to be replaced by an estimate based on $y(t-1)$ and $u(t-1)$ only. Now, since $u(t-1)$ and $e(t-1)$ are independent, it is easy to see that the best estimate of $e(t-1)$, given $y(t-1)$ and $u(t-1)$, is

$$E[e(t-1) \mid y(t-1), u(t-1)] = E[e(t-1) \mid y(t-1)] = \frac{1}{r_0} y(t-1), \quad \checkmark$$

since $u(t-1)$ is independent of $y(t-1)$. The value of the parameter a in the model is thus changed from a_0 to a^* in order to incorporate the contribution from $c_0 \hat{e}(t-1)$.

The values a^*, b^* give a prediction error variance

$$\bar{V}(a^*, b^*) = 1 + c_0^2 - \frac{c_0^2}{r_0}.$$ (4.115)

The "true values" a_0 and b_0 inserted into (4.113) give a higher variance:

$$\bar{V}(a_0, b_0) = 1 + c_0^2.$$ (4.116)

For example, with $b_0 = a_0 = 0$ and $c_0 = 0.9$, we have $\bar{V}(a^*, b^*) = 1.36$ and $\bar{V}(a_0, b_0) = 1.81$.

When we apply the recursive prediction error method (which in this case coincides with RLS) to (4.112), (4.113) the estimates $\hat{a}(t)$ and $\hat{b}(t)$

will, according to theorem 4.3, converge to the values given by (4.114). Since $a^* \neq a_0$ we say that the estimate is "biased." However, it is clear from (4.115)–(4.116) that the bias is beneficial for the prediction performance of the model (4.113). It gives a strictly better predictor for $a = a^*$ than for $a = a_0$. $\square$

The example stresses that the algorithm indeed gives us the best possible predictor, and it uses its parameters as vehicles for that. It is, however, important to keep in mind that what is the best approximate description of a system may depend on the input used when the true system does not belong to the model set. We illustrate this in the next example.

EXAMPLE 4.6 (Effect of Input on Best Approximation)   Consider the system

$$y(t) = b_0 u(t - 1) + v(t) \tag{4.117}$$

where

$$u(t) = d_0 u(t - 1) + w(t) \tag{4.118}$$

and where $\{v(t)\}$ and $\{w(t)\}$ are independent white noise sequences with zero means and unit variances. Let the model set be given by

$$\hat{y}(t \mid b) = bu(t - 2). \tag{4.119}$$

The prediction error variance associated with (4.119) is

$$E[y(t) - bu(t - 2)]^2 = E[b_0 u(t - 1) + v(t) - bu(t - 2)]^2$$

$$= E[(b_0 d_0 - b)u(t - 2) + b_0 w(t - 1)]^2 + Ev^2(t)$$

$$= \frac{(b_0 d_0 - b)^2}{1 - d_0^2} + b_0^2 + 1.$$

Hence, according to theorem 4.3,

$$\hat{b}(t) \to b^* \triangleq b_0 d_0 \text{ w.p.1 as } t \to \infty,$$

since this gives the smallest prediction error variance. Now the model prediction

$$\hat{y}(t \mid b^*) = b_0 d_0 u(t - 2) \tag{4.120}$$

is a fairly reasonable one for the system (4.117) under the input (4.118). It yields the prediction error variance $1 + b_0^2$, which could be compared to the optimal 1 for a correct model and the output variance $1 + b_0^2/(1 - d_0^2)$. Notice, however, that the identified model is heavily dependent upon the input that was used during the identification experiment. For example, suppose that the model (4.120) is used for another input signal, e.g., let the input be white noise with variance 1. Then the model gives the prediction error

$$y(t) - b_0 d_0 u(t - 2),$$

which has variance $1 + b_0^2 + b_0^2 \cdot d_0^2$, which is worse than the output variance itself, which in this case is $1 + b_0^2$. □

The foregoing example shows that the expression "the best possible predictor" should be interpreted with some care. There are, however, other cases where the best approximation does not depend on the particular input even though it does not give an exact description of the system. "Output error methods" form an example of this, as we will now see.

EXAMPLE 4.7 (Box-Jenkins Model Set)   Suppose that the true system is given by [see (3.103)]

$$y(t) = \frac{B_0(q^{-1})}{F_0(q^{-1})} u(t) + \frac{C_0(q^{-1})}{D_0(q^{-1})} e(t), \tag{4.121}$$

where $\{e(t)\}$ is a sequence of independent random variables. Suppose also that the system operates in open loop (no feedback), so that the sequences $\{u(t)\}$ and $\{e(t)\}$ are independent. Let the model set be given by

$$y(t) = \frac{B(q^{-1})}{F(q^{-1})} u(t) + e(t). \tag{4.122}$$

This was called an output error model in section 3.7 (see also figure 3.2b and example 2.8). The predictor corresponding to (4.122) is given by

$$\hat{y}(t \mid \theta) = \frac{B(q^{-1})}{F(q^{-1})} u(t). \tag{4.123}$$

Suppose that the orders of the polynomials B and F are large enough, so that, for some $\theta = \theta_0$,

$$B(q^{-1}) = B_0(q^{-1}), \quad F(q^{-1}) = F_0(q^{-1}). \tag{4.124}$$

The prediction error corresponding to (4.123) is given by

$$\varepsilon(t, \theta) = y(t) - \hat{y}(t \mid \theta) = \left[\frac{B_0(q^{-1})}{F_0(q^{-1})} - \frac{B(q^{-1})}{F(q^{-1})}\right] u(t) + \frac{C_0(q^{-1})}{D_0(q^{-1})} e(t). \quad (4.125)$$

Since $\{u(t)\}$ and $\{e(t)\}$ are independent we note that any reasonable measure of the prediction error is minimized for $\theta = \theta_0$, irrespective of C_0 and D_0. The estimate from a recursive prediction error method will thus converge to the true description of the dynamics B_0 and F_0, despite the fact that the output noise in (4.121) is colored and the true predictor for (4.121) is more complex than (4.123). Output error methods are thus robust with respect to properties of the output noise.

Suppose now we use a more complex model

$$y(t) = \frac{B(q^{-1})}{F(q^{-1})} u(t) + \frac{C(q^{-1})}{D(q^{-1})} e(t), \quad (4.126)$$

with the corresponding predictor [see (3.105)]

$$\hat{y}(t \mid \theta) = \left[1 - \frac{D(q^{-1})}{C(q^{-1})}\right] y(t) + \frac{D(q^{-1})}{C(q^{-1})} \frac{B(q^{-1})}{F(q^{-1})} u(t). \quad (4.127)$$

Suppose also that (4.124) holds, but the true noise properties in (4.121) are more complex than those of the model (4.126), i.e., suppose that the orders of C_0 and D_0 are higher than those of C and D. The prediction error is given by

$$\varepsilon(t, \theta) = y(t) - \hat{y}(t \mid \theta) = \frac{D(q^{-1})}{C(q^{-1})} \left[\frac{B_0(q^{-1})}{F_0(q^{-1})} - \frac{B(q^{-1})}{F(q^{-1})}\right] u(t)$$

$$+ \frac{D(q^{-1})}{C(q^{-1})} \cdot \frac{C_0(q^{-1})}{D_0(q^{-1})} e(t). \quad (4.128)$$

Again the two terms of the right-hand side are independent, and the variance of the first one is always minimized by $B = B_0$, $F = F_0$, no matter what C, D, C_0, and D_0 might be. Therefore also for the model (4.126) we obtain a correct description of the dynamic part, B_0 and F_0, even though C and D may have orders too low to provide a true description of the whole system.

Notice that the key property in (4.126) is that the dynamic part B, F and the noise part C, D are parametrized by independent parameters. If we used an A-polynomial as in (3.104), this would be a common denominator in (4.126) and the results of this example would no longer hold true. □

We have so far discussed the "best possible predictor." The ideal situation is, of course, when this is precisely identical to the "true predictor" so that we have obtained a true system description. There are essentially two conditions associated with this. One is that the model set $\mathcal{M}$ should be large enough so that the system $\mathcal{S}$ actually belongs to it. The other is that the experimental condition $\mathcal{X}$ (the input) should be general enough, so that no other model is equivalent to the system under $\mathcal{X}$. This latter condition is illustrated in the following example.

EXAMPLE 4.8 (Effect of Input)   Suppose that the system is given by

$$y(t) + a^0 y(t - 1) = b_1^0 u(t - 1) + b_1^0 u(t - 2) + e(t) \tag{4.129}$$

and that the input is

$$\mathcal{X} : u(t) \equiv 1. \tag{4.130}$$

Let the model set be given by

$$\mathcal{M} : y(t) + ay(t - 1) = b_1 u(t - 1) + b_2 u(t - 2) + e(t). \tag{4.131}$$

This set is "large enough" to include the true system (4.129). However, under the input (4.130) all models $\mathcal{M}(\theta)$, such that

$$b_1 + b_2 = b_1^0 + b_2^0, \quad a = a^0 \tag{4.132}$$

will give an exact description of the system. All these models will therefore give the best possible predictor, and convergence to the true values $b_1 = b_1^0$, $b_2 = b_2^0$ cannot be guaranteed. The experimental condition (4.130) is not "general enough." ⊔

Let us now turn to a discussion of the covariance matrix of the asymptotic distribution. In the case of the stochastic Gauss-Newton method for a quadratic criterion, (4.106), it is, according to corollary 4.5, given by

$$P = \bar{R}^{-1}, \quad \bar{R} = [\bar{E}\psi(t, \theta_0)\Lambda_0^{-1}\psi^T(t, \theta_0)]. \tag{4.133}$$

As an example, let us evaluate this expression explicitly for a linear regression model.

EXAMPLE 4.9 (Covariance Matrix for a Linear Regression)   Consider the model

$$y(t) + ay(t - 1) = bu(t - 1) + e(t),$$

which, with

$$\varphi^T(t) = (-y(t-1) \ u(t-1)), \quad \theta^T = (a \ b),$$

is written as a linear regression

$$y(t) = \varphi^T(t)\theta + e(t); \quad \hat{y}(t \mid \theta) = \varphi^T(t)\theta.$$

In this case $\psi(t, \theta) = \varphi(t)$. If $\{e(t)\}$ is white noise with variance σ^2, then the inverse of the asymptotic covariance matrix is

$$\bar{R} = \frac{1}{\sigma^2}\bar{E}\varphi(t)\varphi^T(t) = \frac{1}{\sigma^2}\begin{pmatrix} \bar{E}y^2(t-1) & -\bar{E}y(t-1)u(t) \\ -\bar{E}y(t-1)u(t-1) & \bar{E}u^2(t-1) \end{pmatrix}.$$

$$(4.134)$$

This matrix is derived from variances and covariances of the input and output sequences. $\square$

The matrix $\bar{R}$ defined by (4.133) is related to the algorithm (4.106) in the sense that $R(t) \to \bar{R}$ w.p.1 as $t \to \infty$. If $\bar{R}$ is not invertible, this means that the algorithm (4.106) will encounter numerical problems when the gain $\gamma(t)R^{-1}(t)\psi(t)\hat{\Lambda}^{-1}(t)$ is formed. It is therefore of interest to discuss when $\bar{R}$ may be singular. Also, under assumption L1 (page 189), $\bar{R}$ equals the second-derivative matrix of the criterion function $\bar{V}(\theta) = \frac{1}{2}\bar{E}\varepsilon^T(t, \theta)\Lambda_0^{-1}\varepsilon(t, \theta)$, evaluated at $\theta = \theta_0$. Hence, if $\bar{R}$ is singular, the criterion will not typically have a well-defined unique minimum in θ_0. It will have a "valley" (along the null space of $\bar{R}$) and certain linear combinations of θ will not converge (or converge an order of magnitude more slowly).

If $\bar{R}$ is singular, then for some column vector L

$$0 = L^T\bar{R}L = \bar{E}L^T\psi(t, \theta_0)\Lambda_0^{-1}\psi^T(t, \theta_0)L. \quad (4.135)$$

This means that the linear combination

$$L^T\psi(t, \theta_0)$$

is essentially identically zero. Now

$$L^T\psi(t, \theta_0) = L^T\left[\frac{d}{d\theta}\hat{y}(t \mid \theta)\right]^T = \frac{d}{d(L^T\theta)}\hat{y}(t \mid \theta).$$

The last expression is the derivative of the prediction with respect to the linear combination $L^T\theta$ of the parameters. Consequently, *the matrix $\bar{R}$*

is singular if and only if the prediction is (essentially) unaffected by changing certain linear combinations of the model parameters.

This is a perfectly natural result. If a certain parameter (or linear combination of parameters) does not affect the prediction, there is no way to find out the value of this parameter from input-output data.

It is also clear that this situation arises when the model set contains "too many" parameters or when the experimental condition is such that individual parameters do not affect the prediction. In example 4.8, for example, any change of b_1 and b_2 along the line $b_1 + b_2 = $ const does not effect $\hat{y}(t \mid \theta)$, due to the particular input (4.130).

EXAMPLE 4.10 (Overparametrization)   Suppose that the true signal $\{y(t)\}$ is described by the ARMA process

$$y(t) + a_0 y(t - 1) = e(t) + c_0 e(t - 1).$$

Let the model set be given by

$$y(t) + a_1 y(t - 1) + a_2 y(t - 2) = e(t) + c_1 e(t - 1) + c_2 e(t - 2).$$

We see that all a_i, c_i such that

$$1 + a_1 q^{-1} + a_2 q^{-2} = (1 + a_0 q^{-1})(1 + dq^{-1}),$$

$$1 + c_1 q^{-1} + c_2 q^{-2} = (1 + c_0 q^{-1})(1 + dq^{-1}),$$

with an arbitrary number d give the same (and the true) description of the system. Hence the predictor does not depend on the number d (which corresponds to a certain combination of a_i, c_i). In this case, consequently, the matrix $\bar{R}$ is singular. We may say that the model is overparametrized, since it is of unnecessarily high order. □

The concept of *identifiability* is closely related to the problems we have now discussed. Several different definitions of identifiability and identifiable parameters have been given (see, e.g., Gustavsson et al., 1977). The basic idea is that a parameter is said to be identifiable if it can be uniquely determined from the data. A model set is then said to be (parameter) identifiable if all its parameters are identifiable. We have thus seen that lack of identifiability is linked to singularity of the matrix $\bar{R}$ and can be caused by overparametrization or non-exciting inputs.

Finally, we shall derive a useful formula from the expression (4.133) under the assumption L1 (page 189). With

$$\bar{V}(\theta) = \tfrac{1}{2}\bar{E}\varepsilon^{T}(t, \theta)\Lambda_0^{-1}\varepsilon(t, \theta),$$

we can tell how good the model corresponding to $\hat{\theta}(N)$ is by evaluating $\bar{V}(\hat{\theta}(N))$. By Taylor expansion we have

$$\bar{V}(\hat{\theta}(N)) = \bar{V}(\theta_0) + \bar{V}'(\theta_0)(\hat{\theta}(N) - \theta_0)$$

$$+ \tfrac{1}{2}(\hat{\theta}(N) - \theta_0)^{T}\bar{V}''(\theta_0)(\hat{\theta}(N) - \theta_0) + o(|\hat{\theta}(N) - \theta_0|^2).$$

In view of L1, $V'(\theta_0) = 0$ and $\bar{V}''(\theta_0) = \bar{R}$. By deleting the last term we thus have

$$\bar{V}(\hat{\theta}(N)) \approx \bar{V}(\theta_0) + \tfrac{1}{2}\,\mathrm{tr}\,\bar{R} \cdot (\hat{\theta}(N) - \theta_0)(\hat{\theta}(N) - \theta_0)^{T}.$$

Now $\hat{\theta}(N)$ is a random variable, and we may evaluate the expectation of the above expression w.r.t $\hat{\theta}(N)$. Now we use

$$\mathrm{E}(\hat{\theta}(N) - \theta_0)(\hat{\theta}(N) - \theta_0)^{T} \approx \frac{1}{N}P, \quad P = \bar{R}^{-1}$$

from theorem 4.5, and obtain

$$\mathrm{E}\bar{V}(\hat{\theta}(N)) \approx \bar{V}(\theta_0) + \frac{1}{2}\mathrm{tr}\left(\bar{R}\frac{1}{N}P\right)$$

$$= \bar{V}(\theta_0) + \frac{1}{2N}\mathrm{tr}\,\mathrm{I} = \bar{V}(\theta_0) + \frac{d}{2N} = \frac{1}{2}\left(p + \frac{d}{N}\right)$$

or

$$\mathrm{E}\varepsilon(t, \hat{\theta}(N))\Lambda_0^{-1}\varepsilon^{T}(t, \hat{\theta}(N)) \approx \mathrm{E}\varepsilon(t, \theta_0)\Lambda_0^{-1}\varepsilon^{T}(t, \theta_0) + \frac{d}{N}, \tag{4.136}$$

where $d = \dim\theta$.

This result is remarkable in its generality. It tells us that the expected prediction error variance increases with the number of independent parameters in the model (once the model set is large enough to satisfy L1), irrespective of where the parameters enter the model. We used the words "independent parameters" to stress that the derivation assumes $\bar{R}$ to be invertible, so that parameters that do not affect the prediction are not included. The result, which is well known, was first derived by Akaike (1972, 1981). The implications of the result (4.136) for the choice of model set will be discussed in section 5.2.

4.5 Analysis of Pseudolinear Regressions

4.5.1 Introduction

In section 3.7 we studied a single-input/single-output model (3.104),

$$A(q^{-1})y(t) = \frac{B(q^{-1})}{F(q^{-1})}u(t) + \frac{C(q^{-1})}{D(q^{-1})}e(t). \tag{4.137}$$

We found that the predictor for this model could be written as (3.118):

$$\hat{y}(t \mid \theta) = \theta^{T}\varphi(t, \theta), \tag{4.138}$$

where $\varphi(t, \theta)$ is defined by (3.114). The recursive prediction error method applied to the model (4.137) was described in section 3.7.2, and the asymptotic properties of the resulting estimates were given in the analysis in section 4.4.

We also discussed an alternative estimation algorithm for (4.137) in section 3.7.3. This alternative algorithm can be regarded as a recursive prediction error algorithm in which the gradient of $\hat{y}(t \mid \theta)$ is computed from (4.138), neglecting the implicit θ-dependence in $\varphi(t, \theta)$. For this reason we called the algorithm a *pseudolinear regression* (PLR). The algorithm can be written

$$\varepsilon(t) = y(t) - \hat{\theta}^{T}(t - 1)\varphi(t), \tag{4.139a}$$

$$R(t) = R(t - 1) + \gamma(t)[\varphi(t)\varphi^{T}(t) - R(t - 1)], \tag{4.139b}$$

$$\hat{\theta}(t) = \hat{\theta}(t - 1) + \gamma(t)R^{-1}(t)\varphi(t)\varepsilon(t) \tag{4.139c}$$

[see (3.130)], where $\varphi(t)$ is given by (3.124). We also mentioned that it may be of interest to replace $\varphi(t)$ and $\varepsilon(t)$ in (4.139b, c) by filtered versions:

$$\varphi_{F}(t) = S(q^{-1})\varphi(t), \quad \varepsilon_{F}(t) = T(q^{-1})\varepsilon(t). \tag{4.140}$$

Special cases of this algorithm are well known. We treated, e.g., the extended least squares (ELS) method in example 2.7 $\lfloor F(q^{-1}) = D(q^{-1}) = 1 \rfloor$ and Landau's output error method in example 2.8 $[A(q^{-1}) = C(q^{-1}) = D(q^{-1}) = 1]$.

In this section we shall investigate the convergence properties of (4.139)–(4.140). The results will differ from the analysis of the prediction error algorithms in three ways.

(1) We will have to introduce a more precise assumption about how the recorded data is generated. The assumption is that there exists a value θ_0 in the model set such that $\varepsilon(t, \theta_0)$ is uncorrelated with $\varphi(t, \theta)$ for all θ, i.e., the assumption is

$$\bar{E}\varphi(t, \theta)\varepsilon(t, \theta_0) = 0 \quad \forall \theta \in D_{\mathcal{M}}. \tag{4.141}$$

This essentially means that the system belongs to the model set, so that the data has to be generated according to the model set (4.137) for specific values of the polynomials: $A_0(q^{-1})$, $B_0(q^{-1})$, $C_0(q^{-1})$, $D_0(q^{-1})$, and $F_0(q^{-1})$. Hence no results for the case when the system does not belong to the model set will be obtained for PLR.

(2) The results will be more restrictive. Global convergence will be proven only for two special cases, ELS and the output error method. Also, in these cases convergence will depend on positive realness of certain transfer functions associated with the true system. Recall that a transfer function $H(q^{-1})$ is said to be *strictly positive real* if

$$\operatorname{Re} H(e^{i\omega}) > 0, \quad \forall \omega - \pi < \omega \le \pi. \tag{4.142}$$

(3) One step in the derivation of the general recursive prediction error method was motivated by an assumption that the prediction is generated by stable filters. Therefore we introduced projections into this stability region, a measure that indeed proves to be necessary in practice. For PLR it turns out, however, that stability takes care of itself. This was apparently first proved by Solo (1978), and we have the following lemma.

LEMMA 4.2 (Solo, 1978)   Consider the algorithm (4.139) with $\gamma(t) = 1/t$. Suppose that

$$\limsup_{N \to \infty} \frac{1}{N} \sum_1^N y^2(t) < C^*.$$

Then

$$\limsup_{N \to \infty} \frac{1}{N} \sum_1^N \bar{\varepsilon}^2(t) < C^*,$$

where

$$\bar{\varepsilon}(t) = y(t) - \hat{\theta}^{\mathrm{T}}(t)\varphi(t).$$

Remark   Notice that there is no assumption about boundedness of $\{u(t)\}$ or $\{\hat{\theta}(t)\}$.

Proof   Let $\bar{R}(t) = t \cdot R(t)$. Then, from (4.139b),

$$\bar{R}(t) = \bar{R}(t-1) + \varphi(t)\varphi^{\mathrm{T}}(t); \quad ? - R(t-1))$$

and, from (4.139b, c),

$$\hat{\theta}^{\mathrm{T}}(t)\bar{R}(t)\hat{\theta}(t) = \hat{\theta}^{\mathrm{T}}(t-1)\bar{R}(t-1)\hat{\theta}(t-1)$$
$$+ [\hat{\theta}^{\mathrm{T}}(t-1)\varphi(t)]^2 + 2\varepsilon(t)\varphi^{\mathrm{T}}(t)\hat{\theta}(t-1)$$
$$+ \varepsilon^2(t)\varphi^{\mathrm{T}}(t)\bar{R}^{-1}(t)\varphi(t)$$
$$= \hat{\theta}^{\mathrm{T}}(t-1)\bar{R}(t-1)\hat{\theta}(t-1) + y^2(t)$$
$$- \varepsilon^2(t)[1 - \varphi^{\mathrm{T}}(t)\bar{R}^{-1}(t)\varphi(t)],$$

where, in the last step, we used

$$\hat{\theta}^{\mathrm{T}}(t-1)\varphi(t) = y(t) - \varepsilon(t).$$

Summing from $t = 1$ to $t = N$ gives

$$\hat{\theta}^{\mathrm{T}}(N)\bar{R}(N)\hat{\theta}(N) = \hat{\theta}^{\mathrm{T}}(0)\bar{R}(0)\hat{\theta}(0) + \sum_1^N y^2(t)$$

$$- \sum_1^N \varepsilon^2(t)[1 - \varphi^{\mathrm{T}}(t)\bar{R}^{-1}(t)\varphi(t)].$$

Now, use that

$$\bar{\varepsilon}(t) = \frac{\varepsilon(t)}{1 + \varphi^{\mathrm{T}}(t)\bar{R}^{-1}(t-1)\varphi(t)} = \varepsilon(t)[1 - \varphi^{\mathrm{T}}(t)\bar{R}^{-1}(t)\varphi(t)]$$

[see (3.D.10)–(3.D.11)] to obtain

$$\frac{1}{N}\sum_1^N \bar{\varepsilon}^2(t)[1 + \varphi^{\mathrm{T}}(t)\bar{R}^{-1}(t-1)\varphi(t)]$$

$$= \frac{1}{N}\sum_1^N y^2(t) + \frac{1}{N}\hat{\theta}^{\mathrm{T}}(0)R(0)\hat{\theta}(0) - \frac{1}{N}\hat{\theta}^{\mathrm{T}}(N)\bar{R}(N)\hat{\theta}(N),$$

which proves the lemma. ∎

The foregoing lemma has important implications for the stability of the predictor. Consider, e.g., the ELS method. Then the residuals $\bar{\varepsilon}(t)$ are generated by

$$\hat{C}_t(q^{-1})\bar{\varepsilon}(t) = \hat{A}_t(q^{-1})y(t) - \hat{B}_t(q^{-1})u(t). \tag{4.143}$$

Moreover, if we use residuals in the regression vector $\varphi(t)$,

$$\varphi^T(t) = (-y(t-1) \ \ldots \ -y(t-n_a) \ u(t-1) \ \ldots \ -u(t-n_b)$$

$$\bar{\varepsilon}(t-1) \ \ldots \ \bar{\varepsilon}(t-n_c)),$$

then the vectors $\{\varphi(t)\}$ will be mean square bounded according to the lemma. This means that the step size in the algorithm (4.139) will tend to zero and the filter $\hat{C}_t(q^{-1})$ in (4.143) will change more and more slowly with t. Hence if $\hat{C}(q^{-1})$ were to remain outside its stability region or converge to the boundary of this region, the generation of $\bar{\varepsilon}(t)$ in (4.143) would be unstable, thus contradicting the lemma. Hence $\hat{C}_t(q^{-1})$ will be inside the stability region at least infinitely often and it will not converge to its boundary.

After these introductory observations we now turn to the analysis. In section 4.5.2 we shall study a special case for which explicit sufficient conditions for convergence can be given. The differential equation associated with the general pseudolinear regression according to theorem 4.2 is derived in section 4.5.3. Local convergence properties and possible convergence points are discussed in section 4.5.4, and a comment on the asymptotic covariance matrix for these estimates is given in section 4.5.5. A summary is given in section 4.5.6.

4.5.2 Basic Convergence Results for Some Special Cases

The PLRs that we considered in section 3.7.3 were all given by (3.130):

$$\varepsilon(t) = y(t) - \hat{y}(t), \tag{4.144a}$$

$$R(t) = R(t-1) + \gamma(t)[\varphi_F(t)\varphi_F^T(t) - R(t-1)], \tag{4.144b}$$

$$\varepsilon_F(t) = T(q^{-1})\varepsilon(t), \tag{4.144c}$$

$$\hat{\theta}(t) = \hat{\theta}(t-1) + \gamma(t)R^{-1}(t)\varphi_F(t)\varepsilon_F(t), \tag{4.144d}$$

$$\varphi(t+1) = A(\hat{\theta}(t))\varphi(t) + B(\hat{\theta}(t))z(t), \tag{4.144e}$$

$$\hat{y}(t+1) = \hat{\theta}^T(t)\varphi(t), \tag{4.144f}$$

$$\varphi_F(t) = S(q^{-1})\varphi(t). \tag{4.144g}$$

We assume that the invertibility of $R(t)$ is assured by some mechanism such as described by (4.95). Compared to the general algorithm (4.70d, e), the filter (4.144e, f) is simpler, since the gradient approximation $\eta(t) =$

$\varphi(t)$ can itself be used as a state vector $\xi(t)$ (see (3.124)). The algorithm (4.144) is thus a special case of the general algorithm (4.70) for which theorem 4.2 was given, with $\varphi(t)$ corresponding to $\eta(t)$. We can therefore use this theorem to investigate the convergence properties of (4.144). Let us for the moment proceed without verifying the regularity conditions for this theorem. That will be done later, in theorem 4.6. Define $\varphi(t, \theta)$ by

$$\varphi(t + 1, \theta) = A(\theta)\varphi(t, \theta) + B(\theta)z(t); \quad \varphi(0, \theta) = 0. \tag{4.145}$$

Notice that explicit expressions for this vector were given in (3.114). The differential equation (4.80), which according to theorem 4.2 is associated with (4.144), is then given by

$$\frac{d}{d\tau}\theta_D(\tau) - R_D^{-1}(\tau)f_{f_r}(\theta_D(\tau)),$$

$$\tag{4.146}$$

$$\frac{d}{d\tau}R_D(\tau) = G_F(\theta_D(\tau)) - R_D^{-1}(\tau),$$

where we have found the right-hand side, as before, by evaluating the average asymptotic updating direction in (4.144) for a given θ, R:

$$f_F(\theta) = \bar{E}\varphi_F(t, \theta)\varepsilon_F(t, \theta), \tag{4.147a}$$

$$G_F(\theta) = \bar{E}\varphi_F(t, \theta)\varphi_F^T(t, \theta), \tag{4.147b}$$

$$\varepsilon(t, \theta) = y(t) - \theta^T\varphi(t, \theta), \tag{4.148a}$$

$$\varepsilon_F(t, \theta) = T(q^{-1})\varepsilon(t, \theta), \tag{4.148b}$$

$$\varphi_F(t, \theta) = S(q^{-1})\varphi(t, \theta). \tag{4.148c}$$

This and the following two sections will be devoted to a closer study of this differential equation.

Since f_F determines the stability properties of (4.146), it is clear from (4.147a) that the correlation between the "regression vector" $\varphi_F(t, \theta)$ (the gradient approximation) and the prediction error $\varepsilon_F(t, \theta)$ will be crucial for the asymptotic behavior of (4.144). Let us therefore examine this correlation a bit more closely for the ELS method. For ELS we have

$$\varphi^T(t, \theta) = (-y(t - 1) \quad \ldots \quad -y(t - n_a) \; u(t - 1) \quad \ldots \quad u(t - n_b)$$

$$\varepsilon(t - 1, \theta) \quad \ldots \quad \varepsilon(t - n_c, \theta)),$$

$$\theta^T = (a_1 \quad \ldots \quad a_{n_a} \; b_1 \quad \ldots \quad b_{n_b} \; c_1 \quad \ldots \quad c_{n_c}).$$

Here $\varepsilon(t, \theta)$ is recursively defined from z^t by

$$\varepsilon(t, \theta) = y(t) - \theta^\mathrm{T}\varphi(t, \theta). \tag{4.149}$$

Suppose now that the true system is given by

$$A_0(q^{-1})y(t) = B_0(q^{-1})u(t) + C_0(q^{-1})e(t), \tag{4.150}$$

where $\{e(t)\}$ is a sequence of independent random variables. Assume that the orders of the polynomials in (4.150) are less than or equal to the corresponding model orders. Then (4.150) can be rewritten as

$$y(t) = \theta_0^\mathrm{T}\varphi_0(t) + e(t), \tag{4.151}$$

with

$$\theta_0^\mathrm{T} = (a_1^0 \ \ldots \ a_{n_a}^0 \ b_1^0 \ \ldots \ b_{n_b}^0 \ c_1^0 \ \ldots \ c_{n_c}^0),$$

$$\varphi_0^\mathrm{T}(t) = (-y(t-1) \ \ldots \ -y(t-n_a) \ u(t-1) \ \ldots \ u(t-n_b)$$

$$e(t-1) \ \ldots \ e(t-n_c)).$$

Using this expression in (4.149) gives

$$\varepsilon(t, \theta) = \theta_0^\mathrm{T}\varphi_0(t) - \theta^\mathrm{T}\varphi(t, \theta) + e(t)$$

$$= \theta_0^\mathrm{T}[\varphi_0(t) - \varphi(t, \theta)] + (\theta_0 - \theta)^\mathrm{T}\varphi(t, \theta) + e(t).$$

But

$$\theta_0^\mathrm{T}[\varphi_0(t) - \varphi(t, \theta)] = [C_0(q^{-1}) - 1][e(t) - \varepsilon(t, \theta)],$$

according to the definition of C_0 and φ_0. Hence

$$C_0(q^{-1})[\varepsilon(t, \theta) - e(t)] = \varphi^\mathrm{T}(t, \theta)(\theta_0 - \theta), \tag{4.152}$$

which gives us a basic relationship between θ_0, $\varepsilon(t, \theta)$, and $\varphi(t, \theta)$. (See also lemma 4.3 in section 4.5.3.) This relationship motivates the following theorem.

THEOREM 4.6   Consider the algorithm (4.144) subject to conditions M1, M2, R1, and G1 stated in section 4.2.4. Suppose that the data sequence $\{z(t)\}$ is subject to S1. Let $\varphi_F(t, \theta)$ and $\varepsilon_F(t, \theta)$ be defined by (4.145) and (4.148), and suppose that there exists a relationship

$$H(q^{-1})\varepsilon_F(t, \theta) = \varphi_F^\mathrm{T}(t, \theta)(\theta_0 - \theta) + H(q^{-1})e(t) \tag{4.153}$$

for some causal and strictly stable filter $H(q^{-1})$ and some value θ_0, where $\{e(t)\}$ is a sequence of random variables each of zero mean, such that $e(t)$ is independent of $\varphi_F(t, \theta)$ for all $\theta \in D_{\mathcal{M}}$. Assume that

$$\text{Re}\,[[H(e^{i\omega})]^{-1} - \tfrac{1}{2}] > 0 \quad \forall \omega, \quad -\pi < \omega \le \pi, \tag{4.154}$$

and that

$$\bar{E}\varphi_F(t, \theta)\varphi_F^T(t, \theta) = G_F(\theta) \tag{4.155}$$

exists for all θ. Suppose that the algorithm includes a projection that keeps $\hat{\theta}(t)$ in $D_{\mathcal{M}}$ and such that there is a bounded subsequence of $\varphi(t)$. Then, w.p.1,

$$\hat{\theta}(t) \to \bar{D}_c = \{\theta \mid \bar{E}[\varepsilon_F(t, \theta) - e(t)]^2 = 0\} \quad \text{as } t \to \infty \tag{4.156}$$

or

$\hat{\theta}(t)$ converges to the boundary of $D_{\mathcal{M}}$.

Remark   Note that according to (4.152) the relationship (4.153) will hold for ELS with $H = C_0$.

Proof   We shall apply theorem 4.2 to (4.144). All the regularity conditions of this theorem are satisfied, since Cr1 and Cr2 are satisfied for the choice of h and H in (4.144b, d). Moreover, S1 and (4.155) imply that A1 holds w.p.1. (lemma 4.1). The associated differential equation (4.80) is given by (4.146)–(4.148). Using (4.153) we can write

$$\begin{aligned}
f_F(\theta) &= \bar{E}\varphi_F(t, \theta)\varepsilon_F(t, \theta) \\
&= \bar{E}\psi_F(t, \theta)\tilde{\varphi}_F^T(t, \theta)(\theta_0 - \theta) + \bar{E}\varphi_F(t, \theta)e(t) \\
&= \tilde{G}_F(\theta)(\theta_0 - \theta),
\end{aligned}$$

where

$$\tilde{\varphi}_F(t, \theta) = \frac{1}{H(q^{-1})}\varphi_F(t, \theta) \tag{4.157}$$

and

$$\tilde{G}_F(\theta) = E\varphi_F(t, \theta)\tilde{\varphi}_F^T(t, \theta). \tag{4.158}$$

Here we also used the fact that $e(t)$ and $\varphi_F(t, \theta)$ are independent. The d.e. is thus

$$\frac{d}{d\tau}\theta_D(\tau) = R_D^{-1}(\tau)\tilde{G}_F(\theta_D(\tau))[\theta_0 - \theta_D(\tau)],$$

$$(4.159)$$

$$\frac{d}{d\tau}R_D(\tau) = G_F(\theta_D(\tau)) - R_D(\tau).$$

We shall now show that (4.154) implies that the matrix

$$\tilde{G}_F(\theta) + \tilde{G}_F^T(\theta) - G_F(\theta)$$

is positive definite: For any column vector $L \neq 0$, define

$$z(t, \theta) = L^T\varphi_F(t, \theta), \qquad \tilde{z}(t, \theta) = L^T\tilde{\varphi}_F(t, \theta).$$

Then [see (4.157)]

$$\tilde{z}(t, \theta) - \frac{1}{2}z(t, \theta) = \left\{[H(q^{-1})]^{-1} - \frac{1}{2}\right\}z(t, \theta)$$

and [see (4.155), (4.158)]

$$L^T[\tilde{G}_F(\theta) + \tilde{G}_F^T(\theta) - G_F(\theta)]L$$

$$= 2\bar{E}z(t, \theta)\left[\tilde{z}(t, \theta) - \frac{1}{2}z(t, \theta)\right]$$

$$= 2\int_{-\pi}^{\pi}\Phi_z(\omega)\left\{[H(e^{i\omega})]^{-1} - \frac{1}{2}\right\}d\omega$$

$$= 2\int_{-\pi}^{\pi}\Phi_z(\omega)\,\mathrm{Re}\left\{[H(e^{i\omega})]^{-1} - \frac{1}{2}\right\}d\omega \geq 0.$$

Here $\Phi_z(\omega)$ is the spectral density for $z(t, \theta)$, and the second equality follows from (1.A.8). To be exact, $\Phi_z(\omega)$ is the Fourier transform of $r(k) \triangleq \bar{E}z(t, \theta)z(t + k, \theta)$. Equality in the above inequality will hold only for $\Phi_z(\omega) \equiv 0$, i.e., only for

$$\bar{E}[z(t, \theta)]^2 = \bar{E}[L^T\varphi_F(t, \theta)]^2 = 0.$$

We have thus established that

$$L^T(\tilde{G}_F + \tilde{G}_F^T - G_F)L \geq 0 \quad \text{for all } L,$$

and that equality holds only if

$$\bar{E}[L^T\varphi_F(t, \theta)]^2 = 0.$$

With this property we can now show that

$$V(\theta, R) = (\theta - \theta_0)^T R(\theta - \theta_0)$$

is a function that satisfies condition (4.79) of theorem 4.2 when applied to (4.159). We have

$$\frac{d}{d\tau} V(\theta_D(\tau), R_D(\tau))$$

$$= -[\theta_D(\tau) - \theta_0]^T \tilde{G}_F(\theta_D(\tau))[\theta_D(\tau) - \theta_0]$$
$$\quad - [\theta_D(\tau) - \theta_0]^T \tilde{G}_F^T(\theta_D(\tau))[\theta_D(\tau) - \theta_0]$$
$$\quad + [\theta_D(\tau) - \theta_0]^T[G_F(\theta_D(\tau)) - R_D(\tau)][\theta_D(\tau) - \theta_0]$$
$$= -[\theta_D(\tau) - \theta_0]^T[\tilde{G}_F(\theta_D(\tau)) + \tilde{G}_F^T(\theta_D(\tau)) - G_F(\theta_D(\tau))$$
$$\quad + R_D(\tau)][\theta_D(\tau) - \theta_0] \leq 0,$$

with equality only in the set

$$\bar{D}_c = \{\theta \mid \bar{E}[(\theta - \theta_0)^T \varphi_F(t, \theta)]^2 = 0\},$$

where $\theta - \theta_0$ corresponds to L above.

From theorem 4.2 we thus see that $\hat{\theta}(t) \to \bar{D}_c$ w.p.1 as $t \to \infty$. It remains now only to prove that this set $\bar{D}_c$ is the same one as in (4.156). But it follows from (4.153) that $\bar{E}[(\theta - \theta_0)^T \varphi_F(t, \theta)]^2 = 0$ implies that $\bar{E}[\varepsilon_F(t, \theta) - e(t)]^2 = 0$, and hence the theorem is proven. ∎

Remark   Notice that the condition

$$\text{Re}\left\{[H(e^{i\omega})]^{-1} - \frac{1}{2}\right\} > 0$$

is equivalent to

$$|H(e^{i\omega}) - 1| < 1. \tag{4.160}$$

This can be seen as follows:

$$\text{Re}\left\{\frac{1}{z} - \frac{1}{2}\right\} > 0 \Leftrightarrow \frac{2\,\text{Re}\,z - |z|^2}{|z|^2} > 0 \Leftrightarrow 2\,\text{Re}\,z > |z|^2,$$

$$|z - 1| < 1 \Leftrightarrow (z - 1)(\bar{z} - 1) < 1 \Leftrightarrow |z|^2 - 2\,\text{Re}\,z + 1 < 1.$$

Hence the positive realness condition can also be regarded as a condition that $H(q^{-1})$ should be close to the unit transfer function.

COROLLARY 4.6 Consider the gradient version of the algorithm (4.144) in which $R(t)$ is replaced by $r(t) \cdot I$, where

$$r(t) = r(t-1) + \gamma(t)\left[|\varphi_F(t)|^2 - r(t-1)\right].$$

Let the conditions of the theorem hold, with (4.154) weakened to

$$\operatorname{Re} H(e^{i\omega}) > 0 \quad \forall \omega, \quad -\pi < \omega \leq \pi. \tag{4.161}$$

Then the conclusion of the theorem holds.

Proof The associated differential equation is in this case

$$\frac{d}{d\tau}\theta_D(\tau) = \frac{1}{r_D(\tau)}\tilde{G}_F(\theta_D(\tau))\left[\theta_0 - \theta_D(\tau)\right],$$

$$\frac{d}{d\tau}r_D(\tau) = g_F(\theta_D(\tau)) - r_D(\tau), \tag{4.162}$$

where

$$g_F(\theta) = \overline{\mathrm{E}}|\varphi_F(t, \theta)|^2.$$

With $V(\theta) = |\theta - \theta_0|^2$ as a Lyapunov function for (4.162), we see that it is only required that $\tilde{G}_F(\theta)$ be positive definite, and this is implied by (4.153) and (4.161). ∎

Remark In view of the discussion following lemma 4.2 in the section 4.5.1, we see that the projection into the stability region $D_{\mathcal{M}}$ will not be necessary. Nor will the estimate converge to the boundary of the stability region. The assumptions in theorem 4.6 regarding this can therefore be dispensed with, and the conclusion (4.156) still holds for the algorithm (4.144).

Theorem 4.6 can also be proved using the martingale convergence theorem. This latter proof is capable of handling the stability problems in a nicer way. The conditions on the sequence $\{e(t)\}$ in (4.153) are, however, slightly more restrictive. The details of this approach are given in appendix 4.C.

Application to ELS The condition (4.153) of theorem 4.6 imposes a restriction on the model structure, and it prevents the application of the theorem to the general PLR algorithm. Several special cases of interest can, however, be treated. From (4.152) we see that the theorem is applicable for ELS with $H(q^{-1}) = C_0(q^{-1})$. Also, the regularity conditions M1 and M2 are trivially satisfied for the models we are discussing. We shall assume that condition S1 about the true data is satisfied. This is the case, e.g., when they are produced by (4.150) and the input is a deterministic sequence or a stochastic process with rational spectral density or a combination of these. (See the discussion at the end of section 4.3.4.) We shall also assume that R1 and G1 (p. 166) hold. Finally, condition (4.155) will hold provided the limits

$$\bar{E}u(t)u(t-k) \quad \text{and} \quad \bar{E}e(t)e(t-k)$$

exist, since the $\varphi(t, \theta)$-vector consists of elements obtained by filtering u^t and e^t through constant filters. We can consequently apply theorem 4.6 to ELS and find that the estimate $\hat{\theta}(t)$ will converge w.p.1 to the set

$$\bar{D}_c = \{\theta \mid \bar{E}[\varepsilon(t, \theta) - e(t)]^2 = 0\} \tag{4.163}$$

as $t \to \infty$, provided the filter

$$\frac{1}{C_0(q^{-1})} - \frac{1}{2} \tag{4.164}$$

is strictly positive real, and provided the data are described by the system (4.150).

Remark   Notice that we have so far made no assumption that the representation (4.150) should be of minimal order. This gives a certain amount of freedom for the condition (4.164), as pointed out by Shah (1981). Suppose, e.g., that a true, minimal order representation is given by

$$\bar{A}_0(q^{-1})y(t) = \bar{B}_0(q^{-1})u(t) + \bar{C}_0(q^{-1})e(t),$$

where the orders of the polynomials are all $\bar{n}$. Then apply ELS with model orders $n > \bar{n}$. It is then a sufficient condition for convergence to $\bar{D}_c$ that there exists a monic* polynomial $G(q^{-1})$ of order $n - \bar{n}$ such that

$$[G(q^{-1})\bar{C}_0(q^{-1})]^{-1} - \frac{1}{2}$$

is strictly positive real. The reason is that we can regard

$$G(q^{-1})\bar{A}_0(q^{-1})y(t) = G(q^{-1})\bar{B}_0(q^{-1})u(t) + G(q^{-1})\bar{C}_0(q^{-1})e(t)$$

*monic: The leading coefficient is a "1."

as a true (although not minimal) description of the system. Notice, though, that such an increase of the model orders may cause problems in the algorithm (see example 4.10).

All models in the set D_c give the same input-output description of the system, since they give the same prediction errors. If the model orders and the input signal are such that no two different models can give the same description of the true system, we conclude that $D_c = \{\theta_0\}$. Then we obtain convergence of $\hat{\theta}(t)$ to the true parameter values.

Consider now the ELS scheme with filters as in (4.144c, g). The relationship between $\varepsilon_F(t, \theta)$ and $\varphi_F(t, \theta)$ will be given by

$$C_0(q^{-1})\left[\frac{1}{T(q^{-1})}\varepsilon_F(t, \theta)\right] = \frac{1}{S(q^{-1})}\varphi_F(t, \theta)(\theta_0 - \theta) + C_0(q^{-1})e(t),$$

according to (4.152). This can also be written

$$\frac{S(q^{-1})C_0(q^{-1})}{T(q^{-1})}\varepsilon_F(t, \theta) = \varphi_F(t, \theta)(\theta_0 - \theta) + \frac{S(q^{-1})C_0(q^{-1})}{T(q^{-1})}[T(q^{-1})e(t)].$$

Comparing this to (4.153), we see that theorem 4.6 requires $T(q^{-1})e(t)$ to be uncorrelated with $\varphi_F(t, \theta)\forall\theta\in D_{\mathscr{M}}$. [See the sentence following (4.153).] Since $\varphi_F(t, \theta)$ contains $y(t-1)$, this will normally be the case only for $T(q^{-1}) = 1$.

We thus have the following general result.

Consider the ELS method with filtered variables according to (4.144). Suppose that $T(q^{-1}) = 1$. Then $\hat{\theta}(t)$ will converge to D_c, given by (4.163), provided that

$$[S(q^{-1})C_0(q^{-1})]^{-1} - \frac{1}{2} \text{ is strictly positive real} \qquad (4.165)$$

and provided that the true system is given by (4.150), with orders less than or equal to those of the model.

Notice that we could think of the filter $S(q^{-1})$ as a prior estimate of $1/C_0(q^{-1})$, since if these are close enough (4.165) will hold.

Application to Output Error Methods   The relationship (4.153) exists also for Landau's output error method (example 2.8), obtained from the general scheme by setting $A(q^{-1}) = C(q^{-1}) = D(q^{-1}) = 1$. In this case we have

$$\hat{y}(t \mid \theta) = w(t, \theta) = \frac{B(q^{-1})}{F(q^{-1})} u(t) \tag{4.166}$$

[see (3.111) and (3.120)]. With

$$\theta^T = (b_1 \ \ldots \ b_{n_b} \ f_1 \ \ldots \ f_{n_f})$$

and

$$\varphi^T(t, \theta) = (u(t-1) \ \ldots \ u(t-n_b) \ -w(t-1, \theta) \ \ldots \ -w(t-n_f, \theta)),$$

we can write (4.166) as

$$\hat{y}(t \mid \theta) = \theta^T \varphi(t, \theta). \tag{4.167}$$

Now suppose that the true data can be described by

$$y(t) = \frac{B_0(q^{-1})}{F_0(q^{-1})} u(t) + v(t) \tag{4.168}$$

for some polynomials B_0 and F_0 of degrees equal to those of the model (4.166), and a disturbance sequence $\{v(t)\}$. The relation (4.168) can be written

$$y(t) = \theta_0^T \varphi_0(t) + v(t) \tag{4.169}$$

with

$$\theta_0^T = (b_1^0 \ \ldots \ b_{n_b}^0 \ f_1^0 \ \ldots \ f_{n_f}^0),$$

$$\varphi_0^T(t) = (u(t-1) \ \ldots \ u(t-n_h) \ -w_0(t-1) \ \ldots \ -w_0(t-n_f)),$$

where

$$w_0(t) = \frac{B_0(q^{-1})}{F_0(q^{-1})} u(t) = y(t) - v(t).$$

From (4.169) and (4.167) we obtain

$$\varepsilon(t, \theta) = y(t) - \theta^T \varphi(t, \theta)$$

$$= \theta_0^T \varphi_0(t) - \theta^T \varphi(t, \theta) + v(t)$$

$$= \theta_0^T [\varphi_0(t) - \varphi(t, \theta)] + (\theta_0 - \theta)^T \varphi(t, \theta) + v(t)$$

$$= -[F_0(q^{-1}) - 1][w_0(t) - w(t, \theta)] + (\theta_0 - \theta)^T \varphi(t, \theta) + v(t).$$

Using that $w_0(t) = y(t) - v(t)$ and that $w(t, \theta) = y(t) - \varepsilon(t, \theta)$, we can rewrite this expression as

$$F_0(q^{-1})\varepsilon(t, \theta) = \varphi^T(t, \theta)(\theta_0 - \theta) + F_0(q^{-1})v(t). \tag{4.170a}$$

With filtered variables as in (4.144) we obtain

$$\frac{F_0(q^{-1})S(q^{-1})}{T(q^{-1})}\varepsilon_F(t, \theta)$$

$$= \varphi_F^T(t, \theta)(\theta_0 - \theta) + \frac{F_0(q^{-1})S(q^{-1})}{T(q^{-1})}T(q^{-1})v(t). \tag{4.170b}$$

This is exactly of the form (4.153) in theorem 4.6. We must then also require that the noise sequence $T(q^{-1})v(t)$ has zero mean and is independent of $\varphi_F(t, \theta)$ $\forall\theta$. However, $\varphi_F(t, \theta)$ is constructed entirely from the input sequence. Therefore if the measurement noise $\{v(t)\}$ is independent of the input this condition will hold. The regularity conditions of theorem 4.6 will hold just as for ELS. We thus have the following result.

Consider the output-error method obtained from the general PLR with $A(q^{-1}) = C(q^{-1}) = D(q^{-1}) = 1$. Assume that the true system is given by (4.168) with the measurement noise $\{v(t)\}$ independent of the input sequence $\{u(t)\}$. Assume also that

$$\frac{T(q^{-1})}{F_0(q^{-1})S(q^{-1})} - \frac{1}{2} \text{ is strictly positive real.} \tag{4.171}$$

Then $\hat{\theta}(t)$ converges to

$$D_c = \{\theta \mid \bar{E}[\hat{y}(t \mid \theta) - w_0(t)]^2 = 0\} \text{ w.p.1 as } t \to \infty.$$

If the input sequence is not independent of the measurement noise (e.g., due to output feedback) it is also required that $T(q^{-1})v(t)$ be a white noise sequence.

These two schemes, ELS and Landau's output error method, are essentially the only PLRs for which sufficient conditions for convergence are presently known. We can of course use theorem 4.2 to investigate the convergence properties of any PLR algorithm, by studying the associated differential equation (4.146)–(4.148). In section 4.5.3 we give an explicit expression for this equation; however, we will not be able to give an analytic stability treatment of the d.e. in the general case.

4.5.3 The Associated Differential Equation for the General Pseudolinear Regression

The general PLR algorithm (4.144) is associated with the d.e. (4.146)–(4.148) according to theorem 4.2. The expressions for $f_F(\theta)$ in this d.e. can be given a more explicit form by using the particular structure of the algorithm.

We shall start the analysis by deriving an expression for how $\varepsilon(t, \theta)$, given by $y(t) - \theta^T \varphi(t, \theta)$ [see (4.138)], relates to the prediction error for another value of θ, say θ_*. To derive this expression, it is useful to introduce the partitioned structure [see (3.110)–(3.114)]

$$\varphi_y^T(t) = (-y(t-1) \ \ldots \ -y(t-n_a)),$$

$$\theta_y^T = (a_1 \ \ldots \ a_{n_a}),$$

$$\varphi_u^T(t, \theta) = (u(t-1) \ \ldots \ u(t-n_b) \ -w(t-1, \theta) \ \ldots \ -w(t-n_f, \theta)),$$

$$\theta_u^T = (b_1 \ \ldots \ b_{n_b} \ f_1 \ \ldots \ f_{n_f}),$$

$$\varphi_e^T(t, \theta) = (\varepsilon(t-1, \theta) \ \ldots \ \varepsilon(t-n_c, \theta) \ -v(t-1, \theta) \ \ldots \ -v(t-n_d, \theta)),$$

$$\theta_e^T = (c_1 \ \ldots \ c_n \ d_1 \ \ldots \ d_{n_d}).$$

Then

$$\varphi^T(t, \theta) = (\varphi_y^T(t) \ \varphi_u^T(t, \theta) \ \varphi_e^T(t, \theta))$$

and

$$\theta^T = (\theta_y^T \ \theta_u^T \ \theta_e^T).$$

We now have the following result.

LEMMA 4.3 Let the value θ correspond to the model (4.137) and let $\theta_* \in D_{\mathcal{M}}$ correspond to the model

$$A_*(q^{-1})y(t) = \frac{B_*(q^{-1})}{F_*(q^{-1})}u(t) + \frac{C_*(q^{-1})}{D_*(q^{-1})}e(t). \tag{4.172}$$

Then the prediction errors are related by

$$\varepsilon(t, \theta) = (\theta_* - \theta)^T \tilde{\varphi}(t, \theta) + \varepsilon(t, \theta_*), \tag{4.173}$$

where

$$\tilde{\varphi}^{\mathrm{T}}(t, \theta) = (\tilde{\varphi}_y^{\mathrm{T}}(t) \ \tilde{\varphi}_u^{\mathrm{T}}(t, \theta) \ \tilde{\varphi}_e^{\mathrm{T}}(t, \theta)) \tag{4.174}$$

with

$$\tilde{\varphi}_y(t) = \frac{D_*(q^{-1})}{C_*(q^{-1})}\varphi_y(t),$$

$$\tilde{\varphi}_u(t, \theta) = \frac{D_*(q^{-1})}{C_*(q^{-1})F_*(q^{-1})}\varphi_u(t, \theta), \tag{4.175}$$

$$\tilde{\varphi}_e(t, \theta) = \frac{1}{C_*(q^{-1})}\varphi_e(t, \theta).$$

Proof Consider first the expression given by (4.166). In the present notation we have

$$w(t, \theta) = B(q^{-1})u(t) - [F(q^{-1}) - 1]w(t, \theta)$$

$$= b_1 u(t - 1) + \cdots + b_{n_b} u(t - n_b) - f_1 w(t - 1, \theta) - f_{n_f} w(t - n_f, \theta)$$

$$= \theta_u^{\mathrm{T}} \varphi_u(t, \theta). \tag{4.176}$$

Hence

$$w(t, \theta) - w(t, \theta_*)$$

$$= \theta_u^{\mathrm{T}} \varphi_u(t, \theta) - \theta_{*,u}^{\mathrm{T}} \varphi_u(t, \theta_*) \tag{4.177a}$$

$$= (\theta_u^{\mathrm{T}} - \theta_{*,u}^{\mathrm{T}})\varphi_u(t, \theta) - \theta_{*,u}^{\mathrm{T}}[\varphi_u(t, \theta_*) - \varphi_u(t, \theta)].$$

Now by definition of φ_u we have

$$[\varphi_u(t, \theta_*) - \varphi_u(t, \theta)]^{\mathrm{T}}$$

$$= (0 \ \ldots \ 0 \ [w(t - 1, \theta) - w(t - 1, \theta_*)] \ \ldots \tag{4.177b}$$

$$\ldots \ [w(t - n_f, \theta) - w(t - n_f, \theta_*)]).$$

Recall that the *F*-polynomial corresponding to $\theta_{*,u}$ is $F_*(q^{-1})$. Then (from (4.177b) and the definition of θ_*)

$$\theta_{*,u}^{\mathrm{T}}[\varphi_u(t, \theta_*) - \varphi_u(t, \theta)]^{\mathrm{T}} = [F_*(q^{-1}) - 1][w(t, \theta) - w(t, \theta_*)].$$

Inserting this into (4.177a) gives

$$F_*(q^{-1})[w(t, \theta) - w(t, \theta_*)] = (\theta_u - \theta_{*,u})^{\mathrm{T}}\varphi_u(t, \theta). \tag{4.178}$$

Define the vector $\varphi_u^F(t, 0)$ by

$$\varphi_u^F(t, \theta) = \frac{1}{F_*(q^{-1})} \varphi_u(t, \theta), \tag{4.179}$$

i.e., $\varphi_u^F(t, \theta)$ is obtained by filtering the data vector $\varphi_u(t, \theta)$ through $1/F_*(q^{-1})$. Then (4.178) can be written

$$[w(t, \theta) - w(t, \theta_*)] = (\theta_u - \theta_{*,u})^T \varphi_u^F(t, \theta). \tag{4.180}$$

We are now prepared to turn to the expression for the residual. This is handled similarly:

$$\varepsilon(t, \theta) - \varepsilon(t, \theta_*)$$
$$= -\theta^T \varphi(t, \theta) + \theta_*^T \varphi(t, \theta) - \theta_*^T \varphi(t, \theta) + \theta_*^T \varphi(t, \theta_*) \tag{4.181}$$
$$- (\theta_* - \theta)^T \varphi(t, \theta) + \theta_*^T [\psi(t, \theta_*) - \psi(t, \theta)].$$

As above, we find that

$$\theta_*^T [\varphi(t, \theta_*) - \varphi(t, \theta)]$$
$$= [F_*(q^{-1}) - 1][w(t, \theta) - w(t, \theta_*)]$$
$$+ [D_*(q^{-1}) - 1][v(t, \theta) - v(t, \theta_*)] \tag{4.182}$$
$$+ [1 - C_*(q^{-1})][\varepsilon(t, \theta) - \varepsilon(t, \theta_*)].$$

From (3.112) we have

$$v(t, \theta) - v(t, \theta_*) = [A(q^{-1}) - A_*(q^{-1})]y(t) - [w(t, \theta) - w(t, \theta_*)]$$
$$= -(\theta_y - \theta_{*,y})^T \varphi_y(t) - [w(t, \theta) - w(t, \theta_*)]. \tag{4.183}$$

The first term of (4.181) can be written

$$(\theta - \theta_*)^T \varphi(t, \theta)$$
$$= (\theta_y - \theta_{*,y})^T \varphi_y(t) + (\theta_u - \theta_{*,u})^T \varphi_u(t, \theta) + (\theta_e - \theta_{*,e})^T \varphi_e(t, \theta). \tag{4.184}$$

Collecting (4.182)–(4.184) and (4.180) into (4.181) now gives

$$C_*(q^{-1})[\varepsilon(t, \theta) - \varepsilon(t, \theta_*)]$$
$$= (\theta_{*,y} - \theta_y)^T D_*(q^{-1}) \varphi_y(t) + (\theta_{*,u} - \theta_u)^T D_*(q^{-1}) \varphi_u^F(t, \theta) \tag{4.185}$$
$$+ (\theta_{*,e} - \theta_e)^T \varphi_e(t, \theta).$$

Now we introduce $\tilde{\varphi}(t, \theta)$ as in (4.174)–(4.175), and we find that (4.185) transforms into (4.173), and the proof is complete. ∎

Remark   Note that the expression (4.173) holds for any two values θ and θ_* in the model set. Notice also the relationship with the gradient expressions (3.119). Clearly these derivatives could be derived from (4.173) by letting $\theta_* \to \theta$. However, lemma 4.3 is *not* a direct consequence of the mean value theorem.

We also note that when specializing to $F(q^{-1}) = D(q^{-1}) = 1$ we obtain (4.152) and when specializing to $A(q^{-1}) = C(q^{-1}) = D(q^{-1}) = 1$ we obtain (4.170a).

Now assume that there is a value θ_0 in the model set such that a true description of the system is obtained, in the sense that $\varepsilon_F(t, \theta_0)$ $= T(q^{-1})\varepsilon(t, \theta_0)$ is independent of $\varphi_F(t, \theta)$, for all θ, and further assume that it is of zero mean. Except for the output error method case, this holds in general only when $T(q^{-1}) = 1$ and $\{\varepsilon(t, \theta_0)\}$ is a sequence of independent random variables. We then have, from (4.147) and (4.173) with $\theta_* = \theta_0$,

$$f_F(\theta) = \bar{E}\varphi_F(t, \theta)\varepsilon_F(t, \theta)$$

$$= \bar{E}\varphi_F(t, \theta)\tilde{\varphi}_F^{\mathrm{T}}(t, \theta)(\theta_0 - \theta) + \bar{E}\varphi_F(t, \theta)\varepsilon_F(t, \theta_0) \qquad (4.186)$$

$$= \tilde{G}_F(\theta)(\theta_0 - \theta)$$

where

$$\tilde{G}_F(\theta) = \bar{E}\varphi_F(t, \theta)\tilde{\varphi}_F^{\mathrm{T}}(t, \theta) \qquad (4.187a)$$

and

$$\varphi_F(t, \theta) = S(q^{-1})\varphi(t, \theta), \qquad (4.187b)$$

$$\tilde{\varphi}_F(t, \theta) = T(q^{-1})\tilde{\varphi}(t, \theta), \qquad (4.187c)$$

with $\tilde{\varphi}(t, \theta)$ given by (4.174)–(4.175). The d.e. associated with the general PLR algorithm (4.144) is thus

$$\frac{d}{d\tau}\theta_D(\tau) = R_D^{-1}(\tau)\tilde{G}_F(\theta_D(\tau))[\theta_0 - \theta_D(\tau)],$$

$$\qquad (4.188)$$

$$\frac{d}{d\tau}R_D(\tau) = G_F(\theta_D(\tau)) - R_D(\tau),$$

where, as in (4.147),

$$G_F(\theta) = \bar{E}\varphi_F(t, \theta)\varphi_F^{\mathrm{T}}(t, \theta).$$

With theorem 4.2 and the d.e. (4.188) we can investigate the convergence properties of any PLR. However, no Lyapunov function $V(\theta, R)$ for (4.188) is presently known in the general case. Only when $\varphi(t, \theta)$ and $\tilde{\varphi}(t, \theta)$ are related through a single filter is analytic treatment found to be possible, as demonstrated in theorem 4.6.

4.5.4 Possible Convergence Points and Local Convergence

In our study of the properties of the PLR we have been able to give sufficient conditions for convergence to the true parameter values in a number of special cases. These conditions have involved positive realness of certain transfer functions related to the true system description. It is an interesting question whether such conditions indeed are necessary ones, or are merely required in the proof for technical reasons.

Result 4.3 in section 4.3.5 gives us a tool to investigate the necessary conditions for convergence. A desired convergence point for the algorithm must be a stable stationary point of the associated d.e. Otherwise convergence cannot take place. For d.e.s of the kind that we consider here (4.82),

$$\dot{\theta} = R^{-1}f(\theta),$$

$$\dot{R} = G(\theta) - R,$$
(4.189)

we found the condition to be that the matrix (4.85)

$$L^* = [G(\theta^*)]^{-1}H(\theta^*)$$
(4.190)

has all eigenvalues in the left half-plane, where

$$H(\theta^*) = \frac{d}{d\theta}f(\theta)\Big|_{\theta=\theta^*}$$

and where θ^* is the desired convergence point. Let us here use this result to investigate the local convergence properties of the general PLR. Suppose that the true data are generated according to the system

$$A_0(q^{-1})y(t) = \frac{B_0(q^{-1})}{F_0(q^{-1})}u(t) + \frac{C_0(q^{-1})}{D_0(q^{-1})}e(t),$$
(4.191)

where $\{e(t)\}$ is white noise. Let θ_0 denote the parameters of these true polynomials. Suppose that the orders of the model polynomials in (4.137) are greater than or equal to those of (4.191). We then found that the differential equation associated with the general PLR algorithm is given by (4.188). The corresponding matrix L^*, evaluated for $\theta^* = \theta_0$, is then

$$L_0 = -[G_F(\theta_0)]^{-1}\tilde{G}_F(\theta_0), \tag{4.192}$$

where $G_F(\theta_0)$ is given by (4.147) and $\tilde{G}_F(\theta_0)$ by (4.187) and (4.174)–(4.175). We would like to determine the eigenvalues of the matrix L_0 in order to find out if the true value θ_0 is a possible convergence point for the algorithm. In the general case we have to resort to numerical evaluation of L_0. A number of interesting special cases can, however, be treated analytically using the following lemma, which is a development of an earlier result by Holst (1977) [see Stoica et al. (1982)].

LEMMA 4.4   Let

$$\varphi(t) = \frac{1}{F(q^{-1})}(e(t-1) \ \ldots \ e(t-m))^{\mathrm{T}}$$

with $\{e(t)\}$ white noise of zero mean and $F(q^{-1})$ a polynomial of degree less than or equal to m,

$$F(q^{-1}) = 1 + f_1 q^{-1} + \cdots + f_m q^{-m} \quad \text{(possibly } f_m = 0, \text{ etc.)}.$$

Let

$$F^*(z) = z^m + f_1 z^{m-1} + \cdots + f_m,$$

and assume that $F^*(z)$ has all zeros inside the unit circle. Define

$$M = [E\varphi(t)\varphi^{\mathrm{T}}(t)]^{-1}[E\varphi(t) \cdot H(q^{-1})\varphi^{\mathrm{T}}(t)], \tag{4.193}$$

where $H(q^{-1})$ is a rational, asymptotically stable, scalar-valued filter. Then the eigenvalues of the matrix M are given by

$$H(\alpha_i), \quad i = 1, \ldots, m,$$

where $\{\alpha_i\}_{i=1}^m$ are the zeros of the polynomial $F^*(z)$.

Proof   We shall first assume that $F^*(z)$ has m distinct zeros. Then we shall treat the general case of multiple zeros using a perturbation argument.

Therefore, assume that $F^*(z)$ has distinct zeros $\{\alpha_i\}_{i=1}^m$. The eigenvalues of M are given by

$$\det[\lambda I - M] = 0$$

or equivalently

$$\det \Lambda = 0 \quad \text{with} \quad \Lambda = E[\lambda - H(q^{-1})]\varphi(t) \cdot \varphi^{\mathrm{T}}(t). \tag{4.194a}$$

Introduce the noise variance $\sigma^2 = Ee^2(t)$ and the residuals of $1/F(z)F^*(z)$ in the zeros of $F^*(z)$:

$$\beta_k \triangleq \lim_{z \to \alpha_k} \frac{z - \alpha_k}{F(z)F^*(z)}, \quad k = 1, \ldots, m.$$

Then we can write

$$\Lambda = \frac{\sigma^2}{2\pi i} \oint [\lambda - H(z)] \frac{1}{F(z)F^*(z)} \begin{pmatrix} 1 \\ \vdots \\ z^{m-1} \end{pmatrix} (z^{m-1} \; \cdots \; 1) \, dz$$

$$= \sigma^2 \sum_{k=1}^{m} \beta_k [\lambda - H(\alpha_k)] \begin{pmatrix} 1 \\ \alpha_k \\ \vdots \\ \alpha_k^{m-1} \end{pmatrix} (\alpha_k^{m-1} \; \cdots \; \alpha_k \; 1)$$

$$= \sigma^2 \begin{pmatrix} 1 & & 1 \\ \alpha_1 & \cdots & \alpha_m \\ \vdots & & \vdots \\ \alpha_1^{m-1} & \cdots & \alpha_m^{m-1} \end{pmatrix} \begin{pmatrix} \beta_1[\lambda - H(\alpha_1)] & & 0 \\ & \ddots & \\ 0 & & \beta_m[\lambda - H(\alpha_m)] \end{pmatrix}$$

$$\times \begin{pmatrix} \alpha_1^{m-1} & \cdots & \alpha_1 & 1 \\ & \vdots & \\ \alpha_m^{m-1} & \cdots & \alpha_m & 1 \end{pmatrix}. \tag{4.194b}$$

Now, the first matrix in (4.194b) as well as the third one are Vandermonde matrices associated with $F^*(z)$. Since $\{\alpha_i\}$ were assumed to be distinct, these matrices are nonsingular. Furthermore, $\beta_k \neq 0$, $k = 1, \ldots, m$ due to the assumption of distinct zeros. Then it follows easily from (4.194a) and (4.194b) that the eigenvalues of M are given by $\{H(\alpha_k)\}_{k=1}^{m}$.

Assume then that the polynomial $F^*(z)$ has multiple zeros. A small perturbation of order ε of the coefficients of $F^*(z)$ can always be done such that the resulting polynomial, say $F_\varepsilon^*(z)$, has distinct zeros $\{\alpha_k^\varepsilon\}$ (situated inside the unit circle). Denote the matrix (4.193) corresponding to $F_\varepsilon^*(z)$ by M_ε. According to the above analysis, the eigenvalues of M_ε are $\{H(\alpha_k^\varepsilon)\}$. However, since the eigenvalues are continuous functions of the matrix elements it follows that (denoting by $\lambda_k(\cdot)$ the kth eigenvalue of the matrix in question) we formally have

$$\lambda_k(M) = \lambda_k\left(\lim_{\varepsilon \to 0} M_\varepsilon\right) = \lim_{\varepsilon \to 0} H(\alpha_k^\varepsilon) = H(\alpha_k), \quad k = 1, \ldots, m.$$

The proof is thus complete. ∎

COROLLARY Let

$$\Phi(t) = (x(t-1) \ \ldots \ x(t-n_a) \ e(t-1) \ \ldots \ e(t-n_c))^{\mathrm{T}},$$

where $\{e(t)\}$ is white noise and $\{x(t)\}$ is an ARMA process

$$A(q^{-1})x(t) = C(q^{-1})e(t)$$

$$A(z) = a_0 + a_1 z + \cdots + a_{n_a} z^{n_a}$$

$$C(z) = c_0 + c_1 z + \cdots + c_{n_c} z^{n_c}.$$

Assume that $A(z)$ and $C(z)$ are coprime. Let $H(q^{-1})$ be a rational asymptotically stable scalar-valued filter and consider the matrix

$$M = [E\Phi(t) \cdot \Phi^{\mathrm{T}}(t)]^{-1}[E\Phi(t) \cdot H(q^{-1})\Phi^{\mathrm{T}}(t)]. \tag{4.195}$$

This matrix has the eigenvalues:

$H(0)$, of multiplicity n_c

$H(\alpha_k), \quad k = 1, \ldots, n_a,$

where $\{\alpha_k\}_{k=1}^{n_a}$ are the zeros of $A^*(z) = z^{n_a}A(z^{-1})$.

Proof We have, with the notation from the lemma, $m = n_a + n_c$ and

$$\Phi(t) = \mathscr{S}(C, A)\varphi(t),$$

where $F(q^{-1}) = A(q^{-1})$ and the Sylvester matrix $\mathscr{S}(C, A)$ is given by

$$\mathscr{S}(C, A) = \begin{pmatrix} c_0 & c_1 & \cdots & c_{n_c} & & \\ 0 & \ddots & \ddots & & \ddots & 0 \\ & & c_0 & c_1 & \cdots & c_{n_c} \\ \hline a_0 & a_1 & \cdots & a_{n_a} & & \\ 0 & \ddots & \ddots & & \ddots & 0 \\ & & a_0 & a_1 & \cdots & a_{n_a} \end{pmatrix} \begin{matrix} \left.\vphantom{\begin{matrix}1\\1\\1\end{matrix}}\right\} n_a \text{ rows} \\ \\ \left.\vphantom{\begin{matrix}1\\1\\1\end{matrix}}\right\} n_c \text{ rows} \end{matrix}$$

Since the polynomials $A(\cdot)$ and $C(\cdot)$ are coprime, the matrix $\mathscr{S}(C, A)$ is nonsingular (Kailath, 1980). Thus M has the same eigenvalues as

$$[E\varphi(t)\varphi^T(t)]^{-1}[E\varphi(t)\cdot H(q^{-1})\varphi^T(t)].$$

It then follows from lemma 4.4 that the eigenvalues are given by $H(\alpha_k)$, where $\{\alpha_k\}$ are the zeros of $z^{n_a+n_c}A(z^{-1})$. There are n_c eigenvalues in $H(0)$, since $\alpha_1 = \cdots = \alpha_{n_c} = 0$. The remaining ones are

$$H(\alpha_k) \quad \text{with} \quad z^{n_a}A(z^{-1})|_{z=\alpha_k} = 0, \quad k = 1, \ldots, n_a. \quad \blacksquare$$

With the aid of lemma 4.4 we can calculate the eigenvalues of (4.192) corresponding to some particular algorithms. This will be done in three examples.

EXAMPLE 4.11 (ELS Applied to an ARMA Model)   Suppose that we have no input signal and an ARMA model

$$A(q^{-1})y(t) = C(q^{-1})e(t) \tag{4.196}$$

is to be determined. Suppose also that the output signal $y(t)$ has rational spectral density so that it can be described by an ARMA process

$$A_0(q^{-1})y(t) = C_0(q^{-1})e(t), \tag{4.197}$$

where the orders of A_0 and C_0 are less than or equal to those of the model (4.196). We then have

$$\varphi^T(t, \theta_0) = (-y(t-1) \ \ldots \ -y(t-n_a) \ e(t-1) \ \ldots \ e(t-n_c)),$$

$$\tilde{\varphi}(t, \theta_0) = \frac{1}{C_0(q^{-1})}\varphi(t, \theta_0),$$

where the second equality follows from (4.175). We can then apply the corollary of lemma 4.4 directly with $H(q^{-1}) = 1/C_0(q^{-1})$. The eigenvalues of the matrix L_0 (4.192), become

-1, of multiplicity n_c

$-\dfrac{1}{C_0(\alpha_k)}$, $k = 1, \ldots, n_a$, where $\{\alpha_k\}$ are the zeros of the polynomial

$$A_0^*(z) = z^{n_a} + a_1^0 z^{n_a-1} + \cdots + a_{n_a}^0.$$

Therefore, if for some root α_k^* of $A_0^*(z)$, the value $C_0(\alpha_k^*)$ has negative real part, then the matrix (4.192) will have an eigenvalue in the right half-plane. Since this matrix is the system matrix of the corresponding differential equation, linearized around θ_0, this linear d.e. will be unstable.

Hence the true value $\theta_0 = (A_0, C_0)$ is not a possible convergence point for the identification algorithm according to result 4.3.

The foregoing brings out the necessary character of the positive real condition that we encountered as a sufficient condition (4.154) for convergence in section 4.5.2. Clearly if the filter $C_0(q^{-1})$ is positive real, then the matrix L_0 (4.192) will always be stable, no matter what $A_0(q^{-1})$ might be. However, if $C_0(q^{-1})$ is *not* positive real, so that $\text{Re } C_0(z)$ assumes negative values for $z \in D^*$ when D^* is a subset of the unit disc, then there will always exist a polynomial $A_0(z^*)$ with zeros in D^*. Hence as a condition on $C_0(q^{-1})$ *alone*, positive realness is a necessary condition for the ELS algorithm to converge to the true values.

As a numerical example, let us consider the ARMA process

$$y(t) + 0.9y(t-1) + 0.95y(t-2) = e(t) + 1.5e(t-1) + 0.75e(t-2). \tag{4.198}$$

The zeros of the A^*-polynomial $z^2 + 0.9z + 0.95 = 0$ are given by

$$\alpha_{1,2} = -0.45 \pm 0.865i;$$

This gives

$$C_0(\alpha_{1,2}) = -0.0845 \pm 0.713i.$$

The eigenvalues of the corresponding linearized d.e. are $\{-1, -1, 0.162 \pm 1.383i\}$. Hence, when the ELS scheme is applied to identify the system (4.198), the estimates *cannot* converge to the true values. On the other hand, it can also be shown that the true values form the only stationary point of the corresponding d.e. (see Ljung et al., 1975). Therefore, there is *no* possible convergence point for the ELS scheme in this case, and the estimates will continue to oscillate. Simulations of ELS when applied to the identification of (4.198) are given in example 5.15. □

EXAMPLE 4.12 (Landau's Output Error Method with a White Noise Input)   Consider the PLR algorithm (4.144) applied to the model

$$\hat{y}(t \mid \theta) = \frac{B(q^{-1})}{F(q^{-1})} u(t), \tag{4.199}$$

with the filter $S(q^{-1}) = 1$. This is what we have called Landau's output error method. Suppose that the input sequence $\{u(t)\}$ is white noise and that the true system can be described by

$$y(t) = \frac{B_0(q^{-1})}{F_0(q^{-1})} u(t) + v(t), \tag{4.200}$$

where $\{v(t)\}$ is a stochastic process that is independent of $\{u(t)\}$. Assume that the orders of the true system (4.200) are less than or equal to those of the model (4.199). We then have [θ_0 being the parameter vector corresponding to (4.200)]

$$\varphi^T(t, \theta_0) = (-w_0(t-1) \ \ldots \ -w_0(t-n_f) \ u(t-1) \ \ldots \ u(t-n_b)),$$

where

$$w_0(t) = \frac{B_0(q^{-1})}{F_0(q^{-1})} u(t).$$

(We have interchanged the orders between w and u in φ here to be in formal agreement with lemma 4.4.) According to lemma 4.3 and (4.187c), we have

$$\tilde{\varphi}_F(t, \theta) = \frac{T(q^{-1})}{F_0(q^{-1})} \varphi(t, \theta_0).$$

We can now apply lemma 4.4 to calculate the eigenvalues of the matrix

$$L_0 = -[E\varphi(t, \theta_0)\varphi^T(t, \theta_0)]^{-1}[E\varphi(t, \theta_0)\tilde{\varphi}_F^T(t, \theta_0)].$$

It follows directly from the corollary that these eigenvalues are

-1, of multiplicity n_b,

$-\dfrac{T(\alpha_k)}{F_0(\alpha_k)}$, $k = 1, \ldots, n_f$, where $\{\alpha_k\}$ are the zeros of

$$F_0^*(z) = z^{n_f} + f_1^0 z^{n_f-1} + \cdots + f_{n_f}^0.$$

The comments on the necessity of positive realness made in the previous example apply of course also in this case. Similarily, numerical counter-examples to convergence can easily be worked out using the above result. □

EXAMPLE 4.13 (The Box-Jenkins Model with a White Noise Input)   Consider the PLR algorithm (4.144) applied to the model

$$y(t) = \frac{B(q^{-1})}{F(q^{-1})} u(t) + \frac{C(q^{-1})}{D(q^{-1})} e(t) \tag{4.201}$$

with the filters $S(q^{-1}) = T(q^{-1}) = 1$. Suppose that the input sequence $\{u(t)\}$ is white noise and that the true system can be described by

$$y(t) = \frac{B_0(q^{-1})}{F_0(q^{-1})}u(t) + \frac{C_0(q^{-1})}{D_0(q^{-1})}e(t), \qquad (4.202)$$

where $\{e(t)\}$ is a white-noise process that is independent of $\{u(t)\}$. Assume that the orders of the true system (4.202) are less than or equal to those of the model (4.201).

With calculations analogous to those in the previous two examples, we find that the matrix L_0 is block diagonal with eigenvalues

-1, of multiplicity $n_b + n_c$

$-\dfrac{D_0(\alpha_k)}{C_0(\alpha_k)F_0(\alpha_k)}$, $k = 1, \ldots, n_f$, where α_k are the zeros of

$$F_0^*(z) = z^{n_f} + f_1^0 z^{n_f - 1} + \cdots + f_{n_f}^0; \qquad (4.203)$$

$-\dfrac{1}{C_0(\beta_k)}$, $k = 1, \ldots, n_d$, where β_k are the zeros of

$$D_0^*(z) = z^{n_d} + d_1^0 z^{n_d - 1} + \cdots + d_{n_d}^0.$$

The result implies that if the two transfer functions

$$\frac{D_0(q^{-1})}{C_0(q^{-1})F_0(q^{-1})} \quad \text{and} \quad \frac{1}{C_0(q^{-1})} \qquad (4.204)$$

are not both positive real, then the system may be such that the linearized differential equation will have eigenvalues in the right half-plane. For such systems the algorithm will fail to converge to the true parameter values. $\square$

In the foregoing three examples we have given an analytic treatment of the eigenvalues of the matrix L_0 (4.192). Thereby we have also given the exact conditions under which the true parameter values θ_0 form a possible convergence point for the PLR algorithm. If the linearized differential equation has its eigenvalues strictly in the left half-plane, then also the original differential equation will be locally asymptotically stable around θ_0. As we pointed out in section 4.3.2, this will imply local convergence of the algorithm. The three examples, consequently bring out the necessity character of the positive real condition.

4.5.5   Asymptotic Distribution for Pseudolinear Regressions

Unfortunately, derivation of the asymptotic covariance matrix and the asymptotic distribution for the estimates obtained by PLR is not as direct as for the recursive prediction error method. Referring to theorem 4.5, we see that the property that $-\psi(t)$ is approximately the gradient of $\varepsilon(t)$ is crucial for lemma 4.B.4. This lemma allows us to get rid of the last term in (4.B.1). When $\psi(t)$ is replaced by $\varphi(t)$, this is no longer possible. The asymptotic distribution and asymptotic covariance matrix for PLR are therefore not fully understood at the present time. Solo (1978, 1980) has discussed this subject in detail.

For a recursive prediction error method applied to a quadratic criterion we know from theorem 4.5 that the asymptotic covariance matrix is given by

$$P(t) = \frac{1}{t}[\bar{E}\psi(k)\Lambda^{-1}(k)\psi^{T}(k)]^{-1}.$$

Let us stress that the analogous result is *not* true for a PLR. This is shown by a counterexample.

EXAMPLE 4.14 (Covariance Matrix for PLRs)   Consider a second order moving average process

$$y(t) = C_0(q^{-1})e(t) = e(t) + c_1^0 e(t-1) + c_2^0 e(t-2), \tag{4.205}$$

where $\{e(t)\}$ is a sequence of independent random variables each of zero mean and unit variance. If $\{e(t)\}$ has a Gaussian distribution, the Cramér-Rao lower bound for the covariance matrix of any unbiased estimator is given by (see theorem 4.5)

$$\frac{1}{N} \cdot [E\psi(t, \theta_0)\psi^{T}(t, \theta_0)]^{-1} = P_N,$$

where

$$\psi(t, \theta_0) = \frac{1}{C_0(q^{-1})}\begin{pmatrix} e(t-1) \\ e(t-2) \end{pmatrix}$$

and

$$C_0(q^{-1}) = 1 + c_1^0 q^{-1} + c_2^0 q^{-2}.$$

Now we shall show that the matrix Q_N, defined by

$$\frac{1}{N}[\bar{E}\varphi(t, \theta_0)\varphi^T(t, \theta_0)]^{-1} = Q_N,$$

where

$$\varphi(t, \theta_0) = \begin{pmatrix} e(t-1) \\ e(t-2) \end{pmatrix},$$

cannot be the asymptotic covariance matrix for the estimates obtained by applying a PLR to (4.205). If Q_N were the covariance matrix we would have to have

$$Q_N \geq P_N \quad \text{or} \quad P_N^{-1} \geq Q_N^{-1}, \tag{4.206}$$

since P_N is the lower bound. We find after straightforward calculation that

$$Q_N^{-1} = N \cdot I,$$

$$P_N^{-1} = \frac{N}{(1 - c_2^0)[(1 + c_2^0)^2 - (c_1^0)^2]} \begin{pmatrix} 1 + c_2^0 & -c_1^0 \\ -c_1^0 & 1 + c_2^0 \end{pmatrix}.$$

The inequality (4.206) can thus be written

$$\begin{pmatrix} (c_2^0)^2(1 + c_2^0) + (c_1^0)^2(1 - c_2^0) & -c_1^0 \\ -c_1^0 & (c_2^0)^2(1 + c_2^0) + (c_1^0)^2(1 - c_2^0) \end{pmatrix} \geq 0$$

or

$$(c_2^0)^4(1 + c_2^0)^2 + (c_1^0)^4(1 - c_2^0)^2 + (c_1^0)^2[-1 + 2(c_2^0)^2 - 2(c_2^0)^4] \geq 0.$$

This inequality is violated, e.g., for $c_2^0 = 0$, $|c_1^0| < 1$. Thus (4.206) does not hold for these values of c_i^0, and therefore Q_N *cannot* be the asymptotic covariance matrix of the estimates obtained by PLR. $\square$

4.5.6 Summary

We have studied the convergence properties of pseudolinear regressions applied to estimate the parameters of the general linear model

$$A(q^{-1})y(t) = \frac{B(q^{-1})}{F(q^{-1})}u(t) + \frac{C(q^{-1})}{D(q^{-1})}e(t).$$

An explicit expression for the associated d.e. was derived in section 4.5.3 under the assumption that the data are generated by a system

$$A_0(q^{-1})y(t) = \frac{B_0(q^{-1})}{F_0(q^{-1})}u(t) + \frac{C_0(q^{-1})}{D_0(q^{-1})}e(t).$$

This d.e. is given by (4.187)–(4.188).

It is easy to give an analytic treatment of the global stability properties of this d.e. only in the cases

$$F(q^{-1}) = D(q^{-1}) = 1 \quad \text{(ELS)}$$

and

$$A(q^{-1}) = C(q^{-1}) = D(q^{-1}) = 1 \quad \text{(Landau's output error method)}.$$

This has been done in section 4.5.2. Convergence of these algorithms is proved under the condition that

$$\frac{1}{C_0(q^{-1})} - \frac{1}{2} \quad \text{and} \quad \frac{1}{F_0(q^{-1})} - \frac{1}{2},$$

respectively, be positive real.

These conditions can also be written

$$|C_0(e^{i\omega}) - 1| < 1, \quad |F_0(e^{i\omega}) - 1| < 1.$$

When the regression vector and/or the prediction errors are filtered as in (4.144c, g), the above positive reality conditions are modified to (4.165) and (4.171), respectively.

Now

$$\left[\frac{d}{d\theta}\hat{y}(t \mid \theta)\right]^{\mathrm{T}} = \psi(t, \theta) = \frac{1}{C(q^{-1})}\varphi(t, \theta)$$

in the ELS case, and ELS uses φ as a gradient approximation rather than ψ. Thus, the quoted conditions have the natural interpretation that convergence will occur when the C-filter is "close to unity," and hence it is a reasonable approximation to replace ψ by φ.

In section 4.5.4 we investigated the necessity of these conditions on the true system. This was done by examination of the eigenvalues of the linearized d.e. at the true parameter values. Whenever this linear d.e. is unstable, convergence to these values cannot take place. We then found as a condition on $C_0(q^{-1})$ alone in the ELS case, that it is *necessary* for convergence that it be positive real. An explicit counterexample to convergence was also constructed [equation (4.198)]. A corresponding result also holds for Landau's output error method.

4.6 Analysis of Instrumental Variable Methods

4.6.1 Introduction

The general recursive IV algorithm is given by (3.96):

$$\hat{\theta}(t) = \hat{\theta}(t-1) + \gamma(t)R^{-1}(t)\zeta(t)[y_F(t) - \varphi_F^T(t)\hat{\theta}(t-1)], \qquad (4.207a)$$

$$R(t) = R(t-1) + \gamma(t)[\zeta(t)\eta^T(t) - R(t-1)], \qquad (4.207b)$$

$$y_F(t) = T(q^{-1})y(t) \qquad \varphi_F(t) = T(q^{-1})\varphi(t). \qquad (4.207c)$$

For the nonsymmetric version (3.92a) we have

$$\eta(t) = \varphi_F(t), \qquad (4.208)$$

and for the symmetric version (3.92b) we have

$$\eta(t) = \zeta(t). \qquad (4.209)$$

The vector $\varphi(t)$ contains delayed input/output variables, as described in section 3.6.3, and is independent of the parameter estimates. The filter $T(q^{-1})$ as well as the instrumental variables $\zeta(t)$ are to be chosen by the user. In section 3.6.3 we described some common choices of $\zeta(t)$.

This section, together with appendix 4.D, will deal with the asymptotic properties of $\hat{\theta}(t)$ as generated by (4.207). We will discuss how convergence and asymptotic distribution depend on $\zeta(t)$ and $T(q^{-1})$ as well as on the choice between (4.208) and (4.209). Most of the discussion will be confined to single-input/single-output systems and to instrumental variables that are generated according to (3.93). Treatments of more general cases can be found in Stoica and Söderström (1981a, b, 1982a, 1983a, b) and Söderström and Stoica (1983).

For the analysis we shall assume that the data actually are generated by a system

$$y(t) = \varphi^T(t)\theta_0 + v(t), \qquad (4.210)$$

$$v(t) = H(q^{-1})e(t), \qquad (4.211)$$

where $H(q^{-1})$ is an asymptotically stable and inversely stable filter and $\{e(t)\}$ is a white noise sequence with zero mean values and variances σ^2. In difference equation notation, (4.210) is written

$$A_0(q^{-1})y(t) = B_0(q^{-1})u(t) + v(t). \qquad (4.212)$$

We shall also throughout this section assume that

$\{u(t)\}$ and $\{v(t)\}$ are independent.   (4.213)

This means that the generation of $u(t)$ must not depend on past $y(t)$ [see (4.210)]. Hence no output feedback is allowed in the input sequence.

Convergence of the nonsymmetric IV method is studied in section 4.6.2, while the asymptotic distribution of the estimates is calculated in section 4.6.3. The symmetric IV method is examined in appendix 4.D. A natural, adaptive choice of instrumental variables ζ and prefilter T, the so-called refined IVAML method, is described in appendix 4.E, where it is also shown that this method in fact is a recursive prediction error method.

4.6.2   Convergence of the Nonsymmetric IV Method

The convergence analysis will be based on an explicit expression for the estimates, obtained from (4.207).

An Expression for $\hat{\theta}(t)$ Assuming for the moment that $\gamma(t) = 1/t$, and defining

$$\bar{R}(t) = t \cdot R(t),$$

we find from (4.207b) and (4.208) that

$$\bar{R}(t) = \bar{R}(t-1) + \zeta(t)\varphi_F^T(t).$$ (4.214)

Multiplying (4.207a) by $\bar{R}(t)$ now gives

$$\bar{R}(t)\hat{\theta}(t) = [\bar{R}(t-1) + \zeta(t)\varphi_F^T(t)]\theta(t-1) + \zeta(t)y_F(t)$$

$$- \zeta(t)\varphi_F^T(t)\hat{\theta}(t-1) = \bar{R}(t-1)\hat{\theta}(t-1) + \zeta(t)y_F(t).$$

Hence

$$\bar{R}(t)\hat{\theta}(t) = \bar{R}(0)\hat{\theta}(0) + \sum_{k=1}^{t} \zeta(k)y_F(k),$$ (4.215)

where

$$\bar{R}(t) = \bar{R}(0) + \sum_{k=1}^{t} \zeta(k)\varphi_F^T(k).$$ (4.216)

For a general gain sequence $\{\gamma(t)\}$ we obtain analogous results [see (2.116), (2.128)].

If we use the expressions (4.210) and (4.207c) for $y(t)$ and $y_F(t)$ in (4.215), we obtain

$$\bar{R}(t)\hat{\theta}(t) = \bar{R}(0)\hat{\theta}(0) + \sum_{k=1}^{t} \zeta(k)v_F(k) + \left[\sum_{k=1}^{t} \zeta(k)\varphi_F^T(k)\right]\theta_0,$$

where

$$v_F(t) = T(q^{-1})v(t). \tag{4.217}$$

This can also be rewritten as [recall that $\bar{R}(t) = tR(t)$]

$$R(t)[\hat{\theta}(t) - \theta_0] = \frac{1}{t}\bar{R}(0)[\hat{\theta}(0) - \theta_0] + \frac{1}{t}\sum_{k=1}^{t} \zeta(k)v_F(k). \tag{4.218}$$

The convergence analysis is based on this expression, and it will be carried out in three steps:

1. Establish that $R(t)[\hat{\theta}(t) - \theta_0] \to 0$.
2. Establish that $R(t) \to \bar{R}$.
3. Examine when $\bar{R}$ is nonsingular.

Step 1: Convergence of $R(t)[\hat{\theta}(t) - \theta_0]$ From (4.218) it is easy to establish the following result.

LEMMA 4.5   Consider the algorithm (4.207), (4.208) with $\gamma(t) = 1/t$. Suppose that (4.210)–(4.211) hold, and that the instrumental variables $\{\zeta(t)\}$ are of bounded variance and independent of the noise sequence $\{v(t)\}$. Then

$$R(t)[\hat{\theta}(t) - \theta_0] \to 0 \quad \text{w.p.1 as} \to \infty.$$

Proof   Let us consider the second term of the right-hand side of (4.218). According to the assumptions of the lemma we have

$$E\zeta(t)v_F(t) = 0,$$

$$\left|E\zeta(t)v_F(t)[\zeta(s)v_F(s)]^T\right| = \left|E\zeta(t)\zeta^T(s) \cdot Ev_F(t)v_F(s)\right| \le C\lambda^{|t-s|},$$

for $\lambda < 1$. The first equality follows since ζ and v_F are independent; and the second, since $\zeta(t)$ is of bounded variance and since $v_F(t)$ is the output of an exponentially stable filter driven by white noise (see (4.217), (4.211)). Then we can use the convergence result (4.67)–(4.68) to infer that

$$\frac{1}{t}\sum_{1}^{t} \zeta(k)v_F(k) \to 0 \quad \text{w.p.1 as } t \to \infty,$$

which with (4.218) proves the lemma. ∎

Remark The condition that $\{\zeta(t)\}$ and $\{v(t)\}$ should be independent is obviously satisfied if $\zeta(t)$ is constructed entirely from old inputs u^t, provided that the open loop condition (4.213) holds. This case covers most of the common choices of $\zeta(t)$. When current estimates $\hat{\theta}(t)$ are used in the filter that produces $\zeta(t)$ from u^t, as in (3.94), the condition is (slightly) violated, though. The reason is that some dependence of $v(t)$ on $\zeta(t)$ is transferred via $\hat{\theta}$, when $v(t)$ is not white. However, it should be clear that this dependence is negligible asymptotically, that we have $\bar{E}\zeta(t)v_F(t) = 0$, and that the lemma still holds in this case.

Step 2: Convergence of $R(t)$ With the result of lemma 4.5 we can focus our attention on the matrix $R(t)$. If $R(t)$ converges to an invertible matrix, we conclude that $\hat{\theta}(t)$ tends to θ_0 w.p.1 as t approaches infinity. More generally, this result holds as long as the eigenvalues of $R(t)$ are bounded away from zero asymptotically, even if $R(t)$ does not converge.

We confine the study to the case (3.94) where the instrumental variables are generated by a prior chosen and time-invariant filter given by $\bar{A}$ and $\bar{B}$:

$$\zeta(t) = (-x(t-1) \ \ldots \ -x(t-n) \ u(t-1) \ \ldots \ u(t-m))^{\mathrm{T}}, \qquad (4.219a)$$

$$\bar{A}(q^{-1})x(t) = \bar{B}(q^{-1})u(t), \qquad (4.219b)$$

$$\bar{A}(q^{-1}) = 1 + \bar{a}_1 q^{-1} + \cdots + \bar{a}_n q^{-n},$$

$$\bar{B}(q^{-1}) = \bar{b}_1 q^{-1} + \cdots + \bar{b}_m q^{-m}. \qquad (4.219c)$$

As before,

$$\varphi(t) = (-y(t-1) \ \ldots \ -y(t-n) \ u(t-1) \ \ldots \ u(t-m))^{\mathrm{T}}. \qquad (4.220)$$

For the limit of the matrix $R(t)$ we now have the following lemma.

LEMMA 4.6 Suppose that the input sequence $\{u(t)\}$ is subject to (4.213), and asymptotically stationary in the sense that

$$\bar{E}u(t)u(t-j) \triangleq r_u(j) \quad \text{exists for all } j. \qquad (4.221)$$

If $\{u(t)\}$ is regarded as a stochastic process assume also that condition

S1 (section 4.3.4) holds. Let $\varphi_F(t)$ be given by (4.220), (4.207c) and let $\zeta(t)$ be given by (4.219). The data is subject to (4.210)–(4.211). Then

$$\bar{R} = \bar{E}\zeta(t)\varphi_F^T(t) \text{ exists} \tag{4.222}$$

and

$$R(t) = \frac{1}{t}\sum_1^t \zeta(k)\varphi_F^T(k) \to \bar{R} \quad \text{w.p.1 as } t \to \infty. \tag{4.223}$$

Remark Recall that when $\{u(t)\}$ is regarded as a deterministic sequence, then

$$\bar{E}u(t)u(t-j) = \lim_{N\to\infty} \frac{1}{N}\sum_1^N u(t)u(t-j).$$

Proof The existence of $\bar{R}$ in (4.222) is immediate since the entries of this matrix are formed from quantities obtained by stable time-invariant filtering of the (asymptotically) stationary sequences $\{u(t)\}$ and $\{e(t)\}$. The convergence result (4.223) then follows just as in lemma 4.1. ∎

Step 3: Invertibility of $\bar{R}$ We now proceed to study the matrix $\bar{R}$ in (4.222). From lemma 4.5 we know that if

$$\bar{R} \text{ is invertible,} \tag{4.224}$$

then $\hat{\theta}(t)$ will converge to the true value θ_0 w.p.1 as $t \to \infty$.
    It can be shown that

$$
\begin{aligned}
\bar{R} &= \bar{E}\zeta(t)\varphi_F^T(t) \\
&= \bar{E}\begin{pmatrix} -\dfrac{\bar{B}(q^{-1})}{\bar{A}(q^{-1})}u(t-1) \\ \vdots \\ -\dfrac{\bar{B}(q^{-1})}{\bar{A}(q^{-1})}u(t-n) \\ u(t-1) \\ \vdots \\ u(t-m) \end{pmatrix} \cdot
\end{aligned}
\tag{4.225}
$$

$$
\times T(q^{-1})(-y(t-1) \ \ldots \ -y(t-n) \ u(t-1) \ \ldots \ u(t-m))
$$
$$
= \mathscr{S}(-\bar{B}, \bar{A})\mathscr{P}(A_0, \bar{A}, T, u)\mathscr{S}^T(-B_0, A_0),
$$

where $\mathscr{S}^{\mathrm{T}}(-B, A)$ is the Sylvester matrix

$$
\mathscr{S}^{\mathrm{T}}(-B, A) = \begin{pmatrix} 0 & & & 1 & & \\ -b_1 & 0 & & a_1 & & 0 \\ \vdots & \ddots & & \vdots & \ddots & \\ & & \ddots & & & 1 \\ & & -b_1 & a_n & & a_1 \\ & & \vdots & & \ddots & \vdots \\ -b_m & & & & & \\ 0 & \ddots & & 0 & \ddots & \\ & & -b_m & & & a_n \end{pmatrix}, \tag{4.226}
$$

and where $\mathscr{P}$ is given by

$$
\mathscr{P}(A_0, \bar{A}, T, u) = \bar{\mathrm{E}} \frac{1}{\bar{A}(q^{-1})} \begin{pmatrix} u(t-1) \\ \vdots \\ u(t-n-m) \end{pmatrix}
$$

$$
\times \frac{T(q^{-1})}{A_0(q^{-1})} (u(t-1) \ \ldots \ u(t-n-m)). \tag{4.227}
$$

All matrices in (4.225) are square and of order $n + m$. When establishing (4.225) we used the fact that $\{\zeta(t)\}$ is uncorrelated with the disturbance $\{v(t)\}$.

From (4.225) we see that $\bar{R}$ is nonsingular if and only if the matrices $\mathscr{S}(-B_0, A_0)$, $\mathscr{S}(-\bar{B}, \bar{A})$, and $\mathscr{P}(A_0, \bar{A}, T, u)$ all are nonsingular. For the first ones, it is well known (see Kailath, 1980) that $\mathscr{S}(-B, A)$ is nonsingular if and only if the polynomials A and B are coprime. Notice that $\mathscr{P}$ does not depend on $\bar{B}$, so, as long as $\bar{A}$ and $\bar{B}$ are coprime, the choice of $\bar{B}$ does not affect the invertibility of $\bar{R}$.

The condition that A_0 and B_0 are coprime implies that the model is not overparametrized in both A and B. The polynomials A_0 and B_0 are defined through (4.212) for the true system but with degrees n and m given by the chosen model set. If A_0 and B_0 in (4.212) do have a common factor, there exists another representation of the form (4.212) with coprime polynomials such that $\deg A_0 < n$, $\deg B_0 < m$.

Let us now discuss the matrix $\mathscr{P}$ in (4.227). Suppose that the input signal is persistently exciting of order $n + m$, so that with $A_0 = \bar{A} = T = 1$ the matrix is nonsingular (see section 3.2.1). We could then say, heuristically, that it requires some very special relationships between A_0, $\bar{A}$, T, and the properties of u to make (4.227) singular (such an example will be given later: example 4.15). It can in fact be proved that $\mathscr{P}(A_0, \bar{A}, T, u)$ is generically nonsingular (i.e., it is nonsingular outside a null set of A_0, $\bar{A}$, T). The arguments are as follows. Consider $\det[\mathscr{P}(A_0, \bar{A}, T, u)]$ as a function of the coefficients of A_0, $\bar{A}$, and T. This function is analytic and not identical to zero (e.g., for $A_0 = \bar{A} = T = 1$ it is nonzero). Thus by the uniqueness theorem of analytic functions the function is nonzero almost everywhere. This result has been shown by Finigan and Rowe (1974) for the case $T = 1$. It has been extended to general IV schemes by Söderström and Stoica (1983). Here we will give some *sufficient* conditions for the nonsingularity of $\mathscr{P}$:

LEMMA 4.7   Consider the matrix $\mathscr{P}(A, \bar{A}, T, u)$ given by (4.227). The following three cases guarantee the nonsingularity of this matrix:

(i) The matrix $\mathscr{P}(A, A, 1, u)$ is nonsingular if and only if the signal $u(t)$ is persistently exciting of order $n + m$.

(ii) Assume that $u(t)$ is white noise and that the filter $T(q^{-1})$ and its inverse are asymptotically stable. Then $\mathscr{P}(A, \bar{A}, T, u)$ is nonsingular for *all* asymptotically stable A and $\bar{A}$.

(iii) Assume that $\bar{A}(q^{-1})T(q^{-1})/A(q^{-1})$ is strictly positive real. Then $\mathscr{P}(A, \bar{A}, T, u)$ is nonsingular for *all* signals $u(t)$ that are persistently exciting of order $n + m$.

Proof   Let

$$\bar{h} = (h_1 \ \ldots \ h_{n+m})^{\mathrm{T}}$$

be an arbitrary vector, and set $H(q^{-1}) = \sum_{i=1}^{n+m} h_i q^{-i}$. Consider now the expression

$$0 = h^{\mathrm{T}}\mathscr{P}(A, \bar{A}, T, u)h$$

$$= \bar{\mathrm{E}}\left[\frac{H(q^{-1})}{\bar{A}(q^{-1})}u(t)\right]\left[\frac{H(q^{-1})T(q^{-1})}{A(q^{-1})}u(t)\right]$$

$$= \int_{-\pi}^{\pi}\left|\frac{H(e^{i\omega})}{\bar{A}(e^{i\omega})}\right|^2 \Phi_u(\omega)\frac{\bar{A}(e^{i\omega})T(e^{i\omega})}{A(e^{i\omega})}d\omega$$

$$= \int_{-\pi}^{\pi}\left|\frac{H(e^{i\omega})}{\bar{A}(e^{i\omega})}\right|^2 \Phi_u(\omega)\left\{\mathrm{Re}\,\frac{\bar{A}(e^{i\omega})T(e^{i\omega})}{A(e^{i\omega})}\right\}d\omega. \tag{4.228}$$

Here $\Phi_u(\omega)$ denotes the spectral density of $u(t)$ (or the discrete Fourier transform of $\bar{E}u(t + \tau)u(t)$ when $u(t)$ is not stationary). Assume that $u(t)$ is persistently exciting of order $n + m$, but otherwise arbitrary. Then $\Phi_u(\omega) \geq 0$ with strict inequality in at least $n + m$ distinct points. When $\bar{A} \equiv A$ and $T \equiv 1$, equation (4.228) then implies $|H(e^{i\omega})|^2 \Phi_u(\omega) \equiv 0$. However, as $H(z)$ is a polynomial of degree $n + m - 1$ only, it follows that $H(z) \equiv 0$ and $h = 0$. Thus $\mathscr{P}(A, A, 1, u)$ is nonsingular, which proves case (i). Case (iii) follows in the same way from (4.228).

Consider now case (ii). Assume

$$0 = h^{\mathrm{T}} \mathscr{P}(A, \bar{A}, T, u),$$

which implies

$$0 = \bar{E} \left[\frac{H(q^{-1})}{\bar{A}(q^{-1})} u(t) \right] \left[\frac{T(q^{-1})}{A(q^{-1})} u(t - j) \right], \quad 1 \leq j \leq n + m,$$

or, with expectation written as complex integration,

$$0 = \frac{1}{2\pi i} \oint \frac{z^{n+m} H(z^{-1})}{z^{n+m} \bar{A}(z^{-1})} \frac{T(z)}{A(z)} z^j \frac{dz}{z}, \quad 1 \leq j \leq n + m.$$

The integration path is the unit circle. Consider the integrand in the complex integral. Since by assumption $\bar{A}$, T, and T^{-1} are stable, all zeros of $\bar{A}(z)$, $T(z)$, and $T(z)^{-1}$ are outside the unit circle. The poles and zeros of the integrand inside the unit circle are thus given by

$$z^{n+m} \bar{A}(z^{-1}) = 0 : n + m \text{ poles,}$$

and, excluding the origin,

$$z^{n+m} H(z^{-1}) = 0 : n + m - 1 \text{ zeros.}$$

(Recall the definition of $H(q^{-1})$!) Since the integral is zero for $1 \leq j \leq n + m$, it follows from lemma 1 in Åström and Söderström (1974) that the listed poles and zeros all cancel. This is not possible unless $H(z^{-1}) \equiv 0$, which shows that h must be zero and that $\mathscr{P}(A, \bar{A}, T, u)$ is nonsingular also in case (ii). ∎

The following counterexample shows that the matrix $\mathscr{P}(A, \bar{A}, T, u)$ is not always nonsingular.

EXAMPLE 4.15 (A Singular $\mathscr{P}$-matrix) Consider the matrix $\mathscr{P}(A, \bar{A}, T, u)$ (4.227). Let $n = 2$, $m = 1$, $A(z) = (1 - \alpha z)^2$, $\bar{A}(z) = (1 + \alpha z)^2$, $T(z) = 1$, and $u(t) = (1 - \alpha q^{-1})^2(1 + \alpha q^{-1})^2 w(t)$, with $w(t)$ being white noise of zero mean and unit variance. Then

$$\mathscr{P}(A_0, \bar{A}, T, u) = \begin{pmatrix} 1 - 4\alpha^2 + \alpha^4 & -2\alpha(1 - \alpha^2) & \alpha^2 \\ 2\alpha(1 - \alpha)^2 & 1 - 4\alpha^2 + \alpha^4 & -2\alpha(1 - \alpha^2) \\ \alpha^2 & 2\alpha(1 - \alpha^2) & 1 - 4\alpha^2 + \alpha^4 \end{pmatrix},$$

which turns out to be singular if $\alpha = ((3 - \sqrt{5})/2)^{1/2} \approx 0.618$. □

Convergence of $\hat{\theta}(t)$ We can now collect the results that we have obtained in the preceding steps 1–3 into the following theorem:

THEOREM 4.7   Consider the nonsymmetric IV algorithm (4.207), (4.208). Suppose that the true data are given by (4.210), (4.211), where the input is such that (4.213) and (4.221) hold. Assume that the polynomials A_0 and B_0 (of degrees n and m, respectively) have no common factor. Suppose that the instrumental variables $\zeta(t)$ are given by (4.219), where $\bar{A}$ and $\bar{B}$ are coprime. Assume also that the matrix $\mathscr{P}(A_0, \bar{A}, T, u)$ given by (4.227) is nonsingular. Then

$$\hat{\theta}(t) \to \theta_0 \quad \text{w.p.1 as } t \to \infty.$$

Notice that we can use lemma 4.7 to obtain conditions for the non-singularity of $\mathscr{P}$.

4.6.3   Asymptotic Distribution for the Nonsymmetric IV Method

A Basic Result   The basic result about the distribution of the IV estimate (4.207), (4.208) is the following one.

THEOREM 4.8   Consider the nonsymmetric IV estimate (4.207), (4.208). Assume that the system is given by (4.212). Suppose that the assumptions of theorem 4.7 hold. Then the estimate are asymptotically normal:

$$\sqrt{t}[\hat{\theta}(t) - \theta_0] \in \mathrm{AsN}(0, P), \tag{4.229}$$

where

$$P = \sigma^2 [\bar{\mathrm{E}}\zeta(t)\varphi_F^{\mathrm{T}}(t)]^{-1}$$

$$\times \bar{\mathrm{E}}[T(q^{-1})H(q^{-1})\zeta(t) \cdot T(q^{-1})H(q^{-1})\zeta^{\mathrm{T}}(t)] \tag{4.230}$$

$$\times [\bar{\mathrm{E}}\varphi_F(t)\zeta^{\mathrm{T}}(t)]^{-1}.$$

Proof It follows from (4.218), neglecting the initial conditions, that

$$\sqrt{t}[\hat{\theta}(t) - \theta_0] = \left[\frac{1}{t}\sum_{s=1}^{t}\zeta(s)\varphi_F^T(s)\right]^{-1}\left[\frac{1}{\sqrt{t}}\sum_{s=1}^{t}\zeta(s)\cdot T(q^{-1})v(s)\right].$$

According to lemma 4.6 we have

$$\frac{1}{t}\sum_{s=1}^{t}\zeta(s)\varphi_F^T(s) \to \bar{E}\zeta(t)\varphi_F^T(t) \quad \text{w.p.1 as } t \to \infty.$$

Moreover, from a variant of the central limit theorem as given by Ljung (1977c), it follows that

$$\frac{1}{\sqrt{t}}\sum_{s=1}^{t}\zeta(s)\cdot T(q^{-1})v(s) \in \text{AsN}(0, Q_0)$$

with

$$Q_0 \triangleq \lim_{t\to\infty}\frac{1}{t}E\left[\sum_{s=1}^{t}\zeta(s)\cdot T(q^{-1})v(s)\sum_{s'=1}^{t}\zeta^T(s')\cdot T(q^{-1})v(s')\right].$$

It then follows from standard convergence results for random variables that (4.229) is true. It only remains to show that

$$Q_0 = E[T(q^{-1})H(q^{-1})\zeta(t)\cdot T(q^{-1})H(q^{-1})\zeta^T(t)].$$

Introduce for convenience $K(q^{-1}) = \sum_{i=0}^{\infty}k_i q^{-i} \triangleq T(q^{-1})H(q^{-1})$ and put $k_i = 0, i < 0$. Straightforward calculation gives

$$Q_0 = \lim_{t\to\infty}\frac{1}{t}\sum_{s=1}^{t}\sum_{s'=1}^{t}E[\zeta(s)\zeta^T(s')]E[T(q^{-1})v(s)\cdot T(q^{-1})v(s')]$$

$$= \lim_{t\to\infty}\frac{1}{t}\sum_{\tau=-t}^{t}(t-|\tau|)E[\zeta(s)\zeta^T(s+\tau)]E[T(q^{-1})v(s)\cdot T(q^{-1})v(s+\tau)]$$

$$= \sum_{\tau=-\infty}^{\infty}E[\zeta(t)\zeta^T(t+\tau)]E[T(q^{-1})v(t)\cdot T(q^{-1})v(t+\tau)]$$

$$- \lim_{t\to\infty}\frac{1}{t}\sum_{\tau=-t}^{t}|\tau|E[\zeta(s)\zeta^T(s+\tau)]E[T(q^{-1})v(s)\cdot T(q^{-1})v(s+\tau)].$$

Since $K(q^{-1})$ is asymptotically stable it follows that

$$|ET(q^{-1})v(s)\cdot T(q^{-1})v(s+\tau)| \le C\lambda^{|\tau|}$$

for all τ and some constants $C > 0, 0 < \lambda < 1$. Thus the magnitude of the last term is bounded from above by

$$\left| \lim_{t \to \infty} \frac{1}{t} \sum_{\tau = -t}^{t} |\tau| C' \lambda^{|\tau|} \right| \leq \lim_{t \to \infty} \frac{2C'}{t} \sum_{\tau = 0}^{\infty} \tau \lambda^\tau = 0.$$

The last term will thus vanish. The first term can be written as

$$P_0 = \sum_{\tau = -\infty}^{\infty} \mathrm{E}[\zeta(t) \zeta^{\mathrm{T}}(t + \tau)] \left[\sum_{j = -\infty}^{\infty} k_j k_{j+\tau} \right] \sigma^2$$

$$= \sigma^2 \sum_{i=0}^{\infty} \sum_{j=0}^{\infty} k_i k_j \mathrm{E} \zeta(t) \zeta^{\mathrm{T}}(t + i - j)$$

$$= \sigma^2 \mathrm{E} \left[\sum_{i=0}^{\infty} k_i \zeta(t - i) \right] \left[\sum_{j=0}^{\infty} k_j \zeta^{\mathrm{T}}(t - j) \right]$$

$$= \sigma^2 \mathrm{E}[K(q^{-1}) \zeta(t) \cdot K(q^{-1}) \zeta^{\mathrm{T}}(t)].$$

which completes the proof. ∎

Theorem 4.8 shows that the asymptotic accuracy of the estimates is given by the covariance matrix P. While (4.230) gives an explicit expression for P in terms of how it depends on the choice of instrumental variables as well as on the prefilter $T(q^{-1})$, it is not easy to evaluate the effects of these choices. We shall study these effects in the remainder of this section.

Optimal IV Methods   The Cramér-Rao inequality of course imposes a lower bound for P. The following lemma establishes a lower bound for P that is attained within the family of refined IV methods. (It will be shown in example 4.16 that this bound in fact equals the Cramér-Rao bound if the signals are Gaussian.)

LEMMA 4.8   Consider the matrix P given by (4.230). Let $\tilde{\varphi}(t)$ be the "noise-free" part of $\varphi(t)$, i.e., let

$$\tilde{\varphi}(t)$$

$$= \left(-\frac{B_0(q^{-1})}{A_0(q^{-1})} u(t-1) \quad \ldots \quad -\frac{B_0(q^{-1})}{A_0(q^{-1})} u(t-n) \quad u(t-1) \quad \ldots \quad u(t-m) \right)^{\mathrm{T}}.$$

$$(4.231)$$

Then

$$P - \sigma^2 \{ \overline{\mathrm{E}}[H^{-1}(q^{-1})\tilde{\varphi}(t)][H^{-1}(q^{-1})\tilde{\varphi}(t)]^{\mathrm{T}} \}^{-1} \geq 0, \qquad (4.232)$$

i.e., the left-hand side is nonnegative definite. Moreover, strict equality in (4.232) is obtained with the choice

$$\zeta(t) = H^{-1}(q^{-1})\tilde{\varphi}(t), \quad T(q^{-1}) = H^{-1}(q^{-1}). \tag{4.233}$$

Proof   Using again that $K(q^{-1}) = T(q^{-1})H(q^{-1})$ and that $\zeta(t)$ is uncorrelated with the disturbances, we have

$$\bar{E}\zeta(t)\varphi_F^T(t) = \bar{E}[\zeta(t) \cdot T(q^{-1})\tilde{\varphi}^T(t)]$$

$$= \bar{E}[\zeta(t) \sum_{i=0}^{\infty} k_i q^{-i} H^{-1}(q^{-1})\tilde{\varphi}^T(t)]$$

$$= \bar{E}\left\{\left[\sum_{i=0}^{\infty} k_i \zeta(t+i)\right] H^{-1}(q^{-1})\tilde{\varphi}^T(t)\right\}.$$

Moreover,

$$\mathrm{E}[K(q^{-1})\zeta(t) \cdot K(q^{-1})\zeta^T(t)] = \sum_{i=0}^{\infty}\sum_{j=0}^{\infty} k_i \mathrm{E}\zeta(t-i)k_j\zeta^T(t-j)$$

$$= \bar{E}\left[\sum_{i=0}^{\infty} k_i\zeta(t+i)\right]\left[\sum_{j=0}^{\infty} k_j\zeta^T(t+j)\right].$$

Thus the relation (4.232) can be rewritten as

$$\{\bar{E}[\bar{\zeta}(t) \cdot H^{-1}(q^{-1})\tilde{\varphi}^T(t)]\}^{-1}\bar{E}[\bar{\zeta}(t)\bar{\zeta}^T(t)]\{E[H^{-1}(q^{-1})\tilde{\varphi}(t) \cdot \bar{\zeta}^T(t)]\}^{-1}$$

$$- \{\bar{E}[H^{-1}(q^{-1})\tilde{\varphi}(t) \cdot H^{-1}(q^{-1})\tilde{\varphi}^T(t)]\}^{-1} \geq 0,$$

where $\bar{\zeta}(t) = \sum_{i=0}^{\infty} k_i\zeta(t+i)$. This matrix inequality is equivalent to

$$\bar{E}\begin{pmatrix} H^{-1}(q^{-1})\tilde{\varphi}(t) \\ \bar{\zeta}(t) \end{pmatrix}(H^{-1}(q^{-1})\tilde{\varphi}^T(t) \ \bar{\zeta}^T(t)) \geq 0,$$

which is obvious. Thus (4.232) is proved. It is then easy to verify that the choice (4.233) gives strict equality in (4.232). ∎

Comparison with Prediction Error Methods   The important conclusion is that the choice (4.233) of instrumental variables and prefilter gives optimal accuracy. Let us now compare this with the accuracy that will be obtained with a prediction error method.

EXAMPLE 4.16 (Comparison between the Optimal IV Method and Recursive Prediction Error Methods)   The model structure chosen for the recursive prediction error algorithm is important for this comparison. Suppose the model is

$$A(q^{-1})y(t) = B(q^{-1})u(t) + H(q^{-1})e(t), \tag{4.234}$$

where $H(q^{-1})$ is a known filter and $\{e(t)\}$ is white noise. This model is consistent with the true system description (4.210)–(4.211). The asymptotic covariance matrix for the estimates of A and B obtained by a recursive Gauss-Newton method is, according to theorem 4.5, P_{RPEM}/t with

$$P_{\text{RPEM}} = \sigma^2[\bar{\text{E}}\psi(t, \theta_0)\psi^{\text{T}}(t, \theta_0)]^{-1}. \tag{4.235}$$

In this case (4.234) we have

$$\hat{y}(t \mid \theta) = H^{-1}(q^{-1})[H(q^{-1}) - A(q^{-1})]y(t) + H^{-1}(q^{-1})B(q^{-1})u(t),$$

which means that

$$\psi(t, \theta) = H^{-1}(q^{-1})\varphi(t), \tag{4.236}$$

where $\varphi(t)$ is given by (4.220). Hence $\psi(t, \theta) = \varphi_F(t)$ with $T(q^{-1}) = H^{-1}(q^{-1})$. Comparing this to the definition of $\tilde{\varphi}(t)$ in (4.231), we see that

$$\psi(t, \theta_0) = H^{-1}(q^{-1})\varphi(t) = H^{-1}(q^{-1})\tilde{\varphi}(t) + \frac{1}{A_0(q^{-1})}\begin{pmatrix} e(t-1) \\ \vdots \\ e(t-n) \\ 0 \\ \vdots \\ 0 \end{pmatrix}. \tag{4.237}$$

Hence, from (4.235) we have

$$P_{\text{RPEM}} = \left[P_{\text{IV}}^{-1} + \begin{pmatrix} \bar{P} & 0 \\ 0 & 0 \end{pmatrix} \right]^{-1}, \tag{4.238}$$

where P_{IV} is the covariance matrix for the optimal IV method (4.233), i.e.,

$$P_{\text{IV}} = \sigma^2\{\bar{\text{E}}[H^{-1}(q^{-1})\tilde{\varphi}(t)][H^{-1}(q^{-1})\tilde{\varphi}(t)]^{\text{T}}\}^{-1} \tag{4.239a}$$

and $\bar{P}$ is the $n \times n$ positive definite matrix

$$\bar{P} = \frac{1}{\sigma^2}\text{E}\frac{1}{A_0(q^{-1})}\begin{pmatrix} e(t-1) \\ \vdots \\ e(t-n) \end{pmatrix} \cdot \frac{1}{A_0(q^{-1})}(e(t-1) \ \ldots \ e(t-n)). \tag{4.239b}$$

Consequently, the recursive prediction error method for (4.234) is strictly better than the optimal IV method, as far as the asymptotic accuracy is concerned.

The comparison is, however, somewhat unfair. The recursive prediction error method uses the knowledge of H not only to improve accuracy but also to secure unbiased estimates. The optimal IV method uses the knowledge of H only for improving the accuracy. The IV estimates will be unbiased even if H does not correspond to the true noise properties. Moreover, if the filter H in (4.234) was to be estimated by a recursive prediction error method, then the accuracy of the A and B estimates would be worse.

Let us therefore consider the model

$$y(t) = \frac{B(q^{-1})}{F(q^{-1})}u(t) + \bar{H}(q^{-1})e(t), \tag{4.240}$$

where $\bar{H}$ is a known filter and $\{e(t)\}$ is white noise. This model is consistent with the true system description (4.210)–(4.212) with

$$F_0(q^{-1}) = A_0(q^{-1}) \quad \text{and} \quad \bar{H}(q^{-1}) = \frac{1}{A_0(q^{-1})}H(q^{-1}). \tag{4.241}$$

The asymptotic covariance matrix for the recursive Gauss-Newton algorithm is still given by (4.235). This time

$$\hat{y}(t \mid \theta) = \bar{H}^{-1}(q^{-1})\frac{B(q^{-1})}{F(q^{-1})}u(t) + [1 - \bar{H}^{-1}(q^{-1})]y(t), \tag{4.242}$$

which gives [see (3.121)]

$$\psi^T(t, \theta_0) = \frac{\bar{H}^{-1}(q^{-1})}{F_0(q^{-1})}\left(-\frac{B_0(q^{-1})}{F_0(q^{-1})}u(t-1)\right.$$

$$\left. \cdots \quad -\frac{B_0(q^{-1})}{F_0(q^{-1})}u(t-n) \quad u(t-1) \quad \cdots \quad u(t-m)\right).$$

A comparison with (4.231) shows that, using (4.241), we can write

$$\psi(t, \theta_0) = [H(q^{-1})]^{-1}\tilde{\varphi}(t),$$

and hence that

$$P_{\text{RPEM}} = P_{\text{IV}}$$

in this case. The optimal IV method therefore gives an asymptotic covariance matrix that equals the Cramér-Rao lower bound for the estimation problem (4.240), provided the disturbances are Gaussian.

Finally, suppose that $\bar{H}$ in (4.240) is not known so that a model (3.103)

$$y(t) = \frac{B(q^{-1})}{F(q^{-1})} u(t) + \frac{C(q^{-1})}{D(q^{-1})} e(t) \tag{4.243}$$

is used. Then the recursive Gauss-Newton algorithm gives an asymptotic covariance matrix for the B- and F-estimates that still is given by $P_{\mathrm{RPEM}} = P_{\mathrm{IV}}$. (Further details of this are given in appendix 4.E.) $\square$

It follows from the foregoing example that the optimal IV method (4.233) is not only optimal in the IV class, but also gives the best accuracy possible for any estimation method for a reasonably posed problem (i.e., for the model (4.240)). The problem with (4.233) is, of course, that it cannot be exactly implemented, since it requires knowledge both of the system $A_0(q^{-1})$, $B_0(q^{-1})$ and of the filter $H(q^{-1})$. Some ways of including this knowledge by replacing these filters by current estimates are discussed in appendix 4.E.

4.6.4 Summary

In this section we have studied the asymptotic properties of the estimates obtained by the instrumental variable method. Most of the results have been confined to the case where the instruments are generated by a time-invariant filter from the input signal. It has been shown that consistency is assured for "most" choices of filters (theorem 4.7). We have also shown that the estimates have asymptotically normal distributions (theorem 4.8) and have given an explicit expression for the asymptotic covariance matrix (4.230). Examination of this expression shows that optimal accuracy is obtained for a certain choice of instrumental variables, viz., (4.233). The construction of these instrumental variables requires knowledge of the system dynamics and the disturbance properties. Replacing these unknown quantities with their current estimates leads naturally to algorithms of the recursive prediction error class, as explained in appendix 4.E.

4.7 User's Summary

In section 3.9 we described a general family of recursive identification methods. The present chapter has dealt with the analysis of their asymptotic properties. The analysis of most of these algorithms is technically difficult. The chapter has therefore been a long one, and at places filled with technical calculations. At least with the currently available tools, this seems unavoidable. Nevertheless, the results of the analysis are easy to express and understand. Section 4.2 contains a summary of the results; it is necessary to read that summary before continuing with the following chapters.

The basic asymptotic results are given in theorems 4.3–4.8. The significance of the theorems 4.3–4.5 is discussed in section 4.4.4. This section can be read independently of the analysis in the preceding sections.

The tools by which the results have been obtained in this chapter have, as such, no independent interest in this context. Still, let us point out that most of the convergence analysis was carried out in terms of an associated differential equation. We shall use this technique occasionally also in chapter 7. An outline of the idea was given in sections 4.3.1 and 4.3.2, which may provide sufficient insight for understanding its use. We will also (in appendix 4.C) give some details of another approach to convergence analysis, that can handle certain PLR schemes in a very nice fashion.

Finally, we stress that all analytic results in this chapter have been asymptotic. This means that they will apply when the number of processed data has become large. We do not, however, know *how* large it has to be. This caution must be kept in mind when the results of this chapter are applied to practical cases.

4.8 Bibliography

Section 4.1 Analysis for the case when the gain does not decrease to zero and noise is present is difficult. Some interesting studies of this problem are given in Bitmead and Anderson (1980a, b), Polyak and Tsypkin (1979), Kushner and Huang (1981), Macchi and Eweda (1983), and Weiss and Mitra (1979).

Section 4.3 The d.e. approach has also been treated by Kushner and Clark (1978) using a different technique. A simple version of the algorithm (4.52) is treated in detail in Ljung (1978a). Relations, although of a different character, between the recursive algorithm and the d.e. have been derived and used by Khasminskii (1966) and Nevelson and Khasminskii (1973). The martingale convergence approach has been applied to several different adaptive algorithms by Goodwin and his coworkers; see, e.g., Goodwin and Sin (1983).

Section 4.4 Section 4.4.2 is based on Ljung (1981). Theorem 4.5 is a development of a result in Ljung (1980b), which in turn was inspired by Solo (1978, 1981). Solo's technique is slightly different, though. Notice that certain results on asymptotic distribution can also be derived based on stochastic approximation results (Kushner and Huang, 1979).

Section 4.5 Section 4.5.2 is based on Ljung (1977a). Convergence of Landau's method in the noise-free case was proved in Landau (1976). See also Dugard and Landau (1980a) for a discussion of convergence of several PLRs. (Notice, however, that their proof for algorithm 2, i.e., $A = C = 1$, is incomplete. As seen from lemma 4.3, there is no single filter that relates the regression vector to the prediction error in this case. Hence theorem 4.6 cannot be applied.)

Section 4.6   Section 4.6.3 is based on Stoica and Söderström (1981b, 1983a). Extension of the results to the multivariable case is straightforward; see Stoica and Söderström (1982a, 1983b) and Söderström and Stoica (1983) for details. The adaptive implementation of optimal IV methods, as described in appendix 4.E, has been derived by Young (1976), Young and Jakeman (1979), and Jakeman and Young (1979). In these papers, however, the algorithm is derived by approximating the likelihood equations. The results given in section 4.6.3 give a theoretical justification for their approach when the model set (4.240) is used. The relation to RPEM has also been noted by Solo (1978, 1980).

5 Choice of Algorithm

5.1 Introduction

In section 3.9 we summarized the algorithms that we are treating in this book. We listed eight choices that the user has to make. Taken together, these eight choices constitute the actual choice of algorithm. In this chapter we return to the question of how to make these choices. Our discussion will be based partly on the analytical results given in chapter 4 and partly on simulation studies.

Simulations often suffer from a lack of conclusiveness: To what extend does the result depend on the chosen system? It is impossible to recommend nontrivial user choices that are universally applicable. In the present chapter we mostly consider scalar-output systems and models of fairly low order (typically, less than 10 parameters). The conclusions we draw, as summarized at the end of each section, are based on rather extensive simulation studies of such systems. When reading the chapter, it should be kept in mind that the conclusions might be modified for other types of systems.

The reason for cataloging the aforementioned eight choices is so that the user can, in a rational way, find an identification algorithm that is "good" for his application. In this context a "good" identification method is one that gives a "good" model at a low price. In order to evaluate a simulation result we thus need to measure the quality of a model. How can that be done? We can of course use the covariance matrix of the parameter estimates, but it is often more important to look into the application itself. We shall in this chapter use four scalar measures of the validity of a model that relate to its use in prediction, in description of the impulse response of a system, and in control. These measures are illustrated in the following example.

EXAMPLE 5.1 (Measures of Model Validity) Consider a scalar-output stationary system

$$y(t) = G_0(q^{-1})u(t) + H_0(q^{-1})e(t), \tag{5.1}$$

where $H_0(0) = 1$ and where $\{e(t)\}$ is white noise independent of the input $\{u(t)\}$. Use a model

$$y(t) = G(q^{-1}, \theta)u(t) + H(q^{-1}, \theta)e(t), \tag{5.2}$$

where G and H are rational functions of q^{-1}, and where $H(0, \theta) = 1$.

Then one possible measure of how well (5.2) describes (5.1) is to take the one-step prediction error variance. This gives

$$\bar{V}_1^*(\theta) = E\varepsilon^2(t, \theta)$$

$$= E\{H^{-1}(q^{-1}, \theta)[y(t) - G(q^{-1}, \theta)u(t)]\}^2$$

$$= E\{H^{-1}(q^{-1}, \theta)[G_0(q^{-1}) - G(q^{-1}, \theta)]u(t)\}^2 \tag{5.3}$$

$$+ E[H^{-1}(q^{-1}, \theta)H_0(q^{-1})e(t)]^2.$$

Another possibility is to take the sum of squared differences between the impulse responses of the model and the true one. This means that

$$\bar{V}_2^*(\theta) = \frac{1}{2\pi i}\oint [G(z, \theta) - G_0(z)][G(z^{-1}, \theta) - G_0(z^{-1})]\frac{dz}{z}, \tag{5.4}$$

where the path of integration is the unit circle.

Similarly, a measure of how well $H(q^{-1}, \theta)$ approximates $H_0(q^{-1})$ is

$$\bar{V}_3^*(\theta) = \frac{1}{2\pi i}\oint [H(z, \theta) - H_0(z)][H(z^{-1}, \theta) - H_0(z^{-1})]\frac{dz}{z}. \tag{5.5}$$

A further possibility can be constructed as follows. Use the model (5.2) to design a minimum variance controller (Åström, 1970), i.e., one in which the stationary output variance is minimized. If we take

$$G(q^{-1}, \theta) = g_1(\theta)q^{-1} + g_2(\theta)q^{-2} + \cdots$$

with $g_1(\theta) \neq 0$, then the controller becomes

$$u(t) = -\frac{H(q^{-1}, \theta) - 1}{G(q^{-1}, \theta)}y(t).$$

Assume that this controller is applied to the true system (5.2). Then the output variance is

$$\bar{V}_4^*(\theta) = Ey^2(t) = E\left[\frac{H_0(q^{-1})G(q^{-1}, \theta)}{G(q^{-1}, \theta) - G_0(q^{-1}) + G_0(q^{-1})H(q^{-1}, \theta)}e(t)\right]^2, \tag{5.6}$$

which is then taken as the measure of how well the model (5.2) describes the true system.

For the foregoing validity measures the following inequalities apply:

$$\bar{V}_1^*(\theta) \geq \sigma^2, \quad \bar{V}_2^*(\theta) \geq 0, \quad \bar{V}_3^*(\theta) \geq 0, \quad \bar{V}_4^*(\theta) \geq \sigma^2, \tag{5.7}$$

where $\sigma^2 = \mathrm{E}e^2(t)$. Moreover, if the model (5.2) coincides with the system then all the inequalities in (5.7) become equalities. This means that the lower bounds are obtained if the model gives a perfect fit to the system. (The converse is, however, not always true). Note that of these measures only $\bar{V}_1^*(\theta)$ will depend on the input. The criteria $\bar{V}_1^*(\theta)$ and $\bar{V}_4^*(\theta)$ have clear physical meanings.

An important advantage of these measures is that models within different model sets can be conveniently compared. For evaluation of any of these measures it is necessary to know the true system or to make prior assumptions about the properties of the data $\{z(t)\}$. □

As several of the examples in this chapter include simulations it is appropriate to give some general comments on how these were performed.

The disturbance $e(t)$ was simulated using a random number generator giving Gaussian distributed noise. The input $u(t)$ was in some cases simulated as a pseudorandom binary sequence (PRBS). In other cases it was taken as white noise, i.e., a sequence of independent random variables. In all cases the mean values of $e(t)$ and $u(t)$ were kept equal to zero. The variances were selected to obtain a prescribed value of the signal-to-noise ratio

$$S/N = \{\mathrm{E}[G_0(q^{-1})u(t)]^2\}/\{\mathrm{E}[H_0(q^{-1})e(t)]^2\}.$$

Here $G_0(q^{-1})$ and $H_0(q^{-1})$ are the true transfer functions from u and e respectively to y as in (5.1).

The symbol N also denotes generally the number of data generated.

In many examples several runs or realizations were used. For every run a new initial value for the random number generator was used in order to give independent realizations. When results of several runs are presented in a table the arithmetic means and standard deviations evaluated over the number of runs indicated are shown. Plots always represent one particular run.

This chapter is organized as follows. The choice of model set is discussed in general terms in section 5.2. Aspects of how to choose a model set within the SISO family (3.104) are discussed in section 5.3. The choice of input signal is treated in section 5.4, and the choice of criterion function in section 5.5. The step size or gain sequence in the algorithm is studied in section 5.6, and the search direction in section 5.7. Effects of initial conditions are discussed in section 5.8. The choice between RPEM and

PLR (see section 3.7) is illustrated in section 5.9, various IV methods are discussed in section 5.10, and, finally, the choice between prediction errors and residuals in the gradient vector is treated in section 5.11.

5.2 Choice of Model Set

5.2.1 General Considerations

The selection of a suitable model set is no doubt the single most important choice to make for any identification problem. At the same time it is difficult to give general recommendations. The choice is application-dependent, and it may also be influenced by factors such as the availability of particular computer programs.

We list here four factors that should be taken into account when selecting the model set.

1.Flexibility The model set should be capable of describing most of the different system dynamics that can be expected in the application in question. Both the number of parameters and the way they enter the model are important.

2. Parsimony From the important expression (4.136), we have, in the notation of (5.3), that

$$E\bar{V}_1^*(\hat{\theta}(N)) = \bar{V}_1^*(\theta_0)(1 + (\dim \theta)/N), \tag{5.8}$$

once the model set is large enough to contain the true system. There is consequently a strict penalty associated with the use of models with many parameters. The model set should be parsimonious. Results similar to (5.8) apply also to other validity measures (Gustavsson et al., 1977).

3. Algorithm Complexity The model set determines how $\hat{y}(t \mid \theta)$ and $\psi(t, \theta)$ are computed. The complexity of these calculations depend on the dimension of θ as well as on the structure of the model.

4. Properties of the Criterion Function The asymptotic properties of RPEMs depend on the criterion function $V(\theta) = \bar{E}l(t, \theta, \varepsilon(t, \theta))$, as we found in section 4.4. The existence of nonunique global minima as well as that of local minima that are not global is affected by the model parametrization.

We shall illustrate how the foregoing factors can be used as a guide in two special cases: state-space models and, in section 5.3, black box SISO models.

5.2.2 Choice between a General State-Space Model and an Innovations Model

In section 3.8 we discussed the use of state-space models for RPEMs. We pointed out in section 3.8.1 that we could choose between a general model (3.133)

$$x(t + 1) = F(\theta)x(t) + G(\theta)u(t) + v(t),$$
$$y(t) = H(\theta)x(t) + e(t),$$

(5.9)

where $\{v(t)\}$ and $\{e(t)\}$ are white noise sequences with covariance matrices

$$R_1(\theta) = \mathrm{E}v(t)v^{\mathrm{T}}(t), \quad R_2(\theta) = \mathrm{E}e(t)e^{\mathrm{T}}(t), \quad R_{12}(\theta) = \mathrm{E}v(t)e^{\mathrm{T}}(t),$$

and an innovations model (3.139),

$$x(t + 1) = F(\theta)x(t) + G(\theta)u(t) + K(\theta)v(t),$$
$$y(t) = H(\theta)x(t) + v(t).$$

(5.10)

The difference between these two models is that the steady-state Kalman gain is explicitly parametrized in (5.10) while it is computed indirectly from $R_1(\theta)$, $R_2(\theta)$, $R_{12}(\theta)$, $F(\theta)$, and $H(\theta)$ in (5.9). Also, (5.9) can use a time-varying predictor in the transient phase while (5.10) uses the steady-state one all the time.

Both models offer the same *flexibility* for the steady-state predictor. The model (5.9) has an advantage in that it can also handle time-varying predictors. This may be important in applications where the data record is relatively short. If some prior knowledge is gained from physical modeling it would as a rule be in terms of the model (5.9) (or its continuous-time counterpart) rather than in terms of (5.10). The reason is that we might know that one state variable is the derivative of another (e.g., angular velocity and angle). Then no process noise affects this state variable directly, and the corresponding entries in R_1 would be zero. Such structural knowledge is more difficult to incorporate in (5.10).

To consider *parsimony*, we note that the model (5.10) contains *pn* parameters (p = number of outputs, n = number of states) for $K(\theta)$, while $n(n + 1)/2 + np + p(p + 1)/2$ parameters would be required to fully parametrize the noise covariance matrices in (5.9). This would in any case be too many, since only $np + p(p + 1)/2$ (which is equal to the number of independent entries of the steady-state Kalman gain and the prediction error covariance matrix) of them are identifiable. However, with prior structural knowledge as discussed above, the number of parameters in (5.9) may be reduced, while this is less easy to do in (5.10).

The model (5.10) gives a considerably less computationally complex algorithm than (5.9), because the calculations leading to K are avoided in (5.10). This is even more true in an RPEM where the gradient of K with respect to θ is required. The use of (5.9) then requires solution of (3.B.3) to find this gradient, while $dK(\theta)/d\theta$ is immediate in (5.10). Hence from the viewpoint of *algorithmic complexity* (5.10) is to be preferred.

Summary   If the time variation of the predictor in the transient phase is important *or* there is prior structural knowledge about the noise covariance matrices that considerably reduces the number of parameters below pn, then use of (5.9) should be considered. In all other cases use the innovations model (5.10).

5.3 Choice of Model Set within the General Family of SISO Models

In section 3.7 a general family of model sets was given for black box modeling of a SISO system. This family is given by

$$A(q^{-1})y(t) = \frac{B(q^{-1})}{F(q^{-1})}u(t) + \frac{C(q^{-1})}{D(q^{-1})}e(t) \tag{5.11}$$

[see (3.104)]. The choice of model set within this family will now be discussed in terms of the factors listed in section 5.2.1.

5.3.1 Flexibility versus Parsimony

Adequate flexibility can be obtained by either using several of the polynomials in the general structure (5.11) or by taking high degrees of the polynomials included. That means that the absence of some polynomials can, at least to some extent, be compensated for if the polynomials included are given high orders.

With a total number of parameters given, the best flexibility is usually obtained if they are spread out to some different polynomials. This is easy to realize. Consider, e.g., a model with $A(z) \equiv C(z) \equiv D(z) \equiv F(z) \equiv 1$; the model is then $y(t) = B(q^{-1})u(t) + e(t)$, but such a model with a limited number of parameters (a limited degree of $B(z)$) will not be good for systems with a slow impulse response. Similarly, stochastic disturbances can in general be described with fewer parameters as an ARMA process (take $B(z) \equiv 0$, $D(z) \equiv F(z) \equiv 1$) compared to a pure AR description (with $B(z) \equiv 0$, $C(z) \equiv D(z) \equiv F(z) \equiv 1$).

The tradeoff between flexibility and parsimony should thus best be met by using three or more polynomials in the structure (5.11).

If the system is unstable (assuming it is stabilized during the experiment with an appropriate feedback) it is important to note that $F(q^{-1})$ in (5.11) is constrained to be asymptotically stable. Thus if it is known or expected that the system is unstable, then the polynomial $A(q^{-1})$ must be included. It can also be said that it seems to be common practice to use $A(q^{-1})$ or $F(q^{-1})$, but not both, in the model.

The choice of model set within the general structure (5.11) is further illustrated in the following example.

EXAMPLE 5.2 (Achievable Accuracy within Various Model Sets) The purpose of this example is to investigate the role of the model set when it is not large enough to allow an exact modeling of the true system.

A system of the form (5.11) is assumed. The degrees of the polynomials were taken as $n_a = 0$, $n_b = n_f = 7$, $n_c = n_d = 4$. The zeros of the polynomials are given in table 5.1. The static gain $B_0(1)/F_0(1)$ was set to 1.

The impulse responses of $B_0(q^{-1})/F_0(q^{-1})$ and $C_0(q^{-1})/D_0(q^{-1})$ are shown in figures 5.1 and 5.2, respectively. The disturbance sequence $\{e(t)\}$ was assumed to be white noise with zero mean. Two different inputs were considered:

I. $\{u(t)\}$ is filtered white noise so that its covariance function becomes

$$r_u(t) = (1 - 0.25t)r_u(0) \quad t = 1, 2, 3$$

$$r_u(t) = 0 \qquad\qquad t \geq 4$$

This is a good approximation of a PRBS with the clock period equal to 4 sampling intervals.

II. $u(t)$ is white noise.

The variance $r_u(0)$ was chosen so that S/N became equal to 9.95 for case I and 0.68 for case II. A number of different model sets were con-

Table 5.1
Zeros of the polynomials for example 5.2.

Polynomial	$B_0(z)$	$F_0(z)$	$C_0(z)$	$D_0(z)$
Zeros	-0.5	0.4	0.5	0.75
	0.95	0.8	0.7	0.92
	$0.6 \pm 0.2i$	0.9	$0.75 \pm 0.1i$	$0.8 \pm 0.2i$
	$0.7 \pm 0.1i$	$0.8 \pm 0.15i$		
		$0.7 \pm 0.25i$		

sidered, each with $\dim \theta = 8$. For each model set the limit θ^* of RPEM was determined. According to theorem 4.4 θ^* minimizes the validity measure $\bar{V}_1^*(\theta)$ in (5.3). In this example θ^* was determined by numerical minimization of this function, rather than by simulation.

The resulting models were further compared by evaluating all the four validity criteria $\bar{V}_1^*(\theta) - \bar{V}_4^*(\theta)$, (5.3)–(5.6). The numerical results are given in tables 5.2a, b for the two inputs in question.

Some comments to the results are in order. The given values must not be compared to the last digit. Since the minimization of $\bar{V}_1^*(\theta)$ has been performed with a numerical search routine the given values have a limited accuracy.

Another comment is that all optimizations started with all parameter values equal to zero. These initial values caused no problems except for the model set given by $n_a = 0, n_b = 4, n_f = 4, n_c = 0, n_d = 0$ where in case I the algorithm was stuck in a local minimum. Several initial values were tried for this case before the global minimum could be found. This indicates that for a set with $A \equiv C \equiv D \equiv 1$ false local minima may cause convergence problems in practice. The model set with $A \equiv C \equiv D \equiv 1$ has also another interesting property. It corresponds to an output error method. It gives the smallest value of the criterion $\bar{V}_2^*$, but the second largest one of $\bar{V}_1^*$. The reason for this is not difficult to understand. The parameters of this model are essentially determined to minimize $\bar{V}_2^*$. (This is exactly so in case II when $u(t)$ is white noise). On the other hand the criterion $\bar{V}_1^*$ becomes always at least as large as $E[C(q^{-1})/D(q^{-1})e(t)]^2$, which is 2.7707 for this system. The assessment of this model set must thus be strongly influenced by the purpose of the modelling.

The following conclusions can be drawn from this example:

• Many model sets show about the same performance. The difference in performance is not very significant, with a few exceptions.

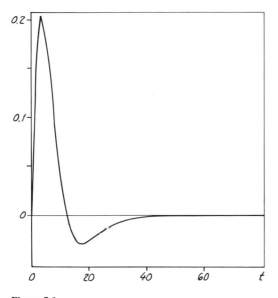

Figure 5.1
Impulse response of $B_0(q^{-1})/F_0(q^{-1})$ for example 5.2.

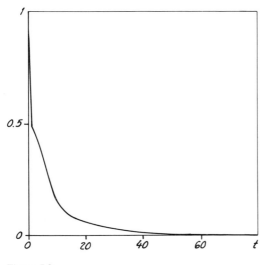

Figure 5.2
Impulse response of $C_0(q^{-1})/D_0(q^{-1})$ for example 5.2.

Table 5.2a
Numerical results for example 5.2, case I. The missing figures for the measure $\bar{V}_4^*$ means that models obtained by minimizing $\bar{V}_1^*(\theta)$ did not lead to stable closed-loop systems when a minimum variance controller was used as feedback.

Model set	Polynomial degrees					$\bar{V}_1^*$	$\bar{V}_2^*$	$\bar{V}_3^*$	$\bar{V}_4^*$
	n_a	n_b	n_f	n_c	n_d				
$y = Bu + e$		8				4.3770	0.7017	1.7707	...
$Ay = Bu + e$	4	4				1.1130	0.0593	0.4069	...
$Ay = Bu + Ce$	3	3		2		1.0943	0.0381	0.4387	1.0841
$Ay = Bu + \dfrac{C}{D}e$	2	2		2	2	1.0213	0.0329	0.0437	1.0151
,,	3	3		1	1	1.0821	0.0306	0.4551	1.4534
$Ay = Bu + \dfrac{1}{D}e$	3	3			2	1.0724	0.0361	0.3818	1.0561
$y = \dfrac{B}{F}u + e$		4	4			3.1954	0.0273	1.7707	...
$y = \dfrac{B}{F}u + \dfrac{C}{D}e$		2	2	2	2	1.0163	0.0333	0.0511	1.0094
,,		3	3	1	1	1.0206	0.0287	0.0227	1.0646
$y = \dfrac{B}{F}u + \dfrac{1}{D}e$		3	3		2	1.0203	0.0310	0.0399	1.0092
$Ay = \dfrac{B}{F}u + \dfrac{C}{D}e$	2	2	2	1	1	1.0858	0.1400	0.4247	1.0326
$Ay = \dfrac{B}{F}u + Ce$	2	2	2	2		1.0644	0.0406	0.3334	1.0653
,,	3	2	2	1		1.0953	0.0471	0.3992	1.0915
$Ay = \dfrac{B}{F}u + \dfrac{1}{D}e$	2	2	2		2	1.2469	0.2052	0.2827	...
,,	2	2	3		1	1.2317	0.1894	0.2225	...

- There is no advantage in using all the five polynomials. On the contrary, most of the other model sets give better performance than this "full structure."

- The purpose of the model, e.g., the choice of criterion $\bar{V}^*$, will strongly influence what model set to select. The model set $y = Bu + e$ is generally inferior to the other ones, though. If it is essential to estimate the dynamics of the system (e.g., if the influence of the input on the output as measured by $\bar{V}_2^*$ should be small) then the model set $y = B/Fu + e$ is the best choice. However, if good prediction (as measured by $\bar{V}_1^*$) is the most important issue, then the model sets $y = B/Fu + C/De$, $y = B/Fu + 1/De$, and to

Table 5.2b
Numerical results for example 5.2, case II.

Model set	Polynomial degrees					$\bar{V}_1^*$	$\bar{V}_2^*$	$\bar{V}_3^*$	$\bar{V}_4^*$
	n_a	n_b	n_f	n_c	n_d				
$y = Bu + e$		8				2.9343	0.1636	1.7707	...
$Ay = Bu + e$	4	4				1.0312	0.2781	0.0440	...
$Ay = Bu + Ce$	3	3		2		1.0140	0.1381	0.1168	1.0285
$Ay = Bu + \dfrac{C}{D}e$	2	2		2	2	1.0146	0.1393	0.1145	1.0387
,,	3	3		1	1	1.0263	0.2517	0.0911	1.0451
$Ay = Bu + \dfrac{1}{D}e$	3	3			2	1.0225	0.1961	0.0852	1.0748
$y = \dfrac{B}{F}u + e$		4	4			2.7981	0.0274	1.7707	2.7689
$y = \dfrac{B}{F}u + \dfrac{C}{D}e$		2	2	2	2	1.0036	0.0321	0.0096	1.0059
,,		3	3	1	1	1.0043	0.0319	0.0137	1.0057
$y = \dfrac{B}{F}u + \dfrac{1}{D}e$		3	3		2	1.0061	0.0316	0.0148	1.0142
$Ay = \dfrac{B}{F}u + \dfrac{C}{D}e$	2	2	2	1	1	1.0209	0.2523	0.0319	1.0880
$Ay = \dfrac{B}{F}u + Ce$	2	2	2	2		1.0225	0.2549	0.0394	1.0749
,,	3	2	2	1		1.0185	0.2339	0.0250	1.0640
$Ay = \dfrac{B}{F}u + \dfrac{1}{D}e$	2	2	2		2	1.0241	0.3305	0.0383	1.0504
,,	2	2	3		1	1.0314	0.3622	0.0484	1.0576

some extent $Ay = Bu + C/De$, are the best. However, these results must be interpreted with care. It is, e.g., clear that the experimental conditions can influence which one of two competitive model sets gives the best performance. □

The signal-to-noise ratio of the application in question also has an important effect on flexibility. For an application with very low noise level, only the dynamic part in (5.11) is of interest. Then we can take $C = D = 1$ and either $F = 1$ or $A = 1$. Since $F = 1$ (which gives the LS method for estimation of A and B) has several advantages, as we shall see, this choice is usually the best one for applications with low noise level.

5.3.2   Algorithm Complexity

The complexity of the equations for updating the estimate $\hat{\theta}(t)$ and the matrix $R(t)$ depends only on the *total* number of parameters; both the number of computations and their complexity are independent of the way the parameters enter into any particular model. To find $\hat{y}(t \mid \theta)$ and $\psi(t, \theta)$ the data must be filtered [see (3.105) and (3.119)]. The number of computations involved in this filtering is proportional to

$$n_c + n_d + n_f.$$

Filtering requires that stability tests be included so that the step length, whenever necessary, can be reduced, and an asymptotically stable filter can be maintained. For the model (5.11) the polynomials $C(z)$ and $F(z)$ must be tested for their stability properties (see section 6.6).

The algorithm becomes especially simple when $C(z) \equiv D(z) \equiv F(z) \equiv 1$. This is nothing but the standard LS method. Then *no* filtering is necessary, since the vector $\psi(t, \theta)$ contains only delayed outputs and inputs.

5.3.3   Properties of the Criterion Function

The properties of the criterion function

$$\bar{V}(\theta) = \bar{\mathrm{E}}\varepsilon^2(t, \theta) = \bar{\mathrm{E}}\left\{\frac{D(q^{-1})}{C(q^{-1})}\left[A(q^{-1})y(t) - \frac{B(q^{-1})}{F(q^{-1})}u(t)\right]\right\}^2 \tag{5.12}$$

will certainly depend on the model set. We shall discuss the existence of local minima and the existence of "valleys" due to overparametrization. Analysis of local minima is important, but harder to perform than a study of the existence of multiple global minima.

Local Minima   Some partial results are known concerning the existence of local minima. Assume that the input is persistently exciting and that the system is operating in open loop, subject to

$$A_0(q^{-1})y(t) = \frac{B_0(q^{-1})}{F_0(q^{-1})}u(t) + \frac{C_0(q^{-1})}{D_0(q^{-1})}e(t). \tag{5.13}$$

Assume further that the polynomials of (5.13) do not have higher degrees than the corresponding polynomials in (5.11). This means that the true system belongs to the model set. The following results are known ("false local minimum" = nonglobal local minimum):

- If $B \equiv 0$ and $D \equiv F \equiv 1$ (i.e., ARMA models), then there are no false local minima. (Åström and Söderström, 1974).

- If $C \equiv D \equiv F \equiv 1$, then there are no false local minima. This is trivial since the criterion function in this case is quadratic in θ.

- If $C \equiv F \equiv 1$, then there are no false local minima if the signal-to-noise ratio is large enough. On the other hand, if it is very small then there are false local minima (Söderström, 1974).

- If $A \equiv 1$, then there are no false local minima if $n_f = 1$. When $n_f > 1$ false local minima can exist in some cases (Söderström, 1973a, 1975).

- If $A \equiv C \equiv D \equiv 1$, then there are no false minima if the input is white noise. For other inputs, however, false local minima can exist (Söderström, 1975; Stearns, 1980).

Overparametrization of the Model Set   We showed in example 4.10 a case where the second derivative matrix $\bar{V}''(\theta^*)$ became singular at the convergence point as a result of using too-high model orders. Such overparametrization leads to problems in the algorithm since the matrix $R(t)$ in (3.158a) will be almost singular. Let us now discuss in more general terms when overparametrization can lead to a singular $\bar{V}''(\theta^*)$.

From the analysis in section 4.4.4 it followed that $\bar{V}''(\theta^*)$ is singular precisely when the prediction $\hat{y}(t \mid \theta)$ is unaffected by changes in certain parameter combinations. For the model (5.11), which also can be written

$$y(t) = \frac{B(q^{-1})}{A(q^{-1})F(q^{-1})}u(t) + \frac{C(q^{-1})}{A(q^{-1})D(q^{-1})}e(t),$$

this is the case when a factor can be cancelled in one of the transfer functions $B/(AF)$ or $C/(AD)$ at the convergence point θ^*. It is easy to see that this in turn happens when any of the following conditions hold:

- There is a factor common to all of A^*, B^*, and C^*,   (5.14a)

- B^* and F^* have a common factor,   (5.14b)

- C^* and D^* have a common factor,   (5.14c)

where the starred polynomials correspond to θ^*. Therefore when the model polynomials have higher degrees than the minimal description at the convergence point, overparametrization may result. Notice, however, that this also depends on the structure of the model. For example, when $F^* = C^* = 1$, neither of the conditions (5.14) can hold.

Independence of Dynamic and Noise Models It was pointed out in example 4.7 that when $A = 1$ and the true transfer function B_0/F_0 can be described within the model set, then the estimation of B/F and of C/D are asymptotically decoupled (provided u and e are independent). This may be a useful advantage when it is more important to have a good model of the dynamic part of a system than of the noise.

5.3.4 Choice of Model Order

The choice of the order of the model (5.11) is a nontrivial problem, that requires a careful tradeoff between good description of the data and model complexity. Most methods of model order selection are developed for the off-line situation. (See Söderström (1977) for a discussion of some different methods.) The basic approach is to compare the performance of models of different orders and test if the higher-order model is worthwhile. For recursive algorithms in on-line applications this would in general require parallel identification of several models.

It should be noted, though, that for the LS method (where φ is a known function of the data) it is easy to afterwards compute models with lower order. This can be seen as follows. The LS estimate is equal to

$$\hat{\theta}(t) = \bar{R}^{-1}(t) \sum_{k=1}^{N} \beta(t, k)\varphi(k)y(k), \tag{5.15a}$$

$$\bar{R}(t) = \sum_{k=1}^{t} \beta(t, k)\varphi(k)\varphi^{\mathrm{T}}(k) \tag{5.15b}$$

(see example 2.10). To derive the estimate for a lower-order model we can write the original model as

$$y(t) = \varphi^{\mathrm{T}}(t)\theta + e(t) = [\varphi_1^{\mathrm{T}}(t) \quad \varphi_2^{\mathrm{T}}(t)]\binom{\theta_1}{\theta_2} + e(t), \tag{5.16}$$

and let the lower-order model be

$$y(t) = \varphi_1^{\mathrm{T}}(t)\theta_1 + e(t). \tag{5.17}$$

Typically $\varphi_2(t)$ then contains u and y with additional delays as compared to $\varphi_1(t)$. We also partition $\bar{R}(t)$ correspondingly, viz.,

$$\bar{R}(t) = \begin{pmatrix} \bar{R}_{11}(t) & \bar{R}_{12}(t) \\ \bar{R}_{12}^{\mathrm{T}}(t) & \bar{R}_{22}(t) \end{pmatrix}.$$

The LS estimate $\hat{\theta}_1^*$ for the lower-order model can then be simply derived:

$$\hat{\theta}_1^*(t) = \bar{R}_{11}^{-1}(t) \sum_{k=1}^{t} \beta(t, k)\varphi_1(t)y(t)$$

$$= \bar{R}_{11}^{-1}(t)[\bar{R}_{11}(t)\hat{\theta}_1(t) + \bar{R}_{12}(t)\hat{\theta}_2(t)] \qquad (5.18)$$

$$= \hat{\theta}_1(t) + \bar{R}_{11}^{-1}(t)\bar{R}_{12}(t)\hat{\theta}_2(t).$$

5.3.5 Summary

The equation error model set

$$A(q^{-1})y(t) = B(q^{-1})u(t) + e(t)$$

(the "LS model") has all advantages except parsimony. It is a good first choice, especially if the noise level is low. A rather high model order might be required when noise is present.

The output error model set

$$y(t) = \frac{B(q^{-1})}{F(q^{-1})}u(t) + e(t)$$

has the advantage that the estimation of the transfer function B/F is independent of the noise properties, as long as the noise is independent of the input.

If more elaborate models are needed one can try either

$$A(q^{-1})y(t) = B(q^{-1})u(t) + C(q^{-1})e(t)$$

or

$$y(t) = \frac{B(q^{-1})}{F(q^{-1})}u(t) + \frac{C(q^{-1})}{D(q^{-1})}e(t),$$

or possibly variants with modified noise models. The first of these is well-tested. It has been used in many applications. The second one has the advantage that the order chosen for $C(q^{-1})/D(q^{-1})$ does not influence the consistency properties for the transfer function estimates. Comparisons of the two model sets have not often been carried out; there is no general result (theoretical or empirical) that clearly favors either model set.

5.4 Choice of Experimental Conditions

The design of an identification experiment involves a number of issues such as choice of input signal, sampling rates, presampling filters, signals to be measured, etc. We shall in this section give some remarks only on the choice of input signal.

The experimental conditions affect the covariance matrix of the estimates. Typically, for a RPEM, the inverse of the asymptotic covariance matrix for a single output system is given by

$$P^{-1} = \kappa^{-1} \mathrm{E} \psi(t, \theta_0) \psi^{\mathrm{T}}(t, \theta_0), \tag{5.19}$$

where κ is some scaling factor [see theorem 4.5 and (5.23b)]. We saw in example 4.8 that with a bad choice of input the parameters of the model may not be identifiable, so that the matrix (5.19) is singular.

In general terms, we may say that the objective of experiment design is to choose an input that enhances interesting parameters and parameter combinations. This means that $\psi(t, \theta) = (d\hat{y}(t \mid \theta)/d\theta)^{\mathrm{T}}$ should be large when the gradient is evaluated with respect to these parameter combinations. There is a rich literature on such experiment design; see, e.g., Mehra (1976, 1981), Zarrop (1979), Gustavsson et al. (1977, 1981), and Goodwin and Payne (1977).

Now, recursive on-line identification is often used during normal plant operation rather than during specifically designed experiments. It is still important, however, to understand how the input choice may affect (5.19). The following example shows how feedback effects may make (5.19) singular.

EXAMPLE 5.3. (Effect of Experimental Conditions on Identifiability)   We take the model set

$$y(t) + ay(t - 1) = bu(t - 1) + e(t),$$

and assume that the true system satisfies

$$y(t) + a_0 y(t - 1) = b_0 u(t - 1) + e(t),$$

where $\{e(t)\}$ is white noise. If identification is performed using the LS method, (5.19) becomes

$$P^{-1} = \kappa^{-1} \mathrm{E} \begin{pmatrix} -y(t) \\ u(t) \end{pmatrix} (-y(t) \ \ u(t)).$$

This matrix is singular precisely when there is a static linear relation between $y(t)$ and $u(t)$. In particular this means that the system is not identifiable if the input is determined as a constant feedback from the output, i.e., if the experimental condition is

$\mathscr{X}_1$ $u(t) = -ky(t)$.

Identifiability can be obtained by modifying the experimental condition to either a feedback law of higher order than in $\mathscr{X}_1$, e.g., by using the condition

$\mathscr{X}_2$ $u(t) = -ky(t) - ly(t-1),$ $l \neq 0,$

or to include also a persistently exciting external signal (e.g., a varying set-point) $v(t)$, to give the condition

$\mathscr{X}_3$ $u(t) = -ky(t) + v(t)$.

It is assumed in this last case that the signal $v(t)$ is uncorrelated with the noise. □

A survey of how an experimental condition with feedback influences the identifiability properties is given by Gustavsson et al. (1977, 1981).

In general terms, it can be said that a persistently exciting signal (see section 3.2) applied to a system in open loop operation will give identifiability. There is no need to use special inputs. If the input is generated in a digital computer it is often convenient to let it be a pseudorandom binary sequence (PRBS), wherein the input can take only two values. It shifts between these values according to certain rules. The minimal time between two shifts, the "clock period," is often chosen as a multiple of the sampling interval. In such a way there will be some long pulses and the slow modes of the system will be well excited. The static properties can then be reasonably well estimated. By varying the clock period, one obtains inputs with different spectral properties.

When the system is too complex to be described within the chosen model set, it is not interesting to discuss identifiability. If the parameter converges to a vector θ^*, then the predictor $\hat{y}(t \mid \theta^*)$ will be the best in the model set for *the experimental condition used under the identification experiment*. It is thus a good practice to choose an input for the identification experiment that as far as possible is similar to inputs to be used for the system at later occasions.

Summary

The design of *optimal* identification experiments is seldom interesting when on-line identification is applied.

If the input is determined as a time-invariant feedback of low order, identifiability of the system can be lost.

The input signal under the identification experiment should if possible be chosen similar to inputs to be used for later control of the system.

The input should excite all modes of the system (i.e., be persistently exciting of a sufficiently high order). PRBSs satisfy this requirement, and can be generated very easily with a digital computer.

5.5 Choice of Criterion Function

In section 3.5 we studied recursive algorithms for minimizing

$$V(\theta) = \sum_{t=1}^{N} l(t, \theta, \varepsilon(t, \theta)). \tag{5.20}$$

The criterion function $l(t, \theta, \varepsilon(t, \theta))$ is at the user's disposal and can be treated as a design variable. It was mentioned in section 3.3 that, for an off-line algorithm minimizing the function $V(\theta)$ in (5.20), optimal accuracy is achieved and the Cramér-Rao lower bound is obtained if the criterion function is chosen as the maximum likelihood one, i.e., if

$$l(t, \theta, \varepsilon(t, \theta)) = -\log \bar{f}(t, \theta, \varepsilon(t, \theta)), \tag{5.21}$$

where $f(\cdot, \cdot, \cdot)$ is the probability density function of the prediction errors. Theorem 4.5 showed that the same accuracy is obtained asymptotically for RPE algorithms if the stochastic Gauss-Newton direction is used and the gain sequence satisfies, at least asymptotically, $\gamma(t) = 1/t$. The expression (5.21) shows in particular that for disturbances with Gaussian distributions it is optimal to let $l(\cdot, \cdot, \cdot)$ be a quadratic function of ε.

In practice it is necessary to be concerned with abnormal data. Abnormality can arise for many reasons, sensor failures being one example. It is clear that some single but large measurements ("outliers") can have a large influence on the identification of a model. If such defects are expected in the data, one should modify the algorithm to make it more robust.

One simple and common way to robustify, at least in process control applications, is to filter all data and to make realiability checks on the data before processing them. However, filtering and checking can also be performed on-line with recursive methods.

Another approach is to choose the criterion function $l(t, \theta, \varepsilon(t, \theta))$ such that the algorithm becomes robust. Assume, e.g., that the prediction errors, excluding some occasions with outliers in the data, are Gaussian. Then (5.21) suggests that $l(\cdot, \cdot, \cdot)$ should be quadratic in ε. However, due to the existence of some outliers the probability for large values of $|\varepsilon(t)|$ is no longer Gaussian. Then according to (5.21) it is better to let $l(\cdot, \cdot, \cdot)$ grow slower than quadratic with $|\varepsilon(t)|$. This means that we base the choice of criterion function on a distribution that gives higher probability for large prediction errors than a Gaussian distribution. This approach for modifying estimation schemes to be less sensitive to large errors in the data has been extensively studied for the static case by Huber (1973), who calls it "robust regression." Applications of the idea to dynamic systems have been discussed by Ljung (1978c) for the off-line case and by Polyak and Tsypkin (1979, 1980) for the recursive case.

Let us now specialize the discussion to scalar-output systems and consider the case when the criterion $l(t, \theta, \varepsilon)$ can be written as a function of the prediction error ε only. This is the far most common case in practice. Then the Gauss-Newton algorithm is given by

$$\hat{\theta}(t) = \hat{\theta}(t - 1) + \gamma(t)R^{-1}(t)\psi(t)\, l_\varepsilon^{\mathrm{T}}(\varepsilon(t)), \tag{5.22a}$$

$$R(t) = R(t - 1) + \gamma(t)[\psi(t)\, l_{\varepsilon\varepsilon}(\varepsilon(t))\psi^{\mathrm{T}}(t) - R(t - 1)] \tag{5.22b}$$

[see (3.72′) and (3.73)]. The quality of the estimates produced by (5.22) will of course depend on the function $l(\varepsilon)$. Using theorem 4.5, we can determine the asymptotic covariance matrix. It is given by

$$P = \kappa(l)[\mathrm{E}\psi(t, \theta_0)\psi^{\mathrm{T}}(t, \theta_0)]^{-1}. \tag{5.23a}$$

This covariance matrix depends on the criterion function *only* through the scalar

$$\kappa(l) = \frac{\mathrm{E}[l_\varepsilon(e(t))]^2}{[\mathrm{E}l_{\varepsilon\varepsilon}(e(t))]^2}. \tag{5.23b}$$

The choice of $l(\cdot)$ should thus be such that this scalar is as small as possible. In view of what we said about the maximum likelihood estimate above, it is clear that the best choice is

$$l(\varepsilon) = -\log \bar{f}(\varepsilon),$$

where $\bar{f}$ is the probability density function of the true prediction errors.

In practice one typically chooses $l(\varepsilon)$ to be a function that is quadratic for small ε but increases more slowly for large ε. This means that the function must contain a parameter, say α, that determines what a "large" prediction error is. The choice of α is a tradeoff, based on the variance of the prediction errors, perhaps recursively determined. Huber (1973) and Polyak and Tsypkin (1979, 1980) contain several examples of such criterion functions. Let us illustrate the behavior for some typical choices of $l(\varepsilon)$.

EXAMPLE 5.4 (Some Criterion Functions)   We simulated the first-order system

$$y(t) - 0.8y(t-1) = 1.0u(t-1) + e(t),$$

where $u(t)$ and $e(t)$ are white noises of zero mean and unit variance, so $S/N = 1$. The system was identified within the model set

$$y(t) + ay(t-1) = bu(t-1) + e(t)$$

using the following methods:

- $\mathscr{I}_1$, given by (5.22) with $l_1(\varepsilon) = \varepsilon^2/2$.

- $\mathscr{I}_2$, given by (5.22) with

$$l_2(\varepsilon) = \begin{cases} -\varepsilon\alpha + \alpha^2/2, & \varepsilon < -\alpha \\ \varepsilon^2/2, & |\varepsilon| \le \alpha \\ \varepsilon\alpha - \alpha^2/2, & \varepsilon > \alpha \end{cases}.$$

Here we have taken $\alpha = 2.1$. Notice that the effect of l_2 in (5.22) is just that a limiter of ± 2.1 is introduced for $\varepsilon(t)$.

- Another way of handling single outliers is $\mathscr{I}_3$, based on filtering of the data. For each t it was tested if the prediction error $\varepsilon(t)$ is large compared to a given limit. When

$$|\varepsilon(t)| < \alpha$$

the algorithm is exactly as for $\mathscr{I}_1$. However, when this condition is not satisfied, the measurement $y(t)$ was considered to be erroneous and $\hat{y}(t \mid \theta) = \psi^T(t)\hat{\theta}(t-1)$ was substituted. The parameter α was taken as 5.

The numerical results are summarized in table 5.3. We also present plots of the parameter estimates in figures 5.3a, b. These plots also show what happened when the identification was repeated with data containing an outlier: The same data as before was used, except that the value of $y(30)$ was changed from -0.416 to 10.00. Table 5.3 and the figures show that the methods give similar results when there are no outliers in the data. However, a single outlier has a drastic influence of the parameter estimates for the method $\mathscr{I}_1$. It takes a long time for the estimates to recover after this single piece of bad data. The method $\mathscr{I}_2$ shows, as expected, better resistance to the outlier. The method $\mathscr{I}_3$ was specifically designed to handle single, large outliers, and gives almost perfect results for this case. □

Summary

When the measured data set contains some values that are abnormal, e.g., due to sensor failures, straightforward use of a quadratic criterion function will give substantial jumps of the parameter estimates. Moreover, a long time elapses before the estimates converge to their previous levels.

A way to make the algorithm robust, is to use a criterion function that grows more slowly with ε than the quadratic one; then large prediction errors will have less influence on the parameter estimates.

Another approach is to test recursively if the data contains outliers. This can be done by comparing the prediction errors with a specified limit. Large prediction errors means that an outlier or a measurement error is probable. The predicted value can then be substituted for the measurement. This approach is applicable when there are only a few outliers in the data.

5.6 Choice of Gain Sequence

The gain sequence $\{\gamma(t)\}$ and tracking were discussed in section 2.6. Two different approaches to manipulate the gain were described.

Table 5.3
Parameter estimates for example 5.4. Ten runs, $N = 100$.

Method	$\hat{a}$	$\hat{b}$
$\mathscr{I}_1$ and $\mathscr{I}_3$	-0.758 ± 0.041	0.982 ± 0.080
$\mathscr{I}_2$	-0.764 ± 0.042	0.992 ± 0.131
True value	-0.8	1.0

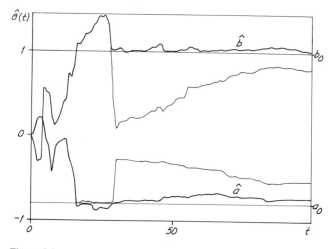

Figure 5.3a
Identification of the system in example 5.4 with methods $\mathscr{I}_1$ and $\mathscr{I}_3$. *Heavy curves:*
Result for $\mathscr{I}_1$ when no outlier was present, as well as the result for $\mathscr{I}_3$ both with and
without outlier data. *Light curves:* Result for $\mathscr{I}_1$ with outlier.

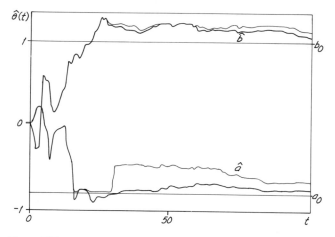

Figure 5.3b
Identification of the system in example 5.4 with method $\mathscr{I}_2$. *Heavy curves:* Without
outlier. *Light curves:* With outlier.

One approach is to use the sequence $\{\gamma(t)\}$. Then the estimates are given by

$$\hat{\theta}(t) = \hat{\theta}(t-1) + L(t)\varepsilon(t), \tag{5.24a}$$

$$L(t) = \gamma(t)R^{-1}(t)\psi(t), \tag{5.24b}$$

$$R(t) = R(t-1) + \gamma(t)[\psi(t)\psi^T(t) - R(t-1)]. \tag{5.24c}$$

It should be noted that the gain sequences in (5.24b) and (5.24c) need not be the same. Usually, though, they are chosen to be equal, and we shall assume that this is the case in the following discussion.

Instead of $\gamma(t)$ and $R(t)$ we may use the matrix $P(t)$ and the forgetting factor $\lambda(t)$, which are defined by

$$\lambda(t) = \frac{\gamma(t-1)}{\gamma(t)}[1 - \gamma(t)], \tag{5.25a}$$

$$P(t) = \gamma(t)R^{-1}(t). \tag{5.25b}$$

The algorithm (5.24) then becomes

$$\hat{\theta}(t) = \hat{\theta}(t-1) + L(t)\varepsilon(t), \tag{5.26a}$$

$$L(t) = P(t)\psi(t) = P(t-1)\psi(t)[\lambda(t)I + \psi^T(t)P(t-1)\psi(t)]^{-1}, \tag{5.26b}$$

$$P(t) = \{P(t-1) - P(t-1)\psi(t)[\lambda(t)I + \psi^T(t)P(t-1)\psi(t)]^{-1} \tag{5.26c}$$
$$\times \psi^T(t)P(t-1)\}/\lambda(t).$$

The design variable for the gain is here the scalar sequence $\{\gamma(t)\}$ or equivalently $\{\lambda(t)\}$.

The other approach is based on the model (2.111) incorporating the behavior of θ and the accuracy of the measurements. The algorithm then becomes

$$\hat{\theta}(t) = \hat{\theta}(t-1) + L(t)\varepsilon(t), \tag{5.27a}$$

$$L(t) = P(t-1)\psi(t)[R_2(t) + \psi^T(t)P(t-1)\psi(t)]^{-1}, \tag{5.27b}$$

$$P(t) = P(t-1) - P(t-1)\psi(t)[R_2(t) + \psi^T(t)P(t-1)\psi(t)]^{-1} \tag{5.27c}$$
$$\times \psi^T(t)P(t-1) + R_1(t).$$

The design variables for the gain are $R_1(t)$ and $R_2(t)$.

We will discuss the choice of gain sequence for time-varying and for time-invariant systems in the following two sections.

5.6.1 Time-Varying Systems

For true real-time identification the purpose is to track time-varying parameters. Then there is obviously a tradeoff between tracking ability and noise sensitivity. It will be impossible to accurately follow parameters which change fast. However, a slow time variation can often be tracked reasonably well.

If the algorithm (5.26) is used a common choice is to take

$$\lambda(t) \equiv \bar{\lambda} < 1. \tag{5.28a}$$

The elements of the corresponding sequence $\{\gamma(t)\}$ are readily found to be

$$\gamma(t) = \frac{1 - \bar{\lambda}}{1 - \bar{\lambda}^t}. \tag{5.28b}$$

A suitable value of the forgetting factor $\bar{\lambda}$ can be determined in the following way. A fixed value of $\lambda(t)$ as in (5.28a) corresponds to a loss function

$$V_N(\theta) = \sum_{t=1}^{N} \bar{\lambda}^{N-t} \varepsilon^2(t, \theta)$$

[see (2.115) and (2.118)]. Old prediction errors thus contribute only marginally to the criterion function. When $\bar{\lambda}$ is close to 1, which is always the case in practice, we have

$$\bar{\lambda}^t = e^{t \ln \bar{\lambda}} = e^{t \ln(\bar{\lambda} - 1 + 1)} \approx e^{t(\bar{\lambda} - 1)}.$$

This gives an exponential-decay time constant of

$$T_0 = \frac{1}{1 - \bar{\lambda}}. \tag{5.29}$$

Hence a prediction error older than T_0 time units has a weight that is less than $e^{-1} \approx 36\%$ of that of the most recent data. We may call T_0 the memory time constant of the criterion.

Note also that for (5.28b) we have

$$\lim_{t \to \infty} \gamma(t) = \frac{1}{T_0}.$$

In the more general case let $\{\gamma(t)\}$ decrease to a positive limit

$$\gamma(t) \to \gamma_0 > 0. \tag{5.30a}$$

Then we get that asymptotically $\gamma(t)$ corresponds to a constant λ:

$$\lambda_0 = 1 - \gamma_0. \tag{5.30b}$$

We can still speak of a memory time constant, in this case given by $T_0 = 1/\gamma_0$. The choice of the time constant T_0 (and hence $\bar{\lambda}$ or γ_0) shall be chosen to match the expected variation of the parameters. These should be "almost constant" over a period of length T_0.

If instead the algorithm (5.27) is used, the matrices $R_1(t)$ and $R_2(t)$ must be selected. If one has prior knowledge about the time variation, this should be used for the choice. If not, the matrices are normally taken as constant diagonal matrices. The diagonal elements of R_1 will describe the supposed rate of change for the different parameters, while the elements of R_2 describe the confidence of the various components of the measured vector.

To illustrate the two approaches consider a simple example.

EXAMPLE 5.5 (Two Approaches to Tracking of Parameters)   The system

$$y(t) + ay(t - 1) = b(t - 1)u(t - 1) + e(t)$$

is of first order. Its gain varies with time while its pole is kept fixed. We simulated the system for $N = 100$ using

$$a = -0.8$$

$$b = \begin{cases} 2, & 0 \le t \le 20, \quad 41 \le t \le 60, \; 81 \le t \le 100 \\ 3, & 21 \le t \le 40, \; 61 \le t \le 80 \end{cases},$$

and where $\{u(t)\}$ and $\{e(t)\}$ were white noise sequences of zero means and unit variances.

In this simulation the parameters behaved quite differently from one another. The parameter a remained constant while the parameter b made some large and quick changes.

The system was identified within the model set

$$y(t) + ay(t - 1) = bu(t - 1) + e(t)$$

using the LS method. First the basic algorithm with $R_1(t) \equiv 0$, $R_2(t) \equiv 1$, $\lambda(t) \equiv 1$ was tried. The estimates so obtained are shown in figure 5.4.

The estimate $\hat{b}$ is, as expected, quite poor; $\hat{b}$ cannot follow the step changes in the true parameter $b(t)$. It merely converges to a mean value of 2.5. Note that the estimate $\hat{a}$ is quite good.

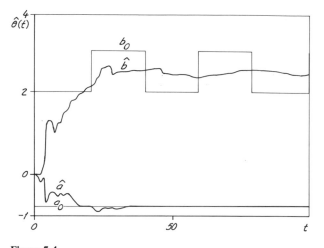

Figure 5.4
Identification results for example 5.5 with $R_1 = 0$, $R_2 = 1$, $\lambda = 1$.

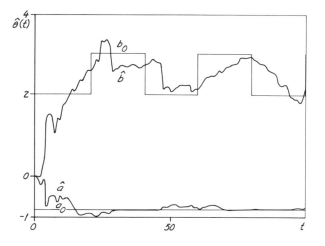

Figure 5.5
Identification results for example 5.5 with $R_1 = 0$, $R_2 = 1$, $\lambda = 0.9$.

Next, the system was identified using $R_1 = 0$, $\lambda(t) = \bar{\lambda} < 1$. With $\bar{\lambda}$ close to 1 the algorithm has a large memory time constant [see (5.29)], and will not be alert enough to follow the changes of $b(t)$. On the other hand, if $\bar{\lambda}$ is small, then the gains $L(t)$ in (5.26b) will be relatively large and the estimate $\hat{a}$ will jump around and not be very accurate. Several $\bar{\lambda}$ values were tried. A reasonable tradeoff between alertness to follow the time variations in $b(t)$ and insensitivity to noise effects was found with $\bar{\lambda} = 0.9$, which corresponds to $T_0 = 10$. The estimates so obtained are shown in figure 5.5. The estimate $\hat{b}(t)$ follows now the time variation of $b(t)$ much better than in figure 5.4, where $\bar{\lambda} = 1$ was used. Note, however, that the estimate $\hat{a}$ is worse since it is now varying more around its true value.

With the approach using the algorithm (5.27) it is possible to obtain a better result. The matrices were taken as

$$R_1(t) = \begin{pmatrix} 0 & 0 \\ 0 & r \end{pmatrix}, \quad R_2(t) = 1.$$

The matrix element r accounts for the time variation of $b(t)$. Again, several values of the design variable were tried to suitably weight alertness against noise sensitivity. It was found that $r = 0.05$ gave a reasonable tradeoff. This corresponds, as we saw in example 2.9, to a model for the parameter b of the form

$$b(t) = b(t-1) + v(t),$$

where the white noise $\{v(t)\}$ has a standard deviation of 0.22. It describes the "average change" of $b(t)$ per time unit. The parameter estimates obtained with this approach are displayed in figure 5.6. It is clear from the figure that this approach is the better one for this example. The estimate $\hat{a}$ is very calm and close to its true value for a long time (i.e., the good property from figure 5.4 is kept). Simultaneously the estimate $\hat{b}(t)$ now follows the time variations in $b(t)$ reasonably well (i.e., the good property from figure 5.5 is also kept). □

It should be noted that although $R_1(t)$, $R_2(t)$, and $\lambda(t)$ have been chosen constant in example 5.5 this is by no means necessary. It can in fact be an advantage to let these quantities be time-varying in a transient phase. This idea applies also to time-invariant systems, as will be discussed in section 5.6.2.

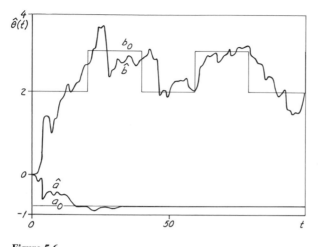

Figure 5.6

Identification results for example 5.5 with $R_1 = \begin{bmatrix} 0 & 0 \\ 0 & 0.05 \end{bmatrix}$, $R_2 = 1$, $\lambda = 1$.

Summary If appropriate prior knowledge about the time variation of the parameters is available, then the algorithm (5.27) will give the best result, especially if different parameters are changing at different rates. The matrices $R_1(t)$ and $R_2(t)$ should be chosen constant and diagonal if no detailed knowledge about the dynamics is at hand.

If there is no prior knowledge of the time variation then the algorithm (5.26) can be used. The forgetting factor $\lambda(t)$ should be chosen as a constant $\bar{\lambda}$ somewhat smaller than 1. The loss function corresponding to the algorithm gives weighting of approximately the last $1/(1 - \bar{\lambda})$ prediction errors. An alternative is to use the algorithm (5.27) with $R_1(t) = r_1 I$, $R_2(t) = r_2 I$ and to choose the scalar r_1 rather small. This choice corresponds to an expected variation of the parameters of approximately $\sqrt{r_1}$ per time unit and a variance r_2 of the measurement noise. Only the ratio r_1/r_2 will influence the parameter estimates.

5.6.2 Time-Invariant Systems

For constant parameters it was shown in section 4.4.3 that the gain should be chosen asymptotically as $\gamma(t) = 1/t$. If the Gauss-Newton direction is used then the estimates will have the minimal achievable variance. However, in the transient phase, i.e., for small and intermediate values of t, the gain $\gamma(t)$ should in some cases be chosen differently. It has in fact turned out in practice that the choice of $\gamma(t)$ [or equivalently $\lambda(t)$] will often have an important influence on the convergence rate. The one situation in which this is not the case is the LS case when the model is a regression

$$y(t) = \theta^T \varphi(t) + e(t),$$

with $\varphi(t)$ exactly known from the measured data. For other cases, the estimates of the variables in the gradient will in the beginning of the recursion be rather poor. Then there is reason to discount these estimates and the corresponding prediction errors in the further processing of the algorithm. This corresponds, to the use of a "weighting profile" $\bar{\beta}(N, t)$ (see example 2.10). The criterion

$$V_N(\theta) = \sum_{t=1}^{N} \bar{\beta}(N, t) \varepsilon^2(t, \theta) \tag{5.31a}$$

is then minimized recursively. As in (2.117), we assume that $\bar{\beta}(N, t)$ has the structure

$$\bar{\beta}(N, t) = \prod_{k=t}^{N-1} \lambda(k), \quad \bar{\beta}(N, N) = 1, \tag{5.31b}$$

which leads to the algorithm (5.26).

When old data are discounted it is required that $\lambda(t) < 1$, which corresponds to $\gamma(t) > 1/t$. On the other hand, it is desirable to let $\lambda(t) \to 1$, i.e., $t\gamma(t) \to 1$ as $t \to \infty$. These objectives can be reached in many ways. In practice it has often been useful to let $\lambda(t)$ grow exponentially with t to 1. This can be written as

$$\lambda(t) = \lambda_0 \lambda(t-1) + (1 - \lambda_0), \tag{5.31c}$$

where the rate λ_0 and the initial value $\lambda(0)$ are design variables. The properties of the gain sequence determined by (5.31) are studied in the following example.

EXAMPLE 5.6 (Gain Sequences)   Consider the forgetting factor $\lambda(t)$ given by (5.31c). The gain sequence $\{\gamma(t)\}$ can be found from (5.25a). It satisfies

$$\gamma(t) = \frac{1}{1 + \dfrac{\lambda(t)}{\gamma(t-1)}}.$$

Plots of $t\gamma(t)$ are shown in figures 5.7a, b, for different values of $\lambda(0)$ and λ_0. In figure 5.8a–c we give plots of the forgetting profile $\bar{\beta}(N, t)$. The product of $t\gamma(t)$ (excluding the case $\lambda_0 = 1$) starts at 1, increases to 2 or 3, and falls again to 1. The parameter $\lambda(0)$ influences mainly the magnitude of the maximum. (A small $\lambda(0)$ gives a large maximum and vice versa). The parameter λ_0 influences both the value of t for which the maximum occurs and the flatness of the maximum. (A large λ_0 gives a late and flat maximum).

The weighting profile $\bar{\beta}(N, t)$ is an increasing function of t. For $N \geq 200$ it typically starts (for $t = 0$) at a small value close to 0. For $t \geq 500$ it is often quite close to 1. □

In practice the exponentially increasing form (5.31c) for $\lambda(t)$ has shown to work quite well. The numerical values

$$\lambda_0 = 0.99, \quad \lambda(0) = 0.95$$

have proven useful in several low-order applications.

The influence of the forgetting factor on the convergence rate will now be illustrated by means of simulations.

EXAMPLE 5.7. (Effect of Forgetting Factor on Convergence Rate) We simulated the first-order system

$$y(t) - 0.8y(t - 1) = 1.0u(t - 1) + e(t) + 0.7e(t - 1).$$

The input was a PRBS with $S/N = 10$ and $S/N = 1$. Identification was by RPEM within the model set

$$y(t) + ay(t - 1) = bu(t - 1) + e(t) + ce(t - 1).$$

Numerical values of the measures $\bar{V}_1^*(\theta) - \bar{V}_4^*(\theta)$ were then computed; these results are shown in tables 5.4a, b. It can be seen from the tables that a time-variable forgetting factor gives better accuracy than $\lambda(t) \equiv 1$. The choice $\lambda_0 = 0.99$, $\lambda(0) = 0.95$ appears to be the best. This is most significant for short data lengths which means that the time-variable forgetting factor improves the convergence rate.

It can be noted that some of the values obtained by simulation are even better than the "ideal values" based on the Cramér-Rao bound. This is no contradiction and happens for a number of realizations. It should be regarded as an illustration of the statistical efficiency of the RPEM.

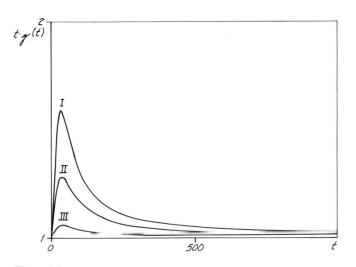

Figure 5.7a
Plots of $t\gamma(t)$ vs. t for $\lambda_0 = 0.95$. I, $\lambda(0) = 0.9$; II, $\lambda(0) = 0.95$; III, $\lambda(0) = 0.99$.

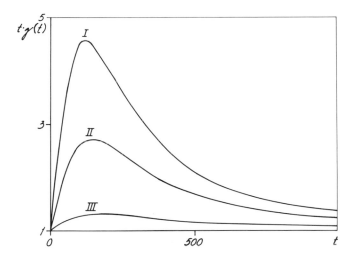

Figure 5.7b
Plots of $t\gamma(t)$ vs. t for $\lambda_0 = 0.99$. I, $\lambda(0) = 0.9$; II, $\lambda(0) = 0.95$; III, $\lambda(0) = 0.99$.

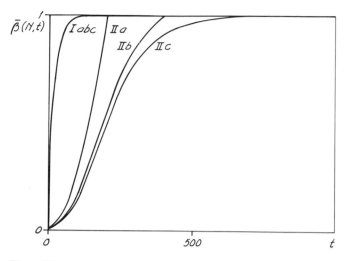

Figure 5.8a
Plots of $\bar{\beta}(N, t)$ vs. t for $N = 200$ (a), $N = 400$ (b), $N = 1,000$ (c). The forgetting factor is given by $\lambda(0) = 0.9$ and $\lambda_0 = 0.95$ (I), $\lambda_0 = 0.99$ (II).

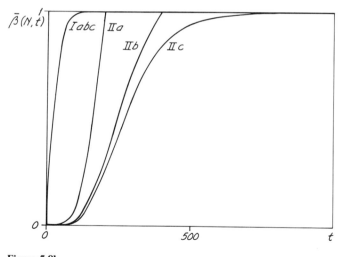

Figure 5.8b
As figure 5.8a, but $\lambda(0) = 0.95$.

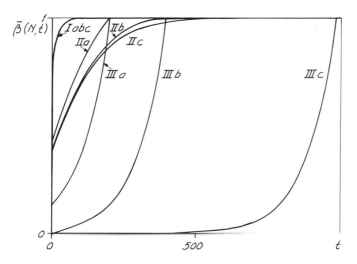

Figure 5.8c
As figure 5.8a, but $\lambda(0) = 0.99$ and $\lambda_0 = 0.95$ (I), $\lambda_0 = 0.99$ (II), $\lambda_0 = 1.0$ (III).

Finally, **PLR** was used on the same runs and with the same values of $\lambda(t)$. The result was again that $\lambda_0 = 0.99$, $\lambda(0) = 0.95$ is the best choice. □

Let us continue with another example illustrating the same type of results.

EXAMPLE 5.8 (Effect of Forgetting Factor on Convergence Rate) Consider a second order system given by

$$y(t) - 1.5y(t - 1) + 0.7y(t - 2)$$
$$= 1.0u(t - 1) + 0.5u(t - 2) + e(t) - 1.0e(t - 1) + 0.2e(t - 2).$$

The input was chosen as a **PRBS** with $S/N = 10$ and $S/N = 1$. Identification was by the **PLR** and the **RPE** methods within the model set

$$y(t) + a_1 y(t - 1) + a_2 y(t - 2)$$
$$= b_1 u(t - 1) + b_2 u(t - 2) + e(t) + c_1 e(t - 1) + c_2 e(t - 2).$$

Numerical results are given in tables 5.5a, b. It is clear from the tables that the results vary a lot from run to run. Nevertheless some general tendencies can be seen.

Table 5.4a
Validity measures for different values of λ_0 and $\lambda(0)$, example 5.7 with $S/N = 10$. Ten runs

λ_0	$\lambda(0)$	N	$\bar{V}_1^*(\theta)$	$\bar{V}_2^*(\theta)$	$\bar{V}_3^*(\theta)$	$\bar{V}_4^*(\theta)$
0.99	0.9	100	1.1951	0.0298	0.0674	...[a]
		500	1.0078	0.0031	0.0133	1.0025
		2000	1.0010	0.0007	0.0018	1.0001
0.99	0.95	100	1.5354	0.0186	0.0597	...
		500	1.0073	0.0024	0.0104	1.0025
		2000	1.0009	0.0007	0.0017	1.0001
0.99	0.99	100	1.1035	0.0452	0.1254	1.1426
		500	1.0082	0.0023	0.0081	1.0035
		2000	1.0011	0.0007	0.0018	1.0002
0.99	0.999	100	1.0901	0.0642	1.0642	1.0665
		500	1.0099	0.0043	0.0128	1.0037
		2000	1.0013	0.0009	0.0002	1.0003
0.999	0.95	100	2.2362	0.0308	0.0767	...
		500	1.0634	0.0181	0.0537	1.0187
		2000	1.0108	0.0044	0.0148	1.0050
0.999	0.99	100	1.1260	0.0394	0.1098	1.3073
		500	1.0104	0.0041	0.0129	1.0031
		2000	1.0029	0.0020	0.0056	1.0010
0.999	0.999	100	1.0910	0.0633	0.1705	1.0687
		500	1.0091	0.0037	0.0110	1.0034
		2000	1.0012	0.0009	0.0022	1.0002
1	1	100	1.094	0.0666	0.1789	1.0626
		500	1.0102	0.0048	0.0137	1.0037
		2000	1.0014	0.0009	0.0024	1.0003
Ideal values based on Cramér-Rao lower bound		100	1.0300	0.0133	0.0429	1.0109
		500	1.0060	0.0027	0.0086	1.0022
		2000	1.0015	0.0007	0.0021	1.0005

[a] Ellipses points ... in this table and in table 5.4b mean that the measure is not defined (due to instability) for at least one run.

Table 5.4b
As table 5.4a with $S/N = 1$.

λ_0	$\lambda(0)$	N	$\bar{V}_1{}^*(\theta)$	$\bar{V}_2{}^*(\theta)$	$\bar{V}_3{}^*(\theta)$	$\bar{V}_4{}^*(\theta)$
0.99	0.9	100	3.2525	0.1096	0.1850	...
		500	1.0065	0.0105	0.0362	1.0047
		2000	1.0007	0.0025	0.0055	1.0004
0.99	0.95	100	8.0698	0.1195	0.3314	...
		500	1.0062	0.0089	0.0290	1.0064
		2000	1.0007	0.0024	0.0053	1.0004
0.99	0.99	100	1.1390	2.0854	5.4485	...
		500	1.0152	0.0219	0.0766	1.0211
		2000	1.0016	0.0055	0.0151	1.0011
0.99	0.999	100	1.1618	0.1438	...	...
		500	1.0352	0.4047	0.9743	1.0317
		2000	1.0047	0.0227	0.0615	1.0027
0.999	0.95	100	2.5892	0.1228	0.2290	...
		500	1.0616	0.0627	0.1384	1.0617
		2000	1.0092	0.0160	0.0342	1.0129
0.999	0.99	100	1.1549	1.3706	3.6263	...
		500	1.0088	0.0145	0.0382	1.0068
		2000	1.0022	0.0063	0.0142	1.0021
0.999	0.999	100	1.1617	0.1429	...	...
		500	1.0306	0.2832	0.6855	1.0275
		2000	1.0026	0.0122	0.0325	1.0014
1	1	100	1.1627	0.1468	...	...
		500	1.0375	0.4964	1.0331	1.1889
		2000	1.0053	0.0258	0.0698	1.0030
Ideal values based on Cramér-Rao lower bound		100	1.0300	0.0790	0.1810	1.0185
		500	1.0060	0.0158	0.0362	1.0037
		2000	1.0015	0.0040	0.0091	1.0009

Table 5.5a
Validity measure $\bar{V}_1^*(\theta)$ for example 5.8 with $S/N = 1$.

Run	Method	$\lambda(t) \equiv 1$			$\lambda(t) = 0.99\lambda(t-1) + 0.01$ $\lambda(0) = 0.95$		
		$N = 200$	$N = 500$	$N = 1000$	$N = 200$	$N = 500$	$N = 1000$
1	PLR	1.1189	1.0398	1.0307	1.0953	1.0518	1.0377
	RPEM	1.3844	1.3212	1.2676	1.2683	1.0811	1.0244
2	PLR	1.1525	1.0988	1.0618	1.2748	1.1005	1.0232
	RPEM	1.0197	1.0093	1.0055	1.0297	1.0169	1.0080
3	PLR	1.1205	1.0497	1.0175	1.0453	1.0253	1.0068
	RPEM	1.0488	1.0072	1.0039	1.0352	1.0033	1.0019
4	PLR	1.0911	1.0251	1.0086	2.3493	1.0643	1.0024
	RPEM	1.4302	1.3884	1.3640	1.2465	1.0314	1.0044
5	PLR	1.0522	1.0379	1.0153	1.0994	1.0303	1.0110
	RPEM	1.5365	1.0039	1.0014	1.0388	1.0094	1.0014
6	PLR	1.0304	1.0322	1.0080	1.1816	1.0169	1.0170
	RPEM	1.4334	1.3999	1.3782	1.0921	1.0037	1.0081
Ideal values based on Cramér-Rao lower bound		1.0300	1.0120	1.0060	1.0300	1.0120	1.0060

- If $\bar{V}_1^*(\theta)$ becomes large, e.g., due to bad parameter estimates in the transient phase, then it decreases quicker for the time-variable forgetting factor as compared to the case $\lambda(t) \equiv 1$. This is natural, since $\lambda(t) < 1$ makes the algorithm more alert. This property is of course an important advantage of the time variable $\lambda(t)$.

- The choice of $\lambda(t)$ is more critical for RPEM then for PLR.

It can be remarked that if prediction errors are used instead of residuals (which there is no reason to do; see section 5.11), then the choice of $\lambda(t)$ becomes more critical than shown in the table.

As a further illustration the parameter estimates are plotted versus time for run no. 6, $S/N = 10$. The RPE method was used. The plots are shown in figures 5.9a, b. They illustrate the effect of using a forgetting factor less than unity in the transient phase. In figure 5.9b, where this is the case, the estimates $\hat{c}_1$, $\hat{c}_2$ converge considerably faster than for

Table 5.5b
As table 5.5a but with $S/N = 10$.

Run	Method	$\lambda(t) \equiv 1$			$\lambda(t) = 0.99\lambda(t-1) + 0.01$ $\lambda(0) = 0.95$		
		$N = 200$	$N = 500$	$N = 1000$	$N = 200$	$N = 500$	$N = 1000$
1	PLR	1.8855	1.0587	1.0537	1.3917	1.0572	1.0594
	RPEM	2.4435	2.1923	1.9456	1.3921	1.0868	1.0259
2	PLR	1.2391	1.1680	1.0937	1.3821	1.0571	1.0112
	RPEM	2.2491	1.9110	1.3779	1.0642	1.0283	1.0104
3	PLR	1.0538	1.0712	1.0323	1.0261	1.0273	1.0122
	RPEM	1.0295	1.0053	1.0033	1.0573	1.0039	1.0028
4	PLR	1.1429	1.0503	1.0252	2.0859	1.0383	1.0097
	RPEM	2.3285	1.7344	1.2195	1.0497	1.0044	1.0003
5	PLR	1.1417	1.0611	1.0409	1.0827	1.0096	1.0110
	RPEM	1.5365	1.2486	1.0617	1.0617	1.0127	1.0013
6	PLR	1.1941	1.1141	1.0505	1.2889	1.0617	1.0201
	RPEM	1.0855	1.0323	1.0157	1.0651	1.0073	1.0096
Ideal values based on Cramér-Rao lower bound		1.0300	1.0120	1.0060	1.0300	1.0120	1.0060

$\lambda(t) \equiv 1$ (see figure 5.9a). It can also be seen that more smooth curves are obtained when $\lambda(t) \equiv 1$, which simply means that the algorithm then is not so alert in tracking parameter changes. □

We stated at the beginning of this section that the matrices R_1 and R_2 can be used as an alternative to $\lambda(t)$ for varying the gain. This is illustrated in the following example where the two approaches are compared.

EXAMPLE 5.9 (Two Approaches for Affecting the Gain)   We simulated the system

$$y(t) - 0.8y(t-1) = 1.0u(t-1) + e(t) + 0.7e(t-1).$$

The input $u(t)$ was a PRBS, $S/N = 1$. The system was identified using PLR within the model set

$$y(t) + ay(t-1) = bu(t-1) + e(t) + ce(t-1).$$

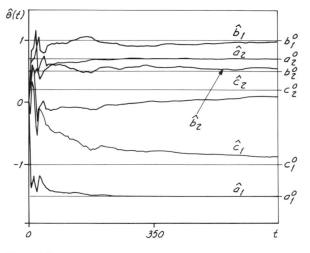

Figure 5.9a
RPEM identification for example 5.8, $S/N = 10$, $\lambda(t) \equiv 1$.

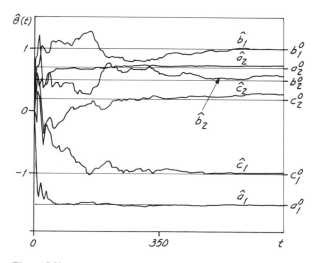

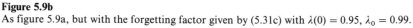

Figure 5.9b
As figure 5.9a, but with the forgetting factor given by (5.31c) with $\lambda(0) = 0.95$, $\lambda_0 = 0.99$.

Three different cases were treated:

(a) the basic algorithm: (5.24) with $\gamma(t) = 1/t$.

(b) the algorithm (5.24), with the gain $\gamma(t)$ given as a forgetting factor $\lambda(t)$ satisfying (5.31c) with $\lambda_0 = 0.99$, $\lambda(0) = 0.95$.

(c) the algorithm (5.27) with

$$
R_1(t) = \begin{pmatrix} 5 & 0 & 0 \\ 0 & 5 & 0 \\ 0 & 0 & 75 \end{pmatrix} \times 10^{-5}, \quad R_2(t) = 1,
$$

where the larger value of the last diagonal element in R_1 is an attempt to increase the convergence rate of the estimate $\hat{c}$.

The system was simulated using 500 data points. The results are shown in figure 5.10a–c. The plots illustrate that roughly the same behavior is obtained in cases (b) and (c). Both these cases show a considerably improved convergence rate over that of case (a). □

Summary The gain sequence has a considerable influence on both the transient behavior and the accuracy for reasonably long data sequences (N a few thousands).

Optimal asymptotic accuracy is obtained if $\lim_{t\to\infty} t\gamma(t) = 1$. For good transient behavior it is required that $\gamma(t) > 1/t$ for moderate values of t. It is often reasonable to let $t\gamma(t)$ have a maximum of 2 to 5, between $t = 50$ and $t = 400$.

One way to obtain the aforementioned gain sequence is to use a forgetting factor $\lambda(t)$ that grows exponentially to 1 [see (5.31c)]. In many cases the numerical values

$$
\lambda(t) = 0.99\lambda(t - 1) + 0.01, \quad \lambda(0) = 0.95
$$

have proved to be useful. When a high-order model is used so that many parameters must be estimated, it may be better to let $\lambda(t)$ grow more slowly to 1.

Another way to obtain similar transient behavior is to add a small matrix R_1 in the updating of $P(t)$ [see (5.27)]. The parameters appearing linearly in $\varepsilon(t)$ should have small values in the corresponding diagonal elements of R_1. An advantage of this approach is that the convergence rate can be manipulated more or less independently for each parameter.

5.7 Choice of Search Direction

The search direction for a recursive algorithm can be chosen in different ways. In this section some possibilities and their consequences will be discussed. As noted in sections 2.4 and 3.5, the two main alternatives are the stochastic Gauss-Newton and the stochastic gradient directions. The stochastic Newton algorithm, for single-output systems, is given by

$$\hat{\theta}(t) = \hat{\theta}(t-1) + \gamma(t) R^{-1}(t) \psi(t) \varepsilon(t), \tag{5.32a}$$

$$R(t) = R(t-1) + \gamma(t) [\psi(t)\psi^{\mathrm{T}}(t) - R(t-1)] \tag{5.32b}$$

[see (3.67)]. The stochastic gradient algorithm is obtained if $R(t)$ in (5.32a) is replaced by $r(t)I$, where (for scalar-output systems) the scalar function $r(t)$ is given by

$$r(t) = r(t-1) + \gamma(t) [\psi^{\mathrm{T}}(t)\psi(t) - r(t-1)] \tag{5.33}$$

[see (3.74)]. There also exist some hybrid variants. Kumar and Moore (1980) discuss a gradient given by

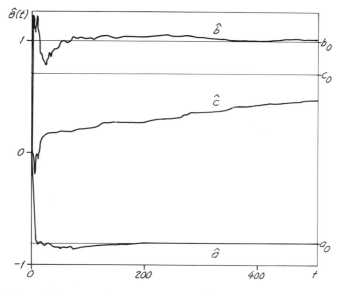

Figure 5.10a
Parameter estimates for the system in example 5.9 identified with PLR. The algorithm (5.24) with $\gamma(t) \equiv 1/t$ was used.

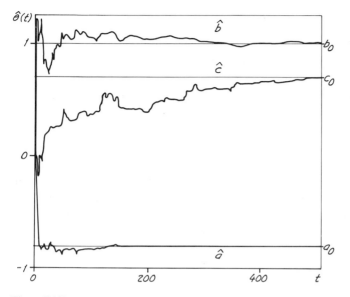

Figure 5.10b
Parameter estimates for the system in example 5.9 identified with PLR. The algorithm
(5.26) was used with a forgetting factor $\lambda(t)$ satisfying $\lambda(t) = 0.99\lambda(t-1) + 0.01$,
$\lambda(0) = 0.95$.

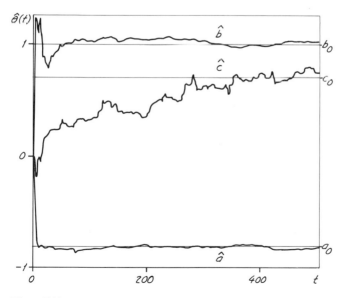

Figure 5.10c
Parameter estimates for the system in example 5.9 identified with PLR. The algorithm
(5.27) was used with the matrices $R_1(t) \equiv \text{diag}(5,\ 5,\ 75)\cdot 10^{-5}$, $R_2(t) \equiv 1$.

$$\hat{\theta}(t) = \hat{\theta}(t-1) + \gamma(t)\psi(t)^{\dagger}\varepsilon(t), \tag{5.34}$$

where $\psi(t)^{\dagger}$ denotes the componentwize pseudoinverse: $\psi_i(t)^{\dagger} = \psi_i(t)^{-1}$ if $\psi_i(t) \neq 0$, and $\psi_i(t)^{\dagger} = 0$ if $\psi_i(t) = 0$, for each component $\psi_i(t)$ of $\psi(t)$. In practice $\psi_i(t)^{\dagger}$ is set to zero if $|\psi_i(t)| < \delta$ for some small $\delta > 0$.

Other schemes have been reported by Hastings-James and Sage (1969) and Young (1976) for specific model sets. In these schemes, the stochastic Newton method has been used, with the parameter vector partitioned into two parts; one part describes the transfer function from $u(t)$ to $y(t)$ while the other part contains the noise parameters. The corresponding $R(t)$ matrix is constrained to be block diagonal. The overall algorithm then consists of two coupled algorithms of lower dimensions due to the partitioning. Such coupled algorithms are illustrated in (4.E.12).

The choice of search direction will influence

- the asymptotic accuracy,
- the convergence rate,
- the algorithm complexity.

It was argued in section 4.4.3 that essentially only the Newton direction will give asymptotically efficient estimates. In general, it can be said that the Newton direction gives quicker convergence than the stochastic gradient direction. A similar statement can be made for numerical minimization algorithms in general. As was discussed in section 5.6, the gain sequence may have a major influence on the convergence rate. It has been found in practice that gradient algorithms are much more sensitive in this respect than Newton algorithms. Since the choice of gain sequence is not trivial, this property argues in favor of choosing Newton directions. It should, however, be pointed out that the modified stochastic gradient algorithm (5.34) can give an improved convergence rate over those of the stochastic gradient algorithms (5.33) (see Kumar and Moore, 1980).

We can quickly compare the complexity of the algorithms. Let d denote the number of parameters; d is assumed to be "large" compared to 1. The basic stochastic Newton algorithm will then require $\approx 4d^2$ arithmetic operations. The storage requirement is $\approx d^2/2$. For the stochastic gradient methods the number of arithmetic operations is between $3d$ and $4d$; and the storage requirement is $\approx 2d$. In these expressions the computation of the prediction errors $\varepsilon(t)$ is not included. The computational requirement for this calculation can differ greatly between different model sets. It should also be mentioned that there are computationally "fast" algorithms for updating the gain $R^{-1}(t)\psi(t)$ (see section 6.3). A more detailed discussion of computational requirements will be given in chapter 6.

The following numerical examples will compare the Newton algorithm (5.32a) and the stochastic gradient algorithm (5.33).

EXAMPLE 5.10 (Comparison of Search Directions)   We simulated the system

$$y(t) - 0.8y(t-1) = 1.0u(t-1) + e(t).$$

The input $u(t)$ was a PRBS, $S/N = 1$. The LS method was applied to the data using the model set

$$y(t) + ay(t-1) = bu(t-1) + e(t).$$

The forgetting factor used obeyed (5.31c):

$$\lambda(t) = \lambda_0\lambda(t-1) + (1-\lambda_0).$$

Different values of λ_0 and $\lambda(0)$ were tried. The results are given in table 5.6. Several features are illustrated by the results. The estimate $\hat{a}$ is insensitive to the choice of method. However, $\hat{b}$ is given with much better accuracy by the Newton than by the gradient method. Finally, the choice of forgetting factor, i.e., the choice of the numbers λ_0 and $\lambda(0)$, has a greater effect on the estimates in the gradient algorithm. □

EXAMPLE 5.11 (Search Directions, Continued)   We simulated the system

Table 5.6
Parameter estimates for Newton and the gradient methods, example 5.10. Ten runs.

Method	λ_0	$\lambda(0)$	$N = 50$		$N = 500$	
			$\hat{a}$	$\hat{b}$	$\hat{a}$	$\hat{b}$
Newton	1.0	1.0	-0.792 ± 0.045	0.937 ± 0.158	-0.796 ± 0.011	0.994 ± 0.054
	0.99	0.95	-0.795 ± 0.034	0.929 ± 0.121	-0.796 ± 0.017	1.005 ± 0.056
Gradient	1.0	1.0	-0.770 ± 0.054	0.916 ± 0.430	-0.792 ± 0.015	0.947 ± 0.300
	0.99	0.95	-0.778 ± 0.032	0.923 ± 0.350	-0.795 ± 0.016	0.971 ± 0.158
	0.9975	0.9	-0.787 ± 0.043	0.932 ± 0.246	-0.785 ± 0.040	1.027 ± 0.056
True values			-0.8	1.0	-0.8	1.0

$$y(t) - 1.8y(t-1) + 1.54y(t-2) - 0.592y(t-3)$$

$$= 1.0u(t-1) - 0.9u(t-1) + 0.196u(t-3) + e(t).$$

The input was a PRBS, $S/N = 1$. The resulting estimates are given in table 5.7. The forgetting factor was again chosen to satisfy (5.31c). As seen from the table, the difference between the algorithms is now much more substantial than for example 5.10. The reason is that in this example there are more parameters to be estimated, which makes the P-matrix more ill-conditioned. As a further illustration, we give in figure 5.11a–d plots of the estimates. □

The difference between gradient and Newton algorithms thus becomes more pronounced as the model order increases. In certain applications, such as adaptive equalization (see section 7.5.2) high-order models are often used. The next example illustrates such a case.

EXAMPLE 5.12 (Comparison of Search Directions for a High-Order Model) (The calculations in this example were performed by F. Soong, 1981.) The system

$$y(t) = \frac{B(q^{-1})}{F(q^{-1})}u(t) + e(t)$$

was simulated, with

$$B(q^{-1}) = 0.2111q^{-1} - 0.1539q^{-2} + 0.9308q^{-3},$$

$$F(q^{-1}) = 1 - 0.2431q^{-1} + 0.4900q^{-2}.$$

The input $\{u(t)\}$ was chosen as white Gaussian noise, and $S/N = 10^4$. The model set was

Table 5.7
Parameter estimates for Newton and gradient methods, example 5.11. One run, $N = 2,000$.

Method	λ_0	$\lambda(0)$	$\hat{a}_1$	$\hat{a}_2$	$\hat{a}_3$	$\hat{b}_1$	$\hat{b}_2$	$\hat{b}_3$
Newton	1.0	1.0	−1.78	1.52	−0.58	0.99	−0.87	0.21
Gradient	1.0	1.0	−0.95	0.40	0.09	1.30	−0.12	−0.07
	0.999	0.8	−1.52	1.15	−0.35	0.99	−0.59	0.04
	0.9995	0.6	−1.68	1.41	−0.52	0.99	−0.87	0.13
True values			−1.8	1.54	−0.592	1.0	−0.9	0.196

$$\hat{y}(t \mid \theta) = \sum_{i=1}^{30} b_i u(t - i).$$

The normalized stochastic gradient algorithm $[(2.82), (2.83b)$, in standard notation$]$,

$$\hat{\theta}(t) = \hat{\theta}(t - 1) + 0.005 \frac{1}{|\varphi(t)|^2} \cdot \varphi(t)\varepsilon(t),$$

as well as the Newton algorithm (5.32) ($=$ RLS in this case) were used. In the latter case we had $\lambda(t) \equiv 0.995$ and $P(0) = 100 \cdot I$. In both cases, $\hat{\theta}(0) = 0$.

Twenty-five runs, each with $N = 1,000$, were performed; for each t the ensemble average

$$\overline{\varepsilon^2(t)} = \frac{1}{25} \sum_{i=1}^{25} [y^i(t) - \hat{\theta}^i(t - 1)^\mathsf{T} \varphi^i(t)]^2$$

was evaluated, where i indicates the ith run. These averages are shown as a function of t in figures 5.12a, b. It is obvious from these plots that for this example the Newton algorithm gives the better convergence.

In this example, the RLS algorithm was, due to the high dimension of θ, implemented as a lattice algorithm. Such an implementation is described in section 6.4. □

Summary

The Gauss-Newton algorithm gives considerably better accuracy than the gradient algorithm. This is to be expected, from the theory developed in section 4.4.3.

The choice of forgetting factor is more crucial for the gradient algorithm than for the Newton algorithm.

The aforementioned differences are more significant for models of high or moderate order than for models of (very) low order. The reason for this is probably that the P-matrix then becomes more ill-conditioned.

The Newton algorithm is more complex computationally than the stochastic gradient algorithm.

If there are no very strong constraints on computer time and storage, the Gauss-Newton direction should thus be chosen.

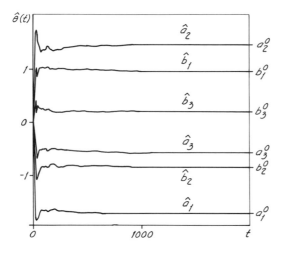

Figure 5.11a
Parameter estimates for example 5.11. The Newton algorithm was applied with $\lambda(0) = 1$, $\lambda_0 = 1$.

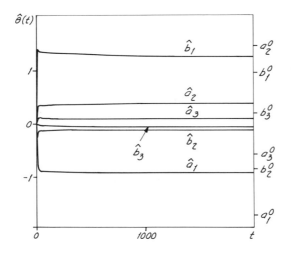

Figure 5.11b
Parameter estimates for example 5.11. The gradient algorithm was applied with $\lambda(0) = 1$, $\lambda_0 = 1$.

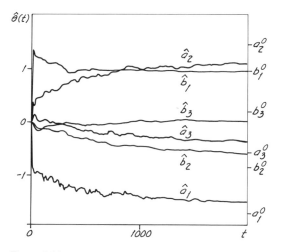

Figure 5.11c
Parameter estimates for example 5.11. The gradient algorithm was applied with $\lambda(0) = 0.8$, $\lambda_0 = 0.999$.

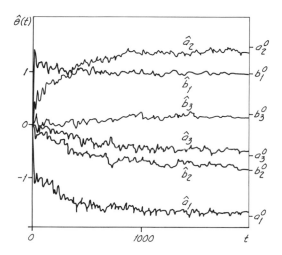

Figure 5.11d
Parameter estimates for example 5.11. The gradient algorithm was applied with $\lambda(0) = 0.6$, $\lambda_0 = 0.9995$.

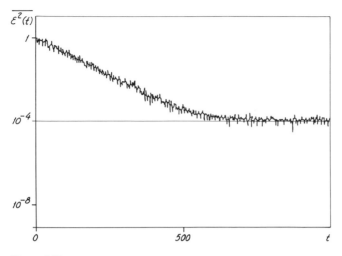

Figure 5.12a
Identification of the system of example 5.12. The stochastic gradient algorithm was used.

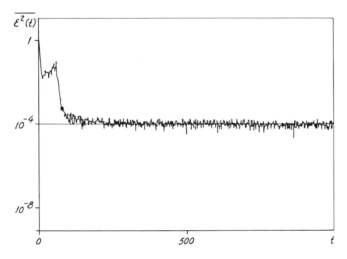

Figure 5.12b
Identification of the system of example 5.12. The Newton algorithm was used.

5.8 Choice of Initial Values

For a recursive algorithm, we require the initial values $\hat{\theta}(0)$, $P(0)$, and $\xi(0)$. The choice of these initial values will be discussed in this section.

In section 2.3 the variables $\hat{\theta}(0)$ and $P(0)$ were interpreted using Bayesian theory. We found that $\hat{\theta}(0)$ can be considered as a prior estimate of the parameter θ. Moreover, for a single-output system the covariance matrix of this estimate is $P(0)$. If, for a single-output system, the algorithm is used with $\Lambda(t) \equiv 1$, then the covariance matrix is $\sigma^2 P(0)$, where σ^2 is the variance of the output innovations. Thus if we have large confidence in the initial value $\hat{\theta}(0)$, the matrix $P(0)$ should be chosen with small elements.

If some prior information about θ is available it should of course be used for determining suitable values of $\hat{\theta}(0)$ and $P(0)$. If σ^2 is not known, a rough estimate of the output variance can be used instead. Then the initial value $P(0)$ can be taken as

$$P(0) = \text{cov}\,[\hat{\theta}(0)]/Ey^2(t). \tag{5.35}$$

If no prior information is available, the most common choice is to take

$$\hat{\theta}(0) = 0, \quad P(0) = \rho \cdot I, \tag{5.36}$$

where ρ is a large number, e.g., $100/Ey^2(t)$. If the input signal and the output signal have significantly different amplitude, this should be taken into account. One way to do this is to let $P(0)$ be a diagonal matrix with different values in the diagonal. An alternative is to use a direct scaling of the signals before starting the estimation.

For the LS method it is possible to make a more explicit statement of the influence of $P(0)$ and $\hat{\theta}(0)$. Let $\hat{\theta}_t$ denote the off-line estimate

$$\hat{\theta}_t = \left[\sum_{k=1}^{t} \varphi(k)\varphi^T(k) \right]^{-1} \left[\sum_{k=1}^{t} \varphi(k)y(k) \right]. \tag{5.37}$$

Then (2.21) with $\alpha_k = 1$ gives easily, with (5.37),

$$\hat{\theta}(t) - \hat{\theta}_t = P(t)P^{-1}(0)[\hat{\theta}(0) - \hat{\theta}_t]. \tag{5.38}$$

Equation (5.38) illustrates how the on-line estimate $\hat{\theta}(t)$ differs from the off-line estimate $\hat{\theta}_t$ due to the initial values. This difference becomes small under two different conditions:

1. When t tends to infinity the matrix $P(t)$ will tend to zero and so will both sides of (5.38).

2. When $P(0)$ is very large the right-hand side of (5.38) will be very small.

The following example illustrates the effect of $P(0)$.

EXAMPLE 5.13 (Effect of $P(0)$)   We simulated the system

$$y(t) - 0.8y(t - 1) = 1.0u(t - 1) + e(t),$$

where $u(t)$ and $e(t)$ were generated as independent Gaussian white noise, $S/N = 1$, $N = 100$. The recursive least squares algorithm was applied with a first-order set,

$$y(t) + ay(t - 1) = bu(t - 1) + e(t).$$

The forgetting factor $\lambda(t)$ was kept equal to 1 for all t.
First, the initial values were taken as

$$\hat{\theta}(0) = 0, \quad P(0) = \rho I.$$

The parameter ρ was varied, and the resulting parameter estimates compared for different values of ρ. The results are shown in figure 5.13. As can be seen, the parameter estimates converge quickly for $\rho \geq 1$.

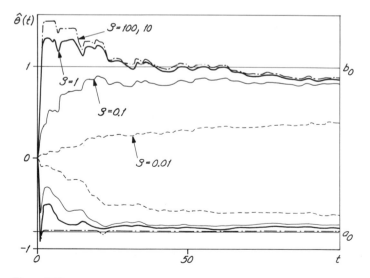

Figure 5.13
Parameter estimates for example 5.13. Initial values were $\hat{\theta}(0) = 0$, $P(0) = \rho I$.

It can also be seen that in the transient phase (say $0 \le t \le 25$) large changes from one measurement to the next one are obtained for large values of ρ. On the other hand, when ρ is small (0.1 or 0.01) the estimates converge slowly. The explanation is that too much confidence is given to the erroneous initial values. In this example, then, $\rho = 1$ seems to be a good choice. It is possible to make an approximate analysis using (5.38) for examining values of ρ for this example. Simple calculations show that the difference between the off-line and the recursive estimates (5.38) decays for the parameter a as $1/(1 + \rho t E y^2(t))$. The same result is true for the parameter b, with $E u^2(t)$ replacing $E y^2(t)$.

Next, the initial values were taken as

$$\hat{\theta}(0) = \begin{pmatrix} -0.8 \\ 1.0 \end{pmatrix}, \quad P(0) = \rho I,$$

i.e., we start with the true values in $\hat{\theta}(0)$. Some runs with different values of ρ are shown in figure 5.14.

A comparison between figures 5.13 and 5.14 shows that for $\rho = 1$ the parameter estimates are practically the same. This means that $\rho = 1$ is too large a value to capitalize on the good initial condition $\hat{\theta}(0)$. □

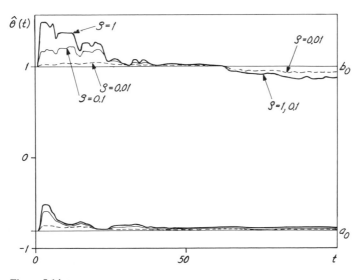

Figure 5.14
Parameter estimates for example 5.13. Initial values were $\hat{\theta}(0) = (-0.8 \;\; 1.0)^{\mathrm{T}}$, $P(0) = \rho I$.

Consider now the choice of initial value $\xi(0)$ in (3.157d). Basically, this value should be determined using data $z(t)$ prior to $t = 0$. Since these are not available, we are left with three possible approaches:

1. Take $\xi(0)$ as an estimate of $E\xi(t)$. If no prior information is available then take $\xi(0) = 0$.

2. Determine $\xi(0)$ based on the assumption that $z(t) = z(0)$ for $t < 0$.

3. Wait to start the updating of $\hat{\theta}$ until $\xi(t)$ has become reliable.

If $\xi(0)$ is badly initialized it can take very long time to compensate for this effect in the estimates. In its effect it therefore resembles an outlier; see the discussion in section 5.5.

In the linear regression case, $\xi(t)$ equals the regression vector $\varphi(t)$ that is determined from data $z(s)$ in a finite time interval $t - M \leq s \leq t - 1$. In this case two of the above choices are known as particular variants of the LS method. The first approach ($\varphi(0) = 0$) is in signal processing known as "prewindowing." For off-line methods one can similarly add zeros in $\varphi(t)$ at the end of the experiment, which is known as "post-windowing." In speech analysis, the LS method with pre- and post-windowing is often referred to as the "autocorrelation method." The third approach, to start updating $\hat{\theta}$ only when $t = M$, is called "non-windowed" in signal processing and the "covariance method" in speech analysis.

In many cases where $\xi(0)$ is reasonably chosen, it has only a quickly decaying influence on the estimates. Then the exact choice is not very crucial. Some care must be used, though, when $\dim \theta$ is not small compared to the number of processed data and/or when $A(\theta_0)$ in (3.157d) has eigenvalues close to the unit circle.

Summary

A large value of $P(0)$ makes the value of $\hat{\theta}(0)$ only marginally important. The parameter estimates will change quickly in the transient phase (for small values of t). A small value of $P(0)$ will give only small corrections of $\hat{\theta}(t)$ since $L(t)$, (3.70d), will be small for all t. The convergence will then be slow unless $\hat{\theta}(0)$ is not close to the convergence limit.

If some prior information is available, then use it to determine $\hat{\theta}(0)$. Furthermore, use (5.35) and the confidence in $\hat{\theta}(0)$ to find an initial value $P(0)$.

If no prior information can be used, take $\hat{\theta}(0) = 0$, $P(0) = \rho I$ with ρ large.

Take $\xi(0)$ as an estimate of $E\xi(t)$.

5.9 Approximation of the Gradient by PLRs

It was shown in sections 2.5.1 and 3.7.3 that the gradient $\psi(t)$ can be approximated using a PLR. When this is done some filtering is omitted and the algorithm complexity is reduced as compared to an RPEM. This section will be devoted to a discussion of the choice between PLR and RPEM. The choice between PLR and RPEM has consequences for

- convergence properties,
- algorithm complexity,
- asymptotic accuracy,
- transient behavior.

5.9.1 Convergence

It should first be noted that the convergence properties are quite different in the two methods. Assume first that the system belongs to the model set. Then for PLR the key condition for convergence is positive realness of appropriate transfer functions and filters (see section 4.5). This condition is satisfied for a great deal of systems but there are also counterexamples. On the other hand, an RPEM gives convergence under very general conditions, which are the same as for the corresponding off-line identification algorithm.

We now give an example which illustrates the different convergence properties of PLR and RPEM.

EXAMPLE 5.14 (Comparison of Convergence Properties using PLR and RPEM). We simulated the system (4.198):

$$y(t) + 0.9y(t-1) + 0.95y(t-2) = e(t) + 1.5e(t-1) + 0.75e(t-2).$$
$$(5.39)$$

[We showed in example 4.11 that the PLR method will not converge when applied to (5.39).] Identification was by PLR and RPEM. The model set was

$$y(t) + a_1 y(t-1) + a_2 y(t-2) = e(t) + c_1 e(t-1) + c_2 e(t-2).$$

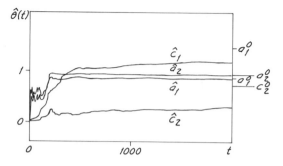

Figure 5.15a
Results of a recursive identification using PLR for the system (5.39) initialized by $P(0) = 100I$, $\hat{\theta} = 0$.

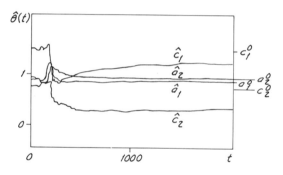

Figure 5.15b
As figure 5.15a, but with $P(0) = 0.0005I$, $\hat{\theta}(0) = \theta_0$.

The results are summarized in figures 5.15a, b, 5.16a, b, and table 5.8. The results clearly demonstrate that for these systems the RPEM converges rapidly, while the PLR does not even converge. □

When the model set is not large enough to include the true system, it is not easy to make comparisons between PLR and RPEM. In section 4.4.4 we saw that RPEM gives an approximation of the true system that is optimal in a certain sense. The PLR, on the other hand, may give a bad approximation of the system when it does not belong to the model set. This is illustrated in the following example.

EXAMPLE 5.15 (PLR and RPEM Approximations in a Small Model Set) In this example, the MA(2) system

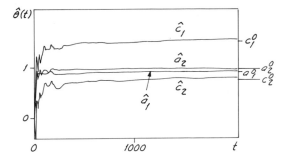

Figure 5.16a
As figure 5.15a, but using RPEM.

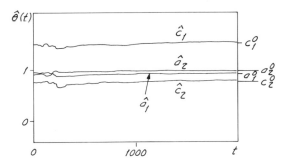

Figure 5.16b
As figure 5.15b, but using RPEM.

Table 5.8
PLR and RPEM parameter estimates and $\bar{V}_1^*(\theta)$ for example 5.14. Ten runs, $N = 2,000$.

Estimate	True value	PLR		RPEM	
		$\theta(0) = 0$ $P(0) = 100\mathrm{I}$	$\theta(0) = \theta_0$ $P(0) = 0.0005\mathrm{I}$	$\theta(0) = 0$ $P(0) = 100\mathrm{I}$	$\theta(0) = \theta_0$ $P(0) = 0.0005\mathrm{I}$
$\hat{a}_1$	0.90	0.898 ± 0.037	0.912 ± 0.049	0.901 ± 0.007	0.901 ± 0.007
$\hat{a}_2$	0.95	0.944 ± 0.041	0.941 ± 0.060	0.953 ± 0.006	0.953 ± 0.006
$\hat{c}_1$	1.50	1.260 ± 0.129	1.338 ± 0.120	1.490 ± 0.021	1.491 ± 0.021
$\hat{c}_2$	0.75	0.407 ± 0.246	0.559 ± 0.153	0.741 ± 0.017	0.742 ± 0.017
$\bar{V}_1^*(\theta)$		1.207	1.206	1.002	1.002

$$y(t) = e(t) + 1.90e(t-1) + 0.95e(t-2), \quad Ee^2(t) = 1,$$

is approximated by the MA(1) model

$$y(t) = e(t) + ce(t-1).$$

The limiting value of the estimates can then be found by solving the equation of $f(\theta) = 0$ (4.24a). This will give $c^*_{PLR} = 0.9987$ for PLR and $c^*_{RPEM} = 0.9744$ for RPEM. The models can be evaluated by examining the variance of the prediction errors. This variance is

$$\bar{V}_1^*(\theta^*) = E\varepsilon^2(t, \theta^*) = E\left[\frac{1 + 1.90q^{-1} + 0.95q^{-2}}{1 + c^*q^{-1}}e(t)\right]^2.$$

With the foregoing values of the estimates we find that the optimal value of $\bar{V}_1(\theta^*)$, which is achieved for θ^*_{RPEM}, is 1.903. Use of PLR will give $\bar{V}_1^*(\theta^*_{PLR}) = 2.765$. In this somewhat artificial example, RPEM gives a much better approximation of the system than PLR. □

5.9.2 Algorithm Complexity

When algorithm complexity is considered it is clear that a RPEM requires more computations than a PLR. The reason for this is twofold. In a RPEM we must include filtering in order to compute the gradient $\psi(t)$. Moreover, the RPEM algorithm must be monitored, i.e., we must, whenever necessary, reduce the updated estimate so that the filter remains asymptotically stable. Although these differences are very important, the requirements on computer time and memory do not differ very much between PLR and RPEM.

5.9.3 Accuracy

The asymptotic accuracy was derived in section 4.4.3. It was shown that under weak assumptions an RPEM gives asymptotically efficient estimates, i.e., that the Cramér-Rao lower bound is achieved. The accuracy of recursive PLR is not yet solved (see section 4.4.5). There is no reason why PLR should be statistically efficient. The RPEM is therefore superior to PLR from the asymptotic accuracy point of view.

We now give numerical illustrations of the differences in accuracy between PLR and RPEM.

EXAMPLE 5.16 (Accuracy of PLR and RPEM)   We simulated the systems

$$y(t) - 1.5y(t-1) + 0.7y(t-2)$$

$$= 1.0u(t-1) + 0.5u(t-2) + e(t) - 1.0e(t-1) + 0.2e(t-2) \tag{5.40a}$$

and

$$y(t) - 1.6y(t-1) + 1.61y(t-2) - 0.776y(t-3)$$

$$= 1.2u(t-1) - 0.95u(t-2) + 0.2u(t-3) + e(t) \tag{5.40b}$$

$$+ 0.1e(t-1) + 0.25e(t-2) + 0.873e(t-3).$$

The input was a PRBS, $S/N = 10$ for (5.40a) and $S/N = 1$ for (5.40b). Identification was by PLR and RPEM in model sets compatible with the system descriptions (5.40). The forgetting factor $\lambda(t)$ was in both cases

$$\lambda(t) = 0.99\lambda(t-1) + 0.01, \quad \lambda(0) = 0.95.$$

The results are given in tables 5.9 and 5.10. It is clear from the tables that for the systems of this example the PLR method is often superior for short data series while RPEM is best for long samples. □

It should be mentioned that when many simulations are made, one occasionally encounters realizations which give outlier performance, i.e., the estimates are nowhere close to the typical behavior. We have found that such outlier behavior is more common for RPEM than for PLR.

5.9.4 Transient Behavior

Concerning the transient behavior of the algorithm, i.e., the parameter estimates based on small data sets, it has been observed in practice, that PLR is often slightly better than RPEM (see example 5.16). A possible explanation of this can be given as follows. When few data are processed the estimate of the noise properties is rather poor in general. Then a bad filtering of the data (using this poor estimate) is obtained for RPEM. It might very well be better to use no filtering at all for short data series. It is thus natural to propose a modified filtering for RPEM such that no or very "weak" filtering take place for the first data. One way to achieve this has been suggested by Friedlander (1982a), as shall be illustrated in the following example.

EXAMPLE 5.17 (Modified Filtering of an RPEM) Consider the model set

$$A(q^{-1})y(t) = B(q^{-1})u(t) + C(q^{-1})e(t).$$

Table 5.9
Parameter estimates and validity measures for the system (5.40a) of example 5.16. 10 runs.
Ellipses ... mean that the measure has no finite value (the corresponding filter is unstable)
at least for one of the runs.

Estimate	True value	PLR			RPEM		
		$N = 100$	$N = 500$	$N = 2,000$	$N = 100$	$N = 500$	$N = 2,000$
$\hat{a}_1$	-1.5	-1.4719	-1.4928	-1.4994	-1.4692	-1.5045	-1.5001
$\hat{a}_2$	0.7	0.6738	0.6865	0.6965	0.6698	0.7038	0.6997
$\hat{b}_1$	1.0	0.9796	1.0046	1.0081	0.9606	1.0046	1.0092
$\hat{b}_2$	0.5	0.6426	0.5047	0.4979	0.6255	0.4804	0.4886
$\hat{c}_1$	-1.0	-0.9048	-0.9558	-0.9883	-0.7233	-0.9855	-0.9920
$\hat{c}_2$	0.2	0.1110	0.1543	0.1754	0.0863	0.2050	0.1954
$\bar{V}_1^*$	1.0	...	1.0415	1.0101	1.1771	1.0096	1.0015
$\bar{V}_2^*$	0.0	0.5126	0.1238	0.0343	0.2761	0.0245	0.0055
$\bar{V}_3^*$	0.0	0.3845	0.0287	0.0041	0.3814	0.0059	0.0004
$\bar{V}_4^*$	1.0	...	1.0149	1.0025	...	1.0043	1.0005

Table 5.10
As table 5.9, for the system (5.40b).

Estimate	True value	PLR			RPEM		
		$N = 100$	$N = 500$	$N = 2,000$	$N = 100$	$N = 500$	$N = 2,000$
$\hat{a}_1$	-1.6	-1.6119	-1.5931	-1.5955	-1.3279	-1.5751	-1.5981
$\hat{a}_2$	1.61	1.5663	1.5997	1.6062	1.2378	1.5826	1.6083
$\hat{a}_3$	-0.776	-0.7751	-0.7670	-0.7722	-0.5369	-0.7489	-0.7741
$\hat{b}_1$	1.2	1.1674	1.2060	1.1939	1.0934	1.2060	1.1867
$\hat{b}_2$	-0.95	-1.0667	-0.9856	-0.9525	-0.6870	-0.9467	-0.9510
$\hat{b}_3$	0.2	0.1456	0.1602	0.1923	0.0774	0.1709	0.1920
$\hat{c}_1$	0.1	0.2599	0.1446	0.1194	0.4952	0.1785	0.1167
$\hat{c}_2$	0.25	0.1494	9.2124	0.2335	0.3182	0.2002	0.2325
$\hat{c}_3$	0.873	0.6482	0.7802	0.8333	0.3742	0.7320	0.8388
V_1^*	(1.0)	...	1.1232	1.0247	1.7202	1.0927	1.0140
V_2^*	(0.0)	8.5201	1.0896	0.1816	12.171	0.7009	0.0343
V_3^*	(0.0)	1.7644	0.4635	0.0799	3.3472	1.1390	0.0997
V_4^*	(1.0)	...	...	1.0375	...	...	1.0370

When the basic RPEM is used, the data and the prediction errors (or the residuals) are filtered through $1/\hat{C}_t(q^{-1})$. To modify the filter, let it have poles closer to the origin for small t. This can be achieved by instead using the filter $1/\hat{C}_t(K_t q^{-1})$. If z_1 is a zero of $z^n \hat{C}_t(z^{-1})$ then $z^n \hat{C}_t(Kz^{-1})$ will have a corresponding zero in Kz_1. The "contraction factor" K_t should be time-varying. For small t it should be close to 0 so that the filtering does not decrease the convergence rate. For large t, however, K_t should be taken as 1 in order to obtain the strong convergence and accuracy results of an RPEM. A simple way to achieve this is to let K_t grow exponentially to 1 according to

$$K_t = \mu K_{t-1} + (1 - \mu), \quad K_0 = 0,$$

where μ is a constant somewhat smaller than 1, e.g., 0.98–0.99. It is clear that, with various choices of K_t, it is possible to obtain any method "between" PLR and RPEM. $\square$

5.9.5 Summary

The estimates obtained with RPEM converge under more general conditions than those obtained with PLR. For use of PLR the key requirement for consistency is that a certain filter is positive real. This condition is not always satisifed.

RPEM gives asymptotically statistically efficient estimates so that the Cramér-Rao lower bound of the covariance matrix is obtained for long data series.

Despite the above results (which are valid for long data series and when the model set is consistent with the data) the PLR method sometimes gives better results than RPEM for short data series.

Important goals like good convergence properties and optimal accuracy are thus obtained if RPEM is used. If "cautious" filtering is applied in the transient phase the transient convergence rate should not be worse than for PLR. The only drawback with RPEM compared to PLR will then be that the algorithm is a bit more complex.

5.10 Approximation of the Gradient by IVs

The possibility of approximating the gradient with instrumental variables (IVs) was discussed in section 3.6.3. In section 4.6 we analysed the IV methods. Here we shall discuss how the IVs should be chosen. We will also make some comparisons between recursive IV methods and RPEMs.

One important difference between RIVs and RPEMs is that for RIVs only the transfer function from the input $u(t)$ to the output $y(t)$ is estimated. For an RPEM, typically, the effect of the disturbance on the system is also included in the identified model.

The asymptotic accuracy can differ considerably between a RIV and a RPEM. This is shown in the following example.

EXAMPLE 5.18 (Asymptotic Accuracy of RIV Methods and RPEMs) Consider a first-order system

$$y(t) + a_0 y(t-1) = b_0 u(t-1) + e(t) + c_0 e(t-1). \tag{5.41}$$

where $\{u(t)\}$ and $\{e(t)\}$ are mutually independent white noises of zero mean and variance σ_u^2 and σ_e^2, respectively. Let $\mu = \sigma_u^2/\sigma_e^2$. Suppose that the system is identified using the IV method (3.93). Then

$$\zeta(t) = (-x(t-1) \ \ u(t-1))^\mathsf{T}, \tag{5.42a}$$

with $x(t)$ being defined by filtering the input as

$$x(t) + \bar{a}x(t-1) = \bar{b}u(t-1). \tag{5.42b}$$

This method will give consistent parameter estimates, i.e., $\hat{\theta}(t)$ will converge to the true value θ_0 as t approaches infinity (see lemma 4.7). The asymptotic covariance matrix P_{RIV} for $\hat{\theta}(t)$ can be found from theorem 4.8. Straightforward but tedious calculation gives

$P_{\mathrm{RIV}}(\bar{a})$

$$= \frac{\mu}{b_0^2(1-\bar{a}^2)} \begin{pmatrix} (1-a_0\bar{a})^2(1+c_0^2-2\bar{a}c_0) & -b_0c_0(1-a_0\bar{a})(1-\bar{a}^2) \\ -b_0c_0(1-a_0\bar{a})(1-\bar{a}^2) & b_0^2(1+c_0^2)(1-\bar{a}^2) \end{pmatrix}. \tag{5.43}$$

The notation $P_{\mathrm{RIV}}(\bar{a})$ is used in order to emphasize the dependence on $\bar{a}$. Note that $\bar{b}$ has no influence on P_{RIV}!

For comparison, consider a RPEM with the Gauss-Newton search direction applied to the model

$$y(t) + ay(t-1) = bu(t-1) + e(t) + ce(t-1).$$

Then the asymptotic covariance matrix of the estimate $\hat{a}(t)$, $\hat{b}(t)$ becomes, according to theorem 4.5,

$$P_{\text{RPEM}} = \frac{\mu}{b_0^2 + \mu(c_0 - a_0)^2}$$

$$\times \begin{pmatrix} (1 - a_0^2)(1 - a_0c_0)^2 & -b_0c_0(1 - a_0^2)(1 - a_0c_0) \\ -b_0c_0(1 - a_0^2)(1 - a_0c_0) & b_0^2(1 - a_0^2c_0^2) + \mu(c_0 - a_0)^2(1 - c_0^2) \end{pmatrix}.$$

$$(5.44)$$

Calculation shows that $P_{\text{RIV}}(\bar{a}) - P_{\text{RPEM}}$ is nonnegative definite for all $\bar{a}$. Thus RPEM is, as expected, generally superior to the RIV method. Moreover, it can be seen from (5.43) that there is no value $\bar{a}^*$ such that $P_{\text{RIV}}(\bar{a}) - P_{\text{RIV}}(\bar{a}^*)$ is nonnegative definite for all $\bar{a}$. To optimize the accuracy within the class of IV methods (5.42) it is thus necessary to have a scalar measure. For this we take the determinant of the covariance matrix, i.e.,

$$W_{\text{RIV}}(\bar{a}) = \det P_{\text{RIV}}(\bar{a})$$

$$= \frac{\mu^2(1 - a_0\bar{a})^2}{b_0^4(1 - \bar{a}^2)}[1 + c_0^2 + c_0^4 - 2c_0\bar{a}(1 - c_0^2) + c_0^2\bar{a}^2]. \qquad (5.45)$$

It can be seen that $W_{\text{RIV}}(\bar{a})$ is not minimized for $\bar{a} = a_0$ unless $c_0 = 0$. This is in contrast to the common conjecture that $\bar{a} = a_0$, $\bar{b} = b_0$ is the best choice of the parameters in (5.42b).

Similarly, (5.44) gives

$$W_{\text{RPFM}} = \det P_{\text{RPEM}} = \frac{\mu^2(1 - a_0^2)(1 - a_0c_0)^2(1 - c_0^2)}{[b^2 + \mu(c_0 - a_0)^2]}. \qquad (5.46)$$

Calculated values of $W_{\text{RIV}}(\bar{a})$ and W_{RPEM} are given in tables 5.11 and 5.12 for two different systems. It is clear from the tables that an RPEM gives far better accuracy than an IV method. Also note that $\bar{a} = 0$ corresponds to an IV method with only delayed inputs as instrumental variables, and is of rather poor accuracy.

A constant prefilter $T(q^{-1}) = 1/(1 + dq^{-1})$ used as in (3.96) can improve the accuracy of the IV estimates considerably, if it is chosen appropriately. For example, consider table 5.11. If a prefilter with $d = 0.7$ is used, the value $W_{\text{IV}}(a_0)$ decreases from 1.3365 to 0.4468. □

We now discuss the relation between the IV variant with only delayed inputs as instrumental variables, [see (3.95a)], and the more general correlation matrix obtained by extending the IV vector.

Table 5.11
W_{RIV} and W_{RPEM} [equations (5.45) and (5.46) of example 5.18] for $a_0 = -0.8$, $b_0 = 1.0$, and $c_0 = 0.7$. $\bar{a}^* = -0.717$ is the value of $\bar{a}$ at which W_{RIV} is a minimum.

	Calculated value for		
Measure	$\mu = 10$	$\mu = 1$	$\mu = 0.1$
$W_{RIV}(a_0)b^2/\mu^2$	1.3365	1.3365	1.3365
$W_{RIV}(0)b^2/\mu^2$	1.7301	1.7301	1.7301
$W_{RIV}(\bar{a}^*)b^2/\mu^2$	1.3013	1.3013	1.3013
$W_{RPEM}b^2/\mu^2$	0.019	0.138	0.365

Table 5.12
As table 5.11, for $a_0 = -0.8$, $b_0 = 1.0$, and $c_0 = -0.8$. $\bar{a}^* = -0.939$. The calculated values are independent of μ.

Measure	Calculated value
$W_{RIV}(a_0)b^2/\mu^2$	0.1296
$W_{RIV}(0)b^2/\mu^2$	2.0496
$W_{RIV}(\bar{a}^*)b^2/\mu^2$	0.0785
$W_{RPEM}b^2/\mu^2$	0.0168

EXAMPLE 5.19. (Comparison of an IV Method and a Correlation Analysis) Consider again the first-order system

$$y(t) + a_0 y(t-1) = b_0 u(t-1) + e(t) + c_0 e(t-1),$$

where $u(t)$ and $e(t)$ are mutually indendent white noises of zero mean and variance σ_u^2, σ_e^2, respectively. Let the IV vector be given by

$$\zeta(t) = (u(t-1) \ u(t-2) \ \ldots \ u(t-n))^T,$$

and assume that for $n > 2$ an overdetermined system of equations is solved in a least squares sense as explained in section 3.6.3. Calculation then gives the normalized covariance matrix

$$P_n = \frac{\sigma_e^2}{\sigma_u^2} \left(\begin{array}{c|c} \dfrac{1+c_0^2}{b_0^2}\dfrac{1-a_0^2}{1-a_0^{2n-2}} - \dfrac{(1-a_0^2)(1-a_0^{2n-4})}{(1-a_0^{2n-2})^2}\dfrac{2a_0c_0}{b_0^2} & -\dfrac{c_0}{b_0}\dfrac{1-a_0^2}{1-a_0^{2n-2}} \\ \hline -\dfrac{c_0}{b_0}\dfrac{1-a_0^2}{1-a_0^{2n-2}} & 1+c_0^2 \end{array} \right).$$

Since the 2, 2 element does not depend on n it is clear that there exists no optimal value n^* of n such that $P_n - P_{n^*}$ is nonnegative definite for all n.

Let us now examine if the scalar $\det P_n$ improves when n is increased. We have

$$\det P_2 = \frac{1 + c_0^2 + c_0^4}{b_0^2} \cdot \frac{\sigma_e^4}{\sigma_u^4},$$

$$\det P_\infty = \frac{1}{b_0^2}(1 - a_0^2)[(1 + c_0^2 - 2a_0 c_0)(1 + c_0^2) - c_0^2(1 - a_0^2)] \cdot \frac{\sigma_e^4}{\sigma_u^4}.$$

For some numerical values, e.g., $a_0 = -0.7$, $c_0 = 0.7$, $\det P_\infty$ will be larger than $\det P_2$, so that it does not generally pay to increase the dimension of $\zeta(t)$. $\square$

The discussion so far has concerned RIV methods where the IV vector $\zeta(t)$ does not depend on previous estimates $\hat\theta(s)$, $s < t$. In practice, however, $\zeta(t)$ is often constructed using previous estimates; an example is (3.94). The analysis of RIV methods given in section 4.6 does not cover such cases. For such a scheme, i.e., one with a built-in adaptive filter, we must require that all models in $D_\mathcal{M}$ have stable polynomials $A(q^{-1})$. To show some properties of such RIV methods we give a numerical example based on simulation studies.

EXAMPLE 5.20 (RIV Method with Time-Varying Filtering vs. RPEM)   We simulated the system

$$y(t) - 1.5y(t - 1) + 0.7y(t - 2)$$

$$= 1.0u(t - 1) + 0.5u(t - 2) + e(t) - 1.0e(t - 1) + 0.2e(t - 2),$$

where $u(t)$ is a PRBS and $S/N = 10$. Identification was by RPEM in the model set

$$y(t) + a_1 y(t - 1) + a_2 y(t - 2)$$

$$= b_1 u(t - 1) + b_2 u(t - 2) + e(t) + c_1 e(t - 1) + c_2 e(t - 2),$$

with the forgetting factor

$$\lambda(t) = 0.99\lambda(t - 1) + 0.01, \quad \lambda(0) = 0.95.$$

The IV method given by

$$\zeta(t) = (-x(t - 1) \ -x(t - 2) \ u(t - 1) \ u(t - 2))^{\mathrm{T}},$$

$$x(t) = \zeta^{\mathrm{T}}(t)\hat\theta(t - 2)$$

Table 5.13
RIV and RPEM parameter estimates and $\bar{V}_2^*(\theta)$ for example 5.20. Ten runs.

Method	N	Parameter				
		a_1	a_2	b_1	b_2	$\bar{V}_2^*(\theta)$
RIV	100	-1.509 ± 0.114	0.722 ± 0.182	1.090 ± 0.159	0.372 ± 0.317	2.352
	500	-1.490 ± 0.023	0.692 ± 0.022	1.025 ± 0.055	0.525 ± 0.083	2.029
	2000	-1.496 ± 0.018	0.697 ± 0.016	1.001 ± 0.044	0.518 ± 0.047	2.036
RPEM	100	-1.265 ± 0.294	0.482 ± 0.280	1.037 ± 0.205	0.588 ± 0.289	2.240
	500	-1.496 ± 0.020	0.699 ± 0.015	1.006 ± 0.079	0.525 ± 0.098	0.054
	2000	-1.498 ± 0.008	0.698 ± 0.007	0.991 ± 0.025	0.515 ± 0.036	0.010
True value		-1.5	0.7	1.0	0.5	0

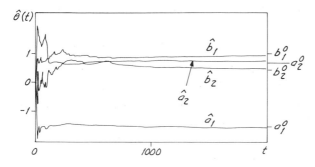

Figure 5.17a
Parameter estimates for one run, example 5.20. The IV method was used.

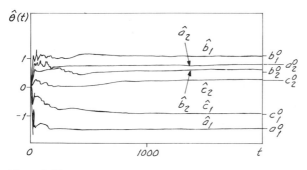

Figure 5.17b
As figure 5.17a, but using RPEM.

was also applied to the simulated data. The reason for using $\hat{\theta}(t-2)$ rather than $\hat{\theta}(t)$ is an attempt to get better stability of the algorithm. Such a delay in $\hat{\theta}(t)$, or, alternatively, use of a low-pass filter, is often used in practice. Note that $x(t)$ can be viewed as an estimate of the noise-free part of the output. The model set used for the IV method is

$$y(t) + a_1 y(t-1) + a_2 y(t-2) - b_1 u(t-1) + b_2 u(t-2) + e(t).$$

The results are given in table 5.13. Besides the parameter estimates, the criterion $\bar{V}_2^*(\theta)$ [see (5.4)] is given. It can be seen from the table that for a small amount of data RIV and RPEM give about the same performance, as measured by the criterion $\bar{V}_2^*(\theta)$, but that the RIV estimates converge faster. Beyond 100 data points the RIV method does not improve the overall accuracy, while the RPEM shows a much improved accuracy for a large amount of data.

The foregoing results are further illustrated in figures 5.17a, b, where plots of the parameter estimates are displayed for one run. It is clear from the plots that the IV method gives slightly more rapid convergence. □

We know from section 4.6 that IV estimates can be improved if properly chosen prefilters are included. The best choice of prefilter and instrumental variables requires knowledge of the true system, and therefore an adaptive approach to the problem is natural. As explained in appendix 4.E, this adaptive refined IV method can be interpreted as an RPEM with a block diagonal Gauss-Newton search direction.

Summary

A basic IV algorithm without filtering of the measured data has low algorithmic complexity and converges rapidly. The accuracy of the estimates is less than that of an RPEM, especially for longer data series.

The IV algorithm estimates only the transfer function from the input to the output. If there is no feedback from output to input, the consistency of the IV estimates is insensitive to the character of the noise affecting the system. On the other hand, the character of the noise can substantially influence the accuracy of the estimates.

When prefiltering of the data is included, considerably improved accuracy can be obtained. The optimal prefilter can only be implemented in approximate ways. Such an approximate algorithm will be as complex as an RPEM.

5.11 Choice between Residuals and Prediction Errors in the Gradient Vector

For many recursive identification algorithms the gradient vector $\psi(t)$ depends in one way or another on the prediction errors. The LS method, wherein only inputs and outputs are used is one exception.

It was shown in sections 2.2.3 and 2.5.1 that instead of the prior prediction errors

$$\varepsilon(t) = y(t) - \hat{y}(t \mid \hat{\theta}^{t-1}) \tag{5.47a}$$

based on estimates up to time $t - 1$, the residuals or posterior prediction errors

$$\bar{\varepsilon}(t) = y(t) - \hat{y}(t \mid \hat{\theta}^t) \tag{5.47b}$$

based on estimates up to time t can be used when computing the elements of the gradient vector.

For PLR we have the following simple relationship between $\bar{\varepsilon}$ and ε:

$$\bar{\varepsilon}(t) = \frac{1}{1 + \varphi^{T}(t)P(t-1)\varphi(t)} \varepsilon(t) \tag{5.48}$$

[see (3.D.11)]. We may expect $\bar{\varepsilon}(t)$ to be a better estimate of the true prediction error $e(t)$ than $\varepsilon(t)$, but the difference is according to (5.48) quite small, except for small t. The effect of this difference on transient convergence behavior is illustrated in the following example.

EXAMPLE 5.21 (Comparison of Residuals and Prediction Errors) We simulated the system

$$y(t) - 0.8y(t-1) = 1.0u(t-1) + e(t) + 0.7e(t-1),$$

with $u(t)$ a PRBS, $S/N = 1$. Identification was by PLR in the model set

$$y(t) + ay(t-1) = bu(t-1) + e(t) + ce(t-1).$$

Both residuals and prediction errors were tried in the gradient. Two different forgetting factors of the form

$$\lambda(t) = \lambda_0\lambda(t-1) + (1 - \lambda_0)$$

were used. The results are summarized in table 5.14. The table shows that

Table 5.14
Parameter estimates for different λ_0 and $\lambda(0)$, using prediction errors and residuals (example 5.21). Ten runs.

N		λ_0	$\lambda(0)$	a	b	c
50	prediction	1.0	1.0	-0.799 ± 0.104	0.958 ± 0.065	0.379 ± 0.140
	errors	0.99	0.95	-0.799 ± 0.043	0.943 ± 0.091	0.474 ± 0.115
	residuals	1.0	1.0	-0.819 ± 0.105	0.978 ± 0.072	0.576 ± 0.157
		0.99	0.95	-0.809 ± 0.056	0.956 ± 0.089	0.638 ± 0.131
500	prediction	1.0	1.0	-0.792 ± 0.019	0.994 ± 0.034	0.606 ± 0.074
	errors	0.99	0.95	-0.794 ± 0.014	1.003 ± 0.036	0.684 ± 0.046
	residuals	1.0	1.0	-0.800 ± 0.023	0.999 ± 0.031	0.665 ± 0.039
		0.99	0.95	-0.796 ± 0.014	1.003 ± 0.036	0.698 ± 0.047
true value				-0.8	1.0	0.7

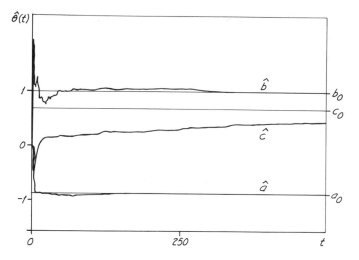

Figure 5.18a
Parameter estimates for one run of example 5.21. Prediction errors were used. The
forgetting factor was $\lambda(t) \equiv 1$.

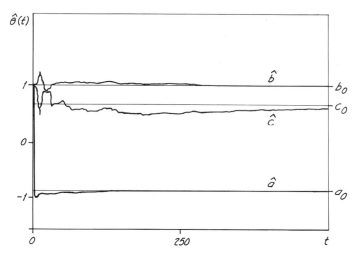

Figure 5.18b
As figure 5.18a, but with residuals.

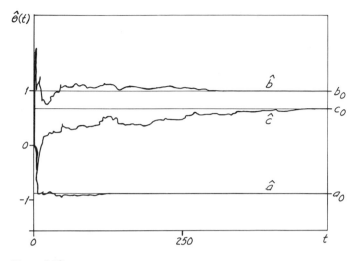

Figure 5.18c
As figure 5.18a, but with the forgetting factor $\lambda(t) = 0.99\lambda(t-1) + 0.01$, $\lambda(0) = 0.95$.

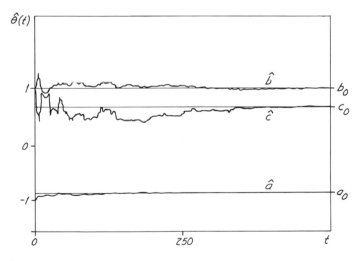

Figure 5.18d
As figure 5.18a, but with residuals and the forgetting factor $\lambda(t) = 0.99\lambda(t-1) + 0.01$, $\lambda(0) = 0.95$.

the estimate $\hat{c}$ is considerably improved when residuals are used instead of prediction errors. The time-variable forgetting factor gives the best performance. A typical run is illustrated in figures 5.18a–d. □

The martingale convergence proof for PLR in appendix 4.C relies upon the use of residuals in the regression vector. Moreover, lemma 4.2 assures a stable PLR without stability monitoring, provided residuals are used in $\varphi(t)$. The use of residuals rather than prediction errors in the gradient vector is thus well motivated.

Summary

The use of residuals (posterior prediction errors) instead of (prior) prediction errors results in more rapid transient convergence and greater accuracy. The effect is most significant for parameters that enter non-linearly into the prediction error. Since residuals require only slightly more computation than prediction errors, it is generally preferable to use residuals in the gradient vector.

5.12 Summary

The choice of a recursive identification algorithm for a particular application may not be easy. The literature offers an abundant supply of candidates. We have in this chapter tried to give a systematic discussion of how to choose a suitable algorithm among the many possible ones. The discussion has been carried out in terms of eight choices within the general family of algorithms (3.157)–(3.158) as listed in section 3.9:

- Choice of model,
- Choice of input,
- Choice of criterion functions,
- Choice of gain sequence,
- Choice of search direction,
- Choice of initial values,
- Choice of gradient approximation: RPEM, PLR, or RIV,
- Choice of residuals or prediction errors in the gradient vector.

The choices have been discussed using the analytical results of chapter 4, calculation, and simulation. The conclusions regarding these choices are summarized at the end of each section. These summaries will certainly be useful for applications to systems similar to the ones used in our examples. But a more important use of this chapter is perhaps that we have shown how to itemize the decision about what algorithm to use, and how to study the effects of the different items.

5.13 Bibliography

Section 5.1 A further discussion on validity criteria can be found in Söderström et al. (1974a, b). Wittenmark and Bar-Shalom (1979) discuss another validity measure related to control applications. It describes how uncertainties in the model parameters influence the closed-loop behavior when the system is used with the controller based on the identified model. Our criterion $\bar{V}_{\ddagger}^{*}(\theta)$ is a special case of this approach.

Section 5.2 An interesting and early discussion on the parsimony principle has been given by Tukcy (1961). See also Box and Jenkins (1970). Equation (5.8) can be used to compare any two model sets. To compare the expected values of other validity measures for two model sets, it is usually necessary to assume that one is a subset of the other; see Gustavsson et al. (1977) and Stoica and Söderström (1982b).

Section 5.4 A discussion of PRBSs and their use in identification can be found in Davies (1970) and Eykhoff (1974).

Section 5.5 Huber (1973) gives a comprehensive discussion of robust regression. The calculations leading to (5.23) can be found in Ljung (1978b).

Section 5.6 The tracking algorithm (5.27) based on an addition of R_1 when $P(t)$ is updated has been used by Bohlin (1976), who gives some interesting case studies.

There are many ways to generate the gain sequence $\{\gamma(t)\}$ such that $\gamma(t) \approx 1/t$ asymptotically. One possibility is to use (5.31c), i.e., to let the forgetting factor grow exponentially to 1. Kumar and Moore (1980) discuss other choices of $\gamma(t)$. In particular the variant

$$\gamma^{-1}(t) = 1 + K(t)\gamma^{-1}(t-1) \tag{5.49a}$$

with

$$K(t) \leq 1, K(t) \to 1 \text{ as } t \to \infty, \tag{5.49b}$$

is recommended. With $K(t)$ a constant less than 1, the gain sequence is given by

$$\gamma(t) = \frac{1 - K}{1 + K^t[(1 - K)/\gamma(0) - 1]}.$$

With some effort it can also be shown that the gain sequence given by (5.49a) will satisfy $\lim_{t \to \infty} t\gamma(t) = 1$, and, with an appropriate initial value, $t\gamma(t) \geq 1$ for all t.

Section 5.8 A discussion of nonwindowing and prewindowing can be found in Morf, Dickinson, Kailath, and Vieira (1977). The application of such algorithms to speech analysis is discussed by Markel and Gray (1976).

Section 5.9 Example 5.14 was first published in Ljung et al. (1975).

Section 5.10 The omitted calculations in example 5.18 can partly be found in Söderström and Stoica (1978). In example 5.19 we use an IV vector of dimension larger than the number of parameters to be estimated. It is shown by Stoica and Söderström (1983a) how theorem 4.8 can be generalized to this case. The result can be written as

$$P = \sigma^2 (R^{\mathrm{T}}R)^{-1} R^{\mathrm{T}} \mathrm{E}\left[T(q^{-1})H(q^{-1})\zeta(t) \cdot T(q^{-1})H(q^{-1})\zeta^{\mathrm{T}}(t) \right] R(R^{\mathrm{T}}R)^{-1},$$

where

$$R = \mathrm{E}\left[\zeta(t) \cdot T(q^{-1})\varphi^{\mathrm{T}}(t) \right].$$

6 Implementation

6.1 Introduction

In chapter 3 we developed the general Gauss-Newton algorithm for quadratic criteria:

$$\varepsilon(t) = y(t) - \hat{y}(t); \tag{6.1}$$

$$\hat{\Lambda}(t) = \hat{\Lambda}(t-1) + \gamma(t)[\varepsilon(t)\varepsilon^{T}(t) - \hat{\Lambda}(t-1)], \tag{6.2a}$$

$$R(t) = R(t-1) + \gamma(t)[\psi(t)\hat{\Lambda}^{-1}(t)\psi^{T}(t) - R(t-1)], \tag{6.2b}$$

$$L(t) = \gamma(t)R^{-1}(t)\psi(t)\hat{\Lambda}^{-1}(t); \tag{6.2c}$$

$$\hat{\theta}(t) = [\hat{\theta}(t-1) + L(t)\varepsilon(t)]_{D_{\mathcal{M}}}; \tag{6.3}$$

$$\xi(t+1) = A(\hat{\theta}(t))\xi(t) + B(\hat{\theta}(t))z(t), \tag{6.4a}$$

$$\begin{pmatrix} \hat{y}(t+1) \\ \mathrm{col}\,\psi(t+1) \end{pmatrix} = C(\hat{\theta}(t))\xi(t+1) \tag{6.4b}$$

[see (3.67)]. We analyzed the properties of (6.1)–(6.4) and its variants in chapter 4, and discussed the user aspects of the choices in chapter 5. We are now faced with the problem of implementing the chosen algorithm.

Even with all of the quantities in (6.1)–(6.4) chosen, there will be many algebraically equivalent ways of performing the calculations. The different ways of organizing the computations may have a substantial influence on the numerical properties of the identification algorithm. By numerical properties we include the following:

- Computing time required by one iteration of (6.1)–(6.4).
- Memory size.
- Numerical accuracy and stability (error propagation) related to round-off and other errors.
- Programming effort.

The algorithm can be regarded as consisting of four parts, (6.1)–(6.4). The first part, (6.1), is then trivial.

The second part, (6.2), gives the gain vector $L(t)$. This is probably the most interesting part from a numerical point of view. The reason is that $R(t)$ is a $d \times d$ ($d = \dim \theta$) matrix that has to be inverted in (6.2c); and this can be done in a number of different ways. We shall devote sections 6.2–6.5 to this problem. In section 6.2, we discuss various ways of implementing (6.2) that are valid for any gradient sequence $\{\psi(t)\}$. Often this sequence is such that elements are shifted down the vector $\psi(t)$ as t increases. When such a structure is at hand, it can be utilized to speed up the calulation of $L(t)$. This is described in sections 6.3 and 6.4. Finally, in section 6.5 we discuss how to cope with the problem that $R(t)$ as defined by (6.2b) may be singular or nearly singular. The problem of making matrices invertible is generally known as regularization.

The third part, (6.3), is fairly straightforward, once $L(t)$ is determined. The problem of how to accomplish the projection into the set $D_{\mathcal{M}}$, is discussed in section 6.6.

The final part, (6.4), depends on the particular model set chosen. Therefore, no general discussion of how to organize this part can be given. In chapter 3 we spelled out these calculations in some detail when we discussed applications to specific model sets [see (3.124) and (3.125) for the general input-output model set, and (3.145e–h) and appendix 3.B for state-space models].

There is a particular way of reorganizing (6.4), together with (6.2) and (6.3), that can be applied to PLRs. This is known as the ladder algorithm (also known as the lattice algorithm), and can be described as a continuous change of basis in the state-space description (6.4) to make the covariance matrix of $\xi(t)$ diagonal. There is a close relationship between such ladder implementations and the fast algorithm for $L(t)$, described in section 6.3. Therefore the ladder algorithms are discussed in the subsequent section 6.4.

6.2 Computation of the Gain Vector for the Gauss-Newton Algorithm

The gain vector is given by

$$L(t) = \gamma(t) R^{-1}(t) \psi(t) \hat{\Lambda}^{-1}(t), \tag{6.5a}$$

$$R(t) = R(t-1) + \gamma(t) [\psi(t) \hat{\Lambda}^{-1}(t) \psi^{\mathrm{T}}(t) - R(t-1)]. \tag{6.5b}$$

These expressions appear in all algorithms. They are not influenced by the model set except the way in which the gradient $\psi(t)$ depends on the measured data $\{z(t)\}$.

In this section we will discuss some approaches for the computation of $L(t)$ for the Gauss-Newton algorithm. A direct solution of (6.5) is not very practical, since (6.5b) contains the matrix $R(t)$, while $R^{-1}(t)$ is needed in (6.5a). This means that a system of linear equations must be solved in each step if (6.5) is to be implemented in a straightforward manner. There are, however, ways to compute $L(t)$ without inverting large matrices in each step. The following three strategies will be described:

1. *Using the Matrix Inversion Lemma.* A difference equation for $P(t) = \gamma(t)R^{-1}(t)$ is used instead of (6.5b) (section 6.2.1).

2. *Using Factorization.* The matrix $P(t)$ is factored into a product of two or three matrices, and these matrices are then updated rather than P itself. The advantage of this approach is that better numerical properties can be obtained (section 6.2.2).

3. *Using Fast Algorithms.* When only $L(t)$ is needed it may be unnecessary to compute $R(t)$, which is of larger dimension than $L(t)$. This idea can be used to derive so-called fast algorithms. With these, the number of operations and the memory requirements can be reduced, in particular for model sets with many independent parameters (section 6.3.2).

We have mentioned earlier, in section 2.3, that recursive identification algorithms are closely linked to Kalman filtering. In fact, much of the results discussed in the present section have been primarily derived for Kalman filters. Since (6.5) is algebraically identical to the gain and the covariance equations for a Kalman filter, those general results can and will be applied here.

6.2.1 Using the Matrix Inversion Lemma

Coping with (6.5) by means of the matrix inversion lemma 2.1 was considered in section 3.4. There it was shown that the substitution

$$P(t) \triangleq \gamma(t)R^{-1}(t) \tag{6.6}$$

gives with (6.5b) the new recursion

$$P(t) = [P(t-1) - P(t-1)\psi(t)S^{-1}(t)\psi^T(t)P(t-1)]/\lambda(t), \tag{6.7a}$$

$$S(t) = \psi^T(t)P(t-1)\psi(t) + \lambda(t)\hat{\Lambda}(t), \tag{6.7b}$$

where the time-varying forgetting factor $\lambda(t)$ is related to the gain sequence $\{\gamma(t)\}$ by

$$\lambda(t) = \frac{\gamma(t-1)}{\gamma(t)}[1 - \gamma(t)] \tag{6.8}$$

[see (3.69)]. Then the gain vector is easily found to be

$$L(t) = P(t)\psi(t)\hat{\Lambda}^{-1}(t) = P(t-1)\psi(t)S^{-1}(t) \tag{6.9}$$

[see (3.70d)]. It is advantageous to use the second of these expressions, since the product $P(t-1)\psi(t)S^{-1}(t)$ must be computed anyway to get $P(t)$ [see (6.7a)].

Unfortunately the recursion (6.7a) is not numerically sound; the equation is sensitive to round-off errors, that can accumulate and make $P(t)$ indefinite. This property of (6.7a) is discussed by Bierman (1977). One might suspect dubious numerical properties of (6.7a) especially when $\lambda(t) = 1$. Then the new P matrix is computed by successive subtractions of correction terms. If the correction terms due to round-off errors become too large, it is easy to imagine that numerical problems can be encountered.

Equation (6.7a) can be rewritten in a form that apparently has better numerical properties. Straightforward calculation gives

$$\begin{aligned} P(t) &= [I - L(t)\psi^{\mathrm{T}}(t)]P(t-1)[I - \psi(t)L^{\mathrm{T}}(t)]/\lambda(t) \\ &\quad + L(t)\hat{\Lambda}(t)L^{\mathrm{T}}(t). \end{aligned} \tag{6.10}$$

This is sometimes referred to as the "stabilized Kalman equation" (Bierman, 1977). With this expression, repeated subtraction is avoided. Nevertheless, this recursion does not guarantee numerical stability, although it has somewhat better numerical properties than the original equation (6.7a). However, it also requires considerably more computation.

One should not exaggerate the numerical problems of (6.7a). When the dimension of $P(t)$ is low, say $d \le 10$, often no problems are encountered. The problems are usually associated with applications for large d and/or ill-conditioned P matrices. As Thornton and Bierman (1977) rightly point out, though, this is no excuse for ignoring the problems altogether.

6.2.2 Using Factorization

For factorization, we start with the equation (6.7a) for $P(t)$ and with $P(t)$ as a product of matrices. Since $P(t)$ is known to be positive definite it can be decomposed, e.g., by using Cholesky decomposition.

We first describe a square root algorithm, in terms of $P(t)$ represented as

$$P(t) = Q(t)Q^{\mathrm{T}}(t), \tag{6.11}$$

where $Q(t)$ is a nonsingular matrix. Equation (6.7a) is then replaced by one for computing $Q(t)$ from $Q(t-1)$.

It is possible to constrain the matrix $Q(t)$ to be triangular. Then the decomposition (6.11) is nothing but the Cholesky decomposition. The algorithm for updating $Q(t)$ will then require more calculations, but it is still relatively simple.

A popular factorization is known as *U-D factorization*; this can be described as a normalized Cholesky decomposition (also called square root free Cholesky decomposition). For it, $P(t)$ is written as

$$P(t) = U(t)D(t)U^{\mathrm{T}}(t), \tag{6.12}$$

where $U(t)$ is an upper triangular matrix with all diagonal elements equal to 1, and where $D(t)$ is a diagonal matrix.

Using factorization for computing $P(t)$ guarantees that $P(t)$ remains positive definite. Extensive experiments for the Kalman filter case have shown that the aforementioned factorizations have good numerical stability, and that the rounding errors do not affect the solution significantly (Bierman, 1977; Thornton and Bierman, 1980).

We shall next describe in some detail a square root algorithm due to Potter (1963) and the *U-D* factorization given by Bierman (1977). We shall consider a system having a scalar output. Then $\hat{\Lambda}(t)$ in (6.7b) is also scalar. If this is chosen to be a constant, then it simply acts as a scale factor, and can as well be chosen as 1, as we remarked in section 3.4. A time-varying scalar weighting $\hat{\Lambda}^{-1}(t) = \alpha_t$ may make sense in some applications. We shall, however, first treat the case $\alpha_t = 1$, remarking on the general case later. We consider the following version of equation (6.7):

$$P(t) = \frac{P(t-1)}{\lambda(t)} - \frac{P(t-1)\psi(t)\psi^{\mathrm{T}}(t)P(t-1)}{\lambda(t) + \psi^{\mathrm{T}}(t)P(t-1)\psi(t)} \cdot \frac{1}{\lambda(t)}. \tag{6.13}$$

The extension to the multivariable case will be discussed later in this section.

Potter's Square Root Algorithm   The square root algorithm due to Potter (1963) is based on the factorization (6.11). The matrix $Q(t)$ is computed by means of the following algorithm.

Potter's Square Root Algorithm	(6.14)

Initialize at time $t = 0$: $Q(0)Q^T(0) = P(0)$.

At time t, update $Q(t-1)$ by performing steps 1–5.

1. $f(t) := Q^T(t-1)\psi(t)$.

2. $\beta(t) := \lambda(t) + f^T(t)f(t)$.

3. $\alpha(t) := 1/[\beta(t) + \sqrt{\beta(t)\lambda(t)}]$.

4. $\bar{L}(t) := Q(t-1)f(t)$.

5. $Q(t) := [Q(t-1) - \alpha(t)\bar{L}(t)f^T(t)]/\sqrt{\lambda(t)}$.

The algorithm becomes a bit simpler when $\lambda(t) = 1$, since the square root and the division in step 5 then can be avoided. The vector $\bar{L}(t)$ is a normalized form of the gain vector $L(t)$. We have

$$L(t) = \bar{L}(t)/\beta(t).$$

Note that it is not necessary to compute $L(t)$ explicitly, since the parameter updating can with advantage be computed by

$$\hat{\theta}(t) = \hat{\theta}(t-1) + \bar{L}(t)[\varepsilon(t)/\beta(t)], \tag{6.15}$$

i.e., the single division $\varepsilon(t)/\beta(t)$ is computed first.

For real-time tracking there is often reason to use the algorithm (2.112) as an alternative to (6.13) (see also section 5.5). Then the equations

$$\tilde{P}(t) = P(t-1) - \frac{P(t-1)\psi(t)\psi^T(t)P(t-1)}{1 + \psi^T(t)P(t-1)\psi(t)}, \tag{6.16a}$$

$$P(t) = \tilde{P}(t) + R_1(t) \tag{6.16b}$$

are used. If a square root approach is taken, the algorithm (6.14) can be used for finding $\tilde{P}(t) = \tilde{Q}(t)\tilde{Q}^T(t)$ from $P(t-1) = Q(t-1)Q^T(t-1)$. It then remains to find $Q(t)$ using (6.16b). One way to do this is as follows. Let $R_1(t)$ be factored as

$$R_1(t) = V(t)V^T(t), \tag{6.17}$$

where $V(t)$ is a $d \times s$-matrix of full rank s. In most cases $R_1(t)$ is a diagonal matrix with some diagonal elements equal to zero. In such cases it is easy to find $V(t)$. Then orthogonal transformations are applied to the rectangular matrix $(\tilde{Q}(t)\ V(t))$. The problem is to find an orthogonal matrix $T(t)$ and a triangular matrix $Q(t)$ such that

$$(\tilde{Q}(t)\ V(t))T(t) = (Q(t)\ 0). \tag{6.18}$$

Then we have

$$\tilde{P}(t) + R_1(t) = \tilde{Q}(t)\tilde{Q}^\mathrm{T}(t) + V(t)V^\mathrm{T}(t)$$

$$= (\tilde{Q}(t) \ \ V(t))T(t)T^\mathrm{T}(t)\begin{pmatrix} \tilde{Q}^\mathrm{T}(t) \\ V^\mathrm{T}(t) \end{pmatrix}$$

$$= (Q(t) \ \ 0)\begin{pmatrix} Q^\mathrm{T}(t) \\ 0 \end{pmatrix} = P(t),$$

as required by (6.16b). The matrices $T(t)$ and $Q(t)$ in (6.18) can be found using a QR factorization or a Gram-Schmidt orthogonalization. Such factorizations are common in numerical linear algebra for solving certain eigenvalue and least squares problems (see Stewart, 1970), and have appeared in many applications.

Bierman's U-D Algorithm for (6.13)   We next turn to the U-D factorization algorithm. Our treatment follows Bierman (1977). Consider again the equation (6.13). Assume that the matrix $P(t-1)$ is factored as

$$P(t-1) = U(t-1)D(t-1)U^\mathrm{T}(t-1),$$

where $U(t-1)$ is upper triangular with all diagonal elements equal to 1, and $D(t-1)$ is a diagonal matrix. We seek a similar factorization of $P(t)$. We have

$$U(t)D(t)U^\mathrm{T}(t)$$

$$= \left[U(t-1)D(t-1)U^\mathrm{T}(t-1) - \frac{U(t-1)g(t)g^\mathrm{T}(t)U(t-1)}{\beta(t)}\right]\bigg/\lambda(t), \tag{6.19a}$$

where

$$f(t) = U^\mathrm{T}(t-1)\psi(t), \tag{6.19b}$$

$$g(t) = D(t-1)f(t), \tag{6.19c}$$

$$\beta(t) = \lambda(t) + \psi^\mathrm{T}(t)P(t-1)\psi(t) = \lambda(t) + f^\mathrm{T}(t)g(t). \tag{6.19d}$$

It follows from (6.19) that

$$U(t)D(t)U^\mathrm{T}(t) = U(t-1)\left[D(t-1) - \frac{g(t)g^\mathrm{T}(t)}{\beta(t)}\right]U^\mathrm{T}(t-1)/\lambda(t).$$

If the part in brackets can be factored as

$$D(t-1) - \frac{g(t)g^{\mathrm{T}}(t)}{\beta(t)} = \bar{U}(t)\bar{D}(t)\bar{U}^{\mathrm{T}}(t), \tag{6.20}$$

we get

$$U(t) = U(t-1)\bar{U}(t), \tag{6.21a}$$

$$D(t) = \bar{D}(t)/\lambda(t). \tag{6.21b}$$

It remains to find the factorization (6.20). To simplify the writing, the time argument will be dropped. Introduce the notation (recall that $d = \dim\theta$)

$$\bar{U}(t) = \begin{pmatrix} \bar{U}_1 & \cdots & \bar{U}_d \end{pmatrix} = \begin{pmatrix} 1 & \bar{U}_{1,2} & \cdots & \bar{U}_{1,d} \\ & 1 & & \vdots \\ & & \ddots & \vdots \\ & & & \bar{U}_{d-1,d} \\ 0 & & & 1 \end{pmatrix},$$

$$\bar{D}(t) = \begin{pmatrix} \bar{D}_1 & & 0 \\ & \ddots & \\ 0 & & \bar{D}_d \end{pmatrix},$$

$$D(t-1) = \begin{pmatrix} D_1 & & 0 \\ & \ddots & \\ 0 & & D_d \end{pmatrix},$$

and let e_i be the ith unit vector. Then (6.20) implies

$$\sum_{i=1}^{d} \bar{U}_i \bar{U}_i^{\mathrm{T}} \bar{D}_i = \sum_{i=1}^{d} D_i e_i e_i^{\mathrm{T}} - \frac{1}{\beta} g g^{\mathrm{T}}. \tag{6.22}$$

Introduce also ($f_i = i$th component of f)

$$\beta_d = \beta = \lambda(t) + \sum_{i=1}^{d} f_i g_i, \quad c_d = 1/\beta, \quad V_d = g.$$

We shall denote the ith component of the column vector V_d by $V_{d,i}$. Then (6.20) can be written

$$\sum_{i=1}^{d} \bar{D}_i \bar{U}_i \bar{U}_i^{\mathrm{T}} = \sum_{i=1}^{d} D_i e_i e_i^{\mathrm{T}} - \frac{1}{\beta_d} V_d V_d^{\mathrm{T}}. \tag{6.23}$$

We shall determine $\bar{D}_i$ and $\bar{U}_i$ (given β, D_i and V_d) from this relation. To do that consider the matrix

$$M_d = \bar{D}_d \bar{U}_d \bar{U}_d^T - D_d e_d e_d^T + \frac{1}{\beta_d} V_d V_d^T.$$

It is easy to verify that the choices

$$\bar{D}_d = D_d - \frac{V_{d,d}^2}{\beta_d}, \tag{6.24a}$$

$$\bar{U}_{d,d} = 1, \tag{6.24b}$$

$$\bar{U}_{i,d} = -\frac{V_{d,d}}{\bar{D}_d \cdot \beta_d} V_{d,i}, \quad i = 1, \ldots, d-1, \tag{6.24c}$$

will make the last row and last column of M_d equal to zero. With

$$V_{d-1} = \begin{pmatrix} V_{d,1} \\ \vdots \\ V_{d,d-1} \\ 0 \end{pmatrix},$$

we thus find that M_d can be written

$$M_d = \left(\frac{V_{d,d}^2}{\bar{D}_d \beta_d^2} + \frac{1}{\beta_d} \right) V_{d-1} V_{d-1}^T. \tag{6.24d}$$

If we introduce

$$\beta_k = \lambda(t) + \sum_{j=1}^{k} f_j g_j \quad (\beta_d = \beta),$$

we find, using (6.19) that

$$\frac{V_{d,d}^2}{\bar{D}_d \beta_d^2} + \frac{1}{\beta_d} = \frac{1}{\beta_{d-1}}$$

and

$$\bar{D}_d = D_d \beta_{d-1} / \beta_d, \tag{6.25a}$$

$$\bar{U}_{i,d} = -(f_d / \beta_{d-1}) g_i, \quad i = 1, \ldots, d-1. \tag{6.25b}$$

Now, returning to (6.23), we find that

$$\sum_{i=1}^{d-1} \bar{D}_i \bar{U}_i \bar{U}_i^{\mathrm{T}} = \sum_{i=1}^{d-1} D_i e_i e_i^{\mathrm{T}} - M_d = \sum_{i=1}^{d-1} D_i e_i e_i^{\mathrm{T}} - \frac{1}{\beta_{d-1}} V_{d-1} V_{d-1}^{\mathrm{T}},$$

provided $\bar{U}_d$ and $\bar{D}_d$ are chosen according to (6.25). This expression is, however, exactly of the form (6.23), except that d has decreased to $d - 1$. Therefore the same procedure can be used again to find numerically $\bar{D}_{d-1}$, $\bar{U}_{d-1}$, etc.

The algorithm to find U and D can be performed together with the multiplication (6.21) to determine the updates $U(t)$, $D(t)$ and the gain vector

$$L(t) = U(t - 1)D(t - 1)U^{\mathrm{T}}(t - 1)\psi(t)/\beta(t) = U(t - 1)g(t)/\beta(t).$$

This leads to the following algorithm, given by Bierman (1977).

Bierman's U-D Factorization Algorithm for (6.13)   (6.26)

Initialize $U(0)$ and $D(0)$ at time $t = 0$, $U(0)D(0)U^{\mathrm{T}}(0) = P(0)$.
At time t, compute $L(t)$ and update $U(t - 1)$ and $D(t - 1)$ by performing steps 1–6.

1. Compute $f := U^{\mathrm{T}}(t - 1)\psi(t)$, $\quad g := D(t - 1)f$, $\quad \beta_0 := \lambda(t)$.

2. For $j = 1, \ldots, d$, go through the steps 3–5.

3. Compute
$$\beta_j := \beta_{j-1} + f_j g_j,$$
$$D(t)_{jj} := \beta_{j-1} D(t - 1)_{jj}/\beta_j \lambda(t),$$
$$v_j := g_j,$$
$$\mu_j := -f_j/\beta_{j-1}.$$

4. For $i = 1, \ldots, j - 1$, go through step 5. (If $j = 1$, skip step 5).

5. Compute
$$U(t)_{ij} := U(t - 1)_{ij} + v_i \mu_j,$$
$$v_i := v_i + U(t - 1)_{ij} v_j.$$

6. $\bar{L}(t) := \begin{pmatrix} v_1 \\ \vdots \\ v_d \end{pmatrix}$; $\quad L(t) := \bar{L}(t)/\beta_d.$

The scalar β_d obtained after the dth cycle of steps 3–5 is the "innovations variance"

$$\beta_d = \lambda(t) + \psi^{\mathrm{T}}(t)P(t-1)\psi(t).$$

Similarly the normalized Kalman gain $\bar{L}(t)$ [see (6.15)] is obtained in step 6 after d cycles of steps 3–5.

U-D Algorithm for (6.16b)   Let us now continue to discuss how to treat the equations (6.16b). The following method, based on Gram-Schmidt orthogonalization, is due to Thornton and Bierman (1977).

Assume again that a factorization (6.17) of $R_1(t)$ is available. Then form a matrix W with row vectors $W_1^{\mathrm{T}}, \ldots, W_d^{\mathrm{T}}$:

$$W = \begin{pmatrix} W_1^{\mathrm{T}} \\ \vdots \\ W_d^{\mathrm{T}} \end{pmatrix} = (\tilde{U}(t) \quad V(t)),$$

and set

$$D = \begin{pmatrix} \tilde{D}(t) & 0 \\ 0 & I_s \end{pmatrix},$$

where I_s is the identity matrix of dimension s (recall $V(t)$ is $d \times s$) and the matrix $\tilde{P}(t)$ of (6.16) is factored as

$$\tilde{P}(t) = \tilde{U}(t)\tilde{D}(t)\tilde{U}^{\mathrm{T}}(t).$$

We can then write (6.16b) as

$$P(t) = \tilde{P}(t) + R_1(t) = \tilde{U}(t)\tilde{D}(t)\tilde{U}^{\mathrm{T}}(t) + V(t)V^{\mathrm{T}}(t) = WDW^{\mathrm{T}}.$$

To obtain a U-D factorization of $P(t)$ we need to rewrite

$$WDW^{\mathrm{T}} = U(t)D(t)U^{\mathrm{T}}(t)$$

for some diagonal matrix $D(t)$ and some upper triangular matrix $U(t)$ with 1's along the diagonal. Applying a Gram-Schmidt orthogonalization to the vectors $W_1, \ldots, W_d$ using the scalar product

$$< W_i, W_j> \ = W_i^{\mathrm{T}}DW_j$$

gives a matrix $\tilde{W} = TW$ with orthogonal column vectors (in the D-norm), i.e.,

$\tilde{W}D\tilde{W}^{\mathrm{T}} = $ diagonal.

Here T is a triangular matrix. It is now easy to verify that the desired decomposition is obtained with $U(t) = T^{-1}$ and $D(t) = \tilde{W}D\tilde{W}^{\mathrm{T}}$. The final algorithm can be organized in the following way.

Updating the U-D Factorization for the Algorithm (6.16b)

At time $t - 1$ $U(t - 1)$ and $D(t - 1)$ are given as well as the factorization $R_1(t) = V(t)V^{\mathrm{T}}(t)$ with $V(t)$ as a full rank $d \times s$ matrix.

1. Compute $\bar{L}(t)$, $\tilde{U}(t)$ and $\tilde{D}(t)$ by performing steps 1–6 of (6.26) ($\tilde{U}(t)$ and $\tilde{D}(t)$ are the matrices called $U(t)$ and $D(t)$ in (6.26)).

2. Define the $(d + s)$-dimensional column vector $W_k^{(0)}$ as the kth column of $U^{\mathrm{T}}(t)$ stacked on top of the kth column of $V^{\mathrm{T}}(t)$; $k = 1, \ldots, d$.

3. Define the $(d + s) \times (d + s)$ diagonal matrix D as the block diagonal matrix formed from $\tilde{D}(t)$ and the $s \times s$ identity matrix.

4. For $j = d, d - 1, \ldots, 2$, go through steps 5–8.

5. Compute
$$D(t)_{jj} := [W_j^{(d-j)}]^{\mathrm{T}}DW_j^{(d-j)}.$$

6. For $i = 1, 2, \ldots, j - 1$, go through step 7.

7. Compute
$$U(t)_{ij} := [W_i^{(d-j)}]^{\mathrm{T}}DW_j^{(d-j)}/D(t)_{jj},$$
$$W_i^{(d-j+1)} := W_i^{(d-j)} - U(t)_{jj}W_j^{(d-j)}.$$

8. Compute
$$D(t)_{11} := [W_1^{(d-1)}]^{\mathrm{T}}DW_1^{(d-1)}.$$

Remark If a general weighting sequence is used for the scalar output case, $\hat{\Lambda}^{-1}(t) = \alpha_t$, the difference in the square root algorithm (6.14) is that (6.14:2) is replaced by

$$\beta(t) := \lambda(t)/\alpha_t + f^{\mathrm{T}}(t)f(t).$$

For the U-D factorization algorithm (6.26) the difference is that the quantity β_0 is initialized as

$$\beta_0 := \lambda(t)/\alpha_t.$$

Multioutput Systems   We shall now discuss the treatment of multioutput systems. This seems to be a topic where there presently is no complete theory. One approach, indicated in Bierman (1977), is to transform the problem into a sequence of scalar problems. To be more specific, consider again (6.7). Assume that a Cholesky decomposition is applied to the estimated covariance matrix $\hat{\Lambda}(t)$. Then

$$\hat{\Lambda}(t) = M(t)M^{\mathrm{T}}(t),$$

where $M(t)$ is a triangular matrix. Its inverse is easy to find. Introduce the notation

$$\bar{\psi}(t) = \psi(t)M^{-\mathrm{T}}(t) = [\bar{\psi}_1(t) \ \ldots \ \bar{\psi}_p(t)],$$

where $M^{-\mathrm{T}}$ denotes $(M^{-1})^{\mathrm{T}}$ and $\bar{\psi}_i(t)$ denotes the ith column vector of $\bar{\psi}(t)$. Then (6.7) becomes

$$P(t) = \{P(t-1) - P(t-1)\bar{\psi}(t)[\bar{\psi}^{\mathrm{T}}(t)P(t-1)\bar{\psi}(t) + \lambda(t)I]^{-1}$$
$$\times \bar{\psi}^{\mathrm{T}}(t)P(t-1)\}/\lambda(t). \tag{6.27}$$

The recursion (6.27) can now be replaced by the sequence of "scalar outputs"

$$P_0(t) = P(t-1), \tag{6.28a}$$

$$P_i(t) = P_{i-1}(t) - P_{i-1}(t)\bar{\psi}_i(t)[\bar{\psi}_i^{\mathrm{T}}(t)P_{i-1}(t)\bar{\psi}_i(t)$$
$$+ \lambda(t)]^{-1}\bar{\psi}_i^{\mathrm{T}}(t)P_{i-1}(t) \tag{6.28b}$$

for $i = 1, \ldots, p$. Then

$$P(t) = P_p(t)/\lambda(t). \tag{6.28c}$$

This follows, e.g., from the interpretation in terms of state-estimation theory. Since there are p different and independent outputs, the optimal state-estimation problem can be solved by adding the effects of one output at a time. Equation (6.28b) looks formally like a time update [see (6.16a)]. We can therefore apply the U-D factorization method in a straightforward way to (6.28b) by using the algorithm (6.26).

6.2.3 Summary

We have now described four methods for determining the gain: equations (6.7a), (6.10), (6.14), and (6.26). A comparison concerning the number of

Table 6.1
Number of arithmetic operations used for updating $P(t)$ once. The number of parameters is d. Adapted from Bierman (1977).

Method	Additions	Multiplications	Divisions	Square roots
Conventional Kalman equation (6.7a)	$1.5d^2 + 3.5d$	$1.5d^2 + 4.5d$	1	0
Stabilized Kalman equation (6.10)	$4.5d^2 + 5.5d$	$4d^2 + 7.5d$	1	0
Potter's square root (6.14)	$3d^2 + 3d$	$3d^2 + 4d$	2	1
U-D factorization (6.26)	$1.5d^2 + 1.5d$	$1.5d^2 + 5.5d$	d	0

arithmetic operations required by these algorithms is given in table 6.1, which is adapted from Bierman (1977). It is clear from the table that the U-D factorization algorithms require roughly the same amount of computation as the Kalman equations. It must be remembered, though, that the updating of $P(t)$ is only a part of the total algorithm, so the total number of operations needed per sample is higher than that given in table 6.1. The computations needed to get $\varepsilon(t)$ and $\psi(t)$ will depend critically on what model set is chosen.

6.3 Fast Algorithms for Gain Computation

The ways of computing the gain vector $L(t)$ in the recursive identification algorithm that we discussed in the previous section were valid for any sequence of "gradient" vectors $\psi(t)$. In this section we shall discuss algorithms that utilize a specific structure of $\{\psi(t)\}$ that is present for many typical model sets. This structure is defined and illustrated in section 6.3.1. In section 6.3.2 it is explained how the structure in question allows a fast way of computing $L(t)$. The computational load per time step is then reduced from being proportional to d^2 ($d = \dim \theta$) for the algorithms in section 6.2 to being proportional to d. Section 6.3.2 is therefore an interesting complement to section 6.2.

It should be noted that the ladder algorithms to be discussed in section 6.4 also use the structure described in section 6.3.1.

6.3.1 A Shift Structure

The gradient vector in the algorithms we have been discussing is of major importance. We have here used the symbol $\psi(t)$ for it. Recall however, that the structure covers also PLRs (ψ replaced by φ) and symmetric recursive IV algorithms (ψ replaced by ζ).

Now, in most cases, the sequence $\{\psi(t)\}$ possesses some structure in the sense that $\psi(t)$ and $\psi(t+1)$ are related. Typically they will have several elements in common. The reason is, loosely speaking, that for dynamical systems the states are closely related at adjacent sampling instants.

We consider the structure

$$\psi(t) = \begin{pmatrix} x(t-1) \\ \vdots \\ x(t-n) \end{pmatrix}, \tag{6.29}$$

where $\{x(\cdot)\}$ is a sequence of α-dimensional column vectors (α-vectors). The dimension of $\psi(t)$ is consequently $\alpha \cdot n = d$. The important thing in (6.29) is that $\psi(t+1)$ and $\psi(t)$ will have a number of elements in common, viz., $x(s)$, $t - n + 1 \le s \le t - 1$.

If we introduce the vector

$$\psi^*(t) \triangleq \begin{pmatrix} x(t) \\ \psi(t) \end{pmatrix}, \tag{6.30}$$

we see that (6.29) can also be expressed as

$$\psi^*(t) = \begin{pmatrix} x(t) \\ \psi(t) \end{pmatrix} = \begin{pmatrix} \psi(t+1) \\ x(t-n) \end{pmatrix}. \tag{6.31}$$

We could also have assumed the slightly more general structure that for some $\psi^*(t)$ we have

$$\begin{pmatrix} x(t) \\ \psi(t) \end{pmatrix} = \mathscr{S}_F \psi^*(t), \quad \begin{pmatrix} \psi(t+1) \\ \tilde{x}(t-n) \end{pmatrix} = \mathscr{S}_B \psi^*(t), \tag{6.32}$$

where $\mathscr{S}_F$ and $\mathscr{S}_B$ are permutation matrices, and $x(t)$ and $\tilde{x}(t)$ are not necessarily the same. Clearly, (6.31) is a special case of (6.32). However, in order not to get involved in too-complex notation, we shall confine ourselves to the simpler assumption (6.29).

We shall now illustrate how the structure (6.29) relates to the model sets we have discussed previously.

EXAMPLE 6.1 (Multivariable Difference Equations)   Consider the linear regression model (3.78) in example 3.8 (with $n = m$):

$$y(t) + A_1 y(t - 1) + \cdots + A_n y(t - n)$$
$$= B_1 u(t - 1) + \cdots + B_n u(t - n) + v(t). \tag{6.33}$$

Here $y(t)$ is a p-dimensional column vector and $u(t)$ an r-dimensional column vector. A_i and B_i are matrices of compatible dimensions. We found in example 3.8 that the predictor for (6.33) could be written

$$\hat{y}(t \mid \theta) = \theta^T \varphi(t), \tag{6.34}$$

where

$$\theta^T = (A_1 \; B_1 \; A_2 \; B_2 \; \ldots \; A_n \; B_n) \tag{6.35}$$

and

$$\varphi(t) = \begin{pmatrix} x(t-1) \\ \vdots \\ x(t-n) \end{pmatrix}, \quad x(t) = \begin{pmatrix} -y(t) \\ u(t) \end{pmatrix}. \tag{6.36}$$

In (6.35) and (6.36) we have reordered the elements compared to (3.79). Due to (6.36), the model (6.33) leads to a shift structure (6.29) for the regression vector $\varphi(t) = \psi(t)$. In this case $\alpha = \dim x = p + r$. □

EXAMPLE 6.2 (A General Linear Input-Output Model)   Consider the general linear black box model (3.104),

$$A(q^{-1}) y(t) = \frac{B(q^{-1})}{F(q^{-1})} u(t) + \frac{C(q^{-1})}{D(q^{-1})} e(t). \tag{6.37}$$

The orders of the polynomials are n_a, n_b, n_f, n_c, and n_d respectively [see (3.97)]. We shall in this example assume that the nonzero orders are all the same, say n. That is,

$$n_a = n \text{ or } 0, \quad n_b = n \text{ or } 0, \quad n_c = n \text{ or } 0, \quad n_d = n \text{ or } 0, \quad n_f = n \text{ or } 0. \tag{6.38}$$

In section 3.7.2 we derived a RPEM for this model. It is given by (3.123)–

(3.125). The gradient vector $\psi(t)$ is given by (3.125):

$$\psi(t) = \begin{pmatrix} \tilde{x}(t-1) \\ \vdots \\ \tilde{x}(t-n) \end{pmatrix}, \tag{6.39}$$

where

$$\tilde{x}(t) = \begin{pmatrix} -\tilde{y}(t) \\ \tilde{u}(t) \\ -\tilde{w}(t) \\ \tilde{\varepsilon}(t) \\ -\tilde{v}(t) \end{pmatrix}. \tag{6.40}$$

Here we have reordered the elements of θ, φ, and ψ, compared to (3.123)–(3.125), so that

$$\theta^{\mathrm{T}} = (a_1 \ b_1 \ f_1 \ c_1 \ d_1 \ a_2 \ b_2 \ f_2 \ c_2 \ d_2 \ \ldots \ a_n \ b_n \ f_n \ c_n \ d_n).$$

Similarly,

$$\varphi(t) = \begin{pmatrix} x(t-1) \\ \vdots \\ x(t-n) \end{pmatrix}, \quad x(t) = \begin{pmatrix} -y(t) \\ u(t) \\ -w(t) \\ \bar{\varepsilon}(t) \\ -v(t) \end{pmatrix}. \tag{6.41}$$

Hence both the RPE algorithm (3.123)–(3.125) and the PLR (3.130) (ψ replaced by φ) are subject to the structure (6.29).

In (6.40) and (6.41) we have assumed that all model orders in (6.37) are nonzero. If a certain polynomial order is zero, then the corresponding entry in $x(t)$ is deleted (e.g., $n_f = 0 \Rightarrow w(t)$ is deleted). Hence the number $\alpha = \dim x$ equals the number of nonzero polynomial orders in (6.37). $\square$

The structure (6.29) or (6.31) is thus quite common for recursive identification algorithms. The restriction that all model polynomials should have the same order is a consequence of the simple structure (6.31). With (6.32) we could have included also the case of different model orders.

6.3.2   Fast Calculation of the Gain Matrix

Here we shall derive an algorithm for computation of the gain $L(t)$, given by (6.2), that utilizes the shift structure (6.29). The section is based on the paper by Ljung, Morf, and Falconer (1978). The derivation is fairly long, and we have broken it into several steps, with details given in appendix 6.B.

The Problem   The problem is to determine $L(t)$, given by (6.2). We shall treat the case when $\psi(t)$ is a column vector (which covers both examples 6.1 and 6.2). Then $\hat{\Lambda}(t)$ is a scalar, and we shall assume that $\hat{\Lambda}(t) \equiv 1$. This means that we consider the algorithm

$$R(t) = R(t - 1) + \gamma(t)[\psi(t)\psi^{\mathrm{T}}(t) - R(t - 1)], \tag{6.42a}$$

$$L(t) = \gamma(t)R^{-1}(t)\psi(t). \tag{6.42b}$$

Assume that the initial condition is

$$R(0) = \delta I \text{ for some } \delta \geq 0. \tag{6.42c}$$

If we introduce

$$\bar{R}(t) = \frac{1}{\gamma(t)}R(t),$$

we know from (2.124b) and (2.126b) that (6.42a) can be rewritten as

$$\bar{R}(t) = \lambda(t)\bar{R}(t - 1) + \psi(t)\psi^{\mathrm{T}}(t), \tag{6.43}$$

where $\lambda(t)$ is given by (6.8).
We shall now make the further restriction that

$$\lambda(t) \equiv \lambda. \tag{6.44}$$

Hence

$$\bar{R}(t) = \sum_{k=1}^{t} \lambda^{t-k}\psi(k)\psi^{\mathrm{T}}(k) + \lambda^{t}\delta \cdot I. \tag{6.45}$$

The problem is thus to find the solution $L(t)$ to

$$\bar{R}(t)L(t) = \psi(t), \tag{6.46}$$

with $\bar{R}(t)$ given by (6.43) and (6.44).

Exploiting the Shift Structure   The idea is to utilize the structure (6.29) or (6.31) in $\psi(t)$ when solving (6.46). This structure will make $L(t)$ and $L(t + 1)$ closely related, as we will now show.

Let $\psi^*(t)$ be defined by (6.30) and define

$$\bar{R}^*(t) = \sum_{k=1}^{t} \lambda^{t-k} \psi^*(k) [\psi^*(k)]^{\mathrm{T}} + \lambda^t \delta \cdot I. \tag{6.47}$$

This is a matrix of dimension $(n + 1)\alpha \times (n + 1)\alpha$. Now, the property (6.31) implies the relationship

$$\bar{R}^*(t) = \begin{pmatrix} ***** \\ * \bar{R}(t) \\ * \end{pmatrix} = \begin{pmatrix} \tilde{R}(t + 1) \\ ********* \end{pmatrix}, \tag{6.48a}$$

where * marks bordering elements, whose exact form does not interest us for the moment. The "width" of these elements is α.

In (6.48a) we have

$$\tilde{R}(t + 1) = \bar{R}(t + 1) - \lambda^t \psi(1) \psi^{\mathrm{T}}(1) + \lambda^t (1 - \lambda) \delta I. \tag{6.48b}$$

Now, if $x(t) = 0$ for $t \le 0$, the second term of the right-hand side will be zero. If either $\lambda = 1$, or $\delta = 0$ the third term will also be zero, in which case we have

$$\tilde{R}(t + 1) = \bar{R}(t + 1). \tag{6.48c}$$

We will in this section as well as in section 6.4 work with relation (6.48a, c). It should be kept in mind that this is exactly correct only when $\lambda = 1$ or $\delta = 0$, but can be regarded as a reasonable approximation of (6.48b) also when λ is close to 1 and δ is small.

The defining relationships for $L(t)$ and $L(t + 1)$, i.e.,

$$\bar{R}(t)L(t) = \psi(t),$$

$$\bar{R}(t + 1)L(t + 1) = \psi(t + 1),$$

can in view of (6.48) also be written

$$\bar{R}^*(t) \begin{pmatrix} 0 \\ L(t) \end{pmatrix} = \begin{pmatrix} * \\ \psi(t) \end{pmatrix}, \tag{6.49}$$

$$\bar{R}^*(t) \begin{pmatrix} L(t + 1) \\ 0 \end{pmatrix} = \begin{pmatrix} \psi(t + 1) \\ * \end{pmatrix}. \tag{6.50}$$

Here again, * denotes an α-vector (an α-dimensional column vector), whose exact expression is unimportant for the definition of $L(t)$.

In view of the relationship (6.31), it seems to be a good idea to determine a quantity $L^*(t)$ as an intermediate step from $L(t)$ to $L(t + 1)$, where

$$\bar{R}^*(t)L^*(t) = \psi^*(t). \tag{6.51}$$

The idea is thus to go from $L(t)$ to $L^*(t)$ and then to $L(t + 1)$.

Some Auxiliary Variables   In order to accomplish the aforementioned steps, we need tools with which we can operate on the *-elements in (6.49) and (6.50). These tools are the $n\alpha \times \alpha$-matrices $A(t)$ and $B(t)$ that satisfy

$$\bar{R}^*(t)\left(\begin{array}{c} I \\ \hline A(t) \end{array}\right) = \left(\begin{array}{c} R^e(t) \\ \hline 0 \end{array}\right) \updownarrow \alpha \text{ rows}, \tag{6.52}$$

$$\bar{R}^*(t)\left(\begin{array}{c} B(t) \\ \hline I \end{array}\right) = \left(\begin{array}{c} 0 \\ \hline R^r(t) \end{array}\right) \updownarrow \alpha \text{ rows}, \tag{6.53}$$

where R^e and R^r are $\alpha \times \alpha$-matrices.

It is useful to interpret these auxiliary variables. The bottom $n\alpha$ rows of (6.52) read

$$\sum_{k=1}^{t} \lambda^{t-k}\psi(k)x^{T}(k) + \bar{R}(t)A(t) = 0$$

or

$$A(t) = -\left[\sum_{1}^{t} \lambda^{t-k}\psi(k)\psi^{T}(k) + \lambda^{t}\delta I\right]^{-1} \sum_{1}^{t} \lambda^{t-k}\psi(k)x^{T}(k). \tag{6.54}$$

A comparison with the least squares formula (2.13), shows that $A(t)$ is the least squares estimate of the matrix A in the regression model

$$x(t) = -A^{T}\psi(t) \tag{6.55}$$

based on data up to time t and using a forgetting factor λ. Moreover, the α first rows of (6.52) read

$$\sum_{1}^{t} \lambda^{t-k}x(k)x^{T}(k) + \lambda^{t}\delta I + \sum_{1}^{t} \lambda^{t-k}x(k)\psi^{T}(k)A(t) = R^e(t).$$

A comparison with the proof of lemma 3.D.1 shows that

$$R^e(t) = \sum_1^t \lambda^{t-k} e^t(k) [e^t(k)]^T + \lambda^t \delta I, \tag{6.56a}$$

where the α-vectors $e^t(k)$ are the residuals from (6.55):

$$e^t(k) = x(k) + A^T(t)\psi(k). \tag{6.56b}$$

The matrix $R^e(t)$ is thus an estimated covariance matrix for the residuals $e^t(k)$.

Similarly, we find that $B(t)$ defined by (6.53) is the least squares estimate of a regression

$$x(t - n) = -B^T \psi(t + 1), \tag{6.57}$$

and that

$$R^r(t) = \sum_1^t \lambda^{t-k} r^t(k) [r^t(k)]^T + \lambda^t \delta I, \tag{6.58a}$$

where

$$r^t(k) = x(k - n) + B^T(t)\psi(k + 1). \tag{6.58b}$$

The interpretations of $A(t)$ and $B(t)$ as *forward and backward predictors* for the sequence $\{x(t)\}$ are quite interesting. For the formal development, however, we need only the algebraic property (6.52)–(6.53).

The formulas for updating A, B, and L are derived in appendix 6.B. The result is summarized as follows.

The Algorithm   Let $\{x(t)\}$ be a sequence of α-vectors such that $x(t) = 0$ for $t \le 0$, and let

$$\psi(t) = \begin{pmatrix} x(t-1) \\ \vdots \\ x(t-n) \end{pmatrix}.$$

Then the gain

$$L(t) = \left[\sum_{k=1}^t \lambda^{t-k} \psi(k)\psi^T(k) + \lambda^t \delta I \right]^{-1} \psi(t)$$

can be recursively computed as follows.

Fast Calculation of the Gain Matrix

1. Initialize:

$A(0) = 0$, $B(0) = 0$, $R^e(0) = \delta \cdot I$, $L(1) = 0$.

2. Given $A(t-1)$, $B(t-1)$, $R^e(t-1)$, and $L(t)$, update:

$$e(t) := x(t) + A^{\mathrm{T}}(t-1)\psi(t), \tag{6.59a}$$

$$A(t) := A(t-1) - L(t)e^{\mathrm{T}}(t), \tag{6.59b}$$

$$\beta(t) := L^{\mathrm{T}}(t)\psi(t), \tag{6.59c}$$

$$\bar{e}(t) := [1 - \beta(t)]e(t), \tag{6.59d}$$

$$R^e(t) := \lambda R^e(t-1) + \bar{e}(t)e^{\mathrm{T}}(t), \tag{6.59e}$$

$$L^*(t) := \begin{pmatrix} [R^e(t)]^{-1}\bar{e}(t) \\ L(t) + A(t)[R^e(t)]^{-1}\bar{e}(t) \end{pmatrix}. \tag{6.59f}$$

Partition $L^*(t)$ as

$$L^*(t) := \left(\begin{array}{c} M(t) \\ \hline \mu(t) \end{array} \right) \begin{array}{l} n\alpha \text{ rows} \\ \alpha \text{ rows} \end{array}. \tag{6.59g}$$

Compute

$$r(t) := x(t-n) + B^{\mathrm{T}}(t-1)\psi(t+1), \tag{6.59h}$$

$$B(t) := [B(t-1) - M(t)r^{\mathrm{T}}(t)][I - \mu(t)r^{\mathrm{T}}(t)]^{-1}, \tag{6.59i}$$

$$L(t+1) := M(t) - B(t)\mu(t). \tag{6.59j}$$

Remark on Initialization   An alternative to step 1 is to take $\delta = 0$ and initialize the algorithm at time $t = t^*$ when the matrix $\bar{R}^*(t^*)$ in (6.47) is of full rank (typically, then, $t^* = (d+1)\alpha$). Then $A(t^*)$, $B(t^*)$, $R^e(t^*)$ and $L(t^*+1)$ are determined from (6.52), (6.53) and (6.50), respectively. In view of (6.48b, c) this may be a better alternative when $\lambda < 1$ and/or $\psi(1) \neq 0$.

The algorithm computes the gain vector $L(t)$. If a linear regression, or a pseudolinear regression is used, this gain is subsequently used to update the parameter estimate:

$$\hat{\theta}(t) = \hat{\theta}(t-1) + L(t)[y(t) - \hat{\theta}^{\mathrm{T}}(t-1)\psi(t)]^{\mathrm{T}} \tag{6.60}$$

[see (3.86) or (3.130)]. Now in several cases the vector $x(t)$ contains $y(t)$ [see (6.36) and (6.41)]. If the first p rows of $x(t)$ equals $-y(t)$ we find by comparison of (6.60) with (6.59a, b) that the first p columns of $A(t)$ will be equal to $\hat{\theta}(t)$. Hence, when we are using a pseudolinear regression for the model (6.37) with $A(q^{-1}) \neq 1$ or a linear regression for the model (6.33), the updating of $\hat{\theta}(t)$ is obtained as a by-product of the algorithm (6.59) to determine the gain.

Computational Complexity   The algorithm (6.59) requires the following elements to be stored at time $t - 1$:

$\psi(t)$: an $n\alpha \times 1$-matrix,

$L(t)$: an $n\alpha \times 1$-matrix,

$A(t - 1)$: an $n\alpha \times \alpha$-matrix,

$B(t - 1)$: an $n\alpha \times \alpha$-matrix,

$R^e(t - 1)$: an $\alpha \times \alpha$ symmetric matrix.

All this requires

$$2n\alpha^2 + 2n\alpha + \tfrac{1}{2}(\alpha^2 + \alpha) \tag{6.61}$$

memory cells. With $d = n\alpha$, we can write (6.61) as

$$2\alpha d + 2d + \tfrac{1}{2}(\alpha^2 + \alpha). \tag{6.61'}$$

In comparison, the algorithms described in section 6.2 need to store the symmetric matrix $P(t)$ (or the triangular matrix $U(t)$ and the diagonal matrix $D(t)$) together with the vector $\psi(t)$. This requires

$$\tfrac{1}{2}(n\alpha)^2 + \tfrac{3}{2}n\alpha \tag{6.62}$$

memory cells, which is an order of magnitude (in n) more.

A count of the number of operations involved in (6.59) gives

$n\alpha^3 + 6n\alpha^2 + n\alpha + \tfrac{2}{3}\alpha^3 + 4\alpha^2 + 2\alpha$ multiplications,

$n\alpha^3 + 7n\alpha^2 + 2n\alpha + \tfrac{7}{2}\alpha^2 + \tfrac{3}{2}\alpha$ additions,

which again is an order of magnitude less (in n) than the algorithms of section 6.2 (see table 6.1).

The algorithm (6.59) has, therefore, a distinct advantage over conventional methods, when the model order n is high: it is much faster and requires much less memory.

Numerical Stability   Most reported numerical studies concern relatively short samples and have shown satisfactory behavior (see Falconer and

Ljung, 1978). The error propagation properties are studied in more detail in Ljung (1983), where it is established that the algorithm is not numerically stable for $\lambda < 1$.

Some Extensions   The general shift structure (6.32) is treated in Ljung, Morf, and Falconer (1978). This reference treats also the nonsymmetric IV method. The case when $\psi(t)$ is a matrix can be treated along the same lines as the vector case; the derivation remains essentially the same.

6.4 Ladder and Lattice Algorithms

For black box models we have worked with difference equations, such as (6.33) or (6.37). From an implementational point of view, these models can be seen as digital filters driven by $\{u(t)\}$ and $\{e(t)\}$ and producing $\{y(t)\}$. There are several ways of implementing a given transfer function as a digital filter, corresponding to the different possible state-space realizations. The difference equation representation itself may not be the best choice, due to sensitivity with respect to round-off errors. This fact has been widely recognized in digital filtering and signal processing (see Oppenheim and Schafer, 1975). A useful form for implementation of digital filters is *ladder* or *lattice* filters. These terms originate from network theory where they denote a certain way of realizing a given transfer function. The name "ladder" or "lattice" then refers to the pattern formed by a diagram of the signal flow in the filter (see figure 6.1 later in this section). Interpreted in terms of the corresponding state-space realization, ladder implementation uses a state vector that has a diagonal covariance matrix. This gives several nice features to the filter, as we shall see.

   In view of the usefulness of the ladder filter, it is natural to look for ways of recursively identifying systems and signals in a ladder representation. Compared to our conventional difference equation models, this means that the model is reparametrized. For a black box model, this should not be much of a disadvantage, since black box parameters are just vehicles for describing the input-output behavior. Recursive identification of ladder representations has been extensively studied by Morf and his coworkers (Morf, 1977; Morf and Lee, 1978; Lee and Morf, 1980). Morf's algorithms are algebraically equivalent to the recursive least squares algorithm. Gradient lattice schemes have also been described and studied (Griffiths, 1977; Makhoul, 1977, 1978).

In this section we shall first describe the rationale for the ladder algorithms (section 6.4.1). We shall then (section 6.4.2, with details in appendix 6.C) develop a basic ladder algorithm for the same setup that we considered in section 6.3, using much the same technique. Finally (section 6.4.3), we shall comment on some variants of the ladder schemes.

6.4.1 Difference Equations and Ladder Forms

Change of Coordinate Basis for Linear Regressions Given a linear regression model

$$\hat{y}(t \mid \theta) = \theta^{\mathrm{T}} \varphi(t),   \tag{6.63}$$

several different representations can be obtained by change of basis in the regressor space (i.e., the space spanned by φ). Ladder forms correspond, as we mentioned, to a particular change of basis. Before discussing ladder forms, we shall first give some general facts on linear transformations of the representation (6.63). Consider a new regression vector

$$\tilde{\varphi}(t) = T\varphi(t)   \tag{6.64}$$

for some invertible matrix T. With

$$\tilde{\theta} = T^{-\mathrm{T}}\theta,   \tag{6.65}$$

we can write the model (6.63) as

$$\hat{y}(t \mid \theta) = \theta^{\mathrm{T}}\varphi(t) = (T^{\mathrm{T}}\tilde{\theta})^{\mathrm{T}} T^{-1}\tilde{\varphi}(t) = \tilde{\theta}^{\mathrm{T}}\tilde{\varphi}(t).   \tag{6.66}$$

The problem of determining the least squares estimate of $\tilde{\theta}$ from (6.66) is of course equivalent to the original problem (6.63), and we obtain

$$\hat{\tilde{\theta}}(t) = [\tilde{R}(t)]^{-1} \sum_{1}^{t} \tilde{\varphi}(k) y(k).   \tag{6.67}$$

Here

$$\tilde{R}(t) = \sum_{1}^{t} \tilde{\varphi}(k)\tilde{\varphi}^{\mathrm{T}}(k) = T\bar{R}(t)T^{\mathrm{T}},   \tag{6.68}$$

where

$$\bar{R}(t) = \sum_{1}^{t} \varphi(k)\varphi^{\mathrm{T}}(k).   \tag{6.69}$$

Obviously, we have

$$\hat{\bar{\theta}}(t) = T^{-\mathrm{T}}\hat{\theta}(t) = T^{-\mathrm{T}}\bar{R}^{-1}(t)\sum_{1}^{t}\varphi(k)y(k). \tag{6.70}$$

A change of basis in the regressor space therefore has some straight-forward consequences for the estimates, as described by (6.65), (6.67), and (6.70).

Increasing the Order of an Autoregression Model   Let us consider a special case of (6.63): an autoregression

$$y(t) + a_1y(t-1) + \cdots + a_ny(t-n) = e(t).$$

Using $y(t-k)$ as regressors in this model is natural, since these are the actual measurements. This choice has, however, the disadvantage that the best (least squares) estimate of a_i actually depends on the order n. We illustrate this by a simple example.

EXAMPLE 6.3   Consider the system

$$y(t) + a_1^0 y(t-1) + a_2^0 y(t-2) = e(t), \tag{6.71}$$

where $\{e(t)\}$ is a sequence of independent random variables each with unit variance and zero mean value. The output sequence generated by (6.71) has a covariance function

$$r_k = \mathrm{E}y(t)y(t-k),$$

where elementary calculation gives

$$r_0 = \frac{1 + a_2^0}{[(1 + a_2^0)^2 - (a_1^0)^2](1 - a_2^0)},$$

$$r_1 = -\frac{a_1^0}{1 + a_2^0}r_0,$$

$$r_2 = -a_2^0 r_0 - a_1^0 r_1.$$

A first-order model

$$\hat{y}(t) = -a_1^{(1)}y(t-1)$$

will give the asymptotic estimate

$$\hat{a}_1^{(1)} = \lim_{t\to\infty}\left[\sum_{k=1}^{t} y^2(k-1)\right]^{-1}\sum_{k=1}^{t} -y(k-1)y(k)$$

$$= -r_1/r_0 = a_1^0/(1 + a_2^0).$$

A second-order model

$$\hat{y}(t) = -a_1^{(2)}y(t-1) - a_2^{(2)}y(t-2)$$

will, however, asymptotically yield the true values

$$\hat{a}_1^{(2)} \to a_1^0, \quad \hat{a}_2^{(2)} \to a_2^0.$$

We notice in particular that the asymptotic values of $\hat{a}_1^{(1)}$ and $\hat{a}_1^{(2)}$ differ if $a_2^0 \neq 0$.

Now, even if the true value of a_2^0 is zero, so the limit of $\hat{a}_1$ does not depend on the model order used, the accuracy is affected. To show this, we proceed as follows: We know from theorem 4.5 that the asymptotic covariance matrix for $\sqrt{N}[\hat{a}_1^{(1)}(N) - a_1^0]$ is

$$[Ey^2(t)]^{-1} = 1/r_0$$

when model order $n = 1$ and $a_2^0 = 0$. For $n = 2$ we find that the covariance matrix for

$$\sqrt{N}\begin{pmatrix} \hat{a}_1^{(2)}(N) - a_1^0 \\ \hat{a}_2^{(2)}(N) \end{pmatrix}$$

is

$$\left(E\begin{pmatrix} -y(t-1) \\ -y(t-2) \end{pmatrix}(-y(t-1) \ -y(t-2))\right)^{-1} = \begin{pmatrix} r_0 & r_1 \\ r_1 & r_0 \end{pmatrix}^{-1},$$

when $a_2^0 = 0$.

We thus obtain that

$$\sqrt{N}[\hat{a}_1^{(2)}(N) - a_1^0] \text{ has the asymptotic variance } \frac{1}{r_0}\left(\frac{1}{1-(a_1^0)^2}\right).$$

The asymptotic accuracy obtained when estimating a_1 in a second-order model is thus *strictly worse* than in a first-order model. There is consequently a penalty for going beyond the true order. (We saw this also in section 4.4.4.) □

The reason why the estimates $\hat{a}_i$, $i \leq n$, change when the model order is increased can be described as follows. Suppose that we have estimated an nth order autoregression, which we write

$$\hat{y}_n(t) = \theta_n^T \varphi_n(t). \tag{6.72}$$

Here the subscript n emphasizes that the model is nth order, and $\varphi_n(t)$ consists of $y(t - i)$, $1 \leq i \leq n$. We now increase the order to $n + 1$ and add the regressor $y(t - n - 1)$:

$$\hat{y}_{n+1}(t) = \theta_n^T \varphi_n(t) - a_{n+1} y(t - n - 1). \tag{6.73}$$

This regressor is typically correlated with $\varphi(t)$, i.e., with previous values of $y(t - i)$, $1 \leq i \leq n$, but it also contains new information that is not present in $\varphi(t)$. We can state this by writing

$$y(t - n - 1) = -B^T \varphi(t) + r_n(t - 1), \tag{6.74}$$

where $r_n(t - 1)$ is the new piece of information in $y(t - n - 1)$, not present in $\varphi(t)$. Hence $r_n(t - 1)$ and $\varphi(t)$ are uncorrelated. With (6.74) inserted into (6.73) we obtain

$$\hat{y}_{n+1}(t) = (\theta_n + a_{n+1} B)^T \varphi(t) - a_{n+1} r_n(t - 1). \tag{6.75}$$

When now the parameter a_{n+1} is adjusted to take the new information in $r_n(t - 1)$ into account, the value of $\hat{\theta}_n$ must also be readjusted compared to (6.72) because of the term $a_{n+1} B$ in (6.75). More specifically, if $\hat{\theta}_n^{(n)}$ denotes the estimate of θ_n in (6.72) and $\hat{\theta}_n^{(n+1)}$ denotes the estimate of θ_n in (6.73) = (6.75), we must have

$$\hat{\theta}_n^{(n+1)} + \hat{a}_{n+1} B = \hat{\theta}_n^{(n)}. \tag{6.76}$$

Changing the Basis of an Autoregression Having understood why the estimates of a_i in an autoregression depend on the model order, it is easy to see how to construct a scheme that does not have this feature. If, when we extended the model to account for the variable $y(t - n - 1)$ we had added only the new information $r_n(t - 1)$ instead of $y(t - n - 1)$, then the estimate of θ_n would have been unchanged. *The idea consequently is to use the variables* $r_k(t - 1)$, $k = 0, \ldots, n$, *as regressors instead of* $y(t - k - 1)$. From (6.74) we see that $r_n(t - 1)$ can be interpreted as the *backward prediction error (or innovation)* associated with predicting $y(t - n - 1)$ from $\varphi^T(t) = (-y(t - 1) \ldots -y(t - n))$.

We also see from (6.74) that $r_k(t - 1)$ is a linear combination of $y(t - k - 1)$ (coefficient 1), $y(t - k)$, $\ldots$, $y(t - 1)$. Hence

$$\tilde{\varphi}(t) \triangleq \begin{pmatrix} r_0(t - 1) \\ \vdots \\ r_{n-1}(t - 1) \end{pmatrix} = T\varphi(t), \tag{6.77}$$

where T is lower triangular with 1's along the diagonal.

The transformation from the original regressors $y(t-k)$ to the backward innovations $r_k(t-1)$ is therefore a linear change of basis in the regressor space as in (6.64). The corresponding change of coefficients is

$$\begin{pmatrix} K_0 \\ \vdots \\ K_{n-1} \end{pmatrix} \triangleq \tilde{\theta} = T^{-T}\theta, \tag{6.78}$$

where the coefficients K_i usually are known as *reflection coefficients* (from an analogy with transmission-line theory). The model description is thus

$$\hat{y}(t) = \tilde{\theta}^T \tilde{\varphi}(t) = K_0 r_0(t-1) + \cdots + K_{n-1} r_{n-1}(t-1). \tag{6.79}$$

To be efficiently used for simulation and prediction the model (6.79) needs to be supplemented with a way of computing $r_i(t-1)$ from y^t. Such a way is described by (6.77), but we shall develop more efficient algorithms in the next subsection. The model (6.79), together with an algorithm to determine $r_i(t)$, is known as a *ladder* or *lattice* form (see figure 6.1, p. 357).

Relation to U-D Factorization   In the foregoing heuristic description of the ladder form we have not been specific about what we formally mean by "uncorrelated" regressors and that $r_n(t-1)$ is "uncorrelated" with $\varphi_n(t)$. A formal treatment is given in the next subsection.

It should be pointed out, though, that the change of variables (6.77) actually depends on time, so that at time t, we use regressors

$$\tilde{\varphi}^t(k) = \begin{pmatrix} r_0^{t-1}(k-1) \\ \vdots \\ r_{n-1}^{t-1}(k-1) \end{pmatrix} = T(t)\varphi(k). \tag{6.80a}$$

The significance of the backward innovations r_i being independent is then that the matrix

$$\tilde{R}(t) = \sum_1^t \tilde{\varphi}^t(k)[\tilde{\varphi}^t(k)]^T \triangleq D(t) \tag{6.80b}$$

is *diagonal*.

Comparing (6.80a) with (6.68) we find that

$$\bar{R}(t) = T^{-1}(t)D(t)T^{-T}(t),$$

or, in terms of the inverse,

$$P(t) = \bar{R}^{-1}(t) = T^{\mathrm{T}}(t)D^{-1}(t)T(t). \tag{6.81}$$

We notice that the change of variables to backward innovations in (6.77) achieves exactly the U-D factorization of the P matrix discussed in section 6.2.2 [see (6.12)].

The fact that the regressors $\tilde{\varphi}$ are uncorrelated has an interesting consequence for the corresponding estimate $\hat{\tilde{\theta}}$ of the reflection coefficients. In the model

$$\hat{y}(t) = \tilde{\theta}^{\mathrm{T}}\tilde{\varphi}(t),$$

the gradient of the prediction with respect to the model parameters $\tilde{\theta}$ is $\tilde{\varphi}(t)$. Hence, according to theorem 4.5, the covariance matrix of the asymptotic distribution for $\sqrt{N}[\hat{\tilde{\theta}}(N) - \tilde{\theta}_0]$ is proportional to

$$\tilde{P} = [\bar{\mathrm{E}}\tilde{\varphi}(t)\tilde{\varphi}^{\mathrm{T}}(t)]^{-1}.$$

This matrix is diagonal by construction. Hence the estimates of the different components of $\tilde{\theta}$ are asymptotically uncorrelated. Neither the value nor the accuracy of the estimate of K_i are therefore affected by the estimation of other reflection coefficient. This is in contrast to the case studied in example 6.3.

As a final comment, we note that if an input signal $\{u(t)\}$ is present in the regression vector $\varphi(t)$, then the diagonalizing transformation $T(t)$ in (6.80) will depend on the properties of this input. This should be kept in mind in applications where the properties of $\{u(t)\}$ may vary.

6.4.2   A Fast Algorithm for Recursive Estimation of the Reflection Coefficients

In this section we shall develop a fast algorithm that both accomplishes the (time-varying) transformation from $\{y(t)\}$ to $\{r_k(t-1)\}$ and estimates the coefficients K_i of the model (6.79). Such an algorithm can be derived in several ways, using various degrees of mathematical sophistication and elegance. Here we choose a straightforward and rather lengthy derivation, that, however, has the advantage of using only elementary matrix manipulations. It is inspired by, but not identical to, the derivations given by Morf and Lee (1978) and by Gevers and Wertz (1983). Details are deferred to appendix 6.C.

The Problem   Consider a linear regression model of order n,

$$\hat{y}_n(t \mid \theta_n) = \theta_n^T \varphi_n(t), \tag{6.82}$$

where $\varphi_n(t)$ is a column vector, subject to the structure (6.29):

$$\varphi_n(t) = \begin{pmatrix} x(t-1) \\ \vdots \\ x(t-n) \end{pmatrix}. \tag{6.83}$$

The elements of the α-vector $x(t)$ may very well contain variables that are constructed using past estimates of θ, as described in example 6.2. Therefore (6.82) includes also pseudolinear regressions. We assume that

$$x(t) = 0 \quad \text{for} \quad t \le 0. \tag{6.84}$$

Remark   Since we shall only treat (pseudo) linear regressions in this section we use the symbol φ for the "gradient" vector in order to be consistent with the rest of the book. Certain formulas from section 6.3 will be used in the derivation here. The reader should keep in mind that then φ replaces ψ, which is only a change of notation. Also, the fact that a more general situation ($x(t)$ being a general α-vector) is considered in (6.83) means that there will be certain departures from the heuristic discussion in section 6.4.1. The "backward innovations" r_i will be α-vectors and the decomposition (6.80) will give a block diagonal ($\alpha \times \alpha$-block) rather than a diagonal matrix.

The problem is to determine the predictions $\hat{y}_n(t)$:

$$\hat{y}_n(t) = \hat{\theta}_n^T(t-1)\varphi_n(t) \tag{6.85}$$

based on the least squares estimate $\hat{\theta}_n(t)$ given by

$$\hat{\theta}_n(t) = \bar{R}_n^{-1}(t) \sum_{t=1}^t \lambda^{t-k} \varphi_n(k) y^T(k), \tag{6.86}$$

where

$$\bar{R}_n(t) = \sum_{k=1}^t \lambda^{t-k} \varphi_n(k) \varphi_n^T(k) + \lambda^t \delta I. \tag{6.87}$$

This should be done for $n = 1$ up to a maximum order $n = M$.

The calculations of $\hat{y}_n(t)$ will not be performed explicitly in terms of

$\hat{\theta}_n(t-1)$. Rather, the problem is to recursively reparametrize (6.82) to a ladder form

$$\hat{y}_n(t) = K_0 r_0(t-1) + \cdots + K_{n-1} r_{n-1}(t-1), \tag{6.88}$$

where $r_i(t-1)$ are uncorrelated α-vector regressors, such that $\{r_0(t-1) \ldots r_{k-1}(t-1)\}$ lie in the span of $\{x(t-1) \ldots x(t-k)\}$. The problem is also to recursively estimate the reflection coefficients $K_i(t)$ of the model (6.88). If the dimension of the output is p, then K_i obviously are $p \times \alpha$-matrices.

The Method   The idea is to construct the $r_k(t-1)$ vectors as *backward prediction errors* of the sequence $\{x(t)\}$, based on n future values, i.e., use a regression

$$x(t-n) = -B_n^T \varphi_n(t+1) + r_n(t). \tag{6.89}$$

The minus sign in (6.89) is used for notational convenience only. The regression problem (6.89) to determine $B_n(t)$ (the estimate based on data up to t) is of the same character as the given one, (6.82), and will be solved simultaneously, using an analogous technique.

When working with models of increasing order n we shall utilize the shift structure (6.83). In the notation of section 6.3.2 we then have (recall that φ replaces ψ in this section)

$$\bar{R}(t) = \bar{R}_n(t), \quad \bar{R}^*(t) = \bar{R}_{n+1}(t+1). \tag{6.90}$$

Consequently the shift structure (6.83) can be expressed as in (6.48a, c):

$$\bar{R}_{k+1}(t) = \begin{pmatrix} ************ \\ * \ \bar{R}_k(t-1) \ \\ * \end{pmatrix} = \begin{pmatrix} * \\ \bar{R}_k(t) \ * \\ ******* \end{pmatrix}. \tag{6.91}$$

Here we have used the assumption (6.84).

When we develop the method the following variables will be crucial.

$\hat{x}_n(t)$: The prediction of $x(t)$ based on the information in $\varphi_n(t)$, and based on data up to $t-1$.

$e_n(t) = x(t) - \hat{x}_n(t)$: The forward prediction errors.

$\check{x}_n(t)$: The backward prediction of $x(t)$ based on the information in $\varphi_n(t+n+1)$ and based on data up to $t+n$.

$r_n(t) = x(t-n) - \check{x}(t-n)$: The backward prediction errors.

$R_n^e(t)$: The sample covariance matrix of the forward prediction errors, based on data up to t.

$R_n^r(t)$: The sample covariance matrix of the backward prediction errors, based on data up to t.

$F_n(t)$: The cross covariance matrix between the forward and the backward prediction errors, based on data up to t.

$$\beta_n(t) = \varphi_n^T(t)\bar{R}_n^{-1}(t)\varphi_n(t).$$

Formal definitions of these variables are given in appendix 6.C. Simple relationships between the variables can be derived using (6.91). The derivation itself is, however, tedious and lengthy. It is given in appendix 6.C.

The Algorithm Collecting the equations in appendix 6.C marked with a † gives the following algorithm.

A Fast Ladder Algorithm (6.92)

---

1. Initialize at time $t = 0$: Set

$R_n^e(0) := \delta I,$

$R_n^r(-1) := \delta I,$

$F_n(0) := 0,$

$r_n(0) := 0, \quad n = 0, \ldots, M - 1.$

2. At time $t - 1$, store

$R_n^e(t - 1), \quad F_n(t - 1), \quad R_n^r(t - 2), \quad r_n(t - 1), \quad n = 0, \ldots, M - 1.$

3. At time $t - 1$, compute, for $n = 0, \ldots, M - 1$,

$K_n(t - 1) := F_n^T(t - 1)[R_n^r(t - 2)]^{-1},$

$K_n^*(t - 1) := F_n(t - 1)[R_n^e(t - 1)]^{-1}.$

4. The predictions $\hat{x}_n(t)$ can now be computed for $n = 1, \ldots, M$ from

$\hat{x}_n(t) := \hat{x}_{n-1}(t) + K_{n-1}(t - 1)r_{n-1}(t - 1),$

$\hat{x}_0(t) := 0.$

5. Compute, for $n = 0, \ldots, M - 1,$

$R_n^r(t - 1) := \lambda R_n^r(t - 2) + [1 - \beta_n(t)]r_n(t - 1)r_n^T(t - 1),$

$\beta_{n+1}(t) := \beta_n(t) + [1 - \beta_n(t)]^2 r_n^T(t - 1)[R_n^r(t - 1)]^{-1}r_n(t - 1),$

$\beta_0(t) := 0.$

6. At time t, $x(t)$ is received.

7. For $n = 0, \ldots, M - 1$, update:

$r_n(t) := r_{n-1}(t - 1) - K^*_{n-1}(t - 1)e_{n-1}(t)$,

$e_n(t) := x(t) - \hat{x}_n(t)$,

$r_0(t) := x(t)$.

8. For $n = 0, \ldots, M - 1$, update:

$R^e_n(t) := \lambda R^e_n(t - 1) + [1 - \beta_n(t)]e_n(t)e^T_n(t)$,

$F_n(t) := \lambda F_n(t - 1) + [1 - \beta_n(t)]r_n(t - 1)e^T_n(t)$.

9. Go to 2.

When $y(t)$ is part of $x(t)$, the recursion (6.92:4) gives the prediction $\hat{y}_n(t)$ sought in (6.85), as an automatic by-product. If $y(t)$ does not belong to $x(t)$, the algorithm (6.92) must be supplemented by the following steps.

$K^y_n(t - 1) := [F^y_n(t - 1)]^T[R^r_n(t - 2)]^{-1}$,   (6.93a)

$\hat{y}_n(t) := \hat{y}_{n-1}(t) + K^y_{n-1}(t - 1)r_{n-1}(t - 1)$,   (6.93b)

$\varepsilon_n(t) := y(t) - \hat{y}_n(t)$,   (6.93c)

$F^y_n(t) := \lambda F^y_n(t - 1) + [1 - \beta_n(t)]r_n(t - 1)\varepsilon^T_n(t)$.   (6.93d)

This follows by entirely analogous arguments.

In the algorithm (6.92) we separated the calculation of $e_n(t)$ into two steps: step 4 and step 7. When the predictions $\hat{x}_n(t)$ are not explicitly required it is natural to skip step 4 and perform (6.C.27):

$e_n(t) = e_{n-1}(t) - K_{n-1}(t - 1)r_{n-1}(t - 1)$,

$r_n(t) = r_{n-1}(t - 1) - K^*_{n-1}(t - 1)e_{n-1}(t)$,

$e_0(t) = r_0(t) = x(t)$.

The signal flow in this equation can be depicted as in figure 6.1. This figure shows why the computations are said to have a "ladder" or "lattice" structure.

For the ELS algorithm we have $x^T(t) = (-y(t) \; u(t) \; \bar{\varepsilon}_M(t))$, and hence the variable $\bar{\varepsilon}_M(t)$ is part of the $x(t)$ vector. Then the sequence of events in the algorithm is as follows:

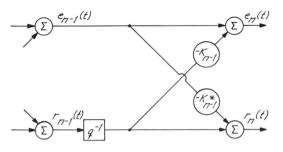

Figure 6.1
Signal flow in the ladder algorithm.

At time $t - 1$ perform steps 2–5.

At time t, $y(t)$ and $u(t)$ are received.

Compute $\varepsilon_M(t) = y(t) - \hat{y}_M(t)$.

Compute $\bar{\varepsilon}_M(t) = [1 - \beta_M(t)]\varepsilon_M(t)$.

$x(t)$ is now complete; continue from step 6.

Some Comments on the Algorithm The algorithm (6.92) requires

$$2(\alpha^2 M + \alpha M) \tag{6.94}$$

memory locations to store its updated variables (which can be reduced to $\frac{3}{2}(\alpha^2 M + \alpha M)$; see next section). A pass through steps 3 to 8 requires approximately

$$3\alpha^3 M + 6\alpha^2 M + 4\alpha M \tag{6.95}$$

multiplications. The ladder algorithm consequently has a computational complexity that is slightly larger than, but of the same order of magnitude as, the fast gain algorithm (6.59). On the other hand, it accomplishes more. For any given maximal order M, we obtain the predictions $\hat{y}_n(t)$ for *all orders* $n \leq M$. These are the *true* nth-order predictions, and they are not equal to what we would have obtained by truncating the estimate $\hat{\theta}_M(t - 1)$ to its $n\alpha$ first rows. The estimates of the reflection coefficients $K_n(t)$ and $K_n^*(t)$ are not affected by the fact that higher-order models are estimated at the same time. In terms of reflection coefficients, our models can thus be truncated to any order $n \leq M$, without affecting the performance of the low-order models. This is in contrast to the case studied in example 6.3.

These facts on models of different orders must be discussed further when the algorithm is applied to PLRs. For ELS we have $x^T(t) = (-y(t)\ u(t)\ \bar{\varepsilon}_M(t))$, where the residual used in $x(t)$ is based on the Mth-order model. Consequently in the $x(t)$ vector itself, there is a decision about what model order is to be estimated. The predictions for other model orders will then only be approximations.

At any desired time, the original parametrization $\hat{\theta}(t)$ can be recovered from the reflection coefficients, using (6.C.18) and (6.C.21). Recall that $\bar{\theta}(t)$ equals the first p columns of $A(t)$, when $-y(t)$ equals the first rows of $x(t)$. Notice, though, that going through (6.C.18) and (6.C.21) from $n = 1$ to $n = M$ requires a number of operations proportional to M^2.

In the algorithm (6.92) we have theoretically that $\bar{e}_n(t) = 0, \bar{r}_n(t-1) = 0$, and $\beta_n(t) = 1$ when $t \leq n$. The reason is that when the number of parameters exceeds the number of data points, then a perfect fit, i.e., zero residual, is possible. Hence in (6.92) one may limit the calculation of $R_n^r(t-1)$, $R_n^e(t)$, $e_n(t)$, and $r_n(t-1)$ to $n \leq \min(t, M)$. (Due to the initialization with $\delta > 0$ in (6.92), we will, however, have a slight deviation from zero in the quantities $\bar{e}_n(t), \bar{r}_n(t-1)$, and $\beta_n(t)$ for $t \leq n$.) The error propagation properties of (6.92) are studied in Ljung (1983). In contrast to the fast algorithm (6.59), the ladder algorithm (6.92) is numerically stable for $\lambda < 1$.

6.4.3   Variants of Ladder Algorithms

In the literature, several different variants of ladder algorithms are suggested and used. They can all be derived using the tools that we developed in the previous subsection. We shall here give a brief account of such variants and how they relate to (6.92).

Order Updates for R^e and R^r   In (6.92) we updated $R_n^e(t)$ from $R_n^e(t-1)$ and $R_n^r(t)$ from $R_n^r(t-1)$ [see (6.92:5, 8)]. As an alternative, we could use the "order updates" (6.C.20) and (6.C.22). The sequence of operations would then be:

1. Store at $t - 1$

$$F_n(t-1), \quad R_n^r(t-2), \quad r_n(t-1), \quad n = 0, \ldots, M-1, \quad R_0^e(t-1).$$

$$(6.96a)$$

(Notice from (6.C.7) and (6.C.5c) that

$$R_0^e(t - 1) = \sum_{k=1}^{t-1} \lambda^{t-1-k} x(k) x^T(k)).$$

2. Determine $R_n^e(t - 1)$ and $K_n(t - 1)$ for $n = 0, \ldots, M - 1$ by

$$K_n(t - 1) := F_n^T(t - 1)[R_n^r(t - 2)]^{-1}, \tag{6.96b}$$

$$R_{n+1}^e(t - 1) := R_n^e(t - 1) - F_n^T(t - 1)K_n^T(t - 1). \tag{6.96c}$$

3. Determine $R_n^r(t - 1)$ and $K_n^*(t - 1)$ for $n = 0, \ldots, M - 1$ by

$$K_n^*(t - 1) := F_n(t - 1)[R_n^e(t - 1)]^{-1}, \tag{6.96d}$$

$$R_{n+1}^r(t - 1) := R_n^r(t - 2) - F_n(t - 1)[K_n^*(t - 1)]^T \tag{6.96e}$$

$$R_0^r(t - 1) := R_0^e(t - 1)$$

4. Perform step 4 of (6.92).
5. Determine $\beta_n(t)$ for $n = 0, \ldots, M - 1$ by

$$\beta_{n+1}(t) := \beta_n(t) + [1 - \beta_n(t)]^2 r_r^T(t - 1)[R_n^r(t - 1)]^{-1} r_n(t - 1), \tag{6.96f}$$

$$\beta_0(t) := 0.$$

6. Perform steps 6 and 7 of (6.92).
7. Update, for $n = 0, \ldots, M - 1$,

$$F_n(t) := \lambda F_n(t - 1) + [1 - \beta_n(t)] r_n(t - 1) e_n^T(t). \tag{6.96g}$$

8. Update

$$R_0^e(t) := \lambda R_0^e(t - 1) + x(t) x^T(t). \tag{6.96h}$$

9. Go to step 1.

This algorithm has the advantage that $R_n^e(t)$ does not have to be stored for $n = 1, \ldots, M - 1$. No major difference in performance between (6.92) and (6.96) has been reported.

Using Residuals Instead of Prediction Errors The quantities $r_n(t)$ and $e_n(t)$ that we have been working with are the "prediction errors," according to (6.C.5a). We obtained the ladder (6.C.25)–(6.C.26) for these variables by first delaying (6.C.18), (6.C.21) one time step before multiplying them by $\varphi_n^T(t)$. If (6.C.18) and (6.C.21) are multiplied by $\varphi_n^T(t)$

and then $x(t)$ and $x(t - n)$, respectively, are subtracted, we obtain the following ladder relationship between the residuals:

$$\bar{e}_n(t) = \bar{e}_{n-1}(t) - K_{n-1}(t)\bar{r}_{n-1}(t - 1), \tag{6.97a}$$

$$\bar{r}_n(t) = \bar{r}_{n-1}(t - 1) - K^*_{n-1}(t)\bar{e}_{n-1}(t). \tag{6.97b}$$

Now, (6.97) can be used in the fast ladder algorithm (6.92) to update the required quantities. The prediction errors used in (6.92) are then transformed to residuals via (6.C.11) and (6.C.13). The resulting algorithm has the disadvantage that the one-step-ahead prediction $\hat{y}_n(t)$ is not automatically available.

With a similar technique, we could also have obtained formulas using k-step-ahead predictions and corresponding prediction errors.

Normalized Equations The matrices $R^e_n(t)$ and $R^r_n(t)$ can be regarded as covariance matrices of $\bar{e}_n(t)$ and $\bar{r}_n(t)$, respectively. The use of their inverses in, i.e., (6.92:3) and (6.92:5) can therefore be interpreted as normalization of these residuals. As an alternative, we could work directly with normalized residuals,

$$\tilde{e}_n(t) \triangleq [R^e_n(t)]^{-1/2}\sqrt{1 - \beta_n(t)}\bar{e}_n(t)$$

and

$$\tilde{r}_n(t) \triangleq [R^r_n(t)]^{-1/2}\sqrt{1 - \beta_n(t + 1)}\bar{r}_n(t).$$

Such an algorithm has been developed by Lee and Morf (1980). It has the advantages that it requires fewer operations per time step and that the matrices R^e_n and R^r_n need not be stored.

Using the Matrix Inversion Lemma The equations

$$R^e_n(t) = \lambda R^e_n(t - 1) + [1 - \beta_n(t)]e_n(t)e^T_n(t),$$

$$F_n(t) = \lambda F_n(t - 1) + [1 - \beta_n(t)]r_n(t - 1)e^T_n(t),$$

$$[K^*_n(t)]^T = [R^e_n(t)]^{-1}F^T_n(t)$$

are entirely analogous to the least squares problem (2.13). We can thus use the matrix inversion lemma (2.16) to develop an expression for updating $K^*_n(t)$ and $[R^e_n(t)]^{-1}$ directly as in (2.19). This could be advantageous when α is not a small number.

"Gradient" Algorithms   If the second-order properties, i.e., the covariance functions, for the sequence $x(t)$ are known a priori, it is possible to derive an expression for the reflection coefficients in the corresponding ladder. That would give, assuming stationarity,

$$K_n = [Ee_n(t)r_n^T(t-1)][Er_n(t-1)r_n^T(t-1)]^{-1},$$

$$K_n^* = [Er_n(t-1)e_n^T(t)][Ee_n(t)e_n^T(t)]^{-1} \tag{6.98}$$

[compare this to (6.92:3)]. When the expected values are not known, a straightforward approach to estimate K_n and K_n^* would be to use the approximations

$$Ee_n(t)r_n^T(t-1) \approx \tilde{F}_n^T(t) = \sum \lambda^{t-k}e_n(k)r_n^T(k-1),$$

and analogously for the other ones. The recursion for $\tilde{F}_n(t)$ will be

$$\tilde{F}_n(t) = \lambda\tilde{F}_n(t-1) + r_n(t-1)e_n^T(t). \tag{6.99}$$

Such an approach will lead to an algorithm that is identical to (6.92), except that $\beta_n(t)$ is always replaced by 0. Such algorithms, known as *gradient lattice algorithms*, have been developed by Griffiths (1977) and Makhoul (1977). Since we have $Er_n(t-1)r_n^T(t-1) = Ee_n(t)e_n^T(t)$, this equality is sometimes enforced when computing R_n^e and R_n^r.

The gradient lattice requires somewhat less computation than (6.92), since $\beta_n(t)$ need not be determined. Moreover, when λ is close to 1, $\beta_n(t)$ is a small value, except in the transient stage. Hence the gradient version appears to be a reasonable approximation. However, it has worse performance than (6.92), as pointed out by Satorius and Pack (1981).

6.5 Regularization

6.5.1   The Problem

In the Gauss-Newton algorithm the gain $L(t)$ is given by

$$R(t) = R(t-1) + \gamma(t)[\psi(t)\hat{\Lambda}^{-1}(t)\psi^T(t) - R(t-1)], \tag{6.100a}$$

$$L(t) = \gamma(t)R^{-1}(t)\psi(t)\hat{\Lambda}^{-1}(t). \tag{6.100b}$$

Now it may happen that the matrix $R(t)$ as defined by (6.100a) is singular or nearly singular. In section 4.4.4 and example 5.3 we discussed some causes for such a situation [see (4.135) and example 4.10]. Typically,

$R(t)$ is almost singular either if the model set contains too many parameters, or if the input signal is not general enough. Let us consider a simple example.

EXAMPLE 6.4.   Consider a first-order difference equation model

$$y(t) + ay(t-1) = bu(t-1) + e(t),$$

to which the recursive least squares procedure is applied. Then

$$\psi(t) = \varphi(t) = \begin{pmatrix} -y(t-1) \\ u(t-1) \end{pmatrix}.$$

With the notation $\bar{R}(t) = R(t)/\gamma(t)$ we have, as in (6.43),

$$\bar{R}(t) = \lambda(t)\bar{R}(t-1) + \varphi(t)\varphi^{T}(t).$$

If $\lambda(t) \equiv \lambda$ and $\bar{R}(0) = 0$, we obtain

$$\bar{R}(t) = \begin{pmatrix} \displaystyle\sum_{k=1}^{t} \lambda^{t-k}y^2(k-1) & \displaystyle -\sum_{k=1}^{t} \lambda^{t-k}y(k-1)u(k-1) \\ \displaystyle -\sum_{k=1}^{t} \lambda^{t-k}y(k-1)u(k-1) & \displaystyle \sum_{k=1}^{t} \lambda^{t-k}u^2(k-1) \end{pmatrix}.$$

Suppose now that at t_0

$$\bar{R}(t_0) = I,$$

and that for $t \geq t_0$ a constant feedback law is used:

$$u(t) = -k_0 y(t). \tag{6.101}$$

We then find that

$$\bar{R}(t) = \begin{pmatrix} \lambda^{t-t_0} + r_0(t) & k_0 r_0(t) \\ k_0 r_0(t) & \lambda^{t-t_0} + k_0^2 r_0(t) \end{pmatrix}, \tag{6.102a}$$

where

$$r_0(t) = \sum_{k=t_0}^{t} \lambda^{t-k}y^2(k-1). \tag{6.102b}$$

This means that

$$\bar{R}^{-1}(t) = \frac{1}{\lambda^{2(t-t_0)} + \lambda^{t-t_0}(1+k_0^2)r_0(t)} \begin{pmatrix} \lambda^{t-t_0} + k_0^2 r_0(t) & -k_0 r_0(t) \\ -k_0 r_0(t) & \lambda^{t-t_0} + r_0(t) \end{pmatrix}.$$

Hence all elements of $\bar{R}^{-1}(t)$ tend to infinity exponentially as $(1/\lambda)^{t-t_0}$ when $t - t_0$ approaches infinity. On the other hand,

$$L(t) = \bar{R}^{-1}(t)\varphi(t)$$

$$= \frac{1}{\lambda^{2(t-t_0)} + \lambda^{t-t_0}(1 + k_0^2)r_0(t)}$$

$$\times \begin{pmatrix} -\lambda^{t-t_0}y(t-1) - k_0^2 r_0(t)y(t-1) + k_0^2 r_0(t)y(t-1) \\ k_0 r_0(t)y(t-1) - k_0 \lambda^{t-t_0}y(t-1) - k_0 r_0(t)y(t-1) \end{pmatrix} \quad (6.103)$$

$$= \frac{1}{\lambda^{t-t_0} + (1 + k_0^2)r_0(t)} \begin{pmatrix} -y(t-1) \\ -k_0 y(t-1) \end{pmatrix},$$

so that the elements of the gain vector are bounded and well-defined. □

The example tells us a number of things. First, that if the input ceases to be general enough (like the proportional feedback law (6.101)), then the elements of $\bar{R}^{-1}(t)$ start to increase exponentially (with rate $1/\lambda$, where λ is the forgetting factor). However, the elements of the gain vector L remain bounded. The latter property obviously holds generally, since the eigenvectors of $\bar{R}^{-1}(t)$ that correspond to eigenvalues that tend to infinity all must lie in the null space of $\psi^T(t)$, $t \geq t_0$. The elements of $L(t)$ are formed as differences of numbers tending to infinity. Therefore numerical problems will eventually occur.

In implementations of recursive identification algorithms for long-term adaptive control or monitoring applications, it is important to have resilience to this sort of numerical problem, since it can never be guaranteed that the input will remain general enough. We must consequently somehow assure ourselves that the elements of $R^{-1}(t)$ will remain bounded, $|R^{-1}(t)| < C$. As a matter of fact, we needed such an assumption also for the theoretical analysis in chapter 4. This was condition R1:

$$R(t) \geq \delta I; \quad \delta > 0. \quad (6.104)$$

Dealing with almost-singular matrices that need to be inverted is generally known as an "ill-posed problem." A standard technique is *regularization*, which can be described as measures to ensure (6.104). In this section we shall discuss how to implement some schemes that achieve (6.104) for the recursive identification problem.

6.5.2   The Levenberg-Marquardt Method

The matrix $R(t)$ in (6.100a) is by construction positive semidefinite. Hence, one way of achieving (6.104) is simply to add δI to (6.100a). We then obtain

$$R(t) = R(t-1) + \gamma(t)[\psi(t)\hat{\Lambda}^{-1}(t)\psi^T(t) + \delta I - R(t-1)]. \tag{6.105}$$

We shall call (6.105) the *Levenberg-Marquardt regularization*, after the techniques suggested by Levenberg (1944) and Marquardt (1963) for a nonlinear least squares minimization problem.

The modification (6.105) is conceptually simple. The positive number δ can be chosen as quite small, say $\delta = 10^{-2}$–10^{-4}, compared to the magnitude of the elements of $\psi(t)$. Therefore the search direction obtained from (6.105) will differ only slightly from the true Gauss-Newton one.

There is, however, an implementational disadvantage to (6.105). We have previously applied the matrix inversion lemma to (6.2b) = (6.100a) to obtain a recursion for $P(t) = \gamma(t)R^{-1}(t)$. This recursion [see (6.7a)] will contain an inverse of a matrix of dimension equal to the rank of the updating quantity $\psi(t)\hat{\Lambda}^{-1}(t)\psi^T(t)$. In (6.7a) this rank was $p = \dim y$, which typically is (much) smaller than $d = \dim \theta$. In (6.105) the updating quantity is $\psi(t)\hat{\Lambda}^{-1}(t)\psi^T(t) + \delta I$, which has rank d. Hence (6.105) is quite costly to use, since it will involve inversion of a $d \times d$-matrix in each time step.

A possible remedy is the following. Instead of adding the $d \times d$-matrix δI at each step, we could add one diagonal element at a time. Let $J_d(t)$ be a matrix whose $t(\bmod d) + 1$ diagonal element is 1 and the other elements are all zero. Then consider

$$R(t) = R(t-1) + \gamma(t)[\psi(t)\hat{\Lambda}^{-1}(t)\psi^T(t) + d\delta J_d(t) - R(t-1)]. \tag{6.106}$$

This expression is virtually identical to (6.105). Over a period of d time steps we have

$$\sum_{t=k+1}^{k+d} d\delta J_d(t) = d\delta I,$$

so a multiple of the identity matrix is being added to $R(t)$ to ensure (6.104). The version (6.106) has the advantage that it can be written

$$R(t) = R(t-1) + \gamma(t)[\psi^*(t)\hat{\Lambda}^*(t)^{-1}\psi^*(t)^T - R(t-1)], \tag{6.107}$$

where $\psi^*(t)$ is the $d \times (p+1)$-matrix

$$\psi^*(t) = \begin{pmatrix} \psi(t) \begin{pmatrix} 0 \\ \vdots \\ 0 \\ 1 \\ \vdots \\ 0 \end{pmatrix} \end{pmatrix} \leftarrow \text{pos } t(\bmod d) + 1 \tag{6.108a}$$

and

$$\hat{\Lambda}^*(t) = \begin{pmatrix} \hat{\Lambda}(t) & 0 \\ 0 & d\delta \end{pmatrix}. \tag{6.108b}$$

The matrix inversion lemma can now be applied to (6.107), giving

$$P(t) = \{P(t-1) - P(t-1)\psi^*(t)[S^*(t)]^{-1}\psi^*(t)^{\mathrm{T}}P(t-1)\}/\lambda(t), \tag{6.109a}$$

$$S^*(t) = \psi^*(t)^{\mathrm{T}}P(t-1)\psi^*(t) + \lambda(t)\hat{\Lambda}^*(t), \tag{6.109b}$$

where

$$P(t) = \gamma(t)R^{-1}(t)$$

and $\lambda(t)$ is given by (6.8). The dimension of S^* has been increased from p in (6.7) to $p+1$ in order to accomodate the Levenberg-Marquardt modification. Otherwise the algorithm for determining the gain vector is left unchanged.

6.5.3 Regularization of U-D Factorization Algorithms

The Levenberg-Marquardt modification (6.105) or (6.109) adds a positive definite matrix to $R(t)$ constantly, whether required or not. It would seem more natural to add such a modification only to an emerging null space, so as to prevent any eigenvalue of $R(t)$ from tending to zero. We described such an algorithm conceptually in (4.95). It apparently requires more work than the straightforward regularization (6.104), since the eigenspaces must also be determined. In the U-D factorization algorithm described in section 6.2.2, this type of information is, however, automatically present. We have, as in (6.12),

$$P(t) = U(t)D(t)U^{\mathrm{T}}(t), \tag{6.110}$$

where $U(t)$ is a normalized upper triangular matrix and $D(t)$ is a diagonal matrix. By introducing limits for the elements of $D(t)$ and $U(t)$ we can consequently ensure that $P(t)$ remains both bounded and positive definite.

For all practical purposes it is, in fact, sufficient to introduce a limit for $D(t)$. The reason is as follows: From (6.110), (3.59), and (3.68) we have

$$\gamma(t)P^{-1}(t) = R(t) = \sum_{k=1}^{t} \beta(t,k)\psi(k)\hat{\Lambda}^{-1}\psi^{\mathrm{T}}(k) = \gamma(t)U^{-\mathrm{T}}(t)D^{-1}(t)U^{-1}(t).$$

Assume that $\gamma(t)$ does not tend to zero and that the ψ vectors do not tend to infinity (if they did, we would have worse problems than numerical ones to worry about.) Then the elements of $U^{-1}(t)$ must remain bounded, and so must the elements of $U(t)$ (since $\det U(t) = 1$). The boundedness of $U(t)$ will thus take care of itself, and it is sufficient to introduce a limit for the elements of $D(t)$. This means that regularization is obtained for the U-D algorithm in (6.26) by modifying step 3 as follows:

3. $\vdots$

$$D(t)_{jj} := \min\left(C, \beta_{j-1}D(t-1)_{jj}/\beta_j\lambda(t)\right). \tag{6.111}$$

$\vdots$

Here C is a positive number that bounds the elements of $D(t)$. It corresponds to $\gamma(t)/\delta$ with δ chosen as in the discussion in the previous section.

The regularization (6.111) is natural and very easy to implement within the U-D algorithm. As long as the input sequence is well-behaved, so that $\psi(t)$ is a full-rank process, the limit in (6.111) is never reached and the update step coincides with the true Gauss-Newton one.

We noted that the ladder algorithms of section 6.4 correspond to block U-D factorization of the P matrix. The matrices $[R_n^e(t)]^{-1}$ and $[R_n^r(t)]^{-1}$ then are the diagonal blocks of $D_B(t)$ with

$$P(t) = U_B(t)D_B(t)U_B^{\mathrm{T}}(t).$$

The corresponding regularization procedure for the ladder algorithm will thus be to ensure the invertibility of those matrices by adding δI to them when necessary.

6.6 Stability Tests and Projection Algorithms

The parameter-updating algorithm (6.3) contains a projection into the subset of stable predictors. This is used to ensure the stability of the filter (6.4a). When we derived the algorithm (section 3.4), we used an argument involving the stability of this filter. Also, for the convergence results in section 4.4 an assumption that $\hat{\theta}(t) \in D_\mathcal{M}$ (at least infinitely often) had to be introduced. These facts as such do not of course imply that the projection in (6.3) is necessary for the practical use of the algorithm. In fact, we saw in lemma 4.2 (section 4.5) that for a PLR the stable behavior of (6.4) is guaranteed even without the projection feature.

For an RPE algorithm without projection, however, experience shows that occasionally the estimate $\hat{\theta}(t)$ steps out of the stability region and makes the algorithm "explode." An RPE algorithm should therefore contain stability monitoring and projection into the stability region, as indicated in (6.3).

Now, the aforementioned stability problem is fairly rare. In the simulations summarized in table 5.10, for example, the projection feature was in action less than ten times in a simulation containing 2,000 data points. This means that the projection does not have to be very sophisticated, and we do not have to worry about the mistreatment of information related to the projection.

In practice, projection is often implemented as a succesive reduction of the correction term for the parameter estimates. This can be done in the following way.

Projection Algorithm (6.112)

1. Choose a factor $0 \leq \mu < 1$.

2. Compute $\tilde{\theta}(t) := \gamma(t) R^{-1}(t) \psi(t) \Lambda^{-1}(t) \varepsilon(t)$.

3. Compute $\hat{\theta}(t) := \hat{\theta}(t-1) + \tilde{\theta}(t)$.

4. Test if $\hat{\theta}(t) \in D_\mathcal{M}$. If yes, go to 6; if no, go to 5.

5. Set $\tilde{\theta}(t) := \mu\tilde{\theta}(t)$ and go to 3.

6. Stop.

In step 4, the testing if $\hat{\theta}(t) \in D_\mathcal{M}$ boils down to ascertaining whether certain polynomials have all zeros inside the unit circle. This is a classical problem in the analysis of linear discrete-time systems. There are well-known algorithms for such tests (see Jury, 1974; Kucera, 1979).

In step 5, the factor μ determines the reduction of the step size. It is our experience that the choice $\mu = 0.5$ works well.

Strictly speaking, the algorithm (6.112) violates the rules of a recursive algorithm as defined in section 1.2, since there is no absolute bound on the number of iterations required. This could be resolved by taking $\mu = 0$. Then a measurement that would take $\hat{\theta}(t)$ out of the stability region is simply ignored.

For the general input-output model (3.104) treated in section 3.7, the difference between the PLR and the RPE algorithms is that the gradient is computed by filtering the regression vector through certain filters. These filters are exactly the ones whose stability is monitored in (6.112: 4). An alternative to step (6.112: 5) would then be to use PLR steps if $\hat{\theta}(t) \notin D_{\mathcal{M}}$ (i.e., skip the filtering when computing the gradient). According to what we said above about PLR, we have then ensured that the estimates will return to $D_{\mathcal{M}}$.

6.7 Summary

The implementation of the general recursive identification algorithm (6.1)–(6.4) has been discussed. The model-independent calculation of the gain vector, i.e., (6.2), has been treated in some detail.

As general advice for implementation of (6.2) we suggest the use of the U-D factorization algorithm described in section 6.2.2. The reason is that this algorithm has better resilience to numerical problems due to ill-conditioned P matrices. It also easily incorporates the regularization feature (6.111). These aspects are especially important for algorithms that are to be used in automatic systems without direct human monitoring.

The complexity of the model-dependent part of the algorithm, i.e., (6.4), very much depends on the character of the model. For difference equation models such as (3.104) or discrete-time state-space models in innovations form, such as (3.138) the computational effort is fairly moderate. As an example, we give in appendix 6.A a FORTRAN subroutine for the RPE algorithm applied to an ARMAX model. It utilizes the U-D algorithm for computation of the gain.

6.8 Bibliography

Section 6.2 The basic reference for the implementation of the Riccati equation (6.7) is Bierman (1977).

Section 6.3 The idea behind the derivation of the fast gain algorithm has its origin in Morf (1974).

Section 6.4 The paper by Itakura and Saito (1971), which deals with speech applications, had an important impact on the use of ladder filters for estimation and signal processing. The gradient algorithm was developed by Griffiths (1977) and Makhoul (1977). The exact equivalent to the least squares algorithm was derived by Morf, Vieira and Lee (1977) and by Lee (1980). A comprehensive presentation of these results is given in Lee, Morf, and Friedlander (1981). A ladder implementation of the RML algorithm is discussed in Friedlander, Ljung, and Morf (1981).

7 Applications of Recursive Identification

7.1 Introduction

At this point we have acquired a certain understanding of how to identify dynamic systems using recursive methods. The question we turn to in this chapter is what this knowledge can be used for.

There are many situations in which a mathematical model of a system is required before we can make certain decisions. Such a model does not necessarily have to be constructed on-line. Quite often, a batch of data is collected from the system, perhaps during an identification experiment which has been specifically designed to yield relevant information about the system's properties. These data are then analyzed by an off-line procedure to infer a model of the system. We called this procedure *off-line identification* in section 1.4, and we gave a few details about it in section 3.3. Even though the algorithms described in this book have been developed for another problem, that of recursive identification, it turns out that they are also quite efficient for off-line identification. The reasons for this, as well as how the algorithms are used in this latter context, are discussed in section 7.2.

There can be any number of reasons why a mathematical model $\mathcal{M}(\theta)$ of a given system would be required. The system could be a plant that has to be controlled by a regulator, and most control design techniques require a model of the plant. Or the system could generate a signal whose properties we need to know in order to predict it, or to design matching filters.

Again, such problems can sometimes be handled by first identifying the system in an off-line fashion and subsequently designing a regulator/predictor/filter based on the resulting model. In such a case, the model is treated as a true description of the system.

In other situations this may be impractical or impossible; e.g., where the system's properties actually are time-varying, or where it is impossible to perform separate identification experiments. Then a natural idea is to use recursive identification in an on-line fashion to infer the properties of the system.

An approach to this problem is to choose a standard model set in terms of which the system can be identified, such as the model sets we discussed in chapter 3 and sections 5.2 and 5.3. Based on the model $\mathcal{M}(\hat{\theta}(t))$ current at time t, a regulator (or predictor or filter) is computed. This regulator/predictor/filter is then used until the next data pair $z(t + 1)$ is collected at time $t + 1$, and the model has been updated, at which time the procedure is repeated using $\mathcal{M}(\hat{\theta}(t + 1))$.

Schemes for on-line determination of suitable regulators/predictors/filters are usually called *adaptive*. We thus speak of adaptive control, adaptive prediction, and adaptive filtering.

The calculation of the regulator from $\mathcal{M}(\theta)$ may be more or less complicated. Given a certain design technique, it is natural to choose the model set $\mathcal{M}$ so that the computations that have to be carried out at each iteration are as simple as possible. The easiest case is when the parameters of $\mathcal{M}$ coincide with those of the regulator/predictor/filter; then the calculations are indeed trivial. We might say that the system is parametrized in terms of the corresponding optimal regulator/predictor/filter. In this chapter we will illustrate how this can be achieved for a number of control and signal-processing problems.

Schemes for which the "decision parameters" (regulator parameters, filter coefficients, etc.) are directly updated by a recursive algorithm are usually not thought of as identification algorithms. In our framework, however, it is useful to regard them as just that, corresponding to a specific choice of model set $\mathcal{M}$, tailored to the particular application. Most of the development and discussion in this book applies to general model sets and hence also to these adaptive control and signal-processing schemes. We shall discuss adaptive control from this point of view in section 7.3. Adaptive prediction and adaptive state estimation are treated in section 7.4, and other adaptive signal-processing problems are discussed in section 7.5.

7.2 Recursive Algorithms for Off-Line Identification

We described the off-line identification problem in section 3.3. Given a model set $\mathcal{M}$, with corresponding predictors $\hat{y}(t \mid \theta)$, we wish to minimize a criterion

$$V_N(\theta, z^N) = \frac{1}{N} \sum_{t=1}^{N} l(t, \theta, \varepsilon(t, \theta)), \tag{7.1}$$

where

$$\varepsilon(t, \theta) = y(t) - \hat{y}(t \mid \theta). \tag{7.2}$$

Let us for ease of notation restrict outselves to

$$l(t, \theta, \varepsilon(t, \theta)) = \tfrac{1}{2}\varepsilon^{\mathrm{T}}(t, \theta)\Lambda^{-1}\varepsilon(t, \theta). \tag{7.3}$$

The off-line estimate $\hat{\theta}_N$ is defined as the parameter that minimizes $V_N(\theta, z^N)$. Except in simple special cases, the minimization of (7.1) must be performed using numerical iterative procedures. Basically, the iterations are determined from the gradient, and perhaps also the Hessian, of $V_N(\theta, z^N)$:

$$\hat{\theta}_N^{(i+1)} = f(\hat{\theta}_N^{(i)}, V_N'(\hat{\theta}_N^{(i)}, z^N), V_N''(\hat{\theta}_N^{(i)}, z^N)). \tag{7.4}$$

Here

$$[V_N'(\theta, z^N)]^{\mathrm{T}} = -\frac{1}{N}\sum_{t=1}^{N}\psi(t, \theta)\Lambda^{-1}\varepsilon(t, \theta), \tag{7.5}$$

$$V_N''(\theta, z^N) = \frac{1}{N}\sum_{t=1}^{N}\psi(t, \theta)\Lambda^{-1}\psi^{\mathrm{T}}(t, \theta) + \frac{1}{N}\sum_{t=1}^{N}\psi'(t, \theta)\Lambda^{-1}\varepsilon(t, \theta), \tag{7.6}$$

where $\varepsilon(t, \theta)$ and $\psi(t, \theta)$ are computed by means of (3.23).

Several minimization schemes (7.4) can be used. For example, we could use a Gauss-Newton scheme

$$\hat{\theta}_N^{(i+1)} = \hat{\theta}_N^{(i)} + \alpha^{(i)}\left[\frac{1}{N}\sum_{1}^{N}\psi(t, \hat{\theta}_N^{(i)})\Lambda^{-1}\psi^{\mathrm{T}}(t, \hat{\theta}_N^{(i)})\right]^{-1}$$

$$\times \frac{1}{N}\sum_{1}^{N}\psi(t, \hat{\theta}_N^{(i)})\Lambda^{-1}\varepsilon(t, \hat{\theta}_N^{(i)}). \tag{7.7}$$

Ideally, $\alpha^{(i)}$ in this equation can be set equal to 1. However, in many cases met in practice, it will be found necessary to assign a value other than 1 to $\alpha^{(i)}$ in order to assure that the criterion decreases.

Another possibility is to provide a standard numerical minimization program with subroutines for the computation of $V_N(\theta, z^N)$ and $V_N'(\theta, z^N)$ for any θ [i.e., (3.23) + (7.1) + (7.5)], and leave the search for the minimizing $\hat{\theta}_N$ to this program.

We may compare the foregoing expressions with the recursive prediction error method (3.157). The programming effort to implement (3.157d, e) and (3.23) is the same. Also, the computing effort to determine the sequence $\{\psi(t), \varepsilon(t)\}$, $t = 1, 2, \ldots, N$ is basically the same as that to find $\{\psi(t, \theta), \varepsilon(t, \theta)\}$ for any given θ. Similarily, the programming and computational effort to determine $\{R(t)\}$ in (3.158a) is comparable to determining the Hessian approximation

$$\frac{1}{N}\sum_1^N \psi(t, \theta)\Lambda^{-1}\psi^{\mathrm{T}}(t, \theta)$$

in (7.7).

In summary, the programming effort spent in implementing the Gauss-Newton algorithm (3.157), (3.158) is comparable to that of implementing (7.7) (not including the choice of $\alpha^{(i)}$). The computational complexity is such that going through the recursive scheme once for the data z^N is comparable to performing one iteration with the off-line scheme (7.4).

We know that except for the recursive least squares procedure, for which the off-line and recursive estimates coincide, the off-line estimate $\hat{\theta}_N$ is a better estimate than the recursive one, $\hat{\theta}(N)$. The reason, briefly, is that in the transient period information is misused in the recursive algorithm. One might then think of improving the recursive estimate by making several passes through the data. The procedure would be the following:

1. Apply the recursive identification method to the data set $z(t)$, $t = 1$, 2, ... N. This gives $\hat{\theta}^{(i)}(N)$ and $P^{(i)}(N) = \gamma(N)R^{-1}(N)$ with $i = 1$.

2. Use the estimate $\hat{\theta}^{(i+1)}(0) = \hat{\theta}^{(i)}(N)$ as an initial condition to apply the recursive scheme once more to the data. For $\xi(0)$, take the initial condition zero. If the eigenvalues of $A(\hat{\theta}^{(i+1)}(0))$ are close to the unit circle let the estimate $\hat{\theta}$ be frozen at its initial condition for that number of steps over which the effects of the initial condition $\xi(0)$ are damped out [this to avoid the influence of biased calculations of $\hat{y}(t)$ and $\psi(t)$]. For $P^{(i+1)}(0)$ use the initial condition $P^{(i)}(N)$ (perhaps somewhat increased to allow for more alert adjustments.) The sequence $\{\gamma(t)\}$ (or $\{\lambda(t)\}$) is, however, not reinitialized.

3. Repeat step 2 until $\hat{\theta}$ has converged.

The foregoing procedure will give estimates $\hat{\theta}^{(i)}(N)$ that converge to $\hat{\theta}_N$ as $i \to \infty$. This limiting value is that which minimizes (locally) the criterion (7.1). This follows from the following lemma.

LEMMA 7.1 Let M be a fixed given value and let z^M be a given data sequence. Construct the infinite sequence $\bar{z}(t)$:

$$\bar{z}(t) = z(j) \quad \text{for} \quad t = k \cdot M + j, \quad k = 0, \ldots.$$

Let $\hat{\bar{\theta}}(t)$ be the estimates that result when the recursive prediction error algorithm is applied to the data sequence $\{\bar{z}(t)\}$. Then $\hat{\bar{\theta}}(t)$ converges to a local minimum of the function

$$V(\theta, z^M) = \frac{1}{M} \sum_{t=1}^{M} \frac{1}{2} \varepsilon^T(t, \theta) \Lambda^{-1} \varepsilon(t, \theta)$$

or to the boundary of $D_{\mathcal{M}}$ (see section 4.3.4) as $t \to \infty$.

Proof Regard z^M as a given deterministic sequence. Then $\bar{z}^t$ will be a deterministic periodic sequence. Since $\bar{z}$ is deterministic, condition S1 of theorem 4.3 will be trivially satisfied (take $z_k^0(t) = z(t)$; this variable is "independent" of z^k); and since the sequence $\bar{z}^t$ is periodic, the conditions A2, A3 will trivially hold. In particular,

$\frac{1}{2}\bar{E}\varepsilon^T(t, \theta)\Lambda^{-1}\varepsilon(t, \theta)$

$$= \lim_{N \to \infty} \frac{1}{N} \sum_{t=1}^{N} \frac{1}{2} \varepsilon^T(t, \theta) \Lambda^{-1} \varepsilon(t, \theta)$$

$$= \lim_{i \to \infty} \frac{1}{i \cdot M} \sum_{t=1}^{i \cdot M} \frac{1}{2} \varepsilon^T(t, \theta) \Lambda^{-1} \varepsilon(t, \theta)$$

$$= \lim_{i \to \infty} \frac{1}{i} \left[\sum_{k=1}^{i} \frac{1}{M} \sum_{t=1}^{M} \frac{1}{2} \varepsilon^T(t, \theta) \Lambda^{-1} \varepsilon(t, \theta) \right]$$

$$= \frac{1}{M} \sum_{t=1}^{M} \frac{1}{2} \varepsilon^T(t, \theta) \Lambda^{-1} \varepsilon(t, \theta).$$

Here, the first equality follows since E can be dispensed with for a deterministic sequence. The third equality follows from the periodicity of $\bar{z}$, which allows us to write $\varepsilon(t, \theta) = \varepsilon(t + k \cdot M, \theta)$. Theorem 4.3 now gives the desired result. ∎

The lemma proves the desired behavior of $\hat{\theta}^{(i)}(M)$, since $\hat{\bar{\theta}}(M \cdot i) = \hat{\theta}^{(i)}(M)$.

The use of recursive identification algorithms for off-line identification has been suggested and extensively used by Young and Jakeman (1979) in connection with the refined instrumental variable method. The technique has several potential advantages:

• There is no need to develop specific software for the off-line problem. As remarked above, the development of such software may involve more work than implementation of (3.157).

• The minimum of the criterion function can be found faster with a recursive algorithm than with the off-line schemes (7.4). The reason for this can be expressed as follows. Once the recursive estimates have come into a neighborhood of the true minimum, a second-order Taylor's expansion becomes very accurate and there is no further loss of information, despite the recursive character of the calculation (see section 4.4.3). This means that if a pass through data has initial values close to the actual minimum, then this pass will essentially bring the estimate to the minimum. We also know from the simulations in chapter 5 that the estimates are already quite close to the true values after one pass through the data when $M \approx 1,000$ and dim $\theta \approx 6$. Hence, relatively few passes will be required. To put it another way, the off-line iterative schemes (7.4) go through all the data to determine the modifications of the first estimates. This is certainly wasteful, since almost as good adjustments could be computed from a fraction of the data.

• The recursive algorithms can be used to detect nonstationarities in the data and the system properties. If a certain pattern in the estimates is repeated during different passes through the data, this is an indication that the dynamics is changing over that period. An off-line method gives only the average behavior and cannot give this more detailed information about the dynamics.

We thus conclude that recursive identification can be a valuable tool in off-line identification. This has been confirmed in practical experience, e.g., by Young and Jakeman (1979).

7.3 Adaptive Control

7.3.1 The Adaptive Control Problem

As mentioned in section 7.1, a typical problem in control design is that the dynamics of the plant to be controlled are not sufficiently well known and/ or are time-varying. The design of a suitable regulator is then difficult. Adaptive control is an approach to this problem that has been widely discussed. For example, see the comprehensive treatments by Landau (1979), Åström et al. (1977), and Goodwin and Sin (1983).

The approach can be described as follows.

(1) Choose a regulator structure. A regulator is a feedback mechanism by which the input is determined based on previous observations $\{z(t)\}$ and possibly on a sequence of reference signal values $\{r(t)\}$. The regulator

contains a number of parameters to be determined. Let these be collected in a parameter vector ρ. We can then describe the regulator symbolically by

$$u(t) = h(\rho; t, y^t, u^{t-1}, r^t).\tag{7.8}$$

For example, a proportional regulator is given by

$$u(t) = K_p[r(t) - y(t)],$$

a simple special case of (7.8), with the gain K_p corresponding to ρ.

(2) Choose a model set $\mathcal{M}$ parametrized by a parameter vector θ.

(3) Choose a design procedure such that the best (or "optimal") values of the regulator parameters ρ can be computed from the model $\mathcal{M}(\theta)$ of the system. In our framework, a design procedure is a mapping from θ to ρ:

$$\rho = k(\theta).\tag{7.9}$$

(4) Choose a recursive identification algorithm that provides estimates $\hat{\theta}(t)$ of θ on-line based on input-output observations z^t.

(5) At time t use the control law corresponding to the regulator parameter

$$\hat{\rho}(t) = k(\hat{\theta}(t)),\tag{7.10}$$

i.e., use

$$u(t) = h(\hat{\rho}(t); t, y^t, u^{t-1}, r^t).\tag{7.11}$$

Now the expressions (7.10) and (7.11), together with the recursive identification algorithm to determine $\hat{\theta}(t)$, define an algorithm for adaptive control. This approach to adaptive control is depicted in figure 7.1.

The adaptive control problem has many interesting aspects. Each of the above five items, as well as the analysis of the resulting scheme, is worthy of a lengthy discussion. In this section, though, we shall concentrate on issues that are directly related to recursive identification.

Two different approaches to the adaptive control problem can be distinguished. These have been called the indirect (or explicit) and the direct (or implicit) approaches, respectively, by Narendra and Valavani (1978) and Landau (1979). In our notation, they can be described as follows:

"Indirect" (*"explicit"*) schemes use "conventional" model sets $\mathcal{M}$ in (2). These schemes are further discussed in section 7.3.2.

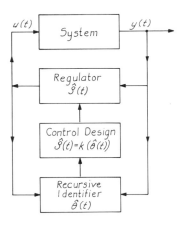

Figure 7.1
An adaptive control scheme.

"Direct" (*"implicit"*) schemes use a model set $\mathscr{M}$ that is tailored to the specific regulator structure and design procedure so that the mapping k in (3) becomes the identity mapping $\rho = \theta$ (or another trivial mapping, perhaps involving scaling of θ to yield ρ). We may then say that the regulator parameters are identified. Such a scheme is discussed in section 7.3.3.

With the control law (7.11) the input-output sequence $\{z(t)\}$ is generated from the reference signal $\{r(t)\}$ (if present) and from the noise sources that affect the system. It is obvious from (7.10) and (7.11) that the data sequence $\{z(t)\}$ will depend on the estimate sequence $\{\hat{\theta}(t)\}$. As we remarked several times in section 4.3, this fact will essentially prevent us from verifying the regularity conditions A1–A3 of section 4.3.3, since these conditions require "asymptotic mean stationarity" of the data sequence. The convergence problem can, however, still be handled by means of the techniques of section 4.3. We give the formal details of this in appendix 7.A. Here the results are outlined at a level that matches section 4.3.2.

Pretend that the regulator parameter ρ is kept fixed in (7.8). This constant feedback law will then give us an input-output sequence that is generated by the reference signal and the disturbance signals (that are independent of ρ) via (7.8) and the system. Denote this (fictitious) data sequence by $\{z(t, \rho)\}$. (It will obviously depend on the regulator parameter ρ.) The data sequence $\{z(t, \rho)\}$ will satisfy "asymptotic mean stationary" conditions provided $\{r(t)\}$ and the noise signals do. Now let $z(t, \rho)$ be the data sequence that drives the generation of model predictions $\hat{y}(t, \theta)$, prediction errors $\varepsilon(t, \theta)$, and gradients (or gradient approximations) $\eta(t, \theta)$ as described in section 4.3.2 (and 4.3.3–4.3.5). To indicate the ρ-dependence explicitly we use the notation $\varepsilon(t, \theta, \rho)$ and $\eta(t, \theta, \rho)$ for these quantities.

We found in section 4.3 that the associated differential equation for the general recursive identification algorithm (4.16) is given by (4.82), (4.64). In the present case, if the identification procedure is actually applied to the data sequence $\{z(t, \rho)\}$, the corresponding differential equation will be

$$\dot{\theta} = R^{-1}\bar{f}(\theta, \rho),$$
$$\dot{R} = \bar{G}(\theta, \rho) - R, \tag{7.12}$$

where

$$\bar{f}(\theta, \rho) = \bar{E}\eta(t, \theta, \rho)\Lambda^{-1}\varepsilon(t, \theta, \rho),$$
$$\bar{G}(\theta, \rho) = \bar{E}\eta(t, \theta, \rho)\Lambda^{-1}\eta^{\mathrm{T}}(t, \theta, \rho). \tag{7.13}$$

In view of the link between θ and ρ expressed in (7.9) and (7.10), the differential equation that is associated with the adaptive control algorithm will be

$$\dot{\theta} = R^{-1}\bar{f}(\theta, k(\theta)),$$
$$\dot{R} = \bar{G}(\theta, k(\theta)) - R. \tag{7.14}$$

A more formal treatment of these arguments is given in appendix 7.A, but the above discussion may be sufficient for an understanding of the convergence studies to follow in the next subsection.

One nontrivial difficulty with the present approach must now be pointed out. The fictitious data sequence $\{z(t, \rho)\}$ will give well-defined limits in (7.13) only if ρ is such that the closed-loop system obtained for the regulator (7.8) is stable. This means that the differential equation (7.14) is defined only for such θ that:

(i) give stable predictors,

(ii) via $\rho = k(\theta)$ give a stable closed-loop system.

The region corresponding to (i) is known to the user, who can always assure that θ belongs to this set. The second region presents a more difficult problem. The stability of the closed-loop system depends on the (unknown) properties of the controlled plant. Therefore a projection into this stability region cannot be used. Hence other techniques must be applied to prove that the closed-loop adaptive control scheme is stable, so that the signals under consideration remain bounded. Knowing that, the differential equation (7.14) can be used to investigate the convergence properties.

The overall stability (or "boundedness") problem for adaptive regulators is an interesting problem, the details of which, however, are beyond the scope of this book. The first complete solutions for special cases appeared in Jeanneau and de Larminat (1975), de Larminat (1979), Egardt (1979a), Goodwin, Ramadge and Caines (1980, 1981), Narendra and Lin (1980), and Fuchs (1980). A comprehensive treatment is given by Goodwin and Sin (1983), but a general theory for the boundedness problem, applicable to the general scheme described above, is still lacking.

7.3.2 Adaptive Control Based on Explicit Identification

The family of "explicit" adaptive control algorithms is a large one. In the recipe described in the previous subsection we may take any combination of model set, recursive identification algorithm, and control design procedure; many specific combinations have been considered in the literature. It may be said that it is not a problem to invent adaptive control algorithms in this family—the problem lies in the evaluation and analysis of the methods.

We shall here not go into details of any specific control laws. Instead, we shall discuss some general convergence aspects of the resulting adaptive control schemes.

Suppose first that we use a recursive prediction error algorithm to estimate the parameters of the model. This means, in the notation of the previous subsection, that we use $\eta(t) = \psi(t)$. Thus we have

$$\psi(t, \theta, \rho) = -\frac{\partial}{\partial \theta}\varepsilon(t, \theta, \rho),\tag{7.15}$$

where the argument ρ indicates, as before, that the algorithm is applied to the data sequence $\{z(t, \rho)\}$. Let us define

$$\bar{V}(\theta, \rho) = \tfrac{1}{2}\bar{E}\varepsilon^{\mathrm{T}}(t, \theta, \rho)\Lambda^{-1}\varepsilon(t, \theta, \rho),$$

with $\bar{f}(\theta, \rho)$ defined as in (7.13). We have, as in section 4.4, that

$$\bar{f}(\theta, \rho) = -\frac{\partial}{\partial\theta} \bar{V}(\theta, \rho).$$ (7.16)

This means that for any given data sequence $\{z(t, \rho)\}$, we can use $\bar{V}(\theta, \rho)$ as a Lyapunov function for (7.12) just as in theorem 4.3. However, when the control loop is closed the associated differential equation is given by (7.14). The function $f(\theta, k(\theta))$ in (7.12) will, however, *not* be the gradient (with respect to θ) of the function $\bar{V}(\theta, k(\theta))$, since the $\rho = k(\theta)$-dependence is not accounted for in (7.16). Therefore $\bar{V}(\theta, k(\theta))$ cannot be used as a Lyapunov function to prove stability of (7.14), and the general convergence result of theorem 4.3 cannot be applied when the recursive prediction error algorithm is used in an adaptive control loop. The differential equation (7.14) still describes the convergence properties, though, and can be analysed in special cases. A general analysis has, however, not yet been performed.

Let us now consider a case wherein a pseudolinear regression (such as ELS) is used. In the notation of section 7.3.1 we then have $\eta(t) = \varphi(t)$ and $\eta(t, \theta, \rho) = \varphi(t, \theta, \rho)$. Suppose now that the system in fact can be described by an ARMAX model:

$$A_0(q^{-1})y(t) = q^{-k}B_0(q^{-1})u(t) + C_0(q^{-1})e(t),$$ (7.17)

where the degrees of A_0, B_0, and C_0 are less than or equal to the model orders. We then still have the relationship between $\varepsilon(t, \theta, \rho)$ and $\varphi(t, \theta, \rho)$ given by (4.152) in section 4.5.2:

$$C_0(q^{-1})\varepsilon(t, \theta, \rho) = \varphi^T(t, \theta, \rho)(\theta_0 - \theta) + C_0(q^{-1})e(t),$$

since this holds for any data sequence z^t. Here θ_0 corresponds to the parameters of (7.17). Thus, as in section 4.5.2, we find

$$\bar{f}(\theta, \rho) = \tilde{G}(\theta, \rho)(\theta_0 - \theta),$$ (7.18a)

where

$$\tilde{G}(\theta, \rho) = \bar{E}\varphi(t, \theta, \rho)\tilde{\varphi}^T(t, \theta, \rho),$$ (7.18b)

$$\tilde{\varphi}(t, \theta, \rho) = \frac{1}{C_0(q^{-1})}\varphi(t, \theta, \rho).$$ (7.18c)

Hence, if the transfer function

$$\frac{1}{C_0(q^{-1})} - \frac{1}{2} \text{ is strictly positive real,}$$ (7.19)

then we have, as in the proof of theorem 4.6, that $\tilde{G}(\theta, \rho) + \tilde{G}^{\mathrm{T}}(\theta, \rho) - G(\theta, \rho)$ is positive semidefinite. This holds for all values of ρ. In particular, it is true for $\rho = k(\theta)$. Therefore we can apply the same convergence analysis to (7.14) as we did in section 4.5.2. This gives that if the ELS method is applied to the system (7.17), where (7.19) holds, then the differential equation (7.14) is globally stable and the trajectories will converge to

$$D_c = \{\theta | \bar{\mathrm{E}}[\varepsilon(t, \theta, k(\theta)) - e(t)]^2 = 0\}. \tag{7.20}$$

This is true for an arbitrary regulator and design technique $\rho = k(\theta)$. To conclude from this that the adaptive control algorithm also converges to D_c requires that the overall stability, or boundedness, condition is satisfied, as explained in section 7.3.1.

In words, the set D_r can be characterized as the set of models that are equivalent, from an input-output point of view, to the true system (7.17) under the feedback law $\rho = k(\theta)$ induced by the model in question. We can summarize this result as follows:

Consider the ELS algorithm applied to an ARMAX model, used in an adaptive control scheme with arbitrary regulator structure and control design technique. Suppose that the true system is given by (7.17) and that (7.19) holds. Then, assuming that the boundedness condition holds, the adaptive control algorithm will converge w.p.1 to θ-values that give a correct input-output description of the system.

7.3.3 Self-Tuning Regulators and Model Reference Adaptive Control Schemes

The schemes described in the previous subsection computed a control law from a current system model. In the adaptive control application we might say that the objective is control, i.e., optimal choice of the regulator parameters. In this case the model is of no interest in itself—it merely serves as a vehicle to arrive at the control law. It would thus seem to be more appealing to develop algorithms that directly update the regulator parameters.

This point of view has been stressed for so-called model reference adaptive systems (MRASs) (see, e.g., Landau, 1979). In these adaptive regulators the difference between the actual output $y(t)$ and that produced by an "ideal" reference model, $r(t)$, are compared. The difference

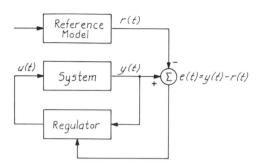

Figure 7.2
A model reference adaptive system.

between these signals is used to update the regulator parameters. The idea is depicted in figure 7.2.

The Self-Tuning Regulator suggested by Åström and Wittenmark (1973) is also based upon the idea that the regulator parameters should be updated directly.

It may seem that the identification step is avoided when the regulator parameters are updated directly. This has sometimes been claimed as an advantage of the model reference approach, compared to the schemes of section 7.3.2. However, from our point of view it is more useful and illuminating to think of the "direct" methods of adaptive control as recursive identification of the parameters of a model set that is specifically chosen to make the mapping (7.9) trivial.

The heuristic idea is the following. There is usually a close relationship between a system model and the regulator that gives optimal behavior (according to a chosen criterion) when applied to the model. Given the model we can compute the optimal regulator. Conversely, given the regulator, we can determine (perhaps up to a scaling factor) the model for which it is optimal. This means that a system can be modeled in terms of its optimal regulator. Hence we can think of a parametrized regulator as a model set for the system. The algorithms described in this book have mostly been developed for general model sets, and nothing prevents their application to the model sets just described. In this sense direct adaptive regulators can be regarded as recursive identification methods. We shall illustrate this interpretation with an example.

EXAMPLE 7.1   Consider a system described by

$$A(q^{-1})y(t) = B(q^{-1})u(t) + C(q^{-1})e(t), \qquad (7.21)$$

where $B(q^{-1})$ starts with the term $b_1 q^{-1}$ and $b_1 = 1$. We want to design a regulator that minimizes the variance of $y(t)$. Simple calculations (see, e.g., Åström, 1970) show that this is accomplished by the control law

$$u(t) = \frac{A(q^{-1}) - C(q^{-1})}{B(q^{-1})} y(t), \qquad (7.22)$$

which gives $y(t) = e(t)$.

In order to implement (7.22) we need to know the polynomials A, B, and C. We could find these adaptively, as in the previous section, by estimating the parameters in (7.21) using RML or ELS and subsequently using these estimates in (7.22). Another approach is to postulate a regulator structure

$$u(t) = \frac{S(q^{-1})}{R(q^{-1})} y(t) \qquad (7.23)$$

and directly update the parameters of the R- and S-polynomials. Let us pursue this idea. The leading coefficient of R is taken to be unity. Then, with

$$\theta^T = (s_1 \; \ldots \; s_n \; r_2 \; \ldots \; r_m),$$

$$\varphi^T(t) = (-y(t-1) \; \ldots \; -y(t-n) \; u(t-2) \; \ldots \; u(t-m+1)),$$

(7.23) can be rewritten as

$$u(t) = -\theta^1 \varphi(t+1). \qquad (7.24)$$

Let us now regard (7.23) as a model of the system. That is, we assume that (7.23) is the true minimum-variance regulator for the system. Then we must have $A - C = q^{-1}S$ and $B = q^{-1}R$. [see (7.22)]. In this case we cannot calculate the model parameters A, B, and C uniquely from S and R. To overcome this lack of uniqueness, we fix C to be C_*. Then $A = C_* + q^{-1}S$, and the system description (7.21) can be written

$$C_*(q^{-1})y(t) = R(q^{-1})u(t-1) - S(q^{-1})y(t-1) + C_*(q^{-1})e(t)$$

$$= u(t-1) + \theta^T \varphi(t) + C_*(q^{-1})e(t).$$

The predictor and its gradient now are

$$\hat{y}(t \mid \theta) = \frac{1}{C_*(q^{-1})} [u(t-1) + \theta^T \varphi(t)],$$

$$\psi(t, \theta) = \frac{1}{C_*(q^{-1})} \varphi(t).$$

With the regulator at time $t - 1$,

$$u(t-1) = -\hat{\theta}^T(t-1)\varphi(t),$$

inserted, we find that

$$\varepsilon(t) = y(t) - \hat{y}(t \mid \hat{\theta}(t-1)) = y(t).$$

Hence the RPEM for estimating θ in the regulator model (7.23) or (7.24) becomes

$$\hat{\theta}(t) = \hat{\theta}(t-1) + \gamma(t)R^{-1}(t)\psi(t)y(t),$$

$$R(t) = R(t-1) + \gamma(t)[\psi(t)\psi^T(t) - R(t-1)],$$

$$\psi(t) = \frac{1}{C_*(q^{-1})} \varphi(t).$$

When we take $C_*(q^{-1}) = 1$, this is exactly the self-tuning regulator suggested by Åström and Wittenmark (1973). Notice, though, that in the framework of our example, it is just a RPEM for the particular model (7.23). □

Adaptive regulators such as the foregoing can be analyzed in the same way we have analyzed the other recursive identification methods in this book. See Ljung (1977b, 1980a), Dugard and Landau (1980b), and Egardt (1980b) for some such results. The martingale approach has been used to prove convergence for a number of similar schemes (Goodwin and Sin, 1983).

7.3.4 Summary

In this section we have sketched the application of recursive identification to the problem of adaptive control. That identification is a key to the understanding of adaptive control schemes that are themselves based upon explicit system identification is obvious. We have, furthermore, pointed out that the recursive estimation techniques treated in this book also provide a key to those algorithms designed to update regulator parameters. Thus recursive identification must be considered an important tool for solving problems in adaptive control and for the design of adaptive control systems.

7.4 Adaptive Estimation

7.4.1 Introduction

Linear estimation concerns the problem of estimating the values of a signal $\{s(t)\}$ based on observations of a related signal $\{y(t)\}$. This subject has attracted extensive interest in the fields of control, communication, and signal processing. See, e.g., the books Kailath (1976) and Anderson and Moore (1979) for comprehensive treatments of the problem.

In the linear estimation literature, the properties of the signal $\{s(t)\}$ are assumed to be known. This information can be given in many ways: as the spectrum of $\{s(t)\}$, as its covariance function, or as a state-space model that generates $\{s(t)\}$ as its output.

A situation that is frequently encountered in practice is that the properties of the signal are not known. By *adaptive estimation* we mean those techniques used to solve the estimation problem when incomplete or no information about $\{s(t)\}$ is available. In the present section we shall comment on this adaptive estimation problem. Some applications to signal processing will be described in section 7.5.

One approach to adaptive estimation is straightforward. Using the techniques described in this book we can recursively build a model of the properties of $\{s(t)\}$. For any such model, we can apply the well-established methods of linear estimation to determine the estimate (prediction) using the current values of the model parameters. This gives a large and obvious family of adaptive estimation methods.

Another approach is to reparametrize the model, so that the computation of the predictor or filter becomes simpler. We may then say that the parameters of the predictor/filter/smoother are directly adapted to the signal. The idea, as well as the technique, is quite similar to the direct adaptive control methods that were described in section 7.3.3.

7.4.2 Adaptive Prediction

Our formulation of the identification problem in this book is closely related to prediction, since we have described a model in terms of its (usually one-step-ahead) prediction. This means that the variable $\hat{y}(t)$, which is present in all our algorithms, indeed is a one-step-ahead *adaptive prediction* of $y(t)$. Adaptive prediction of the next output $y(t)$ is therefore contained in our recursive identification framework. Adaptive prediction of values further into the future can always be achieved by calculations using the current model.

In this section we shall give some details on adaptive prediction of signals described by ARMA models. Such signal models are common in communication and control theory. We consequently consider the model

$$y(t) = \frac{C(q^{-1})}{A(q^{-1})} e(t), \tag{7.25}$$

assuming for simplicity that $y(t)$ is scalar-valued. Here $\{e(t)\}$ is a white-noise process and

$$C(q^{-1}) = 1 + c_1 q^{-1} + \cdots + c_{n_c} q^{-n_c},$$

$$A(q^{-1}) = 1 + a_1 q^{-1} + \cdots + a_{n_a} q^{-n_a}.$$

Fortunately, this is one of the models for which we have a fairly detailed understanding of the convergence properties for the corresponding recursive identification algorithms. Suppose that the signal can indeed be described by

$$y(t) = \frac{C_0(q^{-1})}{A_0(q^{-1})} e(t), \tag{7.26}$$

where the degrees of C_0 and A_0 are less than or equal to those of (7.25).

• If we apply the pseudolinear regression approach (see section 3.7.3) to (7.25), the ELS algorithm results. From section 4.5.2 (or appendix 4.C) we know that a sufficient condition for the convergence w.p.1 of the ELS algorithm to the true description (7.26) is that $1/[C_0(q^{-1})] - 1/2$ is strictly positive real. From section 4.5.4 we know that a necessary condition is that

$$\text{Re } C_0(\alpha_i) > 0, \quad i = 1, \ldots, n_a,$$

where α_i are the roots of

$$z^{n_a} + a_1^0 z^{n_a-1} + \cdots + a_{n_a}^0 = 0.$$

• If we apply the recursive prediction error approach (section 3.7.2) to (7.25) the RML algorithm results. We then know from section 4.4 that, provided the C-polynomial is kept stable, the estimates will converge w.p.1 to the true system, regardless of its properties. This follows since all stationary points of the corresponding criterion function are in fact global minima, a result that was quoted in section 5.3 and proven by Åström and Söderström (1974). Even if the true description of the system may be more complex, the RML algorithm will converge to a value that (locally) minimizes the prediction error variance.

The foregoing paragraphs summarize the situation of recursive identification of the parameters of (7.25). Since both ELS and RML automatically provide a one-step-ahead prediction $\hat{y}(t)$ of the signal, we have also described the properties of the corresponding adaptive predictors. In section 6.4 we described ladder and lattice algorithms to implement the recursive least squares and the ELS schemes. When our prime objective is to determine $\hat{y}(t)$ and the model order is high, these algorithms are quite efficient. The fact that they reparametrize the model set (from difference equation parameters to reflection coefficients) is no disadvantage at all for the adaptive prediction problem.

k-**Step-Ahead Prediction: Indirect Methods**   Let us now turn to the k-step-ahead adaptive prediction problem for the ARMA model. First we shall establish a formula for the k-step-ahead predictor for a given ARMA model (7.25). This predictor will be denoted by $\hat{y}(t + k \mid t; \theta)$, where θ denotes the parameters of (7.25). Conceptually, the k-step-ahead predictor can be obtained by concatenation of k one-step-ahead predictors. We illustrate it for $k = 2$: Use the known formula for $\hat{y}(t + 2 \mid t + 1; \theta)$. Wherever $y(t + 1)$ appears in this expression, replace it by $\hat{y}(t + 1 \mid t; \theta)$. This procedure has some interest for the interpretation of the k-step-ahead predictor. When deriving an explicit expression it is, however, more efficient to go a direct route: From (7.25) we have

$$y(t + k) = \frac{C(q^{-1})}{A(q^{-1})} e(t + k). \tag{7.27}$$

The right-hand side can be expanded in Maclaurin series in powers of q^{-1}. If the series expression is truncated at the power q^{-k} we obtain

$$y(t + k) = F(q^{-1})e(t + k) + \frac{q^{-k}G(q^{-1})}{A(q^{-1})} e(t + k), \tag{7.28}$$

where $F(q^{-1})$ is of degree $k - 1$:

$$F(q^{-1}) = 1 + f_1 g^{-1} + \cdots + f_{k-1} q^{-k+1}.$$

In (7.28) F and G are polynomials that must satisfy

$$C(q^{-1}) \equiv A(q^{-1})F(q^{-1}) + q^{-k}G(q^{-1}) \tag{7.29}$$

for (7.28) to be equivalent to (7.27). It follows from (7.29) that G will be of degree $n - 1$, where $n = \max(n_a, n_c - k + 1, 1)$. In (7.28) the term $F(q^{-1})e(t + k)$ is independent of y^t, since it contains only terms $e(s)$ for $s > t$. Hence

$$\hat{y}(t + k \mid t; \theta) = \frac{G(q^{-1})}{A(q^{-1})} e(t). \tag{7.30}$$

Now using the fact that $e(t) = A(q^{-1})/C(q^{-1})y(t)$, we can rewrite (7.30) as

$$\hat{y}(t + k \mid t; \theta) = \frac{G(q^{-1})}{C(q^{-1})} y(t). \tag{7.31}$$

This is the form given by Åström (1970), and we have basically followed his derivation of it.

From (7.28) we see that the k-step-ahead prediction error will by given by

$$\varepsilon(t \mid t - k; \theta) = y(t) - \hat{y}(t \mid t - k; \theta) = F(q^{-1})e(t). \tag{7.32}$$

To obtain an adaptive k-step-ahead predictor we could simple recursively identify the parameters of the ARMA model (7.25) using either the ELS or the RML method. The current estimates $\hat{A}_t$ and $\hat{C}_t$ are then used in (7.29) to determine $\hat{G}_t$. Finally the predictor $\hat{y}(t + k \mid t)$ is computed from (7.31) using $\hat{G}_t$ and $\hat{C}_t$. Such a procedure could be called an "indirect" adaptive predictor. Its convergence properties coincide with those of the respective recursive identification algorithms, and they were quoted earlier in this section. In particular, the RML algorithm gives a globally convergent adaptive k-step-ahead predictor if the true system is given by (7.26), regardless of the properties of $C_0(q^{-1})$.

k-**Step-Ahead Prediction: Direct Methods** As an alternative to the explicit identification of ARMA parameters we could choose to directly update the predictor coefficients. Introduce the notation, assuming that the degree of the C-polynomial is n,

$$\varphi^{\mathrm{T}}(t + k, \theta) = (y(t) \ \ldots \ y(t - n + 1) \ -\hat{y}(t + k - 1 \mid t - 1; \theta) \ \ldots$$

$$- \hat{y}(t + k - n \mid t - n; \theta)), \tag{7.33}$$

$$\theta^{\mathrm{T}} = (g_0 \ \ldots \ g_{n-1} \ c_1 \ \ldots \ c_n).$$

Then (7.31) can be written

$$\hat{y}(t + k \mid t; \theta) = \theta^{\mathrm{T}}\varphi(t + k, \theta). \tag{7.34}$$

We also find from (7.31) that

$$\frac{d}{d\theta}\hat{y}(t + k \mid t; \theta) = \frac{1}{C(q^{-1})}\varphi(t + k, \theta). \tag{7.35}$$

These expressions look much the same as the corresponding formulas for the ARMA model [see (3.118), (3.122)], and we can easily derive a recursive k-step-ahead prediction error algorithm:

$$\hat{\theta}(t) = \hat{\theta}(t - 1) + \gamma(t)R^{-1}(t)\psi(t)\varepsilon(t), \tag{7.36a}$$

$$\varepsilon(t) = y(t) - \hat{y}(t \mid t - k), \tag{7.36b}$$

$$\hat{y}(t \mid t - k) = \hat{\theta}^{\mathrm{T}}(t - 1)\varphi(t), \tag{7.36c}$$

$$\varphi(t) = (y(t - k) \ \ldots \ y(t - k - n + 1) \ \hat{y}(t - 1 \mid t - k - 1) \ \ldots$$

$$\hat{y}(t - n \mid t - n - k)), \tag{7.36d}$$

$$\psi(t) = \frac{1}{\hat{C}_{t-1}(q^{-1})}\varphi(t), \tag{7.36e}$$

$$R(t) = R(t - 1) + \gamma(t)[\psi(t)\psi^{\mathrm{T}}(t) - R(t - 1)]. \tag{7.36f}$$

Just as in the RML algorithm, it is usually worthwhile to replace the predictions $\hat{y}(t \mid t - k)$ in (7.36d) by their posterior values

$$\hat{\bar{y}}(t \mid t - k) = \hat{\theta}^{\mathrm{T}}(t)\varphi(t) \tag{7.37}$$

to achieve better transient-convergence properties (see section 5.11).

Similarly, by ignoring the implicit θ-dependence in (7.34), we can derive a pseudolinear regression for the estimation of θ:

$$\hat{\theta}(t) = \hat{\theta}(t - 1) + \gamma(t)R^{-1}(t)\varphi(t)\varepsilon(t), \tag{7.38a}$$

$$R(t) = R(t - 1) + \gamma(t)[\varphi(t)\varphi^{\mathrm{T}}(t) - R(t - 1)], \tag{7.38b}$$

$\varepsilon(t)$ and $\varphi(t)$ defined as in (7.36b–d) or (7.37).

The convergence properties of each of these two schemes can be analyzed with the same techniques as in section 4.4 [for (7.36)] and in section 4.5 [for (7.38)]. The first algorithm will converge to a value that minimizes the variance of the k-step-ahead prediction error. The second one will converge to the true predictor, provided that the signal can be described as an ARMA process with compatible orders, and with a C-polynomial such that $[1/C_0(q^{-1})] - \frac{1}{2}$ is positive real. A comprehensive discussion of convergence properties of the algorithm (7.38) is given by Holst (1977) and Goodwin and Sin (1983).

7.4.3 Adaptive State Estimation and Smoothing

Smoothing and filtering of signals to eliminate the influence of additive measurement noise plays a major role in communications and control applications. Optimal enhancement of the signal and suppression of the noise requires knowledge of the properties of the signal. When these are unknown we may use adaptive techniques. Just as for adaptive control and adaptive prediction, this can be achieved by a combination of explicit recursive identification and any established design technique for smoothing and filtering. We may point out that for the algorithms that we developed for state-space models, such as (3.145), the adaptive state estimate $\hat{x}(t)$ is obtained as an automatic by-product of the recursive identification algorithm.

Adaptive state estimation has often been approached through adaptive observer techniques, as described in, e.g., Lüders and Narendra (1974) and Caroll and Lindorff (1973). When the state-space models are obtained by canonical representation of input-output models (as typically is the case for adaptive observers), the state variables are linear combinations of delayed inputs and outputs, with the (unknown) system parameters as coefficients. That, however, means that any recursive identification method that estimates the parameters also yields a natural adaptive observer: one just uses the current parameter estimates in the linear combination of inputs and outputs to obtain the state estimate.

7.5 Adaptive Signal Processing

7.5.1 Introduction

In a broad sense most applications in the control and communication area can be regarded as "signal processing." The recursive identification problem that we have discussed in this book concerns various way of processing a signal $\{z(t)\}$ in order to extract interesting and relevant features.

The term "signal processing" is, however, often used in a more narrow sense. It refers, generally speaking, to filtering used to enhance or modify signals. In this section we shall discuss the use and importance of recursive identification for such applications. For readers with a signal-processing background, the section will also serve as an interface between our problem formulation and notation and that used in signal processing. It is not our intention to give comprehensive treatments of the different problems, or to provide extensive references. Rather, we shall describe a number of important problems in signal processing and explain why they may require adaptive techniques. We shall show how the adaptation process can be described in terms of a recursive identification algorithm. The most common model sets used in the respective applications will be displayed, and the techniques typically applied to estimate their parameters will be outlined.

As a preliminary, we mention that linear regression models are by far the most common ones. This means that of two measured or known signals $\{y(t)\}$ and $\{u(t)\}$, one is typically expressed by a regression on the other:

$$y(t) = B(q^{-1})u(t). \tag{7.39}$$

Such a model is called a finite impulse response (FIR) model; it is, of course, a special case of the linear input-output models described in section 3.7. In signal processing (7.39) is often called a transversal filter or a "tapped delay line," since the output signal $y(t)$ is obtained by linear combination of delayed input signals as shown in figure 7.3.

From our studies in sections 4.4 and 5.3 we know that the model (7.39) has an important property: It is the only model that combines the advantages of being a linear regression and being robust against colored additive noise. The latter property means that if the true data is described by

$$y(t) = B_0(q^{-1})u(t) + v(t), \tag{7.40}$$

then a prediction error method, such as the least squares method, will give consistent estimates of B_0, regardless of the properties of $\{v(t)\}$, as long as it has zero mean and is uncorrelated with $\{u(t)\}$. [This was shown in example 4.7: Here we apply it to the special case $F(q^{-1}) = F_0(q^{-1}) = 1$.] A disadvantage with the model (7.39) is that it may require

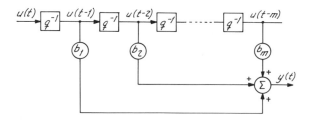

Figure 7.3
A "tapped delay line" model (7.39).

a high-order $B(q^{-1})$-polynomial to give sufficient flexibility to the model. This was illustrated in table 5.2.

When only one signal is involved, it is customary to model that signal as an autoregression

$$A(q^{-1})y(t) = e(t) \tag{7.41}$$

driven by a source $\{e(t)\}$ that is supposed to be white noise, or in some cases an impulse or an impulse train.

A typical feature of models used in signal processing is that the dimension of the parameter vector θ to be estimated is often quite high, i.e., anywhere from about 15 to 250 or so. A reason for this is, of course, that the models (7.39) and (7.41) require high orders to offer sufficient flexibility. At the same time, typical signal-processing applications provide many signal samples, so estimating many parameters is not a problem from the statistical point of view. However, the sampling rates are often high (in most applications, at least several hundred Hz). This means that the calculation per time step must be fast. As a consequence, adaptation techniques have focused on simple gradient schemes, and, in some recent work, on the fast implementations of the least squares method described in sections 6.3 and 6.4.

The well-known LMS algorithm was introduced by Widrow and Hoff (1960) and has been widely used in adaptive signal processing. In our terminology it is a stochastic gradient algorithm for a linear regression model. It is commonly used with a fixed gain γ_0 that is not normalized to the regression vector $\varphi(t)$:

$$\hat{\theta}(t) = \hat{\theta}(t-1) + \gamma_0 \varphi(t)\varepsilon(t). \tag{7.42a}$$

It is, however, suitable to normalize the gain, so that it is invariant to changes in the signal levels:

$$\hat{\theta}(t) = \hat{\theta}(t-1) + \frac{\gamma_0}{|\varphi(t)|^2}\varphi(t)\varepsilon(t) \tag{7.42b}$$

or

$$\hat{\theta}(t) = \hat{\theta}(t-1) + \gamma_0\frac{1}{r(t)}\varphi(t)\varepsilon(t),$$

$$r(t) = r(t-1) + \gamma_0[|\varphi(t)|^2 - r(t-1)]. \tag{7.42c}$$

[See also (2.83b), (2.84)]. The use of fast ladder and lattice methods (see section 6.4) was first suggested by Itakura and Saito (1971), and has since then become widely used, in particular for speech processing.

While the literature on adaptive signal processing has been mostly confined to linear regression models, there is nothing that prevents the use of more sophisticated models. In particular, other members of the general input-output model family (3.104) (allowing one or more of the F, C, and D polynomials to be different from unity) could be used. The advantage of this would be that the same model flexibility could be achieved with fewer parameters, as discussed in section 5.3. This in turn would lead to faster calculations and potentially faster adaptation. Some recent contributions discuss the use of such models in signal processing, e.g., Friedlander (1982b), Goodwin, Doan, and Cantoni (1980), and Johnson (1982). The book by Willsky (1979) also emphasizes the close relationship between identification techniques and signal-processing problems.

7.5.2 Adaptive Equalization

The Problem When a signal $\{y(t)\}$ is transmitted over a communication channel, the received signal $\{u(t)\}$ is always somewhat distorted. This is depicted in figure 7.4. The reason is that the channel acts like a filter which is not a perfect delay. Instead, we may describe the received signal as

$$u(t) = H(q^{-1})y(t) + v(t), \tag{7.43}$$

where $H(q^{-1})$ is a linear filter and $v(t)$ is additive noise. The impulse response of the filter $H(q^{-1})$ has its peak after a delay of k' samples, where typically, k' is something from 10 to 30 samples. In digital signal transmission $\{y(t)\}$ is a sequence of quantized values (like 0's and 1's). But as a consequence of (7.43) their corresponding responses $\{u(t)\}$ will

$u(t) = H(q^{-1}) y(t) + Noise$

Figure 7.4
Transmission of a signal $\{y(t)\}$.

overlap, so that the reading of individual $u(t)$'s will be difficult. Such "smearing out" of the impulse response of (7.43) is known as *intersymbol interference*.

To overcome this problem, the received signal $\{u(t)\}$ is fed into a linear filter whose output will be an estimate $y(t)$ of the transmitted signal:

$$\hat{y}(t - k) = G(q^{-1})u(t). \tag{7.44}$$

Such a filter is known as an *equalizer*, and its characteristics should of course resemble the inverse of (7.43). If in (7.44) we put a delay of k samples to allow for the peak of the response in $y(t)$ to pass before delivering the estimate, the quality of the estimate will be considerably improved. When the input signal $y(t)$ is restricted a finite number of quantized values (like 0's and 1's), it is customary to let the actual estimate be a quantized version $\hat{y}^*(t)$ of $\hat{y}(t)$ as given by (7.44).

If the equalizer filter is a transversal filter [see (7.39)], i.e., when

$$\hat{y}(t - k) = B(q^{-1})u(t), \tag{7.45}$$

where $B(q^{-1})$ is a finite-order polynomial, we talk about a *linear equalizer*.

If previous decisions $\hat{y}^*$ are fed back into the equalizer, as in

$$\hat{y}(t - k) = B(q^{-1})u(t) + [1 - A(q^{-1})]\hat{y}^*(t - k), \tag{7.46}$$

we have a *decision feedback equalizer*. This can also be seen as an implementation of (7.44) with a rational transfer function $G = B/A$.

When the filter $H(q^{-1})$ in (7.43) is known, the design of the equalizer filter $G(q^{-1})$ is conceptually straightforward: we just let it be as good as possible an approximation of $q^{-k}[H(q^{-1})]^{-1}$. A problem arises when the same receiver can be attached to different communication channels, as in a telephone network. Then the equalizer should be able to adapt itself to different channel characteristics. We thus need an *adaptive equalizer*. This is in practice achieved by sending a known "training" sequence

$\{y(t)\}_1^N$ at the beginning of the transmission, during which period the equalizer adapts.

Models   As a model for the communication channel we could use (7.43). It is, however, more suitable to directly parametrize the equalizer. [This resembles the direct adaptive control models, which we have parametrized in terms of the corresponding optimal regulators (section 7.3.3). In the present case, however, our task is easier; we might say that we have parametrized the inverse of the channel characteristics.] This leads to the finite impulse response model

$$y(t - k) = B(q^{-1})u(t) + v(t) \tag{7.47}$$

for the linear equalizer, and a linear difference equation

$$A(q^{-1})y(t - k) = B(q^{-1})u(t) + v(t) \tag{7.48}$$

for the decision feedback equalizer. Notice that during the training period both $\{u(t)\}$ and $\{y(t)\}$ are known sequences, so (7.47) as well as (7.48) can be written as linear regressions. Notice also that the parameters of these models coincide with the parameters of the corresponding equalizers. Hence the parameter estimates obtained by the adaptation procedure can be directly used in the equalizer filter. The polynomials in (7.47) and (7.48) are typically of degree 15–60.

Adaptation Methods   Since the models used here are linear regressions, parameter estimation is easy. The use of gradient methods was suggested by Lucky (1965), and treated also by, e.g., Gersho (1967). The application of the stochastic Newton algorithm for this model, i.e., application of the recursive least squares method, was suggested by Godard (1974). He derived the algorithm by posing the problem as a Kalman filter, as we did in section 2.3. Therefore the recursive least squares scheme is in this application known as the Kalman-Godard algorithm. The application of the fast-update algorithm for the gain (section 6.3.) to the equalizer problem was treated by Falconer and Ljung (1978). Ladder and lattice schemes (section 6.4) are applied to equalizers by Satorius and Pack (1981).

7.5.3   Adaptive Noise Cancellation

The Problem   Measured signals are often corrupted by disturbances of various kinds. It is then desirable to filter the observed signal in order to enhance the useful part of it. We described this problem as a smoothing

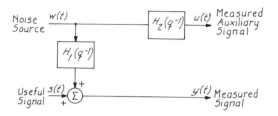

Figure 7.5
A disturbance acting on a useful signal.

problem in section 7.4.3, where we considered the disturbances to be white noise. In many situations, however, we may know what a disturbance source is, and even measure an auxiliary signal that is generated from this source. Think, e.g., of a signal that is corrupted by a 50 or 60Hz periodic signal, which we know arises from the AC voltage mains. Such a case can be depicted as in figure 7.5. The transfer function $H_1(q^{-1})$ is typically not known, however. The relationship between the measured signals can then be written

$$y(t) = \tilde{H}(q^{-1})u(t) + s(t), \tag{7.49}$$

where

$$\tilde{H}(q^{-1}) \triangleq H_1(q^{-1})[H_2(q^{-1})]^{-1}. \tag{7.50}$$

Now the transfer function $\tilde{H}(q^{-1})$ need not be causal. Therefore, we choose k so that the filter

$$H(q^{-1}) \triangleq q^{-k}\tilde{H}(q^{-1})$$

is causal, and rewrite (7.49) as

$$y(t - k) = H(q^{-1})u(t) + s(t - k) + e(t). \tag{7.51}$$

Here, for completeness, we have added a stochastic disturbance term $e(t)$ to account for possible other noise sources affecting the measurements of $y(t)$ and $u(t)$.

Now if the transfer function $H(q^{-1})$ were known, it would be easy to cancel the effect of $\{w(t)\}$ on the measurements $\{y(t)\}$:

$$\hat{s}(t - k) = y(t - k) - H(q^{-1})u(t). \tag{7.52}$$

Then (7.52) would be a *noise-cancelling filter*. The idea is the same as feedforward control in control theory. The problem in practice is that the filter $H(q^{-1})$ is not known, and that it may vary in time. Then *adaptive noise cancellation* is required. This basically means that the transfer function $H(q^{-1})$ in (7.51) is recursively identified, and its estimate is used in the filter (7.52). There have been many successful applications of adaptive noise cancellation. A survey, including several interesting applications, is given by Widrow et al. (1975).

Models In most applications the model (7.51) has been used with $H(q^{-1})$ as a transversal filter, i.e., as a finite-order polynomial

$$y(t - k) = B(q^{-1})u(t) + v(t), \tag{7.53}$$

where the disturbance $v(t)$ corresponds to $s(t - k) + e(t)$. Here $y(t - k)$ and $\{u(t)\}$ are measured signals, so (7.53) is a linear regression model. The fact that the output $y(t - k)$ is delayed k steps causes no problems for the estimation of the B-parameters, but it allows for much more efficient noise reduction in the noise canceller (7.52).

Adaptation Methods The most common recursive identification algorithm for these applications is the LMS algorithm (7.42a). In this case we have

$$\varphi^T(t) = (u(t) \dots u(t - n))$$

and (7.54)

$$\varepsilon(t) = y(t - k) - \hat{\theta}^T(t - 1)\varphi(t).$$

The asymptotic properties of this algorithm follow from our analysis in chapter 4 and were reviewed in section 7.5.1. Let us, however, for the sake of illustration discuss the application of our general convergence results to the described adaptive noise canceller (i.e., using (7.52) with $\hat{H}$ derived from $\hat{\theta}$, which in turn is obtained by using (7.42a), (7.54)).

Suppose that the measured data is actually generated according to (7.51), i.e., by

$$y(t - k) = H(q^{-1})u(t) + s(t - k) + e(t). \tag{7.55}$$

We also assume that the disturbance $w(t)$ is independent of the signal:

$$\bar{E}w(t)s(t - j) = 0 \text{ for all } j, \tag{7.56a}$$

as is the measurement noise:

$$\bar{\mathrm{E}}e(t)s(t-j) = 0 \quad \text{for all } j. \tag{7.56b}$$

Finally, assume that the gain sequence γ in the adaptation mechanism (7.42) tends to zero, so that we can study the convergence properties. This algorithm is then a (gradient) recursive prediction error scheme. Hence theorem 4.3 can be applied to infer that $\hat{\theta}(t)$ will converge to a local (in the present case local = global, since the criterion is quadratic) minimum of the criterion function

$$\bar{V}(\theta) = \bar{\mathrm{E}}[y(t-k) - \theta^{\mathrm{T}}\varphi(t)]^2$$
$$= \bar{\mathrm{E}}[y(t-k) - B(q^{-1})u(t)]^2. \tag{7.57}$$

Inserting the expression for $y(t-k)$ from (7.55) gives

$$\bar{V}(\theta) = \bar{\mathrm{E}}[H(q^{-1})u(t) - B(q^{-1})u(t) + s(t-k) + e(t)]^2$$
$$= \bar{\mathrm{E}}\{[H(q^{-1}) - B(q^{-1})]u(t)\}^2 + \bar{\mathrm{E}}[s(t-k) + e(t)]^2,$$

where the second equality follows from (7.56). The second term is independent of θ, and the first one is quadratic in θ. (Recall that the θ are coefficients of the B-polynomial). Hence $\hat{B}_t$ will converge to a polynomial B^*, corresponding to the global minimum ($=$ the only stationary point) of

$$W(\theta) = \bar{\mathrm{E}}\{[H(q^{-1}) - B(q^{-1})]u(t)\}^2. \tag{7.58}$$

This also gives the best possible predictor in the model set.

Now using the estimate $\hat{B}$ in the noise canceller (7.52) gives

$$\hat{s}(t-k) = y(t-k) - \hat{B}(q^{-1})u(t).$$

With (7.55) we have

$$\hat{s}(t-k) = \{H(q^{-1}) - \hat{B}(q^{-1})\}u(t) + s(t-k) + e(t)$$

and

$$\bar{\mathrm{E}}[\hat{s}(t-k) - s(t-k)]^2 = W(\theta) + \bar{\mathrm{E}}[e(t)]^2. \tag{7.59}$$

Hence the limit estimate B^* that minimizes (7.58) will also give the best signal reconstruction $\hat{s}$ in the mean square sense (7.59). The described adaptive noise canceller therefore asymptotically gives the best possible performance within the given parametrization. The same result would hold for other model choices and other recursive prediction error methods.

7.5.4 Adaptive Spectral Estimation

The spectrum of a signal plays an important role in many signal-processing problems. The spectrum of a stationary stochastic process is defined as the discrete Fourier transform of the covariance function, $r(k) = Ey(t + k)y(t)$, and will thus describe the frequency contents of the signal. The importance of the spectrum is a consequence of the fact that many signal properties are more easily interpreted in the frequency domain than in the time domain.

A common problem is that of finding one or a few low-level sine waves in noise. Finding periodic components in the presence of noise is called the (spectral) *line enhancement* problem. Now, as we have mentioned a number of times, the signal properties are often time-varying. Therefore, the spectrum will also be time-varying, and we must apply adaptive techniques to spectral estimation. For the line enhancement problem the task is to find "the instantaneous frequency" in a rapidly time-varying spectrum. Applications where this is important include vibration measurements, Doppler radar returns, and geophysical processing (Griffiths, 1975). We call this task *adaptive line enhancement* (Widrow et al., 1975).

Several methods of spectral estimation were developed during the 1950s. The idea behind these methods is as follows: estimate the covariance function of the data, $\hat{r}(k)$; multiply this estimate with a "lag window" $w(k)$ i.e., determine $\hat{\hat{r}}(k) = w(k) \cdot \hat{r}(k)$; and then take the discrete Fourier transform of the modified estimate $\hat{\hat{r}}$. The lag window is used to suppress the covariance estimates for large k, which are less reliable. This is the Blackman-Tukey approach (see, e.g., Jenkins and Watts, 1969). With the advent of the fast Fourier tranform (FFT) several spectral estimation algorithms were developed. In these, the spectral estimate is formed by averaging a number of squared transforms constructed from blocks of the original data (see, e.g., Welch, 1967).

The aforementioned approaches have the disadvantage of not being able to give reliable estimates of resonance peaks when the data record is short. The reason for this can be explained as follows. The Blackman-Tukey method typically uses $w(k) = 0$ for $|k| > M$, where M should be less than, say, a tenth of the data record length; otherwise the estimates will be unreliable. This means that the spectral estimate will be a polynomial of order M in $\cos \omega$. The same is true when the FFT method is applied to blocks of length M. To produce a sharp resonance peak, such a function must use a large M, which may require more data than is usually available.

To resolve this problem, Burg (1967) introduced the maximum entropy method (MEM) for spectral analysis. It eventually (van den Bos, 1971), became clear that this method could be regarded as a least squares method for an autoregressive model of the signal: The MEM spectral estimate is given by

$$\hat{\Phi}_{MEM}(\omega) = \frac{\hat{\sigma}^2}{|\hat{A}(e^{-i\omega})|^2}, \tag{7.60}$$

$$\hat{A}(q^{-1}) = 1 + \hat{a}_1 \hat{q}^{-1} + \cdots + \hat{a}_M q^{-M},$$

with $\hat{a}_i$ being the least squares estimates of the parameters in an autoregressive model

$$y(t) + a_1 y(t-1) + \cdots + a_M y(t-M) = e(t), \tag{7.61}$$

and where $\hat{\sigma}^2$ is the estimated variance of $\{e(t)\}$. The number M corresponds to the size of the lag window.

We notice immediately that this approach is well-suited for revealing sharp resonance peaks in the spectrum. Already for $M = 2$, the model (7.61), (7.60) is capable of producing an arbitrarily sharp resonance at an arbitrary frequency.

A number of different variants of the least squares estimates have been discussed for the spectral estimation problem. If we introduce our usual notation

$$y(t) = \theta^T \varphi(t) + e(t)$$

for (7.61), the criterion we have mostly used for the determination of θ is [see (2.12)]

$$\sum_{t=1}^{N} [y(t) - \theta^T \varphi(t)]^2.$$

In the present application, unavailable values of $y(s)$, for $s \leq 0$, entering $\varphi(t)$ are usually taken to be zero. This criterion is sometimes referred to as the "prewindowed case," since a data window is applied to the first data to screen out unavailable unknown values.

We could of course also delay initialization of the criterion until the vector $\varphi(t)$ is filled with known values, i.e., until $t = M + 1$, when the criterion is

$$\sum_{t=M+1}^{N} [y(t) - \theta^T \varphi(t)]^2.$$

All our recursive methods are applicable to minimization of this criterion. The approach is known as the "covariance method" (Makhoul, 1975).

A third criterion results from extending the summation to $N + M$, replacing missing data points by zero:

$$\sum_{t=1}^{N+M} [y(t) - \theta^T \varphi(t)]^2.$$

This makes the autoregressive parameter estimates equal to the Yule-Walker estimates. Makhoul (1975) called the method based on this criterion "the autocorrelation method." It has the advantage of assuring a stable $\hat{A}$-polynomial, but cannot be made into a recursive algorithm as easily as the other criteria.

Finally, Burg (1967) used a fourth criterion

$$\sum_{t=M+1}^{N} \{[y(t) - \theta^T \varphi(t)]^2 + [y(t - M) - \theta^T \bar{\varphi}(t)]^2\},$$

where

$$\bar{\varphi}^T(t) = (y(t - M + 1) \ \ldots \ y(t)),$$

using both forward and backward prediction errors. This criterion also assures a stable $\hat{A}$-polynomial.

The literature on spectral estimation via autoregressive modeling discusses the aforementioned criteria; for some of these aspects, see Makhoul (1975, 1977). Here we note only that as N/M becomes large, the differences between the criteria become insignificant, i.e., the asymptotic properties of the estimates are the same for the four criteria.

The most important aspect of this brief exposé of spectral estimation techniques is that the problem has been reduced to identifying the parameters of the autoregressive model (7.61) and then forming the spectral estimate (7.60). Adaptive spectral estimation and adaptive line enhancement can thus be carried out by applying any recursive identification algorithm to the model (7.61). The use of the LMS algorithm (7.42) has been discussed, e.g., by Widrow et al. (1975), Griffiths (1975), and Treichler (1979). Ladder and lattice forms have been studied in Griffiths (1977) and Reddy et al. (1981). Finally, it must pointed out that another approach is to describe the signal as an ARMA process,

$$A(q^{-1})y(t) = C(q^{-1})e(t), \tag{7.62}$$

and form the spectral estimate

$$\hat{\Phi}(\omega) = \frac{|\hat{C}(e^{-i\omega})|^2}{|\hat{A}(e^{-i\omega})|^2}\sigma^2$$

(van den Bos, 1971). This approach will allow the modeling of sharp peaks, as well as sharp dips, without too many parameters. Adaptive spectral estimation is then achieved by applying a recursive identification method to (7.62).

7.6 Summary

We have discussed a number of important problems in control, communications, and signal processing. Many of these problems require on-line decisions that should be based on a model of the signal or system. We have seen how the adaptive algorithms that solve such problems incorporate recursive identification schemes. These schemes may have been constructed explicitly to yield the desired information about the system or signal. They may also be regarded as implicit ways of gaining that information by adapting an optimal controller/filter to the system. In either case, recursive identification plays a key role in providing the mechanism of adaptation.

7.7 Bibliography

Section 7.3 The literature on adaptive control is extensive. The idea of using recursively identified models for controller design goes back to Kalman (1958). The idea of comparing the controlled plant's output with that of a reference model was first expressed by Whitaker et al. (1958). The role of Lyapunov stability theory for the analysis of these regulators was first noted by Parks (1966). Since then a large number of papers on various aspects of adaptive control have been published. The model reference approach is surveyed in Landau (1974, 1979). The self-tuning approach is surveyed in the paper by Åström et al. (1977). The relationship between these two approaches is clarified by Egardt (1979b, 1980a). A comprehensive study of convergence properties is given in Goodwin and Sin (1983).

Section 7.4 The adaptive prediction problem for ARMA processes is treated, e.g., in

Wittenmark (1974), Holst (1977), and Goodwin and Sin (1983). Adaptive filtering was studied in the late 1960s in relationship to Kalman filter applications for systems with unknown noise properties. This problem is treated, e.g., in Jazwinski (1969) and Mehra (1970). Adaptive smoothing seems to be a problem less studied in the literature. An interesting contribution to this problem is Hagander and Wittenmark (1977).

Section 7.5 The equalizer is a standard tool in data communication. See, e.g., Lucky, Salz, and Weldon (1968). Adaptive equalization using gradient techniques was suggested by Lucky (1965). The technique of Widrow and Hoff (1960), i.e., the LMS algorithm, was applied to equalizer design by Gersho (1967). An adaptive decision feedback equalizer based on stochastic gradient techniques was described in George, Bowen, and Storey (1971).

A good reference for the adaptive noise cancellation is the survey paper by Widrow et al. (1975). This paper gives an account of early applications, as well as a theoretical background pertinent to the use of the LMS algorithm. Friedlander (1982b) discusses the use of other algorithms for this problem. An interesting application of noise cancellation is echo cancellation in telephone networks. The paper by Sondi and Berkley (1980) can serve as a good introduction to this application.

Epilogue

We have now come to the end of the discussion of this book. Some aspects of the approach and of the results deserve to be stressed. The major theme has been to describe the long list of possible recursive identification methods within a unified general framework. We have exposed different particular methods as the result of certain choices to be made by the user within the general framework. In fact, it can be said that the whole discussion of this book has been focused on these user choices: We explained and exposed them in some detail in chapter 3. We provided theory that can be used to make rational decisions regarding some of the choices in chapter 4. The discussion of aspects of the choices was carried out in chapter 5. The problem of how to implement a given algorithm was treated in chapter 6.

We have now a fairly good understanding of the asymptotic properties of the algorithms when the gain tends to zero. Among the more challenging problems for the future is to achieve an equally good understanding of the transient properties, and how they are affected by the user choices, in particular the gain sequence. A good theory for the asymptotic properties of tracking algorithms (the gain not tending to zero) is also highly desirable.

In the book and within our framework we have described a number of different algorithms, ranging from simple least squares schemes to highly sophisticated ones. One may ask which problems can be adequately handled by simple methods and what type of problems necessitate more advanced approaches. The need for sophisticated methods depends very much on the actual problem that has to be solved, and no general advice to this effect can be given. It is an important task for the user to get a good feeling for this question in his or her particular application area.

Appendix 1.A   Some Concepts from Probability Theory

In this appendix we list some concepts from probability theory. Texts that cover the material are Papoulis (1965) or Chung (1968).

An n-dimensional *random variable*, or *random vector*, y is a function from an "event space" or sample space Ω to $\mathbf{R}^n$. The "*outcomes*" or "*realizations*" of y, i.e., the observed values, will be denoted by $y(\omega)$, where $\omega \in \Omega$. However, we shall omit the argument ω when there is no risk of confusion.

There is a probability measure associated with Ω so that certain subsets of Ω are assigned a *probability*. To y there is associated a *probability density function* (pdf) $\bar{f}_y$ from $\mathbf{R}^n$ to $\mathbf{R}$, such that

$$P(y(\omega) \in B) = \int_{x \in B} \bar{f}_y(x) dx, \qquad (1.A.1)$$

where B is a subset of $\mathbf{R}^n$ and "$P(A)$" means "the probability of the event A"; i.e. the probability measure of the set of those ω for which $y(\omega) \in B$.

The *expectation* or *mean value* of y is denoted by

$$Ey = \int_{\mathbf{R}^n} x \bar{f}_y(x) dx. \qquad (1.A.2)$$

The *covariance matrix* of y is

$$\operatorname{cov} y = E(y - m)(y - m)^{\mathrm{T}}, \qquad (1.A.3)$$

where $m = Ey$.

The vector y is said to have *Gaussian* or *normal* distribution if

$$f_y(x) = \frac{1}{(2\pi)^{n/2}} \frac{1}{(\det P)^{1/2}} \exp\left\{-\frac{1}{2}(x - m)^{\mathrm{T}} P^{-1}(x - m)\right\}. \qquad (1.A.4)$$

The mean value is then m and the covariance matrix is P. This is often written as

$$y \in N(m, P). \qquad (1.A.5)$$

Two random vectors y and z are said to be *independent* if

$$P(y(\omega) \in A \quad \text{and} \quad z(\omega) \in B) = P(y(\omega) \in A) \cdot P(z(\omega) \in B)$$

for any subsets A and B for which the probabilities are defined.

A (discrete time) *stochastic process* is a sequence of random vectors $y(t)$, $t = 0, 1, 2, \ldots$. For each ω, the *realization*

$$y(t, \omega), \quad t = 0, 1, 2, \ldots$$

is a sequence of $\mathbf{R}^n$ vectors. If for each $\omega \in \Omega^*$, where Ω^* is a set with measure (probability) one,

$$y(t, \omega) \to y^*(\omega) \quad \text{as} \quad t \to \infty,$$

we say that

$y(t)$ converges to y^* with probability one as $t \to \infty$.   (1.A.6)

"With probability one" is often abbreviated as "w.p.1" or "a.e." or "a.s." (almost everywhere or almost surely). If the associated sequence of probability density functions converges (weakly) to a pdf f^*:

$$f_{y(t)}(x) \to f^*(x) \quad \text{as} \quad t \to \infty,$$

we say that $\{y(t)\}$ *converges in distribution* to the pdf f^*. In the special case when f^* is the Gaussian distribution (1.A.4) we say that $y(t)$ is *asymptotically normal* with mean m and covariance P, and denote it as

$y(t) \in \text{AsN}(m, P)$.   (1.A.7)

For a stochastic process we define the *mean value function*

$$m_y(t) = \text{E}y(t)$$

and the *covariance function*

$$R_y(t, s) = \text{E}[y(t) - m_y(t)][y(s) - m_y(s)]^\text{T}.$$

If $m_y(t) = m_y$ and $R_y(t + \tau, t) = \bar{R}_y(\tau)$, the process is said to be (weakly) *stationary*. We can then define the *spectrum* as

$$\Phi_y(\omega) = \frac{1}{2\pi} \sum_{\tau=-\infty}^{\infty} \bar{R}_y(\tau)e^{-i\omega\tau}.$$

From this a useful formula for the variance can be obtained:

$$R_y(0) = \int_{-\pi}^{\pi} \Phi_y(\omega)d\omega.$$

(See Åström (1970) for efficient schemes for the numerical evaluation.) More generally, if two stochastic processes each of zero mean are related through

$$y(t) = \sum_{k=-\infty}^{t} h_{t-k}u(k),$$

then

$$Ey(t)u^{T}(t) = \int_{-\pi}^{\pi} H(e^{i\omega})\Phi_{u}(\omega)d\omega, \qquad (1.A.8)$$

where

$$H(z) = \sum_{k=0}^{\infty} h_{k}z^{-k}.$$

In the proofs of theorems 4.5 and 4.C.1, we shall use the concepts of *σ-algebras, conditional expectation*, and *martingales*.

A *σ-algebra* $\mathscr{F}$ is a set of subsets of Ω for which probabilities are defined. A random vector y *generates* a σ-algebra by virtue of the fact that all possible outcomes of y define subsets in the event space Ω.

The *conditional expectation* of a random vector z, given the σ-algebra $\mathscr{F}$, is denoted by

$$E(z \mid \mathscr{F}), \qquad (1.A.9)$$

and is itself a random vector. The formal definition is that the probabilities related to $E(z \mid \mathscr{F})$ and to z coincide when evaluated over subsets in $\mathscr{F}$. When $\mathscr{F}$ is the σ-algebra generated by y, (1.A.9) is often written

$$E(z \mid y).$$

Intuitively, $E(z \mid y)$ is the expected value of z, if we happen to know the outcome of the random vector y. It will of course be a function of y.

A *martingale* is a sequence of random variables $y(t)$ and an associated sequence of σ-algebras $\mathscr{F}_{t}$, such that

$$y(t) \in \mathscr{F}_{t}, \qquad (1.A.10a)$$

$$\mathscr{F}_{t-1} \subset \mathscr{F}_{t}, \qquad (1.A.10b)$$

$$E(y(t) \mid \mathscr{F}_{t-1}) = y(t-1). \qquad (1.A.10c)$$

Here (1.A.10a) means that $\mathscr{F}_{t}$ contains the σ-algebra generated by $y(t)$. Often $\mathscr{F}_{t}$ is taken as the σ-algebra generated by $y(0), y(1), \ldots, y(t)$, which clearly satisfies (1.A.10a, b). The intuition in (1.A.10c) is that the increment $y(t) - y(t-1)$ is "unpredictable" even if we know everything that happened up to time $t-1$. An important property of a martingale is that if the second moments $E|y(t)|^{2}$ are bounded in t, then $y(t)$ will converge w.p.1 to a random vector y^{*} as $t \to \infty$.

Appendix 1.B   Some Concepts from Statistics

Identification and recursive identification procedures are in the end nothing but estimation of parameters in specific structures. In this appendix we provide a refresher for some statistical concepts on parameter estimation. For a text on statistical parameter estimation, see Rao (1973).

1.B.1 Samples and Estimators

Consider a random vector in $\mathbf{R}^n$, $y^n = (y(1) \; y(2) \; \ldots \; y(n))$. Let its (joint) probability density function be

$$f(\theta; x_1, \ldots, x_n) = f(\theta; x^n), \tag{1.B.1}$$

which is known up to a finite-dimensional ($\dim \theta = d$) parameter vector θ. The problem we are faced with is to find an *estimate* of θ based on an *observation* (or "a *realization*" or "a *sample*") of y^n.

Realizations of y^n will be denoted by $y^n(\omega)$, where ω is a point in the event space. We shall usually suppress the argument (ω) for realizations, when there is no risk of confusion. We shall also drop the superscript n in y^n when not essential.

Remark   We could also consider the situation when a function of θ, say $r(\theta)$, is to be estimated. Then typically $\dim r < \dim \theta$. This would lead to obvious changes in notation in what follows.

An *estimator* of θ is a function from $\mathbf{R}^n$ to $\mathbf{R}^d$:

$$\hat{\theta}(y). \tag{1.B.2}$$

Since this is a function of a random variable, it is itself a random variable. Its value after $y = y(\omega)$ is observed is $\hat{\theta}(y(\omega))$.

EXAMPLE 1.B.1   Let $y(i)$, $i = 1, \ldots, n$, be independent random variables with normal distribution with means θ (independent of i) and standard deviations σ_i:

$$y(i) \in N(\theta, \sigma_i^2). \tag{1.B.3}$$

The mean θ is to be estimated. Some more or less suitable estimators of θ can be

$$\hat{\theta}_1(y^n) = \frac{1}{n} \sum_{1}^{n} y(i), \tag{1.B.4}$$

$$\hat{\theta}_2(y^n) = y(3), \tag{1.B.5}$$

$$\hat{\theta}_3(y^n) = \sum_1^n \alpha_{i,n} y(i), \tag{1.B.6}$$

where $\alpha_{i,n}$ in (1.B.6) are weighting coefficients. □

1.B.2 Properties of Estimators

Certain properties of estimators are desirable. An estimator is thus said to be *unbiased* if

$$E_\theta \hat{\theta}(y^n) = \theta, \tag{1.B.7}$$

i.e., if its expected value is the true parameter value. Here we have used the subscript θ on the expectation symbol to emphasize that expectation is taken with respect to the probability associated with (1.B.1). That is,

$$E_\theta h(y^n) = \int_{\mathbf{R}^n} h(x^n) f(\theta; x^n) dx^n. \tag{1.B.8}$$

For an unbiased estimator it is of interest to know its variance (covariance matrix if $d > 1$) around the mean:

$$P_s = E[\hat{\theta}_s(y) - \theta][\hat{\theta}_s(y) - \theta]^\mathrm{T}. \tag{1.B.9}$$

An estimator $\hat{\theta}_1(y)$ is said to be *more efficient* than an estimator $\hat{\theta}_2(y)$ if

$$P_1 \le P_2. \tag{1.B.10}$$

(The matrix inequality (1.B.10) means that $P_2 - P_1$ is a positive semi-definite matrix.)

We must sometimes consider the case where the sample n increases to infinity. The dimension of the parameter θ in the joint probability density function (1.B.1) is assumed to remain fixed, even though the dimension of the random vector y^n increases.
The estimator is said to be *consistent* if

$$\hat{\theta}(y^n) \to \theta \text{ as } n \to \infty. \tag{1.B.11}$$

Since $\hat{\theta}(y^n)$ is a random variable, we must specify in what sense (1.B.11) holds. Thus, if (1.B.11) holds w.p.1, we say that $\hat{\theta}(y^n)$ is consistent w.p.1, or that it is *strongly consistent*.

Another question of interest is the *asymptotic distribution* of $\hat{\theta}(y^n)$ as $n \to \infty$. It turns out that the central limit theorem can often be applied to $\hat{\theta}(y^n)$ to infer that $\sqrt{n}[\hat{\theta}(y^n) - \theta]$ is *asymptotically normal* with zero mean as $n \to \infty$.

In fact, much of the analysis of suggested estimators (or *identification methods*, as we call them in this book) concerns questions of consistency and asymptotic normality.

1.B.3 The Cramér-Rao Inequality

We are interested in estimators that are as efficient as possible, i.e., ones that make the covariance matrix in (1.B.9) as small as possible. It is then interesting to note that there is a theoretical lower limit to what variance can be obtained. This is the so called Cramér-Rao inequality.

THEOREM 1.B.1   Suppose that $y(i)$ may take values in interval(s), whose limits do not depend on θ. Suppose that θ is a real scalar and that $f(\theta; \cdot)$ in (1.B.1) is twice continuously differentiable with respect to θ. Let $\hat{\theta}(y^n)$ be an estimator of θ with expected value $E_\theta \hat{\theta}(y^n) = \gamma(\theta)$, which is assumed to be differentiable with respect to θ. Then

$$E_\theta[\hat{\theta}(y^n) - \gamma(\theta)]^2 \geq -\frac{[d\gamma(\theta)/d\theta]^2}{E_\theta[\partial^2 \log f(\theta; y^n)/\partial\theta^2]}. \tag{1.B.12}$$

Proof   By definition

$$E_\theta \hat{\theta}(y^n) = \int \hat{\theta}(x^n) f(\theta; x^n) dx^n = \gamma(\theta).$$

Differentiate w.r.t. θ:

$$\int \hat{\theta}(x^n) \frac{\partial}{\partial\theta} f(\theta; x^n) dx^n$$

$$= \int \hat{\theta}(x^n) \left\{ \frac{\partial}{\partial\theta} \log f(\theta; x^n) \right\} f(\theta; x^n) dx^n$$

$$= E_\theta \hat{\theta}(y^n) \frac{\partial}{\partial\theta} \log f(\theta; y^n) = \frac{d}{d\theta} \gamma(\theta).$$

Since

$$\int f(\theta; x^n)dx^n \equiv 1,$$

by analogous calculation we also have

$$\int \frac{\partial}{\partial \theta} \log f(\theta; x^n) \cdot f(\theta; x^n)dx^n = 0 \qquad (1.B.13)$$

or

$$E_\theta \frac{\partial}{\partial \theta} \log f(\theta; y^n) = 0$$

and

$$E_\theta \gamma(\theta) \frac{\partial}{\partial \theta} \log f(\theta; y^n) = 0.$$

Subtracting this from the previous expression gives

$$E_\theta \left[[\hat{\theta}(y^n) - \gamma(\theta)] \frac{\partial}{\partial \theta} \log f(\theta; y^n) \right] = \frac{d}{d\theta} \gamma(\theta).$$

The Schwartz inequality now gives

$$\left[\frac{d}{d\theta} \gamma(\theta) \right]^2 \leq E_\theta [\hat{\theta}(y^n) - \gamma(\theta)]^2 \cdot E_\theta \left[\frac{\partial}{\partial \theta} \log f(\theta; y^n) \right]^2.$$

It remains only to show that

$$E_\theta \left[\frac{\partial}{\partial \theta} \log f(\theta; y^n) \right]^2 = -E_\theta \frac{\partial^2}{\partial \theta^2} \log f(\theta; y^n). \qquad (1.B.14)$$

But (1.B.14) follows by differentiating (1.B.13):

$$\int \left\{ \frac{\partial^2}{\partial \theta^2} \log f(\theta; x^n) + \left[\frac{\partial}{\partial \theta} \log f(\theta; x^n) \right]^2 \right\} \cdot f(\theta; x^n)dx^n = 0. \quad \blacksquare$$

COROLLARY Let $\hat{\theta}(y^n)$ be a vector-valued unbiased estimator of θ. Then

$$E[\hat{\theta}(y^n) - \theta][\hat{\theta}(y^n) - \theta]^T \geq M^{-1},$$

where

$$M = E_\theta \left[\frac{\partial}{\partial \theta} \log f(\theta; y^n) \right]^T \left[\frac{\partial}{\partial \theta} \log f(\theta; y^n) \right] = -E_\theta \frac{\partial^2}{\partial \theta^2} \log f(\theta; y^n)$$

$$(1.B.15)$$

is the expected value of the second derivative matrix (the Hessian) of the function $-\log f$.

Remark The matrix M in (1.B.15) is known as the *Fisher information matrix*. The corollary shows that it is of fundamental importance in estimation problems.

EXAMPLE 1.B.1 (continued) Let us calculate the Cramér-Rao lower bound for estimators of θ in our example. The random variable $y(i)$ has the probability density function

$$\frac{1}{\sqrt{2\pi}\sigma_i}\exp\left\{-\frac{(x_i-\theta)^2}{2\sigma_i^2}\right\}.$$

Since $y(i)\ i = 1, \ldots, n$ are independent, the joint density function is

$$f(\theta; x_1, \ldots, x_n) = \prod_{i=1}^{n}\frac{1}{\sqrt{2\pi}\sigma_i}\exp\left\{-\frac{(x_i-\theta)^2}{2\sigma_i^2}\right\}, \qquad (1.B.16)$$

and

$$\log f(\theta; x_1, \ldots, x_n) = -\frac{n}{2}\log 2\pi - \sum_{i=1}^{n}\log\sigma_i - \frac{1}{2}\sum_{i=1}^{n}\frac{(x_i-\theta)^2}{\sigma_i^2}.$$

Hence

$$-E_\theta\frac{\partial^2}{\partial\theta^2}\log f(\theta; y(1), \ldots, y(n)) = \sum_{i=1}^{n}\frac{1}{\sigma_i^2}.$$

Therefore all unbiased ($\gamma(\theta) = \theta$) estimators of θ have a variance greater or equal to

$$\frac{1}{\sum\limits_{i=1}^{n}1/\sigma_i^2}. \quad \square \qquad (1.B.17)$$

1.B.4 The Maximum Likelihood Estimator

The joint probability density function for the random vector to be observed is given by (1.B.1). The probability that the realization indeed should take the value y^n is thus proportional to

$$f(\theta; y(1), \ldots, y(n)) = f(\theta; y^n).$$

This is a deterministic function of θ once the numerical values of the observed variables $y(i)$ are inserted. It is called the *likelihood function*. A reasonable estimator of θ could then be to select it so that the observed event becomes as "likely as possible." That is, we seek

$$\max_{\theta} f(\theta; y^n),\qquad\qquad (1.B.18)$$

and let the maximizing vector

$$\hat{\theta}_{ML}(y^n)$$

be our estimator. This is known as the *maximum likelihood estimator* (MLE).

EXAMPLE 1.B.1 (continued) From (1.B.16) we can determine the likelihood function for the estimation problem:

$$f(\theta; y(1) \ldots, y(n)) = \prod_{i=1}^{n} \frac{1}{\sqrt{2\pi}\sigma_i} \exp \left\{ -\frac{(y(i) - \theta)^2}{2\sigma_i^2} \right\}.\qquad (1.B.19)$$

For given observations $y(i)$, we can maximize this function w.r.t. θ by maximizing its logarithm:

$$\max_{\theta} \left[-\frac{n}{2}\log 2\pi - \sum_{i=1}^{n} \log \sigma_i - \frac{1}{2} \sum_{i=1}^{n} \frac{(y(i) - \theta)^2}{\sigma_i^2} \right].$$

We immediately find the maximizing value

$$\hat{\theta}(y^n) = \frac{1}{\sum_{i=1}^{n} 1/\sigma_i^2} \sum_{i=1}^{n} \frac{y(i)}{\sigma_i^2}. \quad \square \qquad (1.B.20)$$

The MLE is a widely used estimator. It has some very attractive asymptotic properties, as the following theorem shows. These properties were first proved by Wald (1949) and Cramér (1946).

THEOREM 1.B.2 Suppose that the random variables $\{y(i)\}$ are independent and are identically distributed, so that

$$f_n(\theta; x_1, \ldots, x_n) = \prod_{i=1}^{n} f_1(\theta, x_i).$$

Then $\hat{\theta}_{ML}(y^n)$ is strongly consistent as n tends to infinity and

$$\sqrt{n}[\hat{\theta}_{ML}(y^n) - \theta]$$

is asymptotically normal, of zero mean, and with a covariance matrix equal to the Cramér-Rao lower bound.

Note that recursive estimators can be constructed with the same nice asymptotic properties for a general identification problem. This is one of the more important analytical results in this book (see theorems 4.3–4.5).

Appendix 1.C Models of Dynamic Stochastic Systems

In this book we use several different types of models for dynamic stochastic systems, as well as for stochastic signals. Much of the discussion on models is developed in examples and sections throughout the book. In this appendix we provide some "initial conditions" for this development. Dynamic stochastic models are treated in detail in Åström (1970), Kwakernaak and Sivan (1972), and Jazwinski (1970).

1.C.1 Input-Output Models

A general linear dynamic model can be described by

$$y(t) = \sum_{k=1}^{\infty} h(k)u(t-k), \tag{1.C.1}$$

where $y(t)$ is the output at time t, $u(t)$ is the input at t, and $h(k)$ is the impulse response of the system. With

$$H(q^{-1}) = \sum_{k=1}^{\infty} h(k)q^{-k}, \tag{1.C.2}$$

where q^{-1} is the delay operator

$$q^{-1}u(t) = u(t-1), \tag{1.C.3}$$

the model (1.C.1) can also be written

$$y(t) = H(q^{-1})u(t). \tag{1.C.4}$$

Now in the real world the output $y(t)$ is always influenced by a number of signals other than the input $u(t)$. Most of them are beyond our control and often also not understood in detail. The effects of such signals can be described as an additive disturbance $v(t)$:

$$y(t) = H(q^{-1})u(t) + v(t). \tag{1.C.5}$$

There are many ways of describing the properties of this disturbance term. For a *stochastic* model it is customary to model $\{v(t)\}$ as a stationary stochastic process with zero mean value and spectrum

$$\Phi_v(\omega). \tag{1.C.6}$$

If, in addition, the process $\{v(t)\}$ is assumed to be Gaussian, the spectrum (1.C.6) will uniquely describe its properties.

Suppose now that $v(t)$ is scalar-valued, so that the spectrum is a real-valued function of ω. Suppose also that it is a *rational* function in $\cos \omega$. It then follows from the *spectral factorization theorem* that

$$\Phi_v(\omega) = \frac{\sigma^2}{2\pi} \frac{C(e^{i\omega})C^*(e^{i\omega})}{D(e^{i\omega})D^*(e^{i\omega})}, \tag{1.C.7}$$

where σ^2 is a positive scalar and where

$$C(z) = 1 + c_1 z + \cdots + c_m z^m,$$

$$C^*(z) = z^m + c_1 z^{m-1} + \cdots + c_m,$$

$$D(z) = 1 + d_1 z + \cdots + d_n z^n, \tag{1.C.8}$$

$$D^*(z) = z^n + d_1 z^{n-1} + \cdots + d_n.$$

Here c_i and d_i are real-valued coefficients. Notice that the zeros of C and C^* (and of D and D^*) are each others' mirror images in the unit circle (i.e., if β is a zero of C, then $1/\bar\beta$ is a zero of C^*). This means that we can always, if desired, select C^* and D^* so that they have all their zeros on or inside the unit circle.

Now the stochastic process defined by

$$v(t) = \frac{C(q^{-1})}{D(q^{-1})} e(t), \tag{1.C.9}$$

where $\{e(t)\}$ is a sequence of independent random variables each of zero mean and variances σ^2, will have the spectrum given by (1.C.7). When the spectrum of v is rational, we can consequently describe the stochastic model (1.C.5) as

$$y(t) = H(q^{-1})u(t) + \frac{C(q^{-1})}{D(q^{-1})} e(t). \tag{1.C.10}$$

Several special cases of (1.C.10) are of interest. A particularly common model is the ARMAX model

$$A(q^{-1})y(t) = B(q^{-1})u(t) + C(q^{-1})e(t), \tag{1.C.11}$$

where A, B and C are polynomials in q^{-1} and $e(t)$ is as above. Clearly (1.C.11) is a special case of (1.C.10) with

$$H(q^{-1}) = \frac{B(q^{-1})}{A(q^{-1})}, \quad D(q^{-1}) = A(q^{-1}).$$

Notice in particular that we can always choose C in (1.C.11) so that $C^*(z)$ has all its zeros inside or on the unit circle.

1.C.2 State-Space Models

Another common way of describing the relationship between the input $u(t)$ and the output $y(t)$ is to use the state-space model

$$x(t + 1) = Fx(t) + Gu(t),$$
$$y(t) = Hx(t),$$
(1.C.12)

where $x(t)$ is an n-dimensional vector, and F, G, and H are matrices of compatible dimensions. In this description we can include stochastic disturbances as follows:

$$x(t + 1) = Fx(t) + Gu(t) + w(t),$$
$$y(t) = Hx(t) + e(t),$$
(1.C.13)

where $\{w(t)\}$ is a sequence of independent random vectors each with zero mean value and covariance matrix

$$Ew(t)w^T(s) = \delta_{t,s}R_1 \quad (\delta_{t,s} \text{ is Kronecker's delta}).$$

This disturbance describes how the process is affected by other signals in addition to the input. We shall therefore use the term *process noise* for $w(t)$.

In (1.C.13) $\{e(t)\}$ is another sequence of independent random variables each with zero mean and covariance matrix

$$Ee(t)e^T(s) = \delta_{t,s}R_2.$$

We shall call it the *measurement noise*. In general it may be correlated with the process noise:

$$Ew(t)e^T(s) = \delta_{t,s}R_{12}.$$

In (1.C.13) the initial condition $x(0)$ is assumed to be a random vector, independent of future noise terms, with mean x_0 and covariance matrix Π_0.

For the model (1.C.13) we may consider the *state estimation problem*, i.e., estimation of the state vector, based on observations of $y(t)$ and $u(t)$. Let

$\hat{x}(t) = E(x(t) \,|\, y(0), u(0), \ldots, y(t-1), u(t-1)).$

If the noise terms $\{e(t)\}$, $\{w(t)\}$, as well as the initial condition $x(0)$, are Gaussian, then $\hat{x}(t)$ is determined by the *Kalman filter:*

$$\hat{x}(t+1) = F\hat{x}(t) + Gu(t) + K(t)[y(t) - H\hat{x}(t)],$$

$$\hat{x}(0) = x_0.$$

(1.C.14)

Here the "Kalman gain" $K(t)$ is determined by

$$K(t) = [FP(t)H^{\mathrm{T}} + R_{12}^{\mathrm{T}}][HP(t)H^{\mathrm{T}} + R_2]^{-1}$$

(1.C.15)

and the $n \times n$ matrix $P(t)$ is calculated from the Riccati equation

$$P(t+1) = FP(t)F^{\mathrm{T}} + R_1 - [FP(t)H^{\mathrm{T}} + R_{12}^{\mathrm{T}}][HP(t)H^{\mathrm{T}} + R_2]^{-1}$$

$$\times [FP(t)H^{\mathrm{T}} + R_{12}^{\mathrm{T}}]^{\mathrm{T}},$$

(1.C.16)

$$P(0) = \Pi_0.$$

In (1.C.13)–(1.C.16) the matrices F, G, H, R_1, R_2, and R_{12} may very well depend on time t, although we did not write out such a dependence explicitly.

When the disturbances are not Gaussian, the estimate $\hat{x}$ given by (1.C.13)–(1.C.16) will not in general be equal to the conditional expectation. However, among all estimates formed by linear operations on y and u, it is always equal to the estimate with the smallest error covariance matrix.

It is sometimes of interest to find the filtered estimate

$\hat{x}(t \,|\, t) = E(x(t) \,|\, y(0), u(0), \ldots, y(t-1), u(t-1), y(t)).$

This is given by (when the noises are Gaussian and $R_{12} = 0$)

$$\hat{x}(t \,|\, t) = \hat{x}(t) + \bar{K}(t)[y(t) - H\hat{x}(t)],$$

(1.C.17)

where

$$\bar{K}(t) = P(t)H^{\mathrm{T}}[HP(t)H^{\mathrm{T}} + R_2]^{-1}$$

(1.C.18)

and where $P(t)$ is given by (1.C.16).

We may also remark that a continuous-time stochastic state-space model can also be used. Formulas for transforming it into a discrete-time one, (1.C.13), are given in Åström (1970). Formulas for computing state estimates for the continuous-time model given discrete-time observations are described in Jazwinski (1970).

1.C.3 Relations between Input-Output and State-Space Models

The state-space model (1.C.13) can be written in input-output form as

$$y(t) = H_1(q^{-1})u(t) + H_2(q^{-1})w(t) + e(t), \tag{1.C.19}$$

where

$$H_1(q^{-1}) = q^{-1}H(I - q^{-1}F)^{-1}G,$$
$$H_2(q^{-1}) = q^{-1}H(I - q^{-1}F)^{-1}. \tag{1.C.20}$$

The input-output model (1.C.11) with a scalar output can be represented in observable canonical state-space form as

$$x(t+1) = \begin{pmatrix} -a_1 & 1 & 0 & \cdots & 0 \\ -a_2 & 0 & 1 & \cdots & 0 \\ \vdots & \vdots & \vdots & & \vdots \\ -a_n & 0 & 0 & \cdots & 0 \end{pmatrix} x(t)$$

$$+ \begin{pmatrix} b_1 \\ \vdots \\ b_n \end{pmatrix} u(t) + \begin{pmatrix} c_1 - a_1 \\ \vdots \\ c_n - a_n \end{pmatrix} v(t), \tag{1.C.21}$$

$$y(t) = (1 \quad 0 \quad \cdots \quad 0)x(t) + v(t),$$

where a_i, b_i and c_i are the coefficients of the polynomials in (1.C.11):

$$A(q^{-1}) = 1 + a_1 q^{-1} + \cdots + a_n q^{-n},$$
$$B(q^{-1}) = b_1 q^{-1} + \cdots + b_n q^{-n}, \tag{1.C.22}$$
$$C(q^{-1}) = 1 + c_1 q^{-1} + \cdots + c_n q^{-n}.$$

Appendix 2.A   The Extended Kalman Filter Algorithm

In this appendix we give the algorithm that results when the extended Kalman filter is applied to the problem of estimating θ in (2.68)–(2.69). The calculations and interpretations are analogous to those in example 2.4. See Ljung (1979a) for further details.

We partition vectors and matrices according to the natural block structure

$$X(t) = \begin{pmatrix} x(t) \\ \theta(t-1) \end{pmatrix}, \quad \bar{K}(t) = \begin{pmatrix} K(t) \\ L(t) \end{pmatrix}, \quad \bar{P}(t) = \begin{pmatrix} P_1(t) & P_2(t) \\ P_2^T(t) & P_3(t) \end{pmatrix},$$

where $\bar{K}$ and $\bar{P}$ are the "Kalman gain" and "covariance" matrix for the extended state. This gives the algorithm:

$$\hat{x}(t+1) = F_t\hat{x}(t) + G_t u(t) + K(t)[y(t) - H_t\hat{x}(t)], \tag{2.A.1}$$

$$\hat{x}(0) = 0;$$

$$\hat{\theta}(t) = \hat{\theta}(t-1) + L(t)[y(t) - H_{t-1}\hat{x}(t)], \tag{2.A.2}$$

$$\hat{\theta}(0) = \theta_0;$$

$$K(t) = [F_t P_1(t) H_t^T + M_t P_2^T(t) H_t^T + F_t P_2(t) D_t^T \tag{2.A.3}$$
$$\qquad + M_t P_2(t) D_t^T + R_{12}]S_t^{-1};$$

$$S_t = H_t P_1(t) H_t^T + H_t P_2(t) D_t^T + D_t P_2^T(t) H_t^T + D_t P_3(t) D_t^T + R_2; \tag{2.A.4}$$

$$L(t) = [P_2^T(t) H_{t-1}^T + P_3(t) D_t^T]S_t^{-1}; \tag{2.A.5}$$

$$P_1(t+1) = F_t P_1(t) F_t^T + F_t P_2(t) M_t^T \tag{2.A.6}$$
$$\qquad + M_t P_2^T(t) F_t^T + M_t P_3(t) M_t^T - K(t)S_t K^T(t) + R_1,$$

$$P_1(0) = \Pi_0(\theta_0);$$

$$P_2(t+1) = F_t P_2(t) + M_t P_3(t) - K(t)S_t L^T(t), \tag{2.A.7}$$

$$P_2(0) = 0;$$

$$P_3(t+1) = P_3(t) - L(t)S_t L^T(t), \tag{2.A.8}$$

$$P_3(0) = P_0.$$

Note that (2.A.7) with the aid of (2.A.5) also can be written as

$$P_2(t+1) = [F_t - K(t)H_t]P_2(t) + [M_t - K(t)D_t]P_3(t). \tag{2.A.7$'$}$$

Here

$$F_t = F(\hat{\theta}(t)),$$
$$G_t = G(\hat{\theta}(t)),$$
$$H_t = H(\hat{\theta}(t)),$$ (2.A.9)
$$M_t = M(\hat{\theta}(t), \hat{x}(t), u(t)),$$

with

$$M(\hat{\theta}, x, u) = \frac{\partial}{\partial\theta}[F(\theta)x + G(\theta)u]\Big|_{\theta=\hat{\theta}} \qquad \text{(an } n \times d\text{-matrix)}, \qquad (2.A.10)$$

and

$$D_t = D(\hat{\theta}(t-1), \hat{x}(t)),$$

with

$$D(\hat{\theta}, x) = \frac{\partial}{\partial\theta}[H(\theta)x]\Big|_{\theta=\hat{\theta}} \qquad \text{(a } p \times d\text{-matrix)}. \qquad (2.A.11)$$

As pointed out in the remark in example 2.4, it may be natural and useful to use the latest available measurements when updating the parameter estimates. To do that, we would update the "filtered" state estimate

$$\hat{x}(t \mid t) = \hat{x}(t) + \bar{K}(t)[y(t) - H_t\hat{x}(t)] \qquad (2.A.12)$$

[see (1.C.17)], where $(R_{12} = 0)$

$$\bar{K}(t) = [P_1(t)H_t^T + P_2(t)D_t^T]S_t^{-1}, \qquad (2.A.13)$$

and define

$$M_t = M(\hat{\theta}(t), \hat{x}(t \mid t), u(t)) \qquad (2.A.14)$$

rather than as in (2.A.9).

Appendix 3.A   An Alternative Gauss-Newton Algorithm

In section 3.4 we derived the Gauss-Newton algorithm (3.70), which is equivalent to (3.67). In this appendix we shall derive an alternative to this algorithm. It is still based on the Gauss-Newton updating direction, but certain filtering operations involved in forming the gain vector $L(t)$ will be done in a different order. The main interest in this modified algorithm is that it will facilitate detailed comparisons with the extended Kalman filter algorithm. See appendix 2.A and appendixes 3.B and 3.C.

In the algorithm (3.70) the gain vector $L(t)$ is given by (3.70d):

$$L(t) = P(t-1)\psi(t)S^{-1}(t), \tag{3.A.1}$$

where $P(t)$ is given by (3.70f):

$$P(t) = [P(t-1) - L(t)S(t)L^{T}(t)]/\lambda(t), \tag{3.A.2}$$

and $\psi(t)$ is obtained from (3.70g, h). By going back to the original equations (3.20) and (3.22) defining $\psi(t, \theta)$, we find that $\psi(t)$ is also given by

$$\zeta(t) = \mathscr{F}_{t-1}\zeta(t-1) + M_{t-1}, \tag{3.A.3a}$$

$$\psi^{T}(t) = \mathscr{H}_{t-1}\zeta(t) + D_{t}, \tag{3.A.3b}$$

where

$$M_{t} = M(\hat{\theta}(t), \varphi(t), z(t)),$$
$$D_{t} = D(\hat{\theta}(t-1), \varphi(t)),$$
$$\mathscr{F}_{t} = \mathscr{F}(\hat{\theta}(t)), \tag{3.A.4}$$
$$\mathscr{H}_{t} = \mathscr{H}(\hat{\theta}(t)),$$

with M and D defined by (3.21). The vector $\varphi(t)$ is obtained by

$$\varphi(t+1) = \mathscr{F}_{t}\varphi(t) + \mathscr{G}_{t}z(t). \tag{3.A.5}$$

The vector $\psi(t)$ can be seen as the output of the linear filter (3.A.3). This filter is driven by the inputs M_{t} and D_{t}. When forming $L(t)$ in (3.A.1), this output $\psi(t)$ is multiplied by $P(t-1)$. As an alternative, we could instead multiply the inputs to the filter by this quantity. This would give

$$\bar{\zeta}(t) = \mathscr{F}_{t-1}\bar{\zeta}(t-1) + M_{t-1}P(t-1), \tag{3.A.6a}$$

$$\bar{\psi}^{T}(t) = \mathscr{H}_{t-1}\bar{\zeta}(t) + D_{t}P(t-1). \tag{3.A.6b}$$

The corresponding gain vector would then be

$$\bar{L}(t) = \bar{\psi}(t)S^{-1}(t).$$ (3.A.7)

The difference between (3.A.1)–(3.A.3) and (3.A.6)–(3.A.7) is that the effect of $P(t)$ passes through the dynamics of the filter (3.A.6) when $\bar{L}(t)$ is formed. If $P(t)$ were a constant, then $L(t)$ and $\bar{L}(t)$ would be identical. The matrix $S(t)$ is given by (3.70c):

$$S(t) = \psi^{\mathrm{T}}(t)P(t-1)\psi(t) + \lambda(t)\hat{\Lambda}(t).$$ (3.A.8)

With the expression (3.A.3b) in (3.A.8) we obtain

$$S(t) = \mathcal{H}_{t-1}\zeta(t)P(t-1)\zeta^{\mathrm{T}}(t)\mathcal{H}_{t-1}^{\mathrm{T}}$$
$$+ \mathcal{H}_{t-1}\zeta(t)P(t-1)D_t^{\mathrm{T}} + D_t P(t-1)\zeta^{\mathrm{T}}(t)\mathcal{H}_{t-1}^{\mathrm{T}}$$
$$+ D_t P(t-1)D_t^{\mathrm{T}} + \lambda(t)\hat{\Lambda}(t).$$

Deleting the first term of the right hand side (which is an ad hoc approximation) and using $\bar{\zeta}(t)$ for $\zeta(t)P(t-1)$ now gives the expression

$$\bar{S}(t) = \mathcal{H}_{t-1}\bar{\zeta}(t)D_t^{\mathrm{T}} + D_t\bar{\zeta}^{\mathrm{T}}(t)\mathcal{H}_{t-1}^{\mathrm{T}} + D_t P(t-1)D_t^{\mathrm{T}} + \lambda(t)\hat{\Lambda}(t).$$

Now, collecting these expressions gives us the alternative Gauss-Newton algorithm:

$$\varepsilon(t) = y(t) - \hat{y}(t),$$ (3.A.9a)

$$\hat{\Lambda}(t) = \hat{\Lambda}(t-1) + \gamma(t)[\varepsilon(t)\varepsilon^{\mathrm{T}}(t) - \hat{\Lambda}(t-1)],$$ (3.A.9b)

$$\bar{S}(t) = \mathcal{H}_{t-1}\bar{\zeta}(t)D_t^{\mathrm{T}} + D_t\bar{\zeta}^{\mathrm{T}}(t)\mathcal{H}_{t-1}^{\mathrm{T}} + D_t\bar{P}(t-1)D_t^{\mathrm{T}} + \lambda(t)\hat{\Lambda}(t),$$ (3.A.9c)

$$\bar{L}(t) = \bar{\psi}(t)\bar{S}^{-1}(t),$$ (3.A.9d)

$$\hat{\theta}(t) = [\hat{\theta}(t-1) + \bar{L}(t)\varepsilon(t)]_{D_{\mathcal{M}}},$$ (3.A.9e)

$$\bar{P}(t) = [\bar{P}(t-1) - \bar{L}(t)\bar{S}(t)\bar{L}^{\mathrm{T}}(t)]/\lambda(t),$$ (3.A.9f)

$$\bar{\zeta}(t+1) = \mathcal{F}_t\bar{\zeta}(t) + M_t\bar{P}(t),$$ (3.A.9g)

$$\bar{\psi}^{\mathrm{T}}(t+1) = \mathcal{H}_t\bar{\zeta}(t+1) + D_{t+1}\bar{P}(t).$$ (3.A.9h)

Here $\hat{y}(t)$ is computed as in (3.70g, h). Notice that using the barred variables in (3.A.9g, h) actually changes $\bar{\zeta}$ compared to (3.A.6).

Remark The matrix $P(t)$ in (3.A.1) and (3.A.2) is guaranteed to be positive semidefinite by construction, since it is given by (3.68). With the approximations made that lead to (3.A.9c, d, f), positive semide-finiteness of $\bar{P}$ may not be automatically guaranteed.

This alternative Gauss-Newton algorithm (3.A.9) is not equivalent to (3.70) since the gains $L(t)$ and $\bar{L}(t)$ differ. They are, however, asymptotically equivalent in the sense that for a constant $\bar{P}$ the vectors L and $\bar{L}$ would be the same. More precisely, it is shown in appendix 4.F that (see also Ljung, 1979a) the algorithms (3.70) and (3.A.9) are associated with the same differential equation, describing their asymptotic properties.

Appendix 3.B An RPE Algorithm for a General State-Space Model

In this appendix we give the formulas for the algorithm described in section 3.8.3.

The predictor is given by

$$\hat{x}(t+1) = F_t\hat{x}(t) + G_t u(t) + K(t)\varepsilon(t), \tag{3.B.1}$$

where, as before

$$F_t = F(\hat{\theta}(t)), \quad G_t = G(\hat{\theta}(t)),$$

and the Kalman gain $K(t)$ is given by (3.135a–c), with θ replaced by the current estimate $\hat{\theta}(t)$:

$$K(t) = [F_t P_1(t) H_t^{\mathrm{T}} + R_{12}(t)] S^{-1}(t), \tag{3.B.2a}$$

$$P_1(t+1) = F_t P_1(t) F_t^{\mathrm{T}} + R_1(t) - K(t)S(t)K^{\mathrm{T}}(t), \tag{3.B.2b}$$

$$S(t) = H_t P_1(t) H_t^{\mathrm{T}} + R_2(t). \tag{3.B.2c}$$

Here

$$R_{12}(t) = R_{12}(\hat{\theta}(t)), \quad R_1(t) = R_1(\hat{\theta}(t)), \quad R_2(t) = R_2(\hat{\theta}(t)),$$

$$H_t = H(\hat{\theta}(t)).$$

The gradient of $K_\theta(t)$ is obtained by differentiating (3.135) and evaluating along $\{\hat{\theta}(k)\}$. If we use the notation

$$\mathcal{K}_t^{(i)} = \frac{d}{d\theta_i} K_\theta(t),$$

$$\sigma_t^{(i)} = \frac{d}{d\theta_i} S_\theta(t),$$

$$\Pi_t^{(i)} = \frac{d}{d\theta_i} P_1(t),$$

we obtain the equations

$$\mathcal{K}_t^{(i)} = \left[\frac{\partial}{\partial\theta_i} F(\theta) P_1(t) H_t^{\mathrm{T}} + F_t \Pi_t^{(i)} H_t^{\mathrm{T}} \right.$$

$$\left. + F_t P_1(t) \frac{\partial}{\partial\theta_i} H^{\mathrm{T}}(\theta) + \frac{\partial}{\partial\theta_i} R_{12}(\theta) \right]\Bigg|_{\theta=\hat{\theta}(t)} \cdot S^{-1}(t) \tag{3.B.3a}$$

$$- K(t)\sigma_t^{(i)} S^{-1}(t),$$

$$\sigma_t^{(i)} = \left[\frac{\partial}{\partial \theta_i} H(\theta) P_1(t) H_t^{\mathrm{T}} + H_t \Pi_t^{(i)}(t) H_t^{\mathrm{T}} \right.$$

$$\left. + H_t P_1(t) \frac{\partial}{\partial \theta_i} H^{\mathrm{T}}(\theta) + \frac{\partial}{\partial \theta_i} R_2(\theta) \right] \Bigg|_{\theta = \hat{\theta}(t)}, \tag{3.B.3b}$$

$$\Pi_{t+1}^{(i)} = \left[\frac{\partial}{\partial \theta_i} F(\theta) P_1(t) F_t^{\mathrm{T}} + F_t \Pi_t^{(i)} F_t^{\mathrm{T}} \right.$$

$$+ F_t P_1(t) \frac{\partial}{\partial \theta_i} F^{\mathrm{T}}(\theta) + \frac{\partial}{\partial \theta_i} R_1(\theta) \tag{3.B.3c}$$

$$\left. - \mathscr{K}_t^{(i)} S(t) K^{\mathrm{T}}(t) - K(t) \sigma_t^{(i)} K^{\mathrm{T}}(t) - K(t) S(t) (\mathscr{K}_t^{(i)})^{\mathrm{T}} \right] \Bigg|_{\theta = \hat{\theta}(t)}.$$

Here $P_1(t)$, $K(t)$, and $S(t)$ are given by (3.B.2). With $\mathscr{K}_t$ defined as above, we can now form M_t^* according to (3.151) and the algorithm reads [see also (3.145)]

$$\varepsilon(t) = y(t) - \hat{y}(t), \tag{3.B.4a}$$

$$R(t) = R(t-1) + \gamma(t) [\psi(t) S^{-1}(t) \psi^{\mathrm{T}}(t) - R(t-1)], \tag{3.B.4b}$$

$$\hat{\theta}(t) = \hat{\theta}(t-1) + \gamma(t) R^{-1}(t) \psi(t) S^{-1}(t) \varepsilon(t), \tag{3.B.4c}$$

$$\hat{x}(t+1) = F_t \hat{x}_t + G_t u(t) + K(t) \varepsilon(t), \tag{3.B.4d}$$

$$\hat{y}(t+1) = H_t \hat{x}(t+1), \tag{3.B.4e}$$

$$M_t^* = M(\hat{\theta}(t), \hat{x}(t), u(t)) + \mathscr{K}_t \varepsilon(t), \tag{3.B.4f}$$

$$W(t+1) = [F_t - K(t) H_t] W(t) + M_t^* - K(t) D_t, \tag{3.B.4g}$$

$$\psi(t+1) = W^{\mathrm{T}}(t+1) H_t^{\mathrm{T}} + D^{\mathrm{T}}(\hat{\theta}(t), \hat{x}(t+1)). \tag{3.B.4h}$$

As before, $D_t = D(\hat{\theta}(t), \hat{x}(t))$. Equations (3.B.2) and (3.B.3) are naturally used between (3.B.4c) and (3.B.4d) to determine $K(t)$ and $\mathscr{K}_t$. We have replaced $\hat{\Lambda}(t)$ in (3.145b) by $S(t)$, defined by (3.B.2), which is an estimate of the prediction error covariance matrix.

If instead the alternative way of calculating the Gauss-Newton direction, described in appendix 3.A, is used, we obtain the algorithm [see (3.A.9)]

$$\varepsilon(t) = y(t) - \hat{y}(t), \tag{3.B.5a}$$

$$\bar{S}(t) = H_{t-1} \bar{W}(t) \bar{D}_t^{\mathrm{T}} + \bar{D}_t \bar{W}(t) H_{t-1}^{\mathrm{T}} + \bar{D}_t \bar{P}(t-1) \bar{D}_t^{\mathrm{T}} + S(t), \tag{3.B.5b}$$

$$\bar{L}(t) = \bar{\psi}(t)\bar{S}^{-1}(t), \tag{3.B.5c}$$

$$\hat{\theta}(t) = [\hat{\theta}(t-1) + \bar{L}(t)\varepsilon(t)]_{D_{\mathscr{M}}}, \tag{3.B.5d}$$

$$\bar{P}(t) = [\bar{P}(t-1) - \bar{L}(t)\bar{S}(t)\bar{L}^{\mathrm{T}}(t)]/\lambda(t), \tag{3.B.5e}$$

$$\overline{W}(t+1) = [F_t - K(t)H_t]\overline{W}(t) + [M_t^* - K(t)D_t]\bar{P}(t), \tag{3.B.5f}$$

$$\bar{\psi}^{\mathrm{T}}(t+1) = H_t\overline{W}(t+1) + \bar{D}_t\bar{P}(t). \tag{3.B.5g}$$

Here $\bar{S}(t)$ is given by (3.B.2) and $\hat{y}(t)$ by (3.B.4). We also introduced the notation

$$\bar{D}_t = D(\hat{\theta}(t-1), \hat{x}(t)).$$

These algorithms for the general state-space model (3.133) are clearly more complex and time-consuming than the algorithm (3.145) for the corresponding innovations model. The main computational burden is in (3.B.3), the calculation of the gradient of the Kalman gain. The number of equations in (3.B.3c) is $n^2 \cdot d$, and that may be forbidding for higher-order problems.

Remark The time indices and the ordering of the equations may seem complicated. The general idea is simple, however, and can be expressed as follows: Always use the latest available parameter estimate for θ and do not update a quantity before it is needed.

Appendix 3.C   Comparison of the EKF and RPEM

We have seen two different algorithms for the general state-space model (3.133): The (EKF) in appendix 2.A and the RPE algorithm in appendix 3.B. Comparing equations (2.A.2)–(2.A.11) with the alternative Gauss-Newton algorithm (3.B.5) shows that the variables are related as shown in table 3.C.1. With these interpretations of corresponding variables, a close look at the algorithms shows that they are indeed identical, with the following two exceptions.

1. In the Kalman filter equations (2.A.3), (2.A.4), (2.A.6), and (3.B.2), respectively, the cross-coupling terms containing $P_2(t)$ and $P_3(t)$ do not appear in (3.B.2). These terms, however, tend to zero as $t \to \infty$, but could of course still be important for the transient behavior of the algorithm. Also, these terms will guarantee that the matrix P_3 is positive semidefinite.

2. M_t in the EKF does not contain the term $\mathcal{K}_t \varepsilon(t)$. This term is included in M_t^*, and corresponds to the coupling between the parameters and the Kalman gain $K(t)$.

Remark   In the RPE algorithm we use the strategy of ordering equations mentioned in the remark of appendix 3.B. In the EKF the time indices are consequences of the general structure. A close look at the equation shows that in (3.B.5f) we use $D(\hat{\theta}(t), \hat{x}(t))$; while in the corresponding expression (2.A.7′), $D(\hat{\theta}(t-1), \hat{x}(t))$ is used. This, however, we consider to be a minor difference.

The only difference of any importance asymptotically between the two algorithms is (2). By deleting the term $\mathcal{K}_t \varepsilon(t)$ a major computational burden, viz., (3.B.3), is eliminated in the EKF. The penalty, however, is the loss of certain convergence properties, as shown in appendix 4.G.

Finally, we comment upon the EKF using filtering state estimates, discussed in appendix 2.A [see (2.A.14)]. We noted that in the original

Table 3.C.1
Variables in EKF and in (3.B.5).

In EKF	In (3.B.5)
$P_2(t)$	$\overline{W}(t)$
M_t	M_t^*
$P_3(t)$	$\overline{P}(t)$
1	$\lambda(t)$
S_t	$\overline{S}(t)$

EKF, (2.A.1)–(2.A.7′), the gain $L(t)$ is formed using only $y(k)$ for k up to and including time $t - 2$. In the filtered version, $y(t - 1)$ is also used. The gain vector $\bar{L}(t)$ in the prediction error algorithm, however, depends on $y(k)$ for $k \leq t - 1$, since M_t^* contains $\varepsilon(t)$ and hence $y(t)$. In fact, the difference between the "filtered" M_t given by (2.A.14) and the "predicted" M_t, given by (2.A.9), can be seen as an approximation of the term $\mathscr{K}_t\varepsilon(t)$:

$$M(\hat{\theta}(t),\, \hat{x}(t \mid t),\, u(t)) - M(\hat{\theta}(t),\, \hat{x}(t),\, u(t)) \approx \mathscr{K}_t\varepsilon(t). \tag{3.C.1}$$

Hence, the filter version (2.A.14) can be regarded as a better approximation of the RPE algorithm. To show (3.C.1) we find that the ith column of the left hand side is given by

$$\frac{\partial}{\partial\theta_i}[F(\theta)\hat{x}(t \mid t) + G(\theta)u(t)] - \frac{\partial}{\partial\theta_i}[F(\theta)\hat{x}(t) + G(\theta)u(t)]$$

$$= \frac{\partial}{\partial\theta_i}F(\theta)[\hat{x}(t \mid t) - \hat{x}(t)] \tag{3.C.2}$$

$$= \frac{\partial}{\partial\theta_i}F(\theta)\bar{K}(t)[y(t) - H_t\hat{x}(t)],$$

where, in the last equality, (2.A.12) was used. With the expression (2.A.13) for $\bar{K}(t)$, suppressing the "small" terms containing $P_2(t)$, we find that (3.C.2) can be written

$$\left[\frac{\partial}{\partial\theta_i}F(\theta)P_1(t)H_t^{\mathrm{T}}S_t^{-1}\right]\varepsilon(t).$$

The factor within brackets is the first term of $\mathscr{K}_t^{(i)}$ in (3.B.3a) which justifies the interpretation (3.C.1). See also Westerlund and Tysső (1980).

Appendix 3.D Some Formulas for Linear and Pseudolinear Regressions

The general linear regression algorithm is given by (3.84) (take $\mu(t) = 0$)

$$\varepsilon(t) = y(t) - \varphi^{\mathrm{T}}(t)\hat{\theta}(t-1), \tag{3.D.1}$$

$$R(t) = R(t-1) + \gamma(t)[\varphi(t)\Lambda^{-1}(t)\varphi^{\mathrm{T}}(t) - R(t-1)], \tag{3.D.2}$$

$$\hat{\theta}(t) = \hat{\theta}(t-1) + \gamma(t)R^{-1}(t)\varphi(t)\Lambda^{-1}(t)\varepsilon(t). \tag{3.D.3}$$

Notice that the PLR (3.130) is given by exactly the same expression for the special case $\Lambda(t) = I$ and $\varepsilon(t)$ a scalar.

In this appendix we shall develop some formulas that are algebraic consequences of the structure (3.D.1)–(3.D.3). Hence, the formulas will apply to both the linear regressions of section 3.6.2 and to the PLRs of section 3.7.3.

First introduce

$$\bar{R}(t) = \frac{1}{\gamma(t)}R(t). \tag{3.D.4}$$

Then (3.D.2) can be written

$$\bar{R}(t) = \lambda(t)\bar{R}(t-1) + \varphi(t)\Lambda^{-1}(t)\varphi^{\mathrm{T}}(t), \tag{3.D.5}$$

where

$$\lambda(t) = \gamma(t-1)[1 - \gamma(t)]/\gamma(t). \tag{3.D.6}$$

If we write

$$P(t) = \bar{R}^{-1}(t), \tag{3.D.7}$$

we know that $P(t)$ will satisfy (3.70c, d, f).
We have, as in the expression preceding (3.70):

$$L(t) = P(t)\varphi(t)\Lambda^{-1}(t) = P(t-1)\varphi(t)[\varphi^{\mathrm{T}}(t)P(t-1)\varphi(t) + \lambda(t)\Lambda(t)]^{-1}$$

or

$$\varphi^{\mathrm{T}}(t)P(t)\varphi(t)\Lambda^{-1}(t)$$

$$= \varphi^{\mathrm{T}}(t)P(t-1)\varphi(t)\Lambda^{-1}(t)[\varphi^{\mathrm{T}}(t)P(t-1)\varphi(t)\Lambda^{-1}(t) + \lambda(t)I]^{-1}.$$

This gives

$$I - \varphi^{\mathrm{T}}(t)P(t)\varphi(t)\Lambda^{-1}(t)$$

$$= [\varphi^{\mathrm{T}}(t)P(t-1)\varphi(t)\Lambda^{-1}(t) + \lambda(t)I - \varphi^{\mathrm{T}}(t)P(t-1)\varphi(t)\Lambda^{-1}(t)]$$
$$\times [\varphi^{\mathrm{T}}(t)P(t-1)\varphi(t)\Lambda^{-1}(t) + \lambda(t)I]^{-1}$$

or

$$I - \varphi^{\mathrm{T}}(t)P(t)\varphi(t)\Lambda^{-1}(t) = \lambda(t)[\varphi^{\mathrm{T}}(t)P(t-1)\varphi(t)\Lambda^{-1}(t) + \lambda(t)I]^{-1},$$
$$(3.\mathrm{D}.8)$$

which is our first basic relationship.

3.D.1 A Relationship between Prediction Errors and Residuals

By definition, the residual is

$$\bar{\varepsilon}(t) = y(t) - \varphi^{\mathrm{T}}(t)\hat{\theta}(t). \tag{3.D.9}$$

Using (3.D.3) in this expression gives

$$\bar{\varepsilon}(t) = y(t) - \varphi^{\mathrm{T}}(t)[\hat{\theta}(t-1) + P(t)\varphi(t)\Lambda^{-1}(t)\varepsilon(t)]$$
$$= \varepsilon(t) - \varphi^{\mathrm{T}}(t)P(t)\varphi(t)\Lambda^{-1}(t)\varepsilon(t)$$

or

$$\bar{\varepsilon}(t) = [I - \varphi^{\mathrm{T}}(t)P(t)\varphi(t)\Lambda^{-1}(t)]\varepsilon(t). \tag{3.D.10}$$

With (3.D.8), this can also be written as

$$\bar{\varepsilon}(t) = \lambda(t)[\lambda(t)I + \varphi^{\mathrm{T}}(t)P(t-1)\varphi(t)\Lambda^{-1}(t)]^{-1}\varepsilon(t). \tag{3.D.11}$$

3.D.2 "Solving" the Recursion

From (3.D.5) we directly obtain

$$\bar{R}(t) = \bar{\beta}(t, 0)\bar{R}(0) + \sum_{k=1}^{t} \bar{\beta}(t, k)\varphi(k)\Lambda^{-1}(k)\varphi^{\mathrm{T}}(k), \tag{3.D.12}$$

where

$$\bar{\beta}(t, k) = \prod_{j=k+1}^{t} \lambda(j), \quad \bar{\beta}(t, t) = 1. \tag{3.D.13}$$

Multiplying (3.D.3) by $\bar{R}(t)$ gives

$$\bar{R}(t)\hat{\theta}(t) = \bar{R}(t)\hat{\theta}(t-1) + \varphi(t)\Lambda^{-1}(t)\varepsilon(t)$$

$$= \lambda(t)\bar{R}(t-1)\hat{\theta}(t-1) + \varphi(t)\Lambda^{-1}(t)\varphi^{\mathrm{T}}(t)\hat{\theta}(t-1)$$

$$+ \varphi(t)\Lambda^{-1}(t)y(t) - \varphi(t)\Lambda^{-1}(t)\varphi^{\mathrm{T}}(t)\hat{\theta}(t-1),$$

where the second equality follows from (3.D.5) and (3.D.1). Hence

$$\bar{R}(t)\hat{\theta}(t) = \lambda(t)\bar{R}(t-1)\hat{\theta}(t-1) + \varphi(t)\Lambda^{-1}(t)y(t), \tag{3.D.14}$$

which can be summed to yield

$$\bar{R}(t)\hat{\theta}(t) = \bar{\beta}(t, 0)\bar{R}(0)\hat{\theta}(0) + \sum_{k=1}^{t} \bar{\beta}(t, k)\varphi(k)\Lambda^{-1}(k)y(k). \tag{3.D.15}$$

This off-line expression coincides, of course, with the least squares formula (2.116). Our derivation shows that it applies also to PLR. It is not useful in practice for PLR, though, since $\{\varphi(k)\}$ has an implicit $\hat{\theta}$-dependence.

3.D.3 Updating the Criterion Function

Let

$$\bar{V}(t) = \bar{\beta}(t, 0)\hat{\theta}^{\mathrm{T}}(t)\bar{R}(0)\hat{\theta}(t)$$

$$+ \sum_{k=1}^{t} \bar{\beta}(t, k)[y(k) - \varphi^{\mathrm{T}}(k)\hat{\theta}(t)]^{\mathrm{T}}$$

$$\times \Lambda^{-1}(k)[y(k) - \varphi^{\mathrm{T}}(k)\hat{\theta}(t)] \tag{3.D.16}$$

The second term is the weighted sum of residuals, computed at time t. The first term is of vanishing importance.

Despite the multiple dependence of t in (3.D.16), there is a remarkably simple updating formula for $\bar{V}(t)$:

LEMMA 3.D.1   Consider the recursion (3.D.1)–(3.D.3). Assume that $\hat{\theta}(0) = 0$. Then

$$\bar{V}(t) = \lambda(t)\bar{V}(t-1) + \varepsilon^{\mathrm{T}}(t)\Lambda^{-1}(t)\bar{\varepsilon}(t), \tag{3.D.17}$$

where $\bar{V}(t)$ is defined by (3.D.16).

Remark   Notice that with the aid of (3.D.11), this can be written as

$$\bar{V}(t) = \lambda(t)\{\bar{V}(t-1) + \varepsilon^{\mathrm{T}}(t)[\lambda(t)\Lambda(t) + \varphi^{\mathrm{T}}(t)P(t-1)\varphi(t)]^{-1}\varepsilon(t)\}. \tag{3.D.18}$$

The scaled quantity $V(t) = \gamma(t)\overline{V}(t)$, which gives a weighted mean of the summands in (3.D.16), obeys

$$V(t) = V(t-1) + \gamma(t)\left\{ \varepsilon^{\mathsf{T}}(t)\left[\Lambda(t) + \varphi^{\mathsf{T}}(t)P(t-1)\frac{\varphi(t)}{\lambda(t)} \right]^{-1} \varepsilon(t) \right.$$

$$\left. - V(t-1) \right\}. \tag{3.D.19}$$

Proof

$$\overline{V}(t) = \sum_{k=1}^{t} \bar{\beta}(t, k)[y(k) - \varphi^{\mathsf{T}}(k)\hat{\theta}(t)]^{\mathsf{T}}\Lambda^{-1}(k)y(k) + Q(t),$$

where

$$Q(t) = -\sum_{k=1}^{t} \{\bar{\beta}(t, k)[y(k) - \varphi^{\mathsf{T}}(k)\hat{\theta}(t)]^{\mathsf{T}}\Lambda^{-1}(k)\varphi^{\mathsf{T}}(k)\hat{\theta}(t)\}$$

$$+ \bar{\beta}(t, 0)\hat{\theta}^{\mathsf{T}}(t)\overline{R}(0)\hat{\theta}(t)$$

$$= \hat{\theta}^{\mathsf{T}}(t)\left[\bar{\beta}(t, 0)\overline{R}(0) + \sum_{k=1}^{t} \bar{\beta}(t, k)\varphi(k)\Lambda^{-1}(k)\varphi^{\mathsf{T}}(k) \right]\hat{\theta}(t)$$

$$- \left[\sum_{k=1}^{t} \bar{\beta}(t, k)y^{\mathsf{T}}(k)\Lambda^{-1}(k)\varphi^{\mathsf{T}}(k) \right]\hat{\theta}(t).$$

A comparison with (3.D.12) and (3.D.15) shows that $Q(t) = 0$. Hence

$$\overline{V}(t) = \sum_{k=1}^{t} \bar{\beta}(t, k)y^{\mathsf{T}}(k)\Lambda^{-1}(k)y(k) - \hat{\theta}^{\mathsf{T}}(t)\sum_{k=1}^{t} \bar{\beta}(t, k)\varphi(k)\Lambda^{-1}(k)y(k)$$

$$= \lambda(t)\sum_{k=1}^{t-1} [\bar{\beta}(t-1, k)y^{\mathsf{T}}(k)\Lambda^{-1}(k)y(k)] + y^{\mathsf{T}}(t)\Lambda^{-1}(t)y(t)$$

$$- \hat{\theta}^{\mathsf{T}}(t-1)\sum_{k=1}^{t-1} [\lambda(t)\bar{\beta}(t-1, k)\varphi(k)\Lambda^{-1}(k)y(k)]$$

$$- \hat{\theta}^{\mathsf{T}}(t-1)\varphi(t)\Lambda^{-1}(t)y(t)$$

$$- \varepsilon^{\mathsf{T}}(t)\Lambda^{-1}(t)\varphi^{\mathsf{T}}(t)\overline{R}^{-1}(t)\sum_{k=1}^{t} \beta(t, k)\varphi(k)\Lambda^{-1}(k)y(k).$$

Here the second equality follows from (3.D.3) and from the definition of $\bar{\beta}(t, k)$. We notice that the sum of the first and third terms is $\lambda(t)\overline{V}(t-1)$. The last term is, according to (3.D.15),

$\varepsilon^{\mathrm{T}}(t)\Lambda^{-1}(t)\varphi^{\mathrm{T}}(t)\hat{\theta}(t).$

Therefore the second, fourth, and last terms add up to

$\varepsilon^{\mathrm{T}}(t)\Lambda^{-1}(t)\bar{\varepsilon}(t).$

Thus

$$\bar{V}(t) = \lambda(t)\bar{V}(t-1) + \varepsilon^{\mathrm{T}}(t)\Lambda^{-1}(t)\bar{\varepsilon}(t),$$

and the lemma is proven. ∎

Appendix 4.A   Proof of Lemma 4.1

Let $z_s^0(t)$ be defined as in S1. Let $\xi_s^0(t, \theta)$ be the corresponding approximation of the vector $\xi(t, \theta)$ defined by (4.63). (We shall henceforth drop the argument θ, since it is a given constant vector in $D_\mathcal{M}$ throughout the proof.) We thus have

$$\xi_s^0(k + 1) = A\xi_s^0(k) + Bz_s^0(k), \quad k \geq s,$$

$$\xi_s^0(s) = 0.$$

Clearly $\xi_s^0(k)$ is independent of anything that happened up to time s. We now have

$$\xi(t) - \xi_s^0(t) = \sum_{k=1}^{s} A^{t-k} Bz(k) + \sum_{k=s+1}^{t} A^{t-k} B[z(k) - z_s^0(k)].$$

Since $\theta \in D_\mathcal{M}$, we have

$$|A^t| < C\lambda_2^t \quad \text{for some} \quad \lambda_2 < 1. \tag{4.A.1}$$

We would like to evaluate the fourth moment of the foregoing expression. If we write $\tilde{z}(k) = z(k) - z_s^0(k)$, with the convention that $z_s^0(k) = 0$ for $k \leq s$, we obtain

$$E|\xi(t) - \xi_s^0(t)|^4$$

$$\leq \sum_{k_1=1}^{t} \sum_{k_2=1}^{t} \sum_{k_3=1}^{t} \sum_{k_4=1}^{t} \{|A^{4t-k_1-k_2-k_3-k_4}| |B|^4 \tag{4.A.2}$$

$$\times E|\tilde{z}(k_1)\tilde{z}(k_2)\tilde{z}(k_3)\tilde{z}(k_4)|\}.$$

From the Schwarz inequality we have

$$E\left|\prod_{1}^{4} \tilde{z}(k_i)\right| \leq \left[\prod_{1}^{4} E|\tilde{z}(k_i)|^4\right]^{1/4}$$

$$\leq C\lambda^{(\tilde{k}_1+\tilde{k}_2+\tilde{k}_3+\tilde{k}_4-4s)/4} \leq C\lambda_1^{\tilde{k}_1+\tilde{k}_2+\tilde{k}_3+\tilde{k}_4-4s},$$

where we used S1 in the second inequality, and used $\lambda_1 = \lambda^{1/4}$ in the last one. Moreover $\tilde{k}_i = \max(k_i, s)$. Now use (4.A.1) and the above expression in (4.A.2) and let $\lambda = \max(\lambda_1, \lambda_2)$. Then we have

$$E|\xi(t) - \xi_s^0(t)|^4 \leq C\lambda^{4(t-s)} \cdot \prod_{i=1}^{4} \sum_{k_i=1}^{t} \lambda^{\tilde{k}_i-k_i} \leq C\lambda^{t-s}. \tag{4.A.3}$$

Now let

$$\begin{pmatrix} \hat{y}_s^0(t) \\ \eta_s^0(t) \end{pmatrix} = C\xi_s^0(t), \quad \varepsilon_s^0(t) = y_s^0(t) - \hat{y}_s^0(t),$$

where $y_s^0(t)$ is the y-component of $z_s^0(t)$, and

$$h_s^0(t) = h(t, \varepsilon_s^0(t), \eta_s^0(t)),$$

$$h(t) = h(t, \varepsilon(t), \eta(t)).$$

Then

$$|h(t) - h_s^0(t)| \le |h_\varepsilon(t)| |\varepsilon(t) - \varepsilon_s^0(t)| + |h_\eta(t)| |\eta(t) - \eta_s^0(t)|$$

$$\le C[1 + |\xi(t)|] \cdot |\xi(t) - \xi_s^0(t)|,$$

where the second inequality follows from Cr1 and the relationship between ε, η, and ξ. We consequently have

$$E|h(t) - h_s^0(t)|^2 \le C[E[1 + |\xi(t)|]^4 \cdot E|\xi(t) - \xi_s^0(t)|^4]^{1/2} \le C\lambda^{t-s}$$

according to (4.A.3). Now let

$$\tilde{h}(t) = h(t) - Eh(t).$$

Notice that $h_s^0(t)$ and $\tilde{h}(k)$ are independent for $k \le s$. Thus, with $\tilde{h}_i$ as the ith component of $\tilde{h}$, we have

$$|E\tilde{h}_i(t) \cdot \tilde{h}_i(s)| = |\mathrm{cov}(h_i(t), h_i(s))|$$

$$= |\mathrm{cov}(h_i(t) - h_{i,s}^0(t) + h_{i,s}^0(t), h_i(s))|$$

$$= |\mathrm{cov}(h_i(t) - h_{i,s}^0(t), h_i(s))|$$

$$\le [E|h(t) - h_s^0(t)|^2 \cdot E|h(s)|^2]^{1/2} \le C \cdot \lambda^{t-s}$$

This means that the sequence $\tilde{h}(t)$ (easily) satisfies (4.67) and thus by (4.68)

$$\frac{1}{N} \sum_{t=1}^{N} \tilde{h}(t) = \frac{1}{N} \sum_{1}^{N} [h(t, \theta, \varepsilon(t, \theta), \eta(t, \theta)) - Eh(t, \theta, \varepsilon(t, \theta), \eta(t, \theta))]$$

tends to zero w.p.1 as $N \to \infty$. With A2 (a) this implies that A1 (a) holds w.p.1. The proof of A1 (b) is of course entirely analogous.

For the proof of A1 (c) we note that

$$E[1 + |z(t)|]^3 < C$$

by the results we already proved. Thus it is sufficient to show that

$$\frac{1}{N}\sum_1^N \{[1 + |z(t)|]^3 - E[1 + |z(t)|]^3\} \to 0 \text{ w.p.1 as } N \to \infty.$$

The proof of this is entirely analogous to the proof of A1 (a). ∎

Appendix 4.B   Proof of Theorem 4.5

In order not to conceal the ideas of the proof with too much notation we first consider the special case of

$$l(t, \theta, \varepsilon) = \frac{1}{2}\varepsilon^{\mathrm{T}}\Lambda^{-1}\varepsilon, \quad r(t) \equiv 0 \quad \text{(see L1)}.$$

The general case is treated later. In the proof, C and λ will denote any positive constants not necessarily the same when appearing in different terms. The constant λ will always be less than one. The constants may depend on the realization of $\{z(t)\}$, which is indicated by the argument (ω).

The Gauss-Newton algorithm for the quadratic case is given by

$$\hat{\theta}(t) = \hat{\theta}(t-1) + \frac{1}{t}R^{-1}(t)\psi(t)\Lambda^{-1}\varepsilon(t),$$

$$R(t) = R(t-1) + \frac{1}{t}[\psi(t)\Lambda^{-1}\psi^{\mathrm{T}}(t) - R(t-1)].$$

With the notation

$$\tilde{\theta}(t) = \hat{\theta}(t) - \theta_0, \quad \bar{R}(t) = t \cdot R(t),$$

we obtain

$$\bar{R}(t)\tilde{\theta}(t) = \bar{R}(t)\tilde{\theta}(t-1) + \psi(t)\Lambda^{-1}\varepsilon(t)$$

$$= \bar{R}(t-1)\tilde{\theta}(t-1) + \psi(t)\Lambda^{-1}\psi^{\mathrm{T}}(t)\tilde{\theta}(t-1) + \psi(t)\Lambda^{-1}\varepsilon(t).$$

This expression can be summed from $t = 0$ to t, giving

$$\bar{R}(t)\tilde{\theta}(t) = \bar{R}(0)\tilde{\theta}(0) + \sum_{k=1}^{t} \psi(k)\Lambda^{-1}[\psi^{\mathrm{T}}(k)\tilde{\theta}(k-1) + \varepsilon(k)].$$

With the notation $\varepsilon(t, \theta_0) = e(t)$, we can rewrite the foregoing expression as

$$\bar{R}(t)\tilde{\theta}(t) = \bar{R}(0)\tilde{\theta}(0) + \sum_{k=1}^{t} \psi(k)\Lambda^{-1}e(k)$$

$$+ \sum_{k=1}^{t} \psi(k)\Lambda^{-1}[\psi^{\mathrm{T}}(k)\tilde{\theta}(k-1) + \varepsilon(k) - e(k)]. \tag{4.B.1}$$

Our analysis will be based on a closer study of each of the terms appearing in (4.B.1). We first have a number of lemmas.

LEMMA 4.B.1   Conditions S2 and M2 along with $\hat{\theta}(t) \to \theta_0$ imply that

$$|\psi(k, \theta)| < C_1 \cdot \alpha(\omega), \quad \left|\frac{d}{d\theta}\psi(k, \theta)\right| < C_1 \cdot \alpha(\omega), \quad |\hat{y}(k \mid \theta)| < C_1 \cdot \alpha(\omega)$$

for all $\theta \in D_{\mathscr{M}}$; and furthermore

$$|\psi(k)| < C(\omega), \quad |\hat{y}(k)| < C(\omega),$$

where $C(\omega)$ is a finite constant that may depend on the realization, but is independent of k. C_1 is a fixed constant and $\alpha(\omega)$ is the random variable defined in S2. Moreover

$$\psi(k) - \psi(k, \theta_0) \to 0 \text{ w.p.1 as } k \to \infty.$$

Proof   We have that $\theta_0 \in D_{\mathscr{M}}$. Hence $|A(\theta_0)^t| \le C \cdot \lambda^t$ for some $\lambda < 1$. Moreover, for θ_k belonging to a small enough neighborhood of θ_0 we also then have

$$\left|\prod_{k=1}^{t} A(\theta_k)\right| \le C \cdot \lambda^t$$

for some (other) $\lambda < 1$. This means, since $\hat{\theta}(t) \to \theta_0$, that $\hat{\theta}(t)$ enters and stays in this neighborhood at some time T (depending on the realization). We thus have

$$\left|\prod_{k=1}^{t} A(\hat{\theta}(k))\right| \le \left|\prod_{k=1}^{T-1} A(\hat{\theta}(k))\right| \cdot \left|\prod_{k=T}^{t} A(\hat{\theta}(k))\right| \le C_T \cdot C \cdot \lambda^{t-T} = C(\omega)\lambda^t,$$

$$(4.B.2)$$

where C_T can be taken, e.g., as C_1^T, where

$$C_1 = \sup_{\theta \in D_{\mathscr{M}}} |A(\theta)|.$$

For each given realization (i.e., for a given T), C_1 is a finite constant. Since from (4.86d) we have

$$\xi(t + 1) = A(\hat{\theta}(t))\xi(t) + B(\hat{\theta}(t))z(t),$$

we obtain

$$\xi(t) = \sum_{k=1}^{t} \left[\prod_{j=k}^{t} A(\hat{\theta}(j))\right] B(\hat{\theta}(k))z(k),$$

which with (4.B.2), and $|B(\theta)| < C$ gives

$$|\xi(t)| \le C \sum_{k=1}^{t} \lambda^{t-k} |z(k)| \le C(\omega).$$

This proves that $|\psi(t)|$ and $|\hat{y}(t)|$ are bounded. The proof that $|\psi(t, \theta)|$, $|d\psi(t, \theta)/d\theta|$, and $|\hat{y}(t \mid \theta)|$ are bounded is immediate, since $A(\theta)$ is exponentially stable for $\theta \in D_{\mathcal{M}}$.

Finally, if we write

$$\tilde{\xi}_k = \xi(k) - \xi(k, \theta_0),$$

$$\tilde{A}_k = A(\hat{\theta}(k)) - A(\theta_0), \quad \tilde{B}_k = B(\hat{\theta}(k)) - B(\theta_0),$$

we have

$$\tilde{\xi}_{k+1} = A(\hat{\theta}(k))\tilde{\xi}_k + \tilde{A}_k \xi(k, \theta_0) + \tilde{B}_k z(k)$$

and

$$\tilde{\xi}_t = \sum_{k=1}^{t} \left[\prod_{j=k}^{t} A(\hat{\theta}(j)) \right] [\tilde{A}_k \xi(k, \theta_0) + \tilde{B}_k z(k)]. \tag{4.B.3}$$

Since $\xi(k, \theta_0)$ and $z(k)$ are bounded and $\tilde{A}_k$, $\tilde{B}_k$ tend to zero, and due to (4.B.2) we find that $\tilde{\xi}_t \to 0$ w.p.1 as $t \to \infty$. Hence $\psi(t) - \psi(t, \theta_0) \to 0$ w.p.1 as $t \to \infty$. ∎

LEMMA 4.B.2

$$R(t) \to \bar{E}\psi(t, \theta_0)\Lambda^{-1}\psi^{\mathrm{T}}(t, \theta_0) = G(\theta_0) \text{ w.p.1 as } t \to \infty.$$

Proof

$$R(t) = \frac{1}{t} \sum_{k=1}^{t} \psi(k)\Lambda^{-1}\psi^{\mathrm{T}}(k) = \frac{1}{t} \sum_{k=1}^{t} \psi(k, \theta_0)\Lambda^{-1}\psi^{\mathrm{T}}(k, \theta_0)$$

$$+ \frac{1}{t} \sum_{k=1}^{t} [\psi(k) - \psi(k, \theta_0)]\Lambda^{-1}\psi^{\mathrm{T}}(k, \theta_0)$$

$$+ \frac{1}{t} \sum_{k=1}^{t} \psi(k)\Lambda^{-1}[\psi(k) - \psi(k, \theta_0)]^{\mathrm{T}}.$$

The last two terms will tend to zero w.p.1 according to lemma 4.B.1. The first one tends to $\bar{E}\psi(t, \theta_0)\Lambda^{-1}\psi^{\mathrm{T}}(t, \theta_0)$ according to lemma 4.1. ∎

LEMMA 4.B.3   For any $\delta > 0$,

$$\frac{1}{t^{1/2+\delta}} \sum_{k=1}^{t} \psi(k)\Lambda^{-1}e(k) \to 0 \quad \text{w.p.1 as } t \to \infty.$$

Proof The proof relies upon the martingale convergence theorem. There is a slight complication in that we know from lemma 4.B.1 only that $|\psi(k)| < C(\omega)$, but we do not know if $C(\omega)$ has finite variance. Therefore introduce the random variable

$$\bar{\psi}(k) = \begin{cases} \psi(k) & \text{if } |\psi(k)| < 2C_1 \cdot \alpha(\omega) \\ 0 & \text{if } |\psi(k)| > 2C_1 \cdot \alpha(\omega) \end{cases}.$$

From lemma 4.B.1 we know that $\psi(k) \to \psi(k, \theta_0)$ and that $\psi(k, \theta_0)$ is bounded by $C_1\alpha(\omega)$. Hence

$$\bar{\psi}(k) = \psi(k) \quad \text{for} \quad k > K(\omega).$$

Now consider

$$s_t = \sum_{k=1}^{t} k^{-(1/2)-\delta}\bar{\psi}(k)\Lambda^{-1}e(k).$$

This random variable is a martingale with respect to the σ-algebra generated by z^{t-1}, since

$$E(s_t|z^{t-1}) = s_{t-1} + E\{t^{-(1/2)-\delta}\bar{\psi}(t)\Lambda^{-1}e(t)|z^{t-1}\}$$

$$= s_{t-1} + t^{-(1/2)-\delta}\bar{\psi}(t)\Lambda^{-1}E\{e(t)|z^{t-1}\} = s_{t-1}.$$

Here the first equality follows since s_{t-1} is z^{t-1}-measurable, the second one since $\bar{\psi}(t)$ is z^{t-1}-measurable, and the last one according to L1. Moreover

$$E|s_t|^2 \le \sum_{k=1}^{t} k^{-1-2\delta}E|\bar{\psi}(k)|^2 \cdot E|e(k)|^2 \cdot |\Lambda^{-1}| \le C \cdot \sum_{k=1}^{\infty} k^{-1-2\delta} < \infty,$$

since $\bar{\psi}(k)$ has bounded variance by construction. Hence s_t is a martingale with bounded variance and it will thus converge w.p.1 to a finite limit

$$s_t \to s_\infty < \infty \text{ w.p.1}$$

(Chung, 1968). Hence, according to Kronecker's lemma (Chung, 1968), we obtain

$$t^{-1/2-\delta} \sum_{k=1}^{t} \bar{\psi}(k)\Lambda^{-1}e(k) \to 0 \text{ w.p.1 as } t \to \infty.$$

To establish the lemma we now only have to prove that $\bar{\psi}$ can be replaced by ψ. But this is true, since $\psi(k)$ and $\bar{\psi}(k)$ coincide from a certain k on. ∎

LEMMA 4.B.4   There is a finite valued constant $C(\omega)$ (dependent on the realization), such that

$$|\psi(k)\Lambda^{-1}[\psi^T(k)\tilde{\theta}(k-1) + \varepsilon(k) - e(k)]| < C(\omega)[|\tilde{\theta}(k-1)|^2 + 1/k].$$

Proof   The intuition for this result is that

$$\varepsilon(k) - e(k) \approx \varepsilon(k, \hat{\theta}(k)) - \varepsilon(k, \theta_0) \approx -\psi^T(k)\tilde{\theta}(k),$$

since ψ is the negative gradient of ε. The formal reasoning goes as follows. In view of lemma 4.B.1, $\psi(t)$ and $\varepsilon(t)$ are bounded; and in view of lemma 4.B.2, $R(t)$ tends to an invertible matrix. Hence from the updating formula for $\hat{\theta}$ we obtain

$$|\hat{\theta}(t) - \hat{\theta}(t-1)| < C(\omega)/t,$$

which implies that

$$|\hat{\theta}(k) - \hat{\theta}(t)| \leq C(\omega) \cdot \log\frac{t}{k} \quad \text{for} \quad t > k. \tag{4.B.4}$$

Introduce

$$\tilde{\xi}(t) = \xi(t) - \xi(t, \hat{\theta}(t)),$$

$$\tilde{A}(k, t) = A(\hat{\theta}(k)) - A(\hat{\theta}(t)),$$

$$\tilde{B}(k, t) = B(\hat{\theta}(k)) - B(\hat{\theta}(t)).$$

According to (4.B.4) and M2, we have

$$|\tilde{A}(k, t)| + |\tilde{B}(k, t)| \leq C(\omega)\log t/k \quad \text{for} \quad t > k. \tag{4.B.5}$$

We also have, as in (4.B.3),

$$\tilde{\xi}(t) = \sum_{k=1}^{t}\left[\prod_{j=k}^{t} A(\hat{\theta}(j))\right][\tilde{A}(k, t)\xi(k, \hat{\theta}(t)) + \tilde{B}(k, t)z(k)].$$

Now, using (4.B.2), (4.B.5), and the boundedness of $\xi(k, \hat{\theta}(t))$ and $z(k)$, we have

$$|\tilde{\xi}(t)| \leq C(\omega)\sum_{k=1}^{t} \lambda^{t-k} \cdot \log t/k \leq C(\omega)/t.$$

This result in particular implies (see definition of $\tilde{\xi}$ and (4.86e))

$$|\varepsilon(t) - \varepsilon(t, \hat{\theta}(t))| < C(\omega)/t,$$

$$|\psi(t) - \psi(t, \hat{\theta}(t))| < C(\omega)/t. \tag{4.B.6}$$

From M3 we have that $\varepsilon(t, \theta)$ is twice differentiable, and Taylor expansion around θ_0 gives

$$\varepsilon(t, \theta) - \varepsilon(t, \theta_0) = \hat{y}(t \mid \theta_0) - \hat{y}(t \mid \theta)$$

$$= -\psi^{\mathrm{T}}(t, \theta)(\theta - \theta_0) + \tfrac{1}{2}(\theta - \theta_0)^{\mathrm{T}}\psi_\theta'(t, \eta)(\theta - \theta_0),$$

where η is a point between θ and θ_0. According to lemma 4.B.1, ψ_θ' is bounded, so when we evaluate the above expression for $\theta = \hat{\theta}(t-1)$ we obtain

$$|\varepsilon(t, \hat{\theta}(t-1)) - e(t) + \psi^{\mathrm{T}}(t, \hat{\theta}(t-1))\tilde{\theta}(t-1)| \le C|\tilde{\theta}(t-1)|^2. \tag{4.B.7}$$

Now use (4.B.6) to replace $\varepsilon(t, \hat{\theta}(t-1))$ by $\varepsilon(t)$ and $\psi(t, \hat{\theta}(t-1))$ by $\psi(t)$. Then (4.B.7) gives the desired expression. ∎

Let us now return to the expression (4.B.1). Using lemmas 4.B.2–4.B.4 for the entities in this expression gives for any $\delta > 0$

$$t \cdot |\tilde{\theta}(t)| \le C \sum_{k=1}^{t} |\tilde{\theta}(k-1)|^2 + C \cdot t^{1/2+\delta} \text{ w.p.1}, \tag{4.B.8}$$

where the constant C may depend on the realization and on δ. To study (4.B.8) we first prove the following result.

LEMMA 4.B.5   Let b_n be a sequence of scalars such that

$$b_n > 0, \ b_n \to 0 \text{ as } n \to \infty$$

and, for some $C > 0$ and $1 > \alpha > 0$,

$$n \cdot b_n < C\left(\sum_{k=1}^{n-1} b_k^2 + n^\alpha\right). \tag{4.B.9}$$

Then

$$\sum_{k=1}^{n-1} b_k^2 < Cn^{\alpha'}, \tag{4.B.10}$$

where

$$\alpha' = \max(0, 2\alpha - 1) \text{ for } \alpha \neq 1/2,$$

$\alpha' = \varepsilon > 0$ (arbitrarily small) for $\alpha = 1/2$.

Proof   Denote

$$T_n = \frac{1}{n^\alpha} \sum_{k=1}^{n-1} b_k^2.$$

We first find that

$$T_{n+1} - T_n = \frac{1}{(n+1)^\alpha} \left[\sum_{k=1}^{n} b_k^2 - \left(1 + \frac{1}{n}\right)^\alpha \sum_{k=1}^{n-1} b_k^2 \right].$$

With the approximation

$$\left(1 + \frac{1}{n}\right)^\alpha \approx 1 + \frac{\alpha}{n},$$

which for sufficiently large n becomes arbitrarily good, we obtain

$$T_{n+1} - T_n \approx \frac{1}{(n+1)^\alpha} \left[b_n^2 - \frac{\alpha}{n} \sum_{k=1}^{n-1} b_k^2 \right]. \tag{4.B.11}$$

Since b_n tends to zero we see that T_n can increase only by a small amount. Hence if T_n is unbounded there must for any $C_1 > 0$ be an infinite sequence n_k such that

$$T_{n_k} > C_1 \quad \text{and} \quad T_{n_k+1} - T_{n_k} > 0. \tag{4.B.12}$$

The second inequality implies, according to (4.B.11) that

$$n_k \cdot b_{n_k}^2 > \alpha \sum_{k=1}^{n_k-1} b_k^2.$$

Using (4.B.9) in this inequality leads to

$$\alpha \cdot \sum_{k=1}^{n_k-1} b_k^2 < n_k \cdot b_{n_k}^2 < C b_{n_k} \sum_{k=1}^{n_k-1} b_k^2 + C b_{n_k} n_k^\alpha$$

or

$$(\alpha - C b_{n_k}) \sum_{k=1}^{n_k-1} b_k^2 < C b_{n_k} n_k^\alpha.$$

Since b_n tends to zero and $\alpha > 0$, this implies that for sufficiently large n_k we can write

$T_{n_k} < Cb_{n_k}$,

which contradicts the first inequality in (4.B.12). Hence T_n is bounded, and (4.B.10) is proven for $\alpha' = \alpha$. Insert this result into (4.B.9), which gives

$b_n < n^{\alpha-1}$.

This in turn implies that

$$\sum_{k=1}^{n-1} b_k^2 \leq C \sum_{k=1}^{n-1} k^{2\alpha-2} \begin{cases} \leq C & \text{if } \alpha < 1/2 \\ \leq C\log n & \text{if } \alpha = 1/2 , \\ \leq Cn^{2\alpha-1} & \text{if } \alpha > 1/2 \end{cases}$$

and the desired result has been proven. ∎

Let us now apply lemma 4.B.5 to (4.B.8) with $b_n = |\tilde{\theta}(n)|$ and $\alpha = (1/2) + \delta$. We find that

$$\sum_{k=1}^{t} |\tilde{\theta}(k-1)|^2 < C \cdot t^{2\delta}. \tag{4.B.13}$$

With (4.B.13) inserted into (4.B.8) we find that

$$t^{(1/2)-\delta'} |\tilde{\theta}(t)| \leq C \cdot t^{-(1/2)+2\delta-\delta'} + Ct^{\delta-\delta'}. \tag{4.B.14}$$

Since δ can be taken as any positive number, we have now established the last statement of the theorem. In particular, returning to (4.B.1) we obtain

$$\sqrt{t}\,\tilde{\theta}(t) - [R(t)]^{-1} \frac{1}{\sqrt{t}} \sum_{k=1}^{t} \psi(k, \theta_0)\Lambda^{-1} e(k) + h(t), \tag{4.B.15}$$

where

$$h(t) = [R(t)]^{-1} \frac{1}{\sqrt{t}} \sum_{k=1}^{t} [\psi(k) - \psi(k, \theta_0)]\Lambda^{-1} e(k) + \frac{1}{\sqrt{t}}[R(t)]^{-1} R(0)\tilde{\theta}(0)$$

$$+ [R(t)]^{-1} \frac{1}{\sqrt{t}} \sum_{k=1}^{t} \psi(k)\Lambda^{-1}[\psi^{\mathrm{T}}(k)\tilde{\theta}(k-1) + \varepsilon(k) - e(k)].$$

LEMMA 4.B.6 $h(t) \to 0$ w.p.1 as $t \to \infty$.

Proof That the second term of $h(t)$ tends to zero is trivial, and that the third term of $h(t)$ tends to zero follows from lemma 4.B.4 and (4.B.13). Consider the variable

$\tilde{\psi}(k) = \psi(k) - \psi(k, \theta_0)$.

Use now (4.B.3) together with the fact that

$|\tilde{A}_k| + |\tilde{B}_k| \le C \cdot k^{-(1/2)+\delta}$

according to M3 and (4.B.14). This gives

$|\tilde{\xi}_k| \le C(\omega) \sum\limits_{j=1}^{k} \lambda^{k-j} j^{-(1/2)+\delta} \le C(\omega) k^{-(1/2)+\delta}$,

and hence

$$|\tilde{\psi}(k)| \le C(\omega) k^{-(1/2)+\delta}. \tag{4.B.16}$$

As in the proof of lemma 4.B.3, we have a complication in that we do not know if $C(\omega)$ and hence $\tilde{\psi}(k)$ has finite variance. (We do know, however, that $C(\omega)$ is finite w.p.1). Therefore introduce

$$\bar{\psi}(k) = \begin{cases} \tilde{\psi}(k) & \text{if } |\tilde{\psi}(k)| < C_1 \cdot k^{-(1/2)+2\delta} \\ 0 & \text{otherwise} \end{cases}$$

Because of (4.B.16), $\bar{\psi}(k) = \tilde{\psi}(k)$ for $k > K(\omega)$. Consider now

$$S_t = \sum\limits_{k=1}^{t} \frac{1}{\sqrt{k}} \bar{\psi}(k) \Lambda^{-1} e(k).$$

As in the proof of lemma 4.B.3 this variable is a martingale, and its variance is subject to

$$\mathrm{E}|S_t|^2 \le \sum\limits_{1}^{\infty} \frac{1}{k} \cdot C_1^2 \cdot \frac{1}{k} \cdot k^{4\delta} \cdot \mathrm{E}|e(k)|^2 \cdot |\Lambda^{-1}| < \infty \quad (\delta < 1/4).$$

Hence S_t converges w.p.1, and, according to Kronecker's lemma, we have

$$\frac{1}{\sqrt{t}} \sum\limits_{1}^{t} \bar{\psi}(k) \Lambda^{-1} e(k) \to 0 \text{ w.p.1 as } t \to \infty.$$

Since $\bar{\psi}(k)$ and $\tilde{\psi}(k)$ coincide for $k > K(\omega)$, the same result holds when $\bar{\psi}$ is replaced by $\tilde{\psi}$. Hence also the first term of $h(t)$ tends to zero w.p.1, and the lemma is proven. ∎

LEMMA 4.B.7

$$\frac{1}{\sqrt{t}} \sum\limits_{k=1}^{t} \psi(k, \theta_0) \Lambda^{-1} e(k) \in \mathrm{AsN}(0, Q),$$

where

$$Q = \bar{E}\psi(t, \theta_0)\Lambda^{-1}\Lambda_0\Lambda^{-1}\psi^T(t, \theta_0).$$

Proof This result is a direct application of Billingsley's central limit theorem for martingales (Billingsley, 1961), which was extended to non-stationary processes by Brown (1971). Brown's theorem is formulated for scalar random variables, and to get the above result for vectors we simply apply it to all linear combinations of the vector components in our case. This is essentially the same as pretending, in the calculations, that e, Λ, and ψ are scalars. To apply Brown's theorem 1, we must verify two conditions:

(1) $\dfrac{V_n^2}{s_n^2} \to 1$ in probability as $n \to \infty$,

where

$$V_n^2 = \sum_{k=1}^{n} E\{[\psi(k, \theta_0)\Lambda^{-1}e(k)]^2 \mid z^{k-1}\}$$

$$= \sum_{k=1}^{n} \psi(k, \theta_0)\Lambda^{-1}\Lambda_0\Lambda^{-1}\psi^T(k, \theta_0)$$

and $s_n^2 = EV_n^2$. But in view of the assumptions of the theorem and lemma 4.1, both s_n^2/n and V_n^2/n tend to $G(\theta_0)$ w.p.1, so this condition is satisfied.

(2) $\dfrac{1}{s_n^2} \sum_{k=1}^{n} E|\psi(k, \theta_0)\Lambda^{-1}|_{\Lambda_0}^2 \cdot P(|\psi(k, \theta_0)\Lambda^{-1}e(k)| > \delta s_n) \to 0$ as $n \to \infty$.

But $s_n \to \infty$ and $\psi(k, \theta_0)\Lambda^{-1}e(k)$ is bounded according to lemma 4.B.1, so this condition is also satisfied. Now Brown's (1971) theorem 1 implies our lemma 4.B.7. ∎

If we now use lemmas 4.B.2, 4.B.6, and 4.B.7 in (4.B.15) we find the final result

$$\sqrt{t}\tilde{\theta}(t) \in AsN(0, P),$$

where

$$P = [G(\theta_0)]^{-1}Q[G(\theta_0)]^{-1},$$

which proves the theorem for l given by the quadratic expression and $r(t) \equiv 0$.

Now suppose that $\varepsilon(t, \theta_0) = e(t) + r(t)$ with $t^{(1/2)+\delta}r(t) \to 0$ w.p.1, as in L1. This gives a fourth term in the expression (4.B.1):

$$D_t = \sum_{k=1}^{t} \psi(k)\Lambda^{-1}r(k) \tag{4.B.17}$$

(In the third term of (4.B.1) we use $\varepsilon(t, \theta_0)$ instead of $e(t)$). Now

$$\left| \frac{1}{\sqrt{t}} D_t \right| \le \frac{C}{\sqrt{t}} \sum_{k=1}^{t} k^{-1/2-\delta} \le C \cdot \frac{t^{1/2-\delta}}{t^{1/2}} \le C \cdot t^{-\delta},$$

where the first inequality follows from lemma 4.B.1 and the properties of $r(k)$. Hence $t^{-1/2}D_t$ tends to zero w.p.1 as $t \to \infty$. Therefore this term will not affect the expression (4.B.8), and it can be included in the term $h(t)$ in (4.B.15).

If we now consider the general Gauss-Newton algorithm

$$\hat{\theta}(t) = \hat{\theta}(t-1) + \frac{1}{t}R^{-1}(t)[-l_\theta^T(t, \hat{\theta}(t-1), \varepsilon(t))$$

$$+ \psi(t)l_\varepsilon^T(t, \hat{\theta}(t-1), \varepsilon(t))],$$

$$R(t) = R(t-1) + \frac{1}{t}[l_{\theta\theta}(t, \hat{\theta}(t-1), \varepsilon(t))$$

$$+ \psi(t)l_{\varepsilon\varepsilon}(t, \hat{\theta}(t-1), \varepsilon(t))\psi^T(t) - R(t-1)],$$

we find with

$$\tilde{\theta}(t) = \hat{\theta}(t) - \theta_0 \quad \text{and} \quad \bar{R}(t) = tR(t)$$

that

$$\bar{R}(t)\tilde{\theta}(t) = \bar{R}(t-1)\tilde{\theta}(t-1) + l_{\theta\theta}(t, \hat{\theta}(t-1), \varepsilon(t))\tilde{\theta}(t-1)$$

$$+ \psi(t)l_{\varepsilon\varepsilon}(t, \hat{\theta}(t-1), \varepsilon(t))\psi^T(t)\tilde{\theta}(t-1)$$

$$- l_\theta^T(t, \hat{\theta}(t-1), \varepsilon(t)) + \psi(t)l_\varepsilon^T(t, \hat{\theta}(t-1), \varepsilon(t)).$$

Summing this expression gives

$$\bar{R}(t)\tilde{\theta}(t)$$

$$= \bar{R}(0)\tilde{\theta}(0) + \sum_{k=1}^{t} [-l_\theta^T(k, \theta_0, e(k)) + \psi(k)l_\varepsilon^T(k, \theta_0, e(k))]$$

$$+ \sum_{k=1}^{t} \{[l_\theta^T(k, \theta_0, e(k)) - l_\theta^T(k, \hat\theta(k-1), \varepsilon(k))$$

$$+ l_{\theta\theta}(k, \hat\theta(k-1), \varepsilon(k))\tilde\theta(k-1)$$

$$+ l_{\theta\varepsilon}(k, \hat\theta(k-1), \varepsilon(k))\psi^T(k)\tilde\theta(k-1)]$$

$$+ \psi(k)[l_\varepsilon^T(k, \hat\theta(k-1), \varepsilon(k)) - l_\varepsilon^T(k, \theta_0, e(k))$$

$$+ l_{\varepsilon\varepsilon}(k, \hat\theta(k-1), \varepsilon(k))\psi^T(k)\tilde\theta(k-1)$$

$$- l_{\varepsilon\theta}(k, \hat\theta(k-1), \varepsilon(k))\tilde\theta(k-1)]\}.$$

This expression is dealt with in the same fashion as (4.B.1). Lemma 4.B.1 is not affected. Lemma 4.B.2 is analogous. Lemma 4.B.3 for the first sum applies, with the analogous proof. Lemma 4.B.4 has a corresponding counterpart, where the terms linear in $\tilde\theta$ in each of the expressions within square brackets in the second sum above cancel when a Taylor expansion around $\hat\theta(k-1)$, $\varepsilon(t)$ is used. The proof is entirely analogous to that for lemma 4.B.4. This brings us to (4.B.8), from where the rest of the proof of the theorem coincides with the one given. ∎

Remark   In the proof, we did not consider the modified algorithm, in which $\hat\theta(t)$ is projected in $D_{\mathcal{M}}$, and the version (4.95) is used for the R-update. However, for $\hat\theta(t) \to \theta_0$, these modifications will be in force only a finite number of times, and will therefore not affect the asymptotic distribution.

Proof of Corollary to Theorem 4.5   We can immediately verify that the proof of the theorem holds if Λ in (4.B.1) is replaced by any sequence $\hat\Lambda(t)$ with $|\hat\Lambda^{-1}(t)| < C$. The algorithm (4.106) produces such a sequence. We only have to prove that

$$G(\theta_0) = \bar{E}\psi(t, \theta_0)\hat\Lambda^{-1}(t)\psi^T(t, \theta_0) = \bar{E}\psi(t, \theta_0)\Lambda_0^{-1}\psi^T(t, \theta_0)$$

and that (in lemma 4.B.7)

$$Q = \bar{E}\psi(t, \theta_0)\hat\Lambda^{-1}(t)\Lambda_0\hat\Lambda^{-1}(t)\psi^T(t, \theta_0)$$

$$= \bar{E}\psi(t, \theta_0)\Lambda_0^{-1}\psi^T(t, \theta_0).$$

But, since $\hat\theta(t) \to \theta_0$, we have that $\varepsilon(t) \to \varepsilon(t, \theta_0)$ (see lemma 4.B.1), and hence

$$\hat{\Lambda}(t) = \frac{1}{t} \sum_1^t \varepsilon(k)\varepsilon^{\mathrm{T}}(k) \to \bar{\mathrm{E}}\varepsilon(t, \theta_0)\varepsilon^{\mathrm{T}}(t, \theta_0) = \Lambda_0 \text{ w.p.1}$$

just as in lemma 4.B.2. Therefore the foregoing expressions for $G(\theta_0)$ and Q hold, which proves the corollary. ∎

Remark   Theorem 4.5 is proven here under the assumption L1 that θ_0 produces asymptotically independent prediction errors. This assumption was used in the proof only in lemmas 4.B.3, 4.B.6, and 4.B.7. The theorem is true also without assumption L1, and can be proven using a central-limit theorem for mixing processes in lemma 4.B.7 (Ljung and Caines, 1979) and using Borel-Cantelli's lemma (Chung, 1968), rather than the martingale convergence theorem in lemmas 4.B.3 and 4.B.6. We must, however, use a full Newton algorithm to obtain the result, since the last three terms of the expression for $\bar{V}''(\theta)$ (p. 190) will not disappear unless L1 holds. This means that in the updating of $R(t)$, the corresponding terms must be included for the counterpart of theorem 4.5 to hold.

Appendix 4.C   A Martingale Convergence Proof of Theorem 4.6

The martingale convergence technique to prove convergence of PLRs was suggested by Moore and Ledwich (1980). A related technique had been used for the least squares case by Ljung (1976b). The technique was carried further by Solo (1978, 1979), and this appendix is based on his work. Goodwin and his coworkers (see the book by Goodwin and Sin, 1983) have developed and applied the technique to a number of interesting problems.

The basic convergence result is the following one, which we will not prove. For the proof, see Neveu (1975), p. 34.

LEMMA 4.C.1   Let $\{T_n\}$ be a sequence of nonnegative random variables and $\{\mathscr{F}_n\}$ a sequence of increasing adapted σ-algebras (i.e., $T_n \in \mathscr{F}_n$). Suppose

$$E(T_n \mid \mathscr{F}_{n-1}) \le T_{n-1} + \alpha_n \qquad (4.C.1)$$

and

$$\sum_1^\infty \alpha_n < \infty \quad \text{w.p.1.}$$

Then T_n converges w.p.1 to a finite nonnegative random variable T as $n \to \infty$.

We now restate this result in a form that suits us better.

LEMMA 4.C.2   Let $\{T_n\}$, $\{\alpha_{n+1}\}$, and $\{\beta_{n+1}\}$ be sequences of nonnegative random variables, adapted to a sequence of increasing σ-algebras $\{\mathscr{F}_n\}$. Suppose that

$$E(T_n \mid \mathscr{F}_{n-1}) \le T_{n-1} + \alpha_n - \beta_n \qquad (4.C.2)$$

and that

$$\sum_1^\infty \alpha_n < \infty \quad \text{w.p.1.}$$

Then $T_n \to T$ w.p.1 as $n \to \infty$, and

$$\sum_1^\infty \beta_n < \infty \quad \text{w.p.1.}$$

Proof   Clearly, (4.C.2) implies (4.C.1), so $T_n \to T$ w.p.1 follows from lemma 4.C.1. Next, introduce

$$T'_n = T_n + \sum_{k=1}^{n} \beta_k;$$

then (4.C.2) can be written as

$$E(T'_n \mid \mathscr{F}_{n-1}) \leq T'_{n-1} + \alpha_n.$$

We can now apply lemma 4.C.1 to T'_n, which proves the lemma. ■

We can now state the counterpart of theorem 4.6.

THEOREM 4.C.1   Let $\{\varepsilon(t)\}$ and $\{\varphi(t)\}$ be sequences of scalars and vectors, respectively, such that

$$\lim_{N \to \infty} \sup \frac{1}{N} \sum_{1}^{N} |\varphi(t)|^2 < \infty. \tag{4.C.3}$$

Define the sequence $\{\hat{\theta}(t)\}$ by

$$\hat{\theta}(t) = \hat{\theta}(t-1) + \frac{1}{t} R^{-1}(t)\varphi(t)\varepsilon(t),$$

$$R(t) = R(t-1) + \frac{1}{t}[\varphi(t)\varphi^{\mathrm{T}}(t) - R(t-1)]. \tag{4.C.4}$$

Let

$$\bar{\varepsilon}(t) = \varepsilon(t)\left[1 - \frac{1}{t}\varphi^{\mathrm{T}}(t)R^{-1}(t)\varphi(t)\right] \tag{4.C.5}$$

and suppose that, for some value θ_0,

$$H(q^{-1})\bar{\varepsilon}(t) = -\varphi^{\mathrm{T}}(t)[\hat{\theta}(t) - \theta_0] + H(q^{-1})e(t), \tag{4.C.6}$$

where

(1) $\{e(t)\}$ is a sequence of random variables such that

$$E(e(t) \mid \mathscr{F}_{t-1}) = 0 \tag{4.C.7a}$$

$$E(e^2(t) \mid \mathscr{F}_{t-1}) = \sigma^2 \tag{4.C.7b}$$

and

$$\varepsilon(t) - e(t) \in \mathscr{F}_{t-1}, \tag{4.C.7c}$$

where $\mathscr{F}_{t-1}$ is the σ-algebra generated by

$e(0), \ldots, e(t-1), \varphi(0), \ldots, \varphi(t)$.

(2) $H(q^{-1})$ is a causal, strictly stable transfer function, such that $\dfrac{1}{H(q^{-1})} - \dfrac{1}{2}$
is strictly positive real, i.e.,

$$\mathrm{Re}\,[[H(e^{i\omega})]^{-1} - \tfrac{1}{2}] > 0 \quad \forall \omega, \quad -\pi < \omega \leq \pi. \tag{4.C.8}$$

Then

$$[\hat{\theta}(t) - \theta_0]^{\mathrm{T}} R(t) [\hat{\theta}(t) - \theta_0] \to 0 \text{ w.p.1 as } t \to \infty \tag{4.C.9a}$$

and

$$\frac{1}{N} \sum_{1}^{N} [\bar{\varepsilon}(t) - e(t)]^2 \to 0 \text{ w.p.1 as } N \to \infty. \tag{4.C.9b}$$

Proof   Let $\bar{R}(t) = t \cdot R(t)$ and $\tilde{\theta}(t) = \hat{\theta}(t) - \theta_0$. Then

$$\frac{1}{t} R^{-1}(t) \varphi(t) \varepsilon(t) = \frac{\bar{R}^{-1}(t) \varphi(t) \bar{\varepsilon}(t)}{1 - \varphi^{\mathrm{T}}(t) \bar{R}^{-1}(t) \varphi(t)}$$

$$= \bar{R}^{-1}(t - 1) \varphi(t) \bar{\varepsilon}(t),$$

and hence from (4.C.4)

$$\tilde{\theta}(t) = \tilde{\theta}(t - 1) + \bar{R}^{-1}(t - 1) \varphi(t) \bar{\varepsilon}(t), \tag{4.C.10a}$$

$$\bar{R}(t) = \bar{R}(t - 1) + \varphi(t) \varphi^{\mathrm{T}}(t). \tag{4.C.10b}$$

Introduce

$$T(t) = \tilde{\theta}^{\mathrm{T}}(t) \bar{R}(t) \tilde{\theta}(t);$$

then the assertion (4.C.9a) can be expressed as $T(t)/t \to 0$ w.p.1 as $t \to \infty$. We shall eventually apply lemma 4.C.2 to $T(t)/t$ and we first seek an expression for how $T(t)$ relates to $T(t - 1)$. We have from (4.C.10a) that

$$\bar{R}(t - 1) \tilde{\theta}(t) = \bar{R}(t - 1) \tilde{\theta}(t - 1) + \varphi(t) \bar{\varepsilon}(t).$$

Add $\varphi(t) \varphi^{\mathrm{T}}(t) \tilde{\theta}(t)$ to both sides and then multiply by $\tilde{\theta}^{\mathrm{T}}(t)$; this gives

$$\tilde{\theta}^{\mathrm{T}}(t) \bar{R}(t) \tilde{\theta}(t)$$

$$= \tilde{\theta}^{\mathrm{T}}(t) \bar{R}(t - 1) \tilde{\theta}(t - 1) + \tilde{\theta}^{\mathrm{T}}(t) \varphi(t) \bar{\varepsilon}(t) + [\tilde{\theta}^{\mathrm{T}}(t) \varphi(t)]^2.$$

Using (4.C.10a) for the first term on the right-hand side gives

$$T(t) = T(t-1) + \varphi^{\mathrm{T}}(t)\bar{\varepsilon}(t)\tilde{\theta}(t-1) + \tilde{\theta}^{\mathrm{T}}(t)\varphi(t)\bar{\varepsilon}(t) + [\tilde{\theta}(t)\varphi(t)]^2$$

$$= T(t-1) + 2\tilde{\theta}^{\mathrm{T}}(t)\varphi(t)\bar{\varepsilon}(t) + [\tilde{\theta}^{\mathrm{T}}(t)\varphi(t)]^2 \qquad (4.C.11)$$

$$- \varphi^{\mathrm{T}}(t)\bar{R}^{-1}(t-1)\varphi(t)\bar{\varepsilon}^2(t).$$

Consider first the two middle terms of the right-hand side:

$$2\tilde{\theta}^{\mathrm{T}}(t)\varphi(t)\bar{\varepsilon}(t) + [\tilde{\theta}^{\mathrm{T}}(t)\varphi(t)]^2$$
$$= \varphi^{\mathrm{T}}(t)\tilde{\theta}(t)\{\varphi^{\mathrm{T}}(t)\tilde{\theta}(t) + 2[\bar{\varepsilon}(t) - e(t)]\} + 2e(t)\varphi^{\mathrm{T}}(t)\tilde{\theta}(t). \qquad (4.C.12)$$

Now we have

$$\mathrm{E}(e(t)\varphi^{\mathrm{T}}(t)\tilde{\theta}(t) \mid \mathscr{F}_{t-1})$$

$$= \mathrm{E}(e(t)\varphi^{\mathrm{T}}(t)\tilde{\theta}(t-1) \mid \mathscr{F}_{t-1}) + \mathrm{E}(e(t)\varphi^{\mathrm{T}}(t)\bar{R}^{-1}(t)\varphi(t)\varepsilon(t) \mid \mathscr{F}_{t-1})$$

$$= \varphi^{\mathrm{T}}(t)\tilde{\theta}(t-1)\mathrm{E}(e(t) \mid \mathscr{F}_{t-1})$$

$$\quad + \varphi^{\mathrm{T}}(t)\bar{R}^{-1}(t)\varphi(t)\mathrm{E}(e^2(t) + e(t)[\varepsilon(t) - e(t)] \mid \mathscr{F}_{t-1})$$

$$= 0 + \varphi^{\mathrm{T}}(t)\bar{R}^{-1}(t)\varphi(t)\sigma^2.$$

In the first equality we used (4.C.4), in the second one that $\varphi(t)$, $\bar{R}(t)$, and $\tilde{\theta}(t-1)$ are $\mathscr{F}_{t-1}$-measurable, and in the third one the properties (4.C.7). For the first term in (4.C.12) we introduce the notation

$$\alpha(t) = -\varphi^{\mathrm{T}}(t)\tilde{\theta}(t),$$
$$\beta(t) = \bar{\varepsilon}(t) - e(t) + \tfrac{1}{2}\varphi^{\mathrm{T}}(t)\tilde{\theta}(t). \qquad (4.C.13)$$

Collecting the expressions (4.C.11)–(4.C.13) now gives

$$\mathrm{E}(T(t) \mid \mathscr{F}_{t-1}) = T(t-1) - 2\mathrm{E}(\alpha(t)\beta(t) \mid \mathscr{F}_{t-1}) + 2\sigma^2\varphi^{\mathrm{T}}(t)\bar{R}^{-1}(t)\varphi(t)$$

$$- \varphi^{\mathrm{T}}(t)\bar{R}^{-1}(t-1)\varphi(t)\mathrm{E}(\bar{\varepsilon}^2(t) \mid \mathscr{F}_{t-1})$$

or

$$\mathrm{E}(T(t) + 2\alpha(t)\beta(t) \mid \mathscr{F}_{t-1}) \le T(t-1) + 2\sigma^2\varphi^{\mathrm{T}}(t)\bar{R}^{-1}(t)\varphi(t).$$

Introduce

$$\tilde{T}(t) = \frac{1}{t}\left[T(t) + 2\sum_1^t \alpha(k)\beta(k)\right] = \tilde{\theta}^{\mathrm{T}}(t)R(t)\tilde{\theta}(t) + \frac{2}{t}\sum_1^t \alpha(k)\beta(k).$$

Then, noting that $\alpha(k) \in \mathscr{F}_{t-1}$ for $k \le t-1$, we have

$$E(t \cdot \tilde{T}(t) \mid \mathscr{F}_{t-1}) \le (t-1)\tilde{T}(t-1) + 2\sigma^2 \varphi^{\mathrm{T}}(t)\bar{R}^{-1}\varphi(t)$$

or

$$E(\tilde{T}(t) \mid \mathscr{F}_{t-1}) \le \tilde{T}(t-1) - \frac{1}{t}\tilde{T}(t-1) + 2\frac{\sigma^2}{t}\varphi^{\mathrm{T}}(t)\bar{R}^{-1}(t)\varphi(t). \quad (4.C.14)$$

We shall now apply lemma 4.C.2 to (4.C.14); in order to do this, we need to establish that $\tilde{T}(t) \ge 0$ and that the last term is summable. These assertions are proven in two lemmas.

LEMMA 4.C.3

$$\sum_{k=1}^{t} \alpha(k)\beta(k) \ge \lambda \cdot \sum_{k=1}^{t} \alpha^2(k),$$

where

$$\lambda = \inf_{-\pi < \omega \le \pi} \left[\operatorname{Re}\left[H(e^{i\omega}) \right]^{-1} - \frac{1}{2} \right].$$

Proof Comparing the definition of α and β in (4.C.13) with (4.C.6) shows that

$$\beta(t) = \frac{1}{H(q^{-1})}\left[-\varphi^{\mathrm{T}}(t)\tilde{\theta}(t) \right] + \frac{1}{2}\varphi^{\mathrm{T}}(t)\tilde{\theta}(t)$$

$$= \left[\frac{1}{H(q^{-1})} - \frac{1}{2} \right]\alpha(t) = \sum_{k=0}^{t} h(t-k)\alpha(k),$$

where

$$\left[\frac{1}{H(q^{-1})} - \frac{1}{2} \right] = \sum_{k=0}^{\infty} h(k)q^{-k} \triangleq \bar{H}(q^{-1}).$$

Thus

$$\sum_{k=1}^{t} \alpha(k)\beta(k) = \sum_{k=1}^{t} \alpha(k) \sum_{s=0}^{k} h(k-s)\alpha(s)$$

$$= \frac{1}{2\pi} \int_{-\pi}^{\pi} \left| \sum_{k=1}^{t} \alpha(k)e^{ik\omega} \right|^2 \cdot \bar{H}(e^{i\omega})d\omega$$

$$= \frac{1}{2\pi} \int_{-\pi}^{\pi} \left| \sum_{k=1}^{t} \alpha(k)e^{ik\omega} \right|^2 \cdot \operatorname{Re}\bar{H}(e^{i\omega})d\omega$$

$$\geq \lambda \cdot \frac{1}{2\pi} \int_{-\pi}^{\pi} \left| \sum_{k=1}^{t} \alpha(k) e^{ik\omega} \right|^2 d\omega$$

$$= \lambda \sum_{k=1}^{t} \alpha(k)^2,$$

which proves the lemma. In the second step, we neglected the exponentially decaying effects from $s \leq 0$. ∎

LEMMA 4.C.4

$$\sum_{n}^{\infty} \frac{1}{t} \varphi^{\mathrm{T}}(t) \bar{R}^{-1}(t) \varphi(t) < \infty \quad \text{w.p.1},$$

where n is such that $\bar{R}(n-1)$ is invertible.

Proof   First note that

$$\bar{R}^{-1}(t)\varphi(t) = \bar{R}^{-1}(t-1)\varphi(t)/[1 + \varphi^{\mathrm{T}}(t)\bar{R}^{-1}(t-1)\varphi(t)]$$

and that

$$\varphi^{\mathrm{T}}(t)\bar{R}^{-1}(t)\bar{R}^{-1}(t-1)\varphi(t) = \mathrm{tr}\, \varphi(t)\varphi^{\mathrm{T}}(t)\bar{R}^{-1}(t)\bar{R}^{-1}(t-1)$$

$$= \mathrm{tr}\,\{[\bar{R}(t) - \bar{R}(t-1)]\bar{R}^{-1}(t)\bar{R}^{-1}(t-1)\} = \mathrm{tr}\,[\bar{R}^{-1}(t-1) - \bar{R}^{-1}(t)].$$

With these, we find

$$\sum_{n}^{\infty} \varphi^{\mathrm{T}}(t)[\bar{R}^{-1}(t)]^2 \varphi(t)$$

$$\leq \sum_{n}^{\infty} \varphi^{\mathrm{T}}(t)\bar{R}^{-1}(t)\bar{R}^{-1}(t-1)\varphi(t)$$

$$= \sum_{n}^{\infty} \mathrm{tr}\,[\bar{R}^{-1}(t-1) - \bar{R}^{-1}(t)]$$ (4.C.15)

$$= \mathrm{tr}\,[\bar{R}^{-1}(n-1) - \bar{R}^{-1}(\infty)] < \infty.$$

We have

$$\varphi^{\mathrm{T}}(t)\bar{R}^{-1}(t)\varphi(t) \leq |\bar{R}(t)| \varphi^{\mathrm{T}}(t)[\bar{R}^{-1}(t)]^2 \varphi(t),$$

where $|\bar{R}(t)|$ is the operator norm of $\bar{R}(t)$. We have

$$|\bar{R}(t)| \leq \mathrm{tr}\, \bar{R}(t) = \sum_{k=1}^{t} |\varphi(k)|^2.$$

Consequently

$$\sum_{n}^{\infty} \frac{1}{t} \varphi^{\mathrm{T}}(t) \bar{R}^{-1}(t) \varphi(t) \le \sum_{n}^{\infty} \frac{\operatorname{tr} \bar{R}(t)}{t} \cdot \varphi^{\mathrm{T}}(t) [\bar{R}^{-1}(t)]^2 \varphi(t) < \infty,$$

according to (4.C.3) and (4.C.15), which proves the lemma. ∎

Lemmas 4.C.3 and 4.C.4 show that lemma 4.C.2 is applicable to (4.C.14) where α_n in (4.C.2) corresponds to $2\sigma^2 \varphi^{\mathrm{T}}(n) \bar{R}^{-1}(n) \varphi(n)/n$, and β_n to $\tilde{T}(n-1)/n$. Hence there exists a nonnegative finite random variable $\tilde{T}$ such that $\tilde{T}(t) \to \tilde{T}$ w.p.1 as $t \to \infty$ and

$$\sum_{1}^{\infty} \frac{1}{t} \tilde{T}(t) < \infty \text{ w.p.1.}$$

The latter condition shows that $\tilde{T} = 0$ w.p.1. Since $\tilde{T}(t)$ is the sum of two positive terms we conclude that

$$\tilde{\theta}^{\mathrm{T}}(t) R(t) \tilde{\theta}(t) \to 0 \text{ w.p.1 as } t \to \infty$$

and

$$\frac{1}{t} \sum_{1}^{t} \alpha(k) \beta(k) \to 0 \text{ w.p.1 as } t \to \infty. \tag{4.C.16}$$

Thus (4.C.9a) has been proven. To see (4.C.9b), we note that lemma 4.C.3 and (4.C.16) imply that

$$\frac{1}{t} \sum_{1}^{t} \alpha^2(k) \to 0 \text{ w.p.1 as } t \to \infty. \tag{4.C.17a}$$

Hence also

$$\frac{1}{t} \sum_{1}^{t} \beta^2(k) \to 0 \text{ w.p.1 as } t \to \infty, \tag{4.C.17b}$$

since $\beta(t)$ is obtained by exponentially stable filtering of $\{\alpha(k)\}$. But

$$\bar{\varepsilon}(t) - e(t) = \beta(t) + \tfrac{1}{2}\alpha(t),$$

and therefore (4.C.9b) follows from (4.C.16) and (4.C.17). Theorem 4.C.1 is now completely proved. ∎

Theorem 4.C.1 applies to general sequences of $\{\varphi(t)\}$ and $\{\varepsilon(t)\}$, related via (4.C.4)–(4.C.7), and is as such more general than theorem 4.6. Let us specialize to the ELS scheme; Landau's output error method will be analogous.

We first note that for ELS

$$\varepsilon(t) = y(t) - \hat{\theta}^{\mathrm{T}}(t-1)\varphi(t), \tag{4.C.18a}$$

$$\varphi^{\mathrm{T}}(t) = (-y(t-1) \;\; \ldots \;\; -y(t-n_a) \quad u(t-1) \;\; \ldots \;\; u(t-n_b),$$
$$\bar{\varepsilon}(t-1) \ldots \bar{\varepsilon}(t-n_c)). \tag{4.C.18b}$$

Then $\bar{\varepsilon}(t)$ given by (4.C.5) is

$$\bar{\varepsilon}(t) = y(t) - \frac{1}{t}\varphi^{\mathrm{T}}(t)R^{-1}(t)\varphi(t)y(t)$$

$$- \varphi^{\mathrm{T}}(t)\hat{\theta}(t-1) + \frac{1}{t}\varphi^{\mathrm{T}}(t)R^{-1}(t)\varphi(t)\hat{\theta}^{\mathrm{T}}(t-1)\varphi(t) \tag{4.C.19}$$

$$= y(t) - \varphi^{\mathrm{T}}(t)\left\{\hat{\theta}(t-1) + \frac{1}{t}R^{-1}(t)\varphi(t)[y(t) - \hat{\theta}^{\mathrm{T}}(t-1)\varphi(t)]\right\}$$

$$= y(t) - \hat{\theta}^{\mathrm{T}}(t)\varphi(t),$$

so $\bar{\varepsilon}(t)$ is the residual, while $\varepsilon(t)$ is the prediction error. We notice that according to lemma 4.2, (4.C.3) will hold when $\varphi(t)$ is given by (4.C.18b), provided

$$\limsup_{N\to\infty}\sum_{1}^{N}[y^2(t) + u^2(t)] < \infty. \tag{4.C.20}$$

We now proceed to verify (4.C.6), which is an analogous, but slightly stronger condition than (4.152)–(4.153). We then assume that the true system indeed can be described by (4.150), i.e., by an ARMAX model of orders less than or equal to those of the model. We can then write [see (4.151)]

$$y(t) = \theta_0^{\mathrm{T}}\varphi_0(t) + e(t),$$

where

$$\varphi_0^{\mathrm{T}}(t) = (-y(t-1) \;\; \ldots \;\; -y(t-n_a) \; u(t-1) \;\; \ldots$$
$$u(t-n_b) \; e(t-1) \;\; \ldots \;\; e(t-n_c)).$$

Then, analogously to the calculations leading to (4.152), we have

$$\bar{\varepsilon}(t) = y(t) - \hat{\theta}^{\mathsf{T}}(t)\varphi(t)$$

$$= \theta_0^{\mathsf{T}}\varphi_0(t) - \hat{\theta}^{\mathsf{T}}(t)\varphi(t) + e(t)$$

$$= \theta_0^{\mathsf{T}}[\varphi_0(t) - \varphi(t)] + [\theta_0 - \hat{\theta}^{\mathsf{T}}(t)]^{\mathsf{T}}\varphi(t) + e(t)$$

$$= [C_0(q^{-1}) - 1][e(t) - \bar{\varepsilon}(t)] + \varphi^{\mathsf{T}}(t)[\theta_0 - \hat{\theta}(t)] + e(t)$$

or

$$C_0(q^{-1})\bar{\varepsilon}(t) = -\varphi^{\mathsf{T}}(t)[\hat{\theta}(t) - \theta_0] + C_0(q^{-1})e(t). \qquad (4.C.21)$$

We have thus established (4.C.6) with $H(q^{-1}) = C_0(q^{-1})$. Conditions (4.C.7a, b) will hold, since $\{e(t)\}$ is a white noise sequence. Finally, condition (4.C.7c) follows from

$$\varepsilon(t) - e(t) = y(t) - \hat{\theta}^{\mathsf{T}}(t-1)\varphi(t) - y(t) + \theta_0^{\mathsf{T}}\varphi_0(t)$$

$$= \theta_0^{\mathsf{T}}\varphi_0(t) - \hat{\theta}^{\mathsf{T}}(t-1)\varphi(t).$$

We now conclude that, provided

$$\mathrm{Re}\left[\frac{1}{C_0(e^{i\omega})} - \frac{1}{2}\right] > 0 \ \forall\omega, \qquad (4.C.22)$$

we have

$$[\hat{\theta}(t) - \theta_0]^{\mathsf{T}} R(t)[\hat{\theta}(t) - \theta_0] \to 0 \text{ w.p.1 as } t \to \infty \qquad (4.C.23a)$$

and

$$\frac{1}{N}\sum_1^N [\bar{\varepsilon}(t) - e(t)]^2 \to 0 \text{ w.p.1 as } N \to \infty. \qquad (4.C.23b)$$

From the latter conclusion it also follows that if

$$R_0(t) = \frac{1}{t}\sum_1^t \varphi_0(k)\varphi_0^{\mathsf{T}}(k),$$

then

$$R(t) - R_0(t) \to 0 \text{ as } t \to \infty. \qquad (4.C.24)$$

Now $R_0(t)$ is defined only in terms of the system variables $\{y(t)\}$, $\{u(t)\}$, and $\{e(t)\}$, and its convergence properties are independent of the estimation procedure. If the model orders and the input are such that no two different models can give the same correct description of the system, then $R_0(t)$ will be bounded from below by a positive definite matrix, and we conclude from (4.C.23a) that $\hat{\theta}(t) \to \theta_0$ w.p.1 as $t \to \infty$.

In the general case we find that under assumption S1 (p. 169) $R_0(t)$ will converge to

$$G(\theta_0) = \bar{E}\varphi_0(t)\varphi_0^T(t),$$

provided this expression exists. Then (4.C.23a) tells us that $\hat{\theta}(t) - \theta_0$ will converge into the null space of $G(\theta_0)$. But

$$\theta \in \{\theta \mid (\theta - \theta_0)^T G(\theta_0) = 0\} \Rightarrow \bar{E}[\varepsilon(t, \theta) - e(t)]^2 = 0, \qquad (4.C.25)$$

which can be seen as follows: Analogously to (4.152) we have

$$C(q^{-1})[\varepsilon(t, \theta) - e(t)] = \varphi_0^T(t)(\theta_0 - \theta),$$

where $C(q^{-1})$ is the polynomial corresponding to the "C-part" of θ. Hence if

$$\bar{E}[\varphi_0^T(t)(\theta_0 - \theta)]^2 = (\theta_0 - \theta)^T G(\theta_0)(\theta_0 - \theta) = 0,$$

we conclude that (4.C.25) holds. We have thus shown that

$$\hat{\theta}(t) \to D_c = \{\theta \mid \bar{E}[\varepsilon(t, \theta) - e(t)]^2 = 0\}. \qquad (4.C.26)$$

Notice the difference between the very similar conclusions (4.C.23b) and (4.C.26)! The result (4.C.26) is the same one as proven by the d.e. approach in section 4.5.2, under essentially the same conditions. Notice, though, that the conclusions (4.C.23) are obtained under weaker assumptions (S1 and the existence of $\bar{E}\varphi_0(t)\varphi_0^T(t)$ are not assumed). We can now summarize the result of the ELS analysis.

Consider the ELS algorithm with $\gamma(t) = 1/t$ and assume that the true data are generated by (4.150), such that (4.C.20) and (4.C.22) hold. Assume that residuals rather than prediction errors are used in the regression vector. Then

$$[\hat{\theta}(t) - \theta_0]^T R_0(t)[\hat{\theta}(t) - \theta_0] \to 0 \text{ w.p.1 as } t \to \infty$$

and

$$\frac{1}{N}\sum_1^N [\bar{\varepsilon}(t) - e(t)]^2 \to 0 \text{ w.p.1 as } N \to \infty.$$

Furthermore, if S1 holds and the limit $\bar{E}\varphi_0(t)\varphi_0^T(t)$ exists, we have that

$$\hat{\theta}(t) \to D_c = \{\theta \mid \bar{E}[\varepsilon(t, \theta) - e(t)]^2 = 0\} \text{ w.p.1 as } t \to \infty.$$

The case where the regression vector is filtered is analogous. Also, the treatment of Landau's output error method will be analogous. However, in this case conditions (4.C.7) limit us to the case where the measurement error $v(t)$ [see (4.168)] is such that $T(q^{-1})v(t)$ is white noise. Thus a less general conclusion is obtained from theorem 4.C.1 than the one derived from theorem 4.6 in section 4.5.

Appendix 4.D Asymptotic Properties of the Symmetric IV Method

We shall in this appendix consider the symmetric IV method, (4.207) and (4.209), only in the case when the generation of $\zeta(t)$ does not depend on the estimates. The associated d.e. is

$$\frac{d}{d\tau}\theta_D(\tau) = R_D^{-1}(\tau)f(\theta_D(\tau)), \tag{4.D.1a}$$

$$\frac{d}{d\tau}R_D(\tau) = G(\theta_D(\tau)) - R_D(\tau), \tag{4.D.1b}$$

where

$$f(\theta) = \bar{E}\zeta(t)[y_F(t) - \varphi_F^T(t)\theta], \tag{4.D.2}$$

$$G(\theta) = \bar{E}\zeta(t)\zeta^T(t).$$

Note that G is in fact independent of θ, since $\zeta(t)$ does not depend on θ. Using the assumption (4.210) about the true system, we can write (4.D.2) as

$$f(\theta) = \bar{E}\zeta(t)[y_F(t) - \varphi_F^T(t)\theta]$$
$$= \bar{E}\zeta(t)\{T(q^{-1})[\varphi^T(t)(\theta_0 - \theta) + v(t)]\}$$
$$= \tilde{G} \times (\theta_0 - \theta) + \bar{E}\zeta(t)v_F(t),$$

where

$$\tilde{G} = \bar{E}\zeta(t)\varphi_F(t). \tag{4.D.3}$$

Now introduce the assumptions

$$\tilde{G} + \tilde{G}^T > 0, \tag{4.D.4a}$$

$$G = \bar{E}\zeta(t)\zeta^T(t) > 0, \tag{4.D.4b}$$

$$\bar{E}\zeta(t)v_F(t) = 0. \tag{4.D.4c}$$

The first assumption is a way of expressing that $\zeta(t)$ should be positively correlated with the gradient $\varphi_F(t)$. (Compare the interpretation of $\zeta(t)$ as an approximate gradient in section 3.6.3!) The last assumption is the usual one, requiring the instrumental variables to be uncorrelated with the noise. The d.e. can then be written

$$\frac{d}{d\tau}\theta_D(\tau) = R_D^{-1}(\tau)\tilde{G}[\theta_0 - \theta_D(\tau)], \tag{4.D.5a}$$

$$\frac{d}{d\tau} R_D(\tau) = G - R_D(\tau). \tag{4.D.5b}$$

To prove stability of this equation, consider the Lyapunov function

$$V(\theta, R) = (\theta - \theta_0)^\mathrm{T} R(\theta - \theta_0).$$

Along trajectories of (4.D.5) we have (see the calculations in the proof of theorem 4.6)

$$\frac{d}{d\tau} V(\theta_D(\tau), R_D(\tau))$$

$$= -[\theta_D(\tau) - \theta_0]^\mathrm{T} [\tilde{G} + \tilde{G}^\mathrm{T} - G + R_D(\tau)][\theta_D(\tau) - \theta_0].$$

From (4.D.5b) we see that $R_D(\tau)$ converges to G independently of θ_D. Hence, in view of (4.D.4a), we have

$$-R_D(\tau) + G < \tilde{G} + \tilde{G}^\mathrm{T}$$

from a certain τ' on. This means that

$$\frac{d}{d\tau} V(\theta_D(\tau), R_D(\tau)) \le 0,$$

with equality only for $\theta_D = \theta_0$. Hence, theorem 4.2 proves that $\hat{\theta}(t)$ converges to θ_0 w.p.1 under the assumptions (4.D.4).

Remark   A perhaps simpler way of seeing the stability of (4.D.5) is to argue as follows. From (4.D.5b) the matrix $R_D(\tau)$ will approach G at an exponential rate (in the τ time scale). Hence the stability of (4.D.5a) is the same as that of the linear time-invariant system

$$\dot{\theta} = G^{-1} \tilde{G}(\theta_0 - \theta). \tag{4.D.6}$$

But G^{-1} and $\tilde{G}$ are positive definite matrices, and the product of two such matrices has its eigenvalues in the right half plane. Hence the stability of (4.D.6) follows.

In the remainder of this appendix we shall confine ourselves to the choice

$$\zeta(t) = T(q^{-1})\tilde{\varphi}(t), \tag{4.D.7}$$

where $\tilde{\varphi}(t)$ is the "noise-free part" of $\varphi(t)$, as defined by (4.231). Recall that this choice of instrumental variables is possible only if we know A_0 and B_0, and should thus be considered as an idealization. We have

$$\varphi(t) = \tilde{\varphi}(t) + \varphi_v(t), \tag{4.D.8}$$

where $\varphi_v(t)$ is a vector depending on v only, and whose last m elements are equal to zero. Since v and $\tilde{\varphi}$ are independent, we obtain

$$\tilde{G} = \bar{\mathrm{E}}\zeta(t)\varphi_F^\mathrm{T}(t)$$

$$= \bar{\mathrm{E}}[T(q^{-1})\tilde{\varphi}(t)][T(q^{-1})\tilde{\varphi}(t)]^\mathrm{T} + \bar{\mathrm{E}}[T(q^{-1})\tilde{\varphi}(t)][T(q^{-1})\varphi_v(t)]^\mathrm{T}$$

$$= \bar{\mathrm{E}}\zeta(t)\zeta^\mathrm{T}(t) = G.$$

Hence in this case the condition (4.D.4a) follows from the much weaker one (4.D.4b). This latter condition can be analyzed in the following way. Let

$$\bar{\alpha} = (\alpha_1 \quad \dots \quad \alpha_{n+m})^\mathrm{T}$$

be an arbitrary vector and consider the quadratic form $\bar{\alpha}^\mathrm{T} G \bar{\alpha}$. By definition G is nonnegative definite. We therefore examine the solutions to

$$0 = \bar{\alpha}^\mathrm{T} G \bar{\alpha} = \bar{\mathrm{E}}[\bar{\alpha}^\mathrm{T}\zeta(t)]^2,$$

which implies that

$$0 = \bar{\alpha}^\mathrm{T}\zeta(t) = \bar{\alpha}^\mathrm{T}\mathscr{S}(-B_0, A_0)\frac{T(q^{-1})}{A(q^{-1})}\begin{pmatrix} u(t-1) \\ \vdots \\ u(t-n-m) \end{pmatrix} \quad \text{w.p.1}$$

[see (4.225), (4.226)]. Provided the polynomials $A_0(z)$, $B_0(z)$ are coprime, the Sylvester matrix $\mathscr{S}(-B_0, A_0)$ is nonsingular. Let $\alpha(q^{-1})$ be defined as

$$\alpha(q^{-1}) = \sum_{i=1}^{n+m} \alpha_i q^{-i}.$$

We then have

$$\frac{\alpha(q^{-1})T(q^{-1})}{A(q^{-1})}u(t) = 0 \quad \text{w.p.1}.$$

If the input is persistently exciting (see lemma 4.7), it follows that $\alpha(q^{-1}) \equiv 0$, i.e., that $\bar{\alpha} = 0$.

These calculations show that the matrix defined in (4.D.4b) is positive definite under mild conditions when $\zeta(t)$ satisfies (4.D.7). It is required only that $A_0(z)$, $B_0(z)$ are coprime and that the input is persistently exciting.

To study the asymptotic distribution of the symmetric IV estimate for the choice (4.D.7) we proceed as follows. Let $\gamma(t) = 1/t$. With

$$\tilde{\theta}(t) = \hat{\theta}(t) - \theta_0, \quad \bar{R}(t) = tR(t),$$

we can rewrite (4.207a, b) as

$$\tilde{\theta}(t) = \tilde{\theta}(t - 1) + \bar{R}^{-1}(t)\zeta(t)[-\varphi_F^{\mathrm{T}}(t)\tilde{\theta}(t - 1) + v_F(t)],$$

$$\bar{R}(t) = \bar{R}(t - 1) + \zeta(t)\zeta^{\mathrm{T}}(t),$$

where we used (4.210). Multiplying the first equation by $\bar{R}(t)$ gives

$$\bar{R}(t)\tilde{\theta}(t) = [\bar{R}(t - 1) + \zeta(t)\zeta^{\mathrm{T}}(t)]\tilde{\theta}(t - 1)$$

$$- \zeta(t)\varphi_F^{\mathrm{T}}(t)\tilde{\theta}(t - 1) + \zeta(t)v_F(t),$$

and hence

$$R(t)\tilde{\theta}(t) = \frac{1}{t}\bar{R}(0)\tilde{\theta}(0) + \frac{1}{t}\sum_1^t \zeta(k)v_F(k)$$

$$+ \frac{1}{t}\sum_1^t \zeta(k)[\zeta(k) - \varphi_F(k)]^{\mathrm{T}}\tilde{\theta}(k - 1).$$

(4.D.9)

This expression is like (4.218), except for the last sum (and the fact that the Rs are different). However, for the choice (4.D.7) we have, according to (4.D.8), that

$$\zeta(k) - \varphi_F(k) = T(q^{-1})\varphi_v(k).$$

The terms of the second sum in (4.D.9) are therefore of the same character as those of the first sum, except that they are multiplied by $\tilde{\theta}(k - 1)$, which we know tends to zero as $k \to \infty$. It may consequently be argued that the first sum of the right-hand side of (4.D.9) dominates the second sum. We leave the question at this heuristic level and tentatively conclude that

$$\tilde{\theta}(t) \approx R^{-1}(t)\frac{1}{t}\sum_1^t \zeta(k)v_F(t)$$

(4.D.10)

for large t. We also have

$$R(t) = \frac{1}{t}\sum_1^t \zeta(k)\zeta^T(k)$$

$$= \frac{1}{t}\sum_1^t \zeta(k)\varphi_F^T(k) + \frac{1}{t}\sum_1^t \zeta(k)[T(q^{-1})\varphi_v(k)]^T,$$

where the second sum tends to zero since $\zeta(k)$ and $\varphi_v(k)$ are independent. Therefore the R matrix in the symmetric IV method will asymptotically coincide with that of the nonsymmetric method (again, assuming (4.D.7)). This, together with (4.D.10) indicates that the asymptotic distribution for the symmetric IV method will coincide with that of the nonsymmetric method for the instrumental variables (4.D.7). This means in particular that the optimal IV accuracy, discussed in section 4.6.3, can also be obtained for a symmetric IV method.

Appendix 4.E   Refined IV Methods and RPE Methods

In this appendix we shall discuss recursive estimation of the parameters of the input-output model (3.103):

$$y(t) = \frac{B(q^{-1})}{F(q^{-1})}u(t) + \frac{C(q^{-1})}{D(q^{-1})}e(t). \tag{4.E.1}$$

In section 3.7 we described the RPE method and a PLR approach to the identification of (4.E.1). The IV method suggests still another approach: To estimate the dynamic part B/F using an IV algorithm and to determine C/D by modeling $y - (B/F)u$ as an ARMA processes. Such a procedure has been suggested and extensively used by Young (1976), who calls it IVAML [AML (approximate maximum likelihood) is a synonym for ELS].

From the analysis in chapter 4 we know that the accuracy of the IV estimates can be improved by carefully selecting the instruments and the prefilter T. Young (1976) and Young and Jakeman (1979) have described a recursive algorithm with prefilter, under the name *refined IVAML*. This algorithm is based on the equation for the stationary point of the likelihood function (assuming Gaussian disturbances). It leads to the prefilter $T(q^{-1}) = H^{-1}(q^{-1})$. We shall first describe this method and then compare it with the RPE method for (4.E.1), as given in section 3.7.2.

The dynamic part of (4.E.1) can be written

$$F(q^{-1})y(t) = B(q^{-1})u(t) + F(q^{-1})v(t), \tag{4.E.2}$$

$$v(t) = \frac{C(q^{-1})}{D(q^{-1})}e(t). \tag{4.E.3}$$

If B, F, C, and D were known polynomials, we could estimate the dynamic part in (4.E.2) using the optimal IV method derived and described in section 4.6.3 and appendix 4.D. This method is described by

$$w(t) = \frac{B(q^{-1})}{F(q^{-1})}u(t), \tag{4.E.4a}$$

$$\eta(t) = (-y(t-1) \ \ldots \ -y(t-n_f) \ u(t-1) \ \ldots \ u(t-n_b))^{\mathrm{T}}, \tag{4.E.4b}$$

$$\varphi_1(t) = (-w(t-1) \ \ldots \ -w(t-n_f) \ u(t-1) \ \ldots \ u(t-n_b))^{\mathrm{T}}, \tag{4.E.4c}$$

$$\zeta(t) = \frac{D(q^{-1})}{C(q^{-1})F(q^{-1})}\varphi_1(t), \tag{4.E.4d}$$

$$\eta_F(t) = \frac{D(q^{-1})}{C(q^{-1})F(q^{-1})}\eta(t), \tag{4.E.4e}$$

$$y_F(t) = \frac{D(q^{-1})}{C(q^{-1})F(q^{-1})} y(t), \tag{4.E.4f}$$

$$\theta_1 = (f_1 \ \ldots \ f_{n_f} \ b_1 \ \ldots \ b_{n_b})^{\mathrm{T}}, \tag{4.E.4g}$$

$$\hat{\theta}_1(t) = \hat{\theta}_1(t-1) + \gamma(t) R_1^{-1}(t)\zeta(t)[y_F(t) - \eta_F^{\mathrm{T}}(t)\hat{\theta}_1(t-1)], \tag{4.E.5a}$$

$$R_1(t) = R_1(t-1) + \gamma(t)[\zeta(t)\zeta^{\mathrm{T}}(t) - R_1(t-1)]. \tag{4.E.5b}$$

Here we have chosen the symmetric IV variant. According to the discussion in appendix 4.D, this will not affect the asymptotic properties of the estimates. Notice that the filter $T = H^{-1}$ is D/CF in this case, according to the assumptions (4.E.2) and (4.E.3).

Similarily, if F and B were known polynomials, we could compute $v(t)$ as

$$v(t) = y(t) - \frac{B(q^{-1})}{F(q^{-1})} u(t), \tag{4.E.6}$$

and apply the RML method to estimate the parameters of the ARMA model (4.E.3) (see sections 2.2.3 and 3.7.2):

$$D(q^{-1})v(t) = C(q^{-1})e(t).$$

Let

$$\theta_2 = (d_1 \ \ldots \ d_{n_d} \ c_1 \ \ldots \ c_{n_c})^{\mathrm{T}}, \tag{4.E.7a}$$

$$\varphi_2(t) = (-v(t-1) \ \ldots \ -v(t-n_d) \ \bar{\varepsilon}(t-1) \ \ldots \ \bar{\varepsilon}(t-n_c))^{\mathrm{T}}, \tag{4.E.7b}$$

$$\bar{\varepsilon}(t) = v(t) - \varphi_2^{\mathrm{T}}(t)\hat{\theta}_2(t), \tag{4.E.7c}$$

$$\psi_2(t) = \frac{1}{\hat{C}_t(q^{-1})} \varphi_2(t). \tag{4.E.7d}$$

Then the RML algorithm is

$$\hat{\theta}_2(t) = \hat{\theta}_2(t-1) + \gamma(t) R_2^{-1}(t)\psi_2(t)[v(t) - \varphi_2^{\mathrm{T}}(t)\hat{\theta}_2(t-1)], \tag{4.E.8a}$$

$$R_2(t) = R_2(t-1) + \gamma(t)[\psi_2(t)\psi_2^{\mathrm{T}}(t) - R_2(t-1)]. \tag{4.E.8b}$$

Now the algorithm as described cannot be implemented, since the polynomials B, C, D, and F in (4.E.4a, d–f) and (4.E.6) are unknown. The obvious solution to this problem is:

Replace F, B, C, and D in (4.E.4) and (4.E.6) by their current estimates according to $\hat{\theta}_1(t)$ and $\hat{\theta}_2(t)$. (4.E.9)

The algorithm (4.E.4)–(4.E.9) is now the refined IVAML algorithm as described by Young and Jakeman (1979). In their paper, they also discuss a variant of (4.E.8) where $\psi_2^{\mathrm{T}}(t)$ in (4.E.8b) is replaced by $\varphi_2^{\mathrm{T}}(t)$, which they call refined AML.

Let us now discuss the RPE method applied to (4.E.1). It is given by (3.123)–(3.125). We can summarize it as follows: Let

$$\theta = \begin{pmatrix} \theta_1 \\ \theta_2 \end{pmatrix}, \quad \varphi(t) = \begin{pmatrix} \varphi_1(t) \\ \varphi_2(t) \end{pmatrix}, \quad \psi(t) = \begin{pmatrix} \zeta(t) \\ \psi_2(t) \end{pmatrix},$$

where θ_i, φ_i, ζ, and ψ_2 are given by (4.E.4)–(4.E.8). [To see that $\zeta(t)$ is indeed the gradient of $\hat{y}(t \mid \theta)$ with respect to θ_1, check (3.119b, c).] The algorithm then is

$$\hat{\theta}(t) = \hat{\theta}(t-1) + \gamma(t)R^{-1}(t)\psi(t)[y(t) - \varphi^{\mathrm{T}}(t)\hat{\theta}(t-1)], \tag{4.E.10a}$$

$$R(t) = R(t-1) + \gamma(t)[\psi(t)\psi^{\mathrm{T}}(t) - R(t-1)]. \tag{4.E.10b}$$

The matrix $R(t)$ will have a block structure:

$$R(t) = \begin{pmatrix} \sum_1^t \beta(t,k)\zeta(k)\zeta^{\mathrm{T}}(k) & \sum_1^t \beta(t,k)\zeta(k)\psi_2^{\mathrm{T}}(k) \\ \sum_1^t \beta(t,k)\psi_2(k)\zeta^{\mathrm{T}}(k) & \sum_1^t \beta(t,k)\psi_2(k)\psi_2^{\mathrm{T}}(k) \end{pmatrix}, \tag{4.E.11}$$

where $\zeta(k)$ is determined entirely from the input u^k (neglecting the vanishing influence of y that leaks over via $\hat{C}$, $\hat{D}$, and $\hat{F}$, as discussed in section 4.6.2). When B and F in (4.E.6) are equal to the true values, φ_2 and ψ_2 will depend entirely on the noise term e. Hence $\zeta(k)$ and $\psi_2(k)$ are asymptotically uncorrelated as the algorithm converges to the true parameter values, provided the input sequence is independent of the noise. This means according to (4.E.11) that the matrix $R(t)$ converges to a block-diagonal matrix. It is therefore reasonable to replace $R(t)$ in (4.E.10) by a block-diagonal version:

$$\bar{R}(t) = \begin{pmatrix} R_1(t) & 0 \\ 0 & R_2(t) \end{pmatrix}.$$

This will, among other things, reduce the computational complexity of (4.E.10). With this modification, the algorithm (4.E.10) can be written

$$\hat{\theta}_1(t) = \hat{\theta}_1(t-1) + \gamma(t)R_1^{-1}(t)\zeta(t)[y(t) - \varphi^{\mathrm{T}}(t)\hat{\theta}(t-1)], \tag{4.E.12a}$$

$$R_1(t) = R_1(t-1) + \gamma(t)[\zeta(t)\zeta^T(t) - R_1(t-1)], \tag{4.E.12b}$$

$$\hat{\theta}_2(t) = \hat{\theta}_2(t-1) + \gamma(t)R_2^{-1}(t)\psi_2(t)[y(t) - \varphi^T(t)\hat{\theta}(t-1)], \tag{4.E.12c}$$

$$R_2(t) = R_2(t-1) + \gamma(t)[\psi_2(t)\psi_2^T(t) - R_2(t-1)]. \tag{4.E.12d}$$

This algorithm very much resembles the refined IVAML scheme (4.E.4)–(4.E.9). In fact, if

$$y(t) - \varphi^T(t)\hat{\theta}(t-1) = y_F(t) - \eta_F^T(t)\hat{\theta}_1(t-1) \tag{4.E.13}$$

and

$$y(t) - \varphi^T(t)\hat{\theta}(t-1) = v(t) - \varphi_2^T(t)\hat{\theta}_2(t-1) \tag{4.E.14}$$

were to hold, the algorithm (4.E.12) would indeed be indentical to (4.E.5), (4.E.8).

Let us study these relationships. From (3.110) and (3.117), we find that

$$y(t) - \varphi^T(t)\hat{\theta}(t-1)$$

$$= \frac{\hat{D}_{t-1}(q^{-1})}{\hat{C}_{t-1}(q^{-1})}\left[y(t) - \frac{\hat{B}_{t-1}(q^{-1})}{\hat{F}_{t-1}(q^{-1})}u(t)\right] \tag{4.E.15}$$

and that

$$v(t) - \varphi_2^T(t)\hat{\theta}_2(t-1) = \frac{\hat{D}_{t-1}(q^{-1})}{\hat{C}_{t-1}(q^{-1})}v(t). \tag{4.E.16}$$

Now using the expressions (4.E.6) and (4.E.9) for $v(t)$, we see that (4.E.14) holds. Moreover, from (4.E.4b, g) we have

$$y(t) - \eta^T(t)\hat{\theta}_1(t-1) = \hat{F}_{t-1}(q^{-1})y(t) - \hat{B}_{t-1}(q^{-1})u(t).$$

Hence

$$y_F(t) - \eta_F^T(t)\hat{\theta}_1(t-1)$$

$$= \frac{\hat{D}_{t-1}(q^{-1})}{\hat{C}_{t-1}(q^{-1})\hat{F}_{t-1}(q^{-1})}[\hat{F}_{t-1}(q^{-1})y(t) - \hat{B}_{t-1}(q^{-1})u(t)]$$

$$= \frac{\hat{D}_{t-1}(q^{-1})}{\hat{C}_{t-1}(q^{-1})}\left[y(t) - \frac{\hat{B}_{t-1}(q^{-1})}{\hat{F}_{t-1}(q^{-1})}u(t)\right],$$

so that (4.E.13) also holds. Therefore, our conclusion is that the refined IVAML algorithm (4.E.4)–(4.E.9) is identical to the RPEM method (4.E.12) with a block-diagonal R-matrix.

It should be said that there may be differences of transient nature in (4.E.13) and (4.E.14), depending on how the time-varying filter operations are ordered.

Our conclusion implies in particular that the convergence result of theorem 4.3 holds for the refined IVAML algorithm. Also, since the block-diagonal approximation of R becomes exact asymptotically, the algorithm is (asymptotically) a true Gauss-Newton scheme, so that the asymptotic-distribution result, theorem 4.5, can also be applied. The expression for the asymptotic covariance matrix of the θ_1 estimate is, according to this theorem and (4.E.11),

$$P_1 = \sigma^2 [\bar{E}\zeta(t)\zeta^T(t)]^{-1}.$$

This result of course coincides with the expression (4.239a) for the optimal IV estimate.

Appendix 4.F   The Associated Differential Equation for the Alternative Gauss-Newton Direction

In the derivation of the Gauss-Newton algorithm in section 3.4, we gave the basic form (3.67) and an algebraically equivalent form (3.70). We also pointed out a variant (3.A.9) that we claimed to be asymptotically equivalent. This claim will be verified in this appendix.

In Theorem 4.2 we determined the d.e. associated with (3.67). When $\hat{\Lambda}(t) = \Lambda$ (constant), it is

$$\dot{\theta} = R^{-1}f(\theta),$$
$$\dot{R} = G(\theta) - R,$$

(4.F.1)

where

$$f(\theta) = \bar{E}\psi(t, \theta)\Lambda^{-1}\varepsilon(t, \theta),$$
$$G(\theta) = \bar{E}\psi(t, \theta)\Lambda^{-1}\psi^{T}(t, \theta).$$

(4.F.2)

Here $\psi(t, \theta)$ is determined by

$$\zeta(t + 1, \theta) = \mathscr{F}(\theta)\zeta(t, \theta) + M(\theta, \varphi(t, \theta), z(t)),$$
$$\psi^{T}(t, \theta) = \mathscr{H}(\theta)\zeta(t, \theta) + D(\theta, \varphi(t, \theta))$$

(4.F.3)

[see (3.22)]. Let us now consider (3.70) with $\lambda(t) \equiv 1$ and $\hat{\Lambda}(t) \equiv \Lambda$. Recall that $P(t) = R^{-1}(t)/t$ [see (3.68)], so the elements of P and L decay as $1/t$. Therefore introduce

$$\tilde{P}(t) = t \cdot P(t), \quad \tilde{L}(t) = t \cdot L(t).$$

(4.F.4)

Now (3.70c–f) can be rewritten as

$$S(t) = \Lambda + O(1/t),$$

(4.F.5a)

$$\tilde{L}(t) = \tilde{P}(t - 1)\psi(t)S^{-1}(t) + O(1/t),$$

(4.F.5b)

$$\tilde{P}(t) = \tilde{P}(t - 1) + \frac{1}{t}[-\tilde{P}(t - 1)\psi(t)S^{-1}(t)\psi^{T}(t)\tilde{P}(t - 1)$$

$$+ \tilde{P}(t - 1)] + O(1/t^{2})$$

(4.F.5c)

$$\hat{\theta}(t) = \hat{\theta}(t - 1) + \frac{1}{t}\tilde{P}(t - 1)\psi(t)S^{-1}(t)\varepsilon(t) + O(1/t^{2}),$$

(4.F.5d)

where $O(\alpha)$ denotes a term that behaves like $|\alpha|$ when $\alpha \to 0$. Comparing with the general algorithm (4.52), x corresponds to $\hat{\theta}$, col $\tilde{P}$. Keeping these variables constant and evaluating the resulting average updating direction in (4.F.5c, d) gives the d.e.

$$\dot{\theta} = \tilde{P}f(\theta),$$

$$\dot{\tilde{P}} = \tilde{P} - \tilde{P}G(\theta)\tilde{P}, \tag{4.F.6}$$

where f and G are given by (4.F.2). Obviously (4.F.6) is equivalent to (4.F.1) with the change of variables $\tilde{P} = R^{-1}$. This is also trivial, since we know that (3.67) and (3.70) were algebraically equivalent to begin with.

Let us now consider the alternative Gauss-Newton algorithm (3.A.9) for $\hat{\Lambda}(t) \equiv \Lambda$ and $\lambda(t) \equiv 1$. As in (3.70) the elements of $\bar{L}$ and $\bar{P}$ will tend to zero as $1/t$. Therefore introduce

$$\tilde{L}(t) = t \cdot \bar{L}(t), \quad \tilde{P}(t) = t \cdot \bar{P}(t), \quad \tilde{\zeta}(t) = t \cdot \bar{\zeta}(t), \quad \tilde{\psi}(t) = t \cdot \bar{\psi}(t).$$

Equations (3.A.9c–h) can now be rewritten as

$$\bar{S}(t) = \Lambda + O(1/t), \tag{4.F.7a}$$

$$\tilde{\zeta}(t+1) = \mathscr{F}_t\tilde{\zeta}(t) + M_t\tilde{P}(t) + O(1/t), \tag{4.F.7b}$$

$$\tilde{\psi}^{\mathsf{T}}(t) = \mathscr{H}_{t-1}\tilde{\zeta}(t) + D_t\tilde{P}(t-1) + O(1/t), \tag{4.F.7c}$$

$$\tilde{L}(t) = \tilde{\psi}(t)\Lambda^{-1} + O(1/t), \tag{4.F.7d}$$

$$\tilde{P}(t) = \tilde{P}(t-1) + \frac{1}{t}[-\tilde{\psi}(t)\Lambda^{-1}\tilde{\psi}^{\mathsf{T}}(t) + \tilde{P}(t-1)] + O(1/t^2), \tag{4.F.7e}$$

$$\hat{\theta}(t) = \hat{\theta}(t-1) + \frac{1}{t}\tilde{\psi}(t)\Lambda^{-1}\varepsilon(t), \tag{4.F.7f}$$

Comparing with the general algorithm (4.52) we find that the estimate x corresponds to $\hat{\theta}$, col $\tilde{P}$. Keeping these variables constant and evaluating the resulting average updating direction in (4.F.7e, f) gives

$$\dot{\theta} = \tilde{f}(\theta, \tilde{P}),$$

$$\dot{\tilde{P}} = \tilde{P} - \tilde{G}(\theta, \tilde{P}), \tag{4.F.8}$$

where

$$\tilde{f}(\theta, \tilde{P}) = \bar{E}\tilde{\psi}(t, \theta, \tilde{P})\Lambda^{-1}\varepsilon(t, \theta) \tag{4.F.9}$$

and

$$\tilde{G}(\theta, \tilde{P}) = \bar{E}\tilde{\psi}(t, \theta, \tilde{P})\Lambda^{-1}\tilde{\psi}(t, \theta, \tilde{P}). \tag{4.F.10}$$

Here $\tilde{\psi}(t, \theta, \tilde{P})$ is the variable that (4.F.7b, c) produces when θ and $\tilde{P}$ are kept constant, i.e.,

$$\tilde{\tilde{\zeta}}(t + 1, \theta, \tilde{\tilde{P}}) = \mathscr{F}(\theta)\tilde{\tilde{\zeta}}(t, \theta, \tilde{\tilde{P}}) + M(\theta, \varphi(t, \theta), z(t))\tilde{\tilde{P}},$$

$$\tilde{\tilde{\psi}}^{\mathrm{T}}(t, \theta, \tilde{\tilde{P}}) = \mathscr{H}(\theta)\tilde{\tilde{\zeta}}(t, \theta, \tilde{\tilde{P}}) + D(\theta, \varphi(t, \theta))\tilde{\tilde{P}}. \qquad (4.\text{F}.11)$$

Comparing this to (4.F.3), we see that

$$\tilde{\tilde{\psi}}(t, \theta, \tilde{\tilde{P}}) = \tilde{\tilde{P}}\psi(t, \theta). \qquad (4.\text{F}.12)$$

Hence

$$\tilde{f}(\theta, \tilde{\tilde{P}}) = \tilde{\tilde{P}}f(\theta),$$

$$\tilde{G}(\theta, \tilde{\tilde{P}}) = \tilde{\tilde{P}}G(\theta)\tilde{\tilde{P}}, \qquad (4.\text{F}.13)$$

where f and G are given by (4.F.2). This means that (4.F.6) and (4.F.8) coincide, and our claim has been proved.

Appendix 4.G The Differential Equation Associated with the Extended Kalman Filter

Let us derive the d.e. associated with the RPE algorithm (3.B.4) (or (3.B.5), which according to appendix 4.F is the same) and the EKF (2.A.1)–(2.A.11).

In both of these algorithms there is a technical complication in the application of theorem 4.1. The filters generating $\hat{x}(t)$ and $W(t)$ in (3.B.4d, g) are not entirely determined by $\hat{\theta}(t)$, as required in (4.52b) ($\hat{\theta}$ corresponds to x; $\hat{x}$ and w to φ). When we ask our question, "What would the average updating directions be, if θ is held constant?" we must therefore first evaluate to what limit, $\bar{K}_\theta$, $K(t)$ in (3.B.2) would tend and then use this constant value for K in (3.B.4d, g). The technical formalities, allowing this slightly more general structure, are justified in Ljung (1979a).

The d.e. associated with (3.B.4) will thus be

$$\dot{\theta} = R^{-1}f(\theta),$$

$$\dot{R} = G(\theta) - R,$$
(4.G.1)

where

$$f(\theta) = \bar{E}\psi(t, \theta)\bar{S}_\theta^{-1}\varepsilon(t, \theta),$$

$$G(\theta) = \bar{E}\psi(t, \theta)\bar{S}_\theta^{-1}\psi^{\mathrm{T}}(t, \theta).$$
(4.G.2)

Here $\varepsilon(t, \theta)$ is defined from (3.B.4a, d, e) in the usual manner, and $\psi(t, \theta)$ is obtained from (3.B.4f h) as

$$W(t + 1, \theta) = [F(\theta) - \bar{K}_\theta H(\theta)] W(t, \theta)$$

$$+ M(\theta, \hat{x}(t, \theta), u(t)) + \mathcal{K}_\theta\varepsilon(t, \theta),$$
(4.G.3a)

$$\psi(t, \theta) = W^{\mathrm{T}}(t, \theta)H^{\mathrm{T}}(\theta) + D^{\mathrm{T}}(\theta, \hat{x}(t, \theta)),$$
(4.G.3b)

and $\bar{S}_\theta$ and $\bar{K}_\theta$ are the steady-state solutions defined in (3.136), and $\mathcal{K}_\theta$ is the limit to which $\mathcal{K}_t$ in (3.B.3) converges for a fixed $\hat{\theta}(t) = \theta$. By construction,

$$\mathcal{K}_\theta\varepsilon = \frac{\partial}{\partial\theta}\bar{K}_\theta\varepsilon.$$
(4.G.4)

Therefore $\psi(t, \theta)$ will obey

$$\psi^{\mathrm{T}}(t, \theta) = \frac{d}{d\theta}\hat{y}(t \mid \theta),$$
(4.G.5)

which, of course was the objective of the algorithm (3.B.4). This also means that the d.e. (4.G.1) have the stability properties of theorem 4.3. The general convergence results on RPE algorithms consequently apply also to (3.B.1)–(3.B.4). Notice that we can alternatively use (3.B.5).

Now, as we found in Appendix 3.C, the only, nontransient difference between (3.B.5) and the EKF (2.A.1)–(2.A.11) is the missing term $\mathcal{K}_t \varepsilon(t)$. Hence, the d.e. associated with the EKF will be given by (4.G.1)–(4.G.3) *with the exception that the last term in (4.G.3a) is deleted.* Exclusion of this term may, however, result in loss of the nice convergence properties, as shown in example 4.1 in Ljung (1979a).

Appendix 6.A   A FORTRAN Subroutine for RPE Identification

In this appendix we give the code for a FORTRAN subroutine which performs RPE identification. In order to not complicate the code unnecessarily, we choose a SISO model set

$$A(q^{-1})y(t) = B(q^{-1})u(t) + C(q^{-1})e(t), \qquad (6.A.1)$$

with

$$A(q^{-1}) = 1 + a_1 q^{-1} + \cdots + a_n q^{-n},$$
$$B(q^{-1}) = b_1 q^{-1} + \cdots + b_n q^{-n}, \qquad (6.A.2)$$
$$C(q^{-1}) = 1 + c_1 q^{-1} + \cdots + c_n q^{-n}.$$

Extensions to cases with different degrees of the polynomials or the more general model structure, (3.104) are straightforward. However, more administration will be needed in the subroutine, since the amount of old data necessary to save for the filtering procedures, is then no longer identical to the vectors $\varphi(t)$, (3.124d), and $\psi(t)$, (3.125f). With the model set (6.A.1), the administrative parts of the program become very simple.

The subroutine is written so that the basic algorithm (6.1)–(6.4) is iterated one step. This means that a new call to the subroutine must be done at every new sampling point. The U-D algorithm described in section 6.2 is used.

It should be noted that the algorithm can be made still more efficient. To make the description clear both the normalized gain vector $\bar{L}(t)$ [see (6.15)] and the usual gain vector $L(t)$ are computed. This is, of course, not necessary.

In table 6.A.1 we give a list of the variables involved in RPEM and the corresponding notations used in the book.

Notice that the choice $K = 0$ gives the PLR algorithm for (6.A.1), i.e., ELS.

The variable V shown in table 6.A.1 is used to estimate the loss function. It is taken as

$$V(t) = \sum_{s=1}^{t} \frac{\varepsilon^2(s)}{\lambda(s) + \varphi^T(s)P(s-1)\varphi(s)}. \qquad (6.A.3)$$

For $\lambda(s) \equiv 1$ this expression gives the exact minimal loss for the least squares case (see lemma 3.D.1).

The subroutine has been implemented on a NORD100 computer. The CPU time for floating-point addition is 8 μsec. Typical CPU times

Table 6.A.1
Notation in subroutine RPEM and in the book.

Variable in RPEM	Notation in the book
THETA	$\hat{\theta}$
P	P
N	n
U	$u(t)$
Y	$y(t)$
LAMBDA	$\lambda(t)$
K	K see ex. 5.17
C	C see (6.111)
V	—
EPS	$\varepsilon(t)$
EPS1	$\bar{\varepsilon}(t)$
FI	$\varphi(t)$
PSI	$\psi(t)$
L	$L(t)$
AMY	μ see (6.112)
Y1	$\tilde{y}(t)$ see (3.125a)
U1	$\tilde{u}(t)$ see (3.125b)
E1	$\tilde{\varepsilon}(t)$ see (3.125d)

for one call to the subroutine is 4.6 msec for a first-order model and 8.7 msec for a second-order model.

The subroutine contains the following seven steps.

1. *Initialization*, lines 81–94. If the integer INIT is given a nonzero value (typically at the first call to the subroutine), then the variables are initialized as

$(t = 0)$

$\varphi(0) = 0 \quad \psi(0) = 0$

$\hat{\theta}(0) = 0 \quad L(0) = 0 \quad P(0) = 0$

$V = 0.$

2. *Computation of the prediction error*, lines 96–100. This computation is given by (3.123a) and (3.124e).

3. *Updating the parameter estimates*, lines 102–126. When the integer ISTAB1 is nonzero, monitoring is performed. Then the step length is

reduced so that $C(z)$ has all zeros outside the unit circle. The algorithm (6.112) for this is contained in the lines 108–121. The stability test on line 117 is given in a separate subroutin NSTABL. This routine is based on the Schur-Cohn algorithm as given by Kucera (1980). The updating (6.3) of the parameter estimates are then performed in lines 125, 126.

4. *Computation of the residuals*, lines 128–132. The equation (3.124c) is used here.

5. *Computation of filtered signals*, lines 134–144. Here the equation (3.125a, b, d) are implemented.

6. *Updating the vectors $\varphi(t)$ and $\psi(t)$*, lines 146–164. These computations are quite straightforward.

7. *Computation of the gain vector and updating of $P(t)$ and V*, lines 166–203. Here the FORTRAN mechanization, given by Thornton and Bierman (1980), of the *U-D* algorithm is implemented. The computations are described in section 6.2.

```
 1*          SUBROUTINE  RPEM(THETA,P,N,U,Y,LAMBDA,K,C,ISTAB1,
 2*         +          ISTAB2,V,EPS,EPS1,INIT,P0,IDIM)
 3*   C
 4*   C     RECURSIVE PREDICTION ERROR   METHOD
 5*   C
 6*   C     THE SUBROUTINE PERFORMS THE MODIFICATION OF THE PARAMETER
 7*   C     ESTIMATES THETA FOR ONE SAMPLING INTERVAL.
 8*   C     A NEW  CALL TO RPEM MUST BE MADE FOR EVERY
 9*   C     NEW SAMPLING INTERVAL
10*   C
11*   C
12*   C
13*   C     MODEL STRUCTURE USED
14*   C
15*   C       -1          -1          -1
16*   C     A(Q )Y(T) =  B(Q )U(T) + C(Q )E(T)
17*   C
18*   C     ***  DESCRIPTION OF PARAMETERS  ***
19*   C
20*   C     THETA - VECTOR OF ORDER (3 * N) CONTAINING
21*   C             THE PARAMETER  ESTIMATES
22*   C             THETA=  (A(1) ... A(N),B(1) ... B(N),C(1) ... C(N))
23*   C             THETA IS CHANGED IN THE SUBROUTINE
24*   C     P     - SYMMETRIC MATRIX OF ORDER  (3 * N)
25*   C             P IS USED IN THE U-D FORM
26*   C             P = U*D*U(TRANSPOSED)
27*   C             WITH D DIAGONAL AND U UPPER TRIANGULAR
28*   C             THE ELEMENTS OF D ARE STORED IN THE DIAGONAL OF P
29*   C             THE ELEMENTS OF U ARE STORED IN
30*   C             THE UPPER TRIANGULAR PART OF P
31*   C             P IS CHANGED IN THE SUBROUTINE
32*   C     N     - MODEL ORDER    (MIN 1, MAX 10)
33*   C     U     - THE LAST INPUT VALUE
34*   C     Y     - THE LAST OUTPUT VALUE
35*   C     LAMBDA- THE FORGETTING FACTOR   (TO BE ENTERED)
36*   C     K     - THE CONTRACTION FACTOR USED FOR FILTERING OF
37*   C             THE DATA  (TO BE ENTERED)
38*   C             COMMENTS:
39*   C             FOR REASONABLE RESULTS
40*   C             0.LT. LAMBDA .LE.1    LAMBDA CLOSE TO 1 AFTER MANY
41*   C             CALLS TO RPEM
42*   C             0.LT. K .LE.1   K CLOSE TO 1 AFTER MANY CALLS TO RPEM
43*   C     C     - PARAMETER USED FOR THE REGULARISATION
44*   C             C SHOULD BE CHOSEN RATHER LARGE
45*   C     ISTAB1- FLAG (TO BE ENTERED) FOR STABILITY TESTS OF C(Z).
46*   C             IF  ISTAB1=0  NO MONITORING (STABILITY TEST
47*   C             AND STEP SIZE REDUCTIONS) IS PERFORMED
48*   C             IF  ISTAB1 .NE.0  MONITORING (STABILITY TEST AND
49*   C             POSSIBLY STEP SIZE REDUCTION) IS PERFORMED
50*   C     ISTAB2- INTEGER AT RETURN GIVING THE NUMBER OF STEP SIZE
51*   C             REDUCTIONS PERFORMED.  IF ISTAB1=0
52*   C             THEN ISTAB2 IS NOT SIGNIFICANT
53*   C     V     - LOSS FUNCTION - SUM OF SQUARED PREDICTION ERRORS.
54*   C             MODIFICATION DUE TO UNCERTAINTIES IN THE
55*   C             TRANSIENT PHASE IS INCLUDED
56*   C             V IS CHANGED IN THE SUBROUTINE
57*   C     EPS   - THE PREDICTION ERROR (GIVEN AT RETURN)
58*   C     EPS1  - THE RESIDUAL       (GIVEN AT RETURN)
```

```
59*  C      INIT  - FLAG TO BE USED FOR STARTING THE RECURSION
60*  C              IF  INIT=0  ALL PARAMETERS ARE UPDATED
61*  C              IF  INIT.NE.0  APPROPRIATE INITIAL VALUES ARE
62*  C                            FIRST SET. THEN THE PARAMETERS ARE
63*  C                            UPDATED USING THE AVAILABLE DATA U,Y
64*  C      PO    - SCALAR PARAMETER USED TO GIVE P AN INITIAL
65*  C              VALUE  (TO BE ENTERED WHEN INIT.NE.0).
66*  C              IF  INIT.NE.0  P= PO*UNIT MATRIX
67*  C      IDIM  - DIMENSION PARAMETER
68*  C
69*  C
70*  C      SUBROUTINE REQUIRED
71*  C              NSTABL
72*  C
73*  C
74*         DIMENSION  THETA(1),P(IDIM,1),R(1)
75*         DIMENSION  FI(30),PSI(30),TSTAB(11),WORK(22),F(30),G(30)
76*         REAL  LAMBDA,K,L(30)
77*  C
78*         IF (N.LT.1 .OR. N.GT.10) RETURN
79*         NN= N*3
80*  C
81*  C      *****  TEST FOR INITIALIZATION  *****
82*  C
83*         IF (INIT.EQ.0)  GO TO 100
84*  C
85*         V=0.
86*         DO 10 I=1,NN
87*         DO 10 J=1,NN
88*      10 P(I,J)= 0.
89*         DO 20 I=1,NN
90*         P(I,I)= PO
91*         THETA(I)= 0.
92*         L(I)= 0.
93*         FI(I)= 0.
94*      20 PSI(I)= 0.
95*  C
96*  C      *****  COMPUTE PREDICTION ERROR  *****
97*  C
98*     100 EPS= Y
99*         DO 110 I=1,NN
100*    110 EPS= EPS - FI(I)*THETA(I)
101* C
102* C      *****  COMPUTE NEW PARAMETER ESTIMATES  *****
103* C
104*        AMY=1.
105* C
106* C      *** TEST FOR NEED OF MONITORING ***
107* C
108*        IF(ISTAB1.EQ.0) GO TO 200
109*        ISTAB2=0
110*    120 DO 130 I=1,N
111*        NI=2*N+I
112*    130 TSTAB(I+1)= THETA(NI)+L(NI)*EPS*AMY
113*        TSTAB(1)=1.
114* C
```

```
115*    C       *** TEST FOR STABILITY OF C(Z) ***
116*    C
117*            CALL NSTABL(TSTAB,N,WORK,IST)
118*            IF(IST.EQ.0) GO TO 200
119*            AMY=AMY/2.
120*            ISTAB2=ISTAB2+1
121*            GO TO 120
122*    C
123*    C       *** UPDATE PARAMETER ESTIMATES ***
124*    C
125*      200 DO 210 I=1,NN
126*      210 THETA(I)= THETA(I) + L(I)*EPS*AMY
127*    C
128*    C       *****   COMPUTE RESIDUALS   *****
129*    C
130*            EPS1= Y
131*            DO 220 I=1,NN
132*      220 EPS1= EPS1 - FI(I)*THETA(I)
133*    C
134*    C       ***** COMPUTE FILTERED SIGNALS Y1,U1,E1 *****
135*    C
136*            Y1=Y
137*            U1=U
138*            E1=EPS1
139*    C
140*            DO 620 I=1,N
141*            CI=THETA(2*N+I)*K**I
142*            Y1=Y1+CI*PSI(I)
143*            U1=U1-CI*PSI(N+I)
144*      620 E1=E1-CI*PSI(N*2+I)
145*    C
146*    C       ***** UPDATE VECTORS FI AND PSI *****
147*    C
148*            IF(N.EQ.1) GO TO 720
149*            DO 700 J=2,N
150*            I=N+2-J
151*            FI(I)=FI(I-1)
152*            PSI(I)=PSI(I-1)
153*            I=2*N+2-J
154*            FI(I)=FI(I-1)
155*            PSI(I)=PSI(I-1)
156*            I=3*N+2-J
157*            FI(I)=FI(I-1)
158*      700 PSI(I)=PSI(I-1)
159*      720 FI(1)=-Y
160*            PSI(1)=-Y1
161*            FI(N+1)=U
162*            PSI(N+1)=U1
163*            FI(2*N+1)=EPS1
164*            PSI(2*N+1)=E1
165*    C
```

```
166*   C       *****   COMPUTE GAIN VECTOR L, UPDATE P AND V *****
167*   C
168*           DO 810 I=2,NN
169*           J=NN+2-I
170*           ALFA=PSI(J)
171*           J1=J-1
172*           DO 800 KK=1,J1
173*     800 ALFA=ALFA+P(KK,J)*PSI(KK)
174*           F(J)=ALFA
175*     810 G(J)=P(J,J)*ALFA
176*           G(1)=P(1,1)*PSI(1)
177*           F(1)=PSI(1)
178*   C
179*           ALFA=LAMBDA+F(1)*G(1)
180*           GAMMA=0.
181*           IF(ALFA.GT.0.) GAMMA=1./ALFA
182*           IF(G(1).NE.0.) P(1,1)=GAMMA*P(1,1)
183*   C
184*           DO 830 J=2,NN
185*           BETA=ALFA
186*           DD=G(J)
187*           ALFA=ALFA+DD*F(J)
188*           IF(ALFA.EQ.0.) GO TO 830
189*           AL=-F(J)*GAMMA
190*   C
191*           J1=J-1
192*           DO 820 I=1,J1
193*           S=P(I,J)
194*           P(I,J)=S+AL*G(I)
195*     820 G(I)=G(I)+DD*S
196*           GAMMA=1./ALFA
197*           P(J,J)=BETA*GAMMA*P(J,J)/LAMBDA
198*           P(J,J)=AMIN1(P(J,J),C)
199*     830 CONTINUE
200*   C
201*           V=V+EPS**2/ALFA
202*           DO 840 I=1,NN
203*     840 L(I)=G(I)/ALFA
204*   C
205*   C       *****   END OF COMPUTATIONS   *****
206*   C
207*           RETURN
208*           END
```

```
 1*          SUBROUTINE NSTABL(A,N,W,IST)
 2*    C
 3*    C     TEST FOR STABILITY
 4*    C
 5*    C
 6*    C     REFERENCE V. KUCERA: DISCRETE LINEAR CONTROL, 1980, P. 153
 7*    C
 8*    C     A  - VECTOR OF ORDER N+1 CORRESPONDING TO THE POLYNOMIAL
 9*    C          A(Z)=A(1)*Z**N+A(2)**(N-1)+ ... +A(N+1)
10*    C     N  - ORDER OF THE POLYNOMIAL (MIN 0, NO MAX)
11*    C     W  - WORKING ARRAY OF ORDER 2*N+2
12*    C     IST- INTEGER AT RETURN SHOWING THE STABILITY OF A(Z)
13*    C          IF IST=0 THEN A(Z) HAS ALL ZEROS STRICTY INSIDE THE UNIT CIRCLE
14*    C          IF IST=1 THEN A(Z) HAS AT LEAST ONE ZERO
15*    C          ON OR OUTSIDE THE UNIT CIRCLE
16*    C
17*    C     SUBROUTINES REQUIRED
18*    C             NONE
19*    C
20*    C
21*          DIMENSION A(1).W(1)
22*    C
23*    C
24*          IST=1
25*          N1=N+1
26*          DO 1 I=1,N1
27*          W(I)=A(I)
28*        1 W(N1+I)=0.
29*    C
30*          K=0
31*       10 IF (K.EQ.N) GO TO 99
32*          NK1=N-K+1
33*          DO 11 J=1,NK1
34*       11 W(N1+J)=W(NK1-J+1)
35*          IF(W(N1+NK1).EQ.0.) GO TO 98
36*          AL=W(NK1)/W(N1+NK1)
37*          IF(ABS(AL).GE.1.0) GO TO 98
38*          NK=N-K
39*          DO 12 J=1,NK
40*       12 W(J)=W(J)-AL*W(N1+J)
41*          K=K+1
42*          GO TO 10
43*    C
44*       98 RETURN
45*       99 IST=0
46*          RETURN
47*          END
```

Appendix 6.B   Derivation of a Fast Algorithm for Gain Calculation

In this appendix we derive the algorithm (6.59). For easy reference we have marked those equations with an asterisk * that are important for the final algorithm, and not just intermediate results. In addition, the equations marked with a dagger [†] will be used in the derivation of the fast ladder algorithm in appendix 6.C. The variables introduced in (6.47)–(6.58) will be crucial for the derivation.

6.B.1 Updating $L(t)$

We now suppose that we know $L(t)$. As in (6.49) we find that

$$\bar{R}^*(t)\begin{pmatrix} 0 \\ L(t) \end{pmatrix} = \begin{pmatrix} \hat{x}(t) \\ \psi(t) \end{pmatrix}, \tag{6.B.1}$$

where the α-vector $\hat{x}(t)$ is given by

$$\hat{x}(t) = \left[\sum_{k=1}^{t} \lambda^{t-k} x(k)\psi^{\mathrm{T}}(k) \right] L(t).$$

Using that

$$L(t) = \bar{R}^{-1}(t)\psi(t)$$

and comparing with (6.54), we find that

$$\hat{x}(t) = -A^{\mathrm{T}}(t)\psi(t). \tag{6.B.2}$$

This expression, can, according to (6.55) be interpreted as a prediction of $x(t)$, which explains the notation.

Now in order to determine $L^*(t)$ in (6.51), i.e.,

$$\bar{R}^*(t)L^*(t) = \psi^*(t) = \begin{pmatrix} x(t) \\ \psi(t) \end{pmatrix}, \tag{6.B.3}$$

we need just replace $\hat{x}(t)$ in (6.B.1) by $x(t)$. This we can accomplish with

$$\begin{pmatrix} I \\ A(t) \end{pmatrix},$$

that according to (6.52) can operate on the first α rows on the right-hand side, without affecting the $n\alpha$ bottom rows. In fact, with

$$\bar{e}(t) \triangleq e'(t) = x(t) - \hat{x}(t), \tag{6.B.4}$$

we have

$$\bar{R}^*(t)\left[\begin{pmatrix} 0 \\ L(t) \end{pmatrix} + \begin{pmatrix} I \\ A(t) \end{pmatrix}[R^e(t)]^{-1}\bar{e}(t)\right]$$

$$= \begin{pmatrix} \hat{x}(t) \\ \psi(t) \end{pmatrix} + \begin{pmatrix} R^e(t) \\ 0 \end{pmatrix}[R^e(t)]^{-1}\bar{e}(t)$$

$$= \begin{pmatrix} \hat{x}(t) + \bar{e}(t) \\ \psi(t) \end{pmatrix} = \begin{pmatrix} x(t) \\ \psi(t) \end{pmatrix} = \psi^*(t).$$

A comparison with (6.B.3) now shows that we must have

$$L^*(t) = \begin{pmatrix} [R^e(t)]^{-1}\bar{e}(t) \\ L(t) + A(t)[R^e(t)]^{-1}\bar{e}(t) \end{pmatrix}. \qquad (6.B.5)^*$$

We have now gone from $L(t)$ to $L^*(t)$. In order to go from $L^*(t)$ to $L(t+1)$, we first make another decomposition of $L^*(t)$, where we single out the last α elements:

$$L^*(t) = \begin{pmatrix} M(t) \\ \hline \mu(t) \end{pmatrix} \updownarrow \alpha \text{ rows}. \qquad (6.B.6)^*$$

We have

$$\bar{R}^*(t)\begin{pmatrix} M(t) \\ \mu(t) \end{pmatrix} = \begin{pmatrix} \psi(t+1) \\ x(t-n) \end{pmatrix}. \qquad (6.B.7)$$

We would like to obtain (6.50); i.e., we would like to obtain

$$\bar{R}^*(t)\begin{pmatrix} L(t+1) \\ 0 \end{pmatrix} = \begin{pmatrix} \psi(t+1) \\ * \end{pmatrix} \qquad (6.B.8)$$

for some α-vector denoted here by $*$. In order to achieve this we have to remove the last α rows of the left-hand side of (6.B.7) without affecting the upper $n\alpha$ rows of the right-hand side. This can be done using

$$\begin{pmatrix} B(t) \\ I \end{pmatrix}.$$

We have

$$\bar{R}^*(t)\left[\begin{pmatrix} M(t) \\ \mu(t) \end{pmatrix} - \begin{pmatrix} B(t) \\ I \end{pmatrix}\mu(t)\right]$$

$$= \begin{pmatrix} \psi(t+1) \\ x(t-n) \end{pmatrix} - \begin{pmatrix} 0 \\ R^r(t) \end{pmatrix} \mu(t) = \begin{pmatrix} \psi(t+1) \\ x(t-n) - R^r(t)\mu(t) \end{pmatrix}.$$

Comparing this to (6.B.8) we find that

$$L(t+1) = M(t) - B(t)\mu(t). \tag{6.B.9}*$$

With (6.B.5), (6.B.8), and (6.B.9) we have now gone from $L(t)$ to $L(t+1)$.

6.B.2 Updating $A(t)$ and $R^e(t)$

Suppose that we have $A(t-1)$ available, i.e., suppose we have

$$\bar{R}^*(t-1)\begin{pmatrix} I \\ A(t-1) \end{pmatrix} = \begin{pmatrix} R^e(t-1) \\ 0 \end{pmatrix}. \tag{6.B.10}$$

Since from (6.47),

$$\bar{R}^*(t) = \lambda \bar{R}^*(t-1) + \psi^*(t)(\psi^*(t))^T, \tag{6.B.11}$$

we also have

$$[\psi^*(t)]^T\begin{pmatrix} I \\ A(t-1) \end{pmatrix} = x^T(t) + \psi^T(t)A(t-1).$$

Hence,

$$\bar{R}^*(t)\begin{pmatrix} I \\ A(t-1) \end{pmatrix}$$

$$= \lambda \bar{R}^*(t-1)\begin{pmatrix} I \\ A(t-1) \end{pmatrix} + \psi^*(t)[x^T(t) + \psi^T(t)A(t-1)] \tag{6.B.12}$$

$$= \begin{pmatrix} \lambda R^e(t-1) \\ 0 \end{pmatrix} + \psi^*(t)e^T(t),$$

where we have introduced the notation

$$e(t) \triangleq e^{t-1}(t) = x(t) + A^T(t-1)\psi(t). \tag{6.B.13}*$$

We may view $e(t)$ as the prediction error associated with the linear regression (6.55) [see (6.56b)].

From (6.B.12) we see that we need to remove the last $n\alpha$ rows of the last term. This can be achieved with the matrix

$$\begin{pmatrix} 0 \\ L(t) \end{pmatrix}.$$

In fact, using (6.B.1) we have

$$\bar{R}^*(t) \begin{pmatrix} 0 \\ L(t) \end{pmatrix} e^T(t) = \begin{pmatrix} \hat{x}(t) \\ \psi(t) \end{pmatrix} e^T(t). \tag{6.B.14}$$

Consequently, from (6.B.12) and (6.B.14) we obtain

$$\bar{R}^*(t) \left[\begin{pmatrix} I \\ A(t-1) \end{pmatrix} - \begin{pmatrix} 0 \\ L(t) \end{pmatrix} e^T(t) \right]$$

$$= \begin{pmatrix} \lambda R^e(t-1) \\ 0 \end{pmatrix} + \begin{pmatrix} x(t) \\ \psi(t) \end{pmatrix} e^T(t) - \begin{pmatrix} \hat{x}(t) \\ \psi(t) \end{pmatrix} e^T(t)$$

$$= \begin{pmatrix} \lambda R^e(t-1) + [x(t) - \hat{x}(t)]e^T(t) \\ 0 \end{pmatrix}.$$

Comparing this to (6.52), we find that we have

$$A(t) = A(t-1) - L(t)e^T(t) \tag{6.B.15}*$$

and

$$R^e(t) = \lambda R^e(t-1) + \bar{e}(t)e^T(t), \tag{6.B.16}^\dagger$$

where we have used the definition (6.B.4) of $\bar{e}(t)$.

There is a useful simple relationship between $\bar{e}(t)$ and $e(t)$. We have from (6.B.15)

$$\bar{e}(t) = x(t) + A^T(t)\psi(t)$$

$$= x(t) + [A(t-1) - L(t)e^T(t)]^T\psi(t)$$

$$= x(t) + A^T(t-1)\psi(t) - e(t)L^T(t)\psi(t)$$

$$= e(t)[1 - L^T(t)\psi(t)].$$

Let us introduce the notation

$$\beta(t) = L^T(t)\psi(t). \tag{6.B.17}*^\dagger$$

Using this scalar, we then have

$$\bar{e}(t) = e(t)[1 - \beta(t)] \tag{6.B.18}*^\dagger$$

and

$$R^e(t) = \lambda R^e(t-1) + [1 - \beta(t)]e(t)e^T(t). \tag{6.B.19}*$$

With (6.B.15), (6.B.17), and (6.B.19) we have thus updated $A(t)$ and $R^e(t)$.

6.B.3 Updating $B(t)$

Assume that $B(t-1)$ is known, so that we can compute

$$\bar{R}^*(t-1)\begin{pmatrix} B(t-1) \\ I \end{pmatrix} = \begin{pmatrix} 0 \\ R^r(t-1) \end{pmatrix}. \tag{6.B.20}$$

We then have from (6.B.11)

$$\bar{R}^*(t)\begin{pmatrix} B(t-1) \\ I \end{pmatrix}$$

$$= \begin{pmatrix} 0 \\ \lambda R^r(t-1) \end{pmatrix} + \psi^*(t)[\psi^T(t+1)B(t-1) + x^T(t-n)].$$

We introduce the notation

$$r(t) = r^{t-1}(t) = x(t-n) + B^T(t-1)\psi(t+1). \tag{6.B.21}*†$$

Then we have

$$\bar{R}^*(t)\begin{pmatrix} B(t-1) \\ I \end{pmatrix} = \begin{pmatrix} 0 \\ \lambda R^r(t-1) \end{pmatrix} + \begin{pmatrix} \psi(t+1) \\ x(t-n) \end{pmatrix} r^T(t). \tag{6.B.22a}$$

We would like to eliminate the first $n\alpha$ rows of the right-hand side. We first note that

$$\bar{R}^*(t)\begin{pmatrix} L(t+1) \\ 0 \end{pmatrix} = \begin{pmatrix} \psi(t+1) \\ \check{x}(t-n) \end{pmatrix} \tag{6.B.22b}$$

[see (6.50)], where

$$\check{x}(t-n) = \left[\sum_{k=1}^{t} \lambda^{t-k}x(k-n)\psi^T(k+1)\right]L(t+1). \tag{6.B.22c}$$

Now using that

$$L(t+1) = R^*(t+1)^{-1}\psi(t+1)$$

and recalling the interpretation of $B(t)$, we find that

$$\check{x}(t - n) = -B^T(t)\psi(t + 1),$$

which is the backward prediction of $x(t - n)$, given $\psi(t + 1)$. Also introduce

$$\bar{r}(t) = x(t - n) - \check{x}(t - n) = x(t - n) + B^T(t)\psi(t + 1).$$

We then obtain

$$\bar{R}^*(t)\left[\begin{pmatrix} B(t - 1) \\ I \end{pmatrix} - \begin{pmatrix} L(t + 1) \\ 0 \end{pmatrix} r^T(t)\right]$$

$$= \begin{pmatrix} 0 \\ \lambda R^r(t - 1) \end{pmatrix} + \begin{pmatrix} \psi(t + 1) \\ x(t - n) \end{pmatrix} r^T(t) - \begin{pmatrix} \psi(t + 1) \\ \check{x}(t - n) \end{pmatrix} r^T(t)$$

$$= \begin{pmatrix} 0 \\ \lambda R^r(t - 1) + [x(t - n) - \check{x}(t - n)]r^T(t) \end{pmatrix}.$$

Comparing this to (6.53), we find that

$$B(t) = B(t - 1) - L(t + 1)r^T(t) \qquad\qquad\qquad (6.B.23)$$

and

$$R^r(t) = \lambda R^r(t - 1) + \bar{r}(t)r^T(t). \qquad\qquad\qquad (6.B.24)^\dagger$$

Now the equations (6.B.9) and (6.B.23) are in fact a system of equations for solving for $L(t + 1)$ and $B(t)$. Substituting (6.B.9) into (6.B.23) gives

$$B(t)[I - \mu(t)r^T(t)] = B(t - 1) - M(t)r^T(t)$$

or

$$B(t) = [B(t - 1) - M(t)r^T(t)][I - \mu(t)r^T(t)]^{-1}. \qquad (6.B.25)^*$$

With (6.B.25) we have now also updated $B(t)$.

With this the derivation is completed. The equations marked with an asterisk * constitute the algorithm as summarized in (6.59).

Appendix 6.C   Derivation of the Fast Ladder Algorithm

In this appendix we shall derive the algorithm (6.92). We mark equations that will be part of the algorithm with a dagger † in order to separate them from all the intermediate relationships.

6.C.1 Auxiliary Variables

We use $B_n(t)$ as the least squares estimate of B_n in the regression (6.89) based on $x(j)$, $0 \le j \le t$, i.e., we use

$$B_n(t) = \left[\sum_{j=1}^{t} \lambda^{t-j} \varphi_n(j+1) \varphi_n^{\mathrm{T}}(j+1) + \lambda^t \delta I \right]^{-1}$$

$$\times \sum_{j=1}^{t} \lambda^{t-j} \varphi_n(j+1)[-x^{\mathrm{T}}(j-n)].$$

As in (6.53), this relationship can also be written [recall (6.90)]

$$\bar{R}_{n+1}(t+1)\left(\frac{B_n(t)}{I}\right) = \left(\frac{0}{R_n^r(t)}\right) \updownarrow \alpha \text{ rows} . \tag{6.C.1}$$

We shall also use $A_k(t)$ as the least squares estimate of A_k in the regression

$$x(t) = -A_k^{\mathrm{T}} \varphi_k(t)$$

based on $x(j)$, $0 \le j \le t$. It is defined, as in (6.52)–(6.55), by

$$\bar{R}_{n+1}(t+1)\left(\frac{I}{A_n(t)}\right) = \left(\frac{R_n^e(t)}{0}\right) \updownarrow \alpha \text{ rows} . \tag{6.C.2}$$

Notice that when $-y(t)$ is part of the $x(t)$-vector, say equal to its first p rows, then the first p columns of the $n\alpha \times \alpha$-matrix $A_n(t)$ will be equal to $-\hat{\theta}_n(t)$ as defined by (6.86).

We shall also use the $n\alpha$-vector $L_n(t)$ as an auxiliary variable. It is defined by

$$\bar{R}_n(t) L_n(t) = \varphi_n(t) \tag{6.C.3}$$

[see also (6.46)].

6.C.2 Additional Notation

We use the residuals

$$e_n^t(k) = x(k) + A_n^{\mathrm{T}}(t) \varphi_n(k), \tag{6.C.4a}$$

Table 6.C.1
Matrix dimensions.

scalar	$\beta_n(t)$
α-vectors	$x(t),\ \hat{x}_n(t),\ e_n(t),\ \bar{e}_n(t),\ r_n(t),\ \bar{r}_n(t)$
$n\alpha$-vectors	$\varphi_n(t),\ L_n(t)$
$n\alpha \times \alpha$-matrices	$A_n(t),\ B_n(t)$
$\alpha \times \alpha$-matrices	$R_n^r(t),\ R_n^e(t),\ F_n(t),\ K_n(t),\ K_n^*(t)$
p-vectors	$y(t),\ \hat{y}_n(t),\ \varepsilon_n(t),\ \bar{\varepsilon}_n(t)$
$p \times \alpha$-matrices	$K_n^y(t),\ F_n^y(t)^{\mathrm{T}}$
$n\alpha \times p$-matrix	$\hat{\theta}_n(t)$

$$r_n^t(k) = x(k - n) + B_n^{\mathrm{T}}(t)\varphi_n(k + 1), \tag{6.C.4b}$$

$$\varepsilon_n^t(k) = y(k) - \hat{\theta}_n^{\mathrm{T}}(t)\varphi_n(k). \tag{6.C.4c}$$

In particular, we use

$$e_n(t) = e_n^{t-1}(t);\ r_n(t) = r_n^{t-1}(t);\ \varepsilon_n(t) = \varepsilon_n^{t-1}(t) \tag{6.C.5a}$$

and

$$\bar{e}_n(t) = e_n^t(t);\ \bar{r}_n(t) = r_n^t(t);\ \bar{\varepsilon}_n(t) = \varepsilon_n^t(t). \tag{6.C.5b}$$

For notational convenience we introduce

$$e_0(t) = r_0(t) = \bar{e}_0(t) = \bar{r}_0(t) = x(t). \tag{6.C.5c}$$

We also use the one-step-ahead prediction of $x(t)$:

$$\hat{x}_n(t) = -A_n^{\mathrm{T}}(t - 1)\varphi_n(t). \tag{6.C.6}$$

We thus have

$$e_n(t) = x(t) - \hat{x}_n(t).$$

We give the matrix dimensions for important quantities in table 6.C.1.

6.C.3 Useful Results from Section 6.3.2

The definitions (6.C.1) and (6.C.2) coincide with (6.52) and (6.53). Therefore (6.56a) implies that

$$R_n^e(t) = \sum_{k=1}^{t} \lambda^{t-k} e_n^t(k)\left[e_n^t(k)\right]^{\mathrm{T}} + \lambda^t \delta I, \tag{6.C.7}$$

and (6.58a) implies that

$$R_n^r(t) = \sum_{k=1}^{t} \lambda^{t-k} r_n^t(k) [r_n^t(k)]^\mathrm{T} + \lambda^t \delta I. \tag{6.C.8}$$

Moreover from (6.B.16) and (6.B.24) we know that

$$R_n^e(t) = \lambda R_n^e(t-1) + \bar{e}_n(t) e_n^\mathrm{T}(t), \tag{6.C.9}$$

$$R_n^r(t) = \lambda R_n^r(t-1) + \bar{r}_n(t) r_n^\mathrm{T}(t). \tag{6.C.10}$$

We also established the following relation between $\bar{e}(t)$ and $e(t)$ [see (6.B.17) and (6.B.18)]:

$$\bar{e}_n(t) = e_n(t)[1 - \beta_n(t)], \tag{6.C.11}$$

where

$$\beta_n(t) = L_n^\mathrm{T}(t) \varphi_n(t). \tag{6.C.12}$$

Analogously, from (6.B.21) and (6.B.23) it follows that

$$\bar{r}_n(t) = r_n(t)[1 - \beta_n(t+1)]. \tag{6.C.13}$$

6.C.4 Order Updates for A_n, B_n, R_n^e, and R_n^r

Suppose that $A_{n-1}(t)$ and $B_n(t-1)$ are known at time t; from these, we wish to determine $A_n(t)$ and $B_n(t)$. From the defining relation (6.C.2) and from the second equality in (6.91) we have

$$
\bar{R}_{n+1}(t+1)
\begin{pmatrix}
I \\
\hline
A_{n-1}(t) \\
\hline
0
\end{pmatrix} \updownarrow \alpha \text{ rows}
$$

$$
= \begin{pmatrix}
\bar{R}_n(t+1) \quad \vdots \\
\end{pmatrix}
\begin{pmatrix}
I \\
\hline
A_{n-1}(t) \\
0
\end{pmatrix} \tag{6.C.14}
$$

$$
= \begin{pmatrix}
\bar{R}_n(t+1)\left(\dfrac{I}{A_{n-1}(t)}\right) \\
\hline
F_{n-1}(t)
\end{pmatrix}
= \begin{pmatrix}
R_{n-1}^e(t) \\
\hline
0 \\
\hline
F_{n-1}(t)
\end{pmatrix}
\begin{matrix}
\updownarrow \alpha \text{ rows} \\
\\
\updownarrow \alpha \text{ rows}
\end{matrix} .
$$

$$\underset{\alpha \text{ columns}}{\longleftrightarrow}$$

Here $F_{n-1}(t)$ is the product of the last (starred) α rows of $\bar{R}_{n+1}(t+1)$, and

$$\left(\begin{array}{c} I \\ \hline A_{n-1}(t) \\ \hline 0 \end{array}\right).$$

Spelling this out gives

$$F_{n-1}(t) = \sum_{k=1}^{t+1} \lambda^{t+1-k} x(k-n-1)\varphi_{n+1}^{\mathrm{T}}(k) \left(\begin{array}{c} I \\ \hline A_{n-1}(t) \\ \hline 0 \end{array}\right)$$

$$= \sum_{k=1}^{t} \lambda^{t-k} x(k-n)[x(k) + A_{n-1}^{\mathrm{T}}(t)\varphi_{n-1}(k)]^{\mathrm{T}} \qquad (6.\mathrm{C}.15)$$

$$= \sum_{k=1}^{t} \lambda^{t-k} x(k-n)[e_{n-1}^{t}(k)]^{\mathrm{T}}.$$

Similarly we have

$$\bar{R}_{n+1}(t+1) \left(\begin{array}{c} 0 \\ \hline B_{n-1}(t-1) \\ \hline I \end{array}\right) = \left(\begin{array}{c} {*******} \\ \vdots \\ \bar{R}_n(t) \\ \vdots \\ {} \end{array}\right) \left(\begin{array}{c} 0 \\ \hline B_{n-1}(t-1) \\ \hline I \end{array}\right)$$

$$= \left(\begin{array}{c} F_{n-1}^{*}(t) \\ \hline 0 \\ \hline R_{n-1}^{r}(t-1) \end{array}\right) \begin{array}{c} \updownarrow \alpha \text{ rows} \\ {} \\ \updownarrow \alpha \text{ rows} \end{array}. \qquad (6.\mathrm{C}.16)$$

An explicit expression for $F_{n-1}^{*}(t)$ can be derived as in (6.C.15), but there is simple relationship between F^{*} and F. To find that, we multiply (6.C.14) from the left by

$$\left(\begin{array}{c} 0 \\ \hline B_{n-1}(t-1) \\ \hline I \end{array}\right)^{\mathrm{T}},$$

and (6.C.16) from the left by

$$\left(\begin{array}{c} I \\ \hline A_{n-1}(t) \\ \hline 0 \end{array}\right)^{\mathrm{T}}.$$

The left-hand sides of the resulting expression are then each other's transposes by construction, while the right-hand sides yield

$$F^*_{n-1}(t) = F^T_{n-1}(t). \tag{6.C.17}$$

Now the matrices $A_n(t)$ and $B_n(t)$ are defined by (6.C.1) and (6.C.2). In order to go from (6.C.14) to (6.C.2) we need to remove the last α rows from the right-hand side of (6.C.14). This can be accomplished using (6.C.16). We obtain

$$
\bar{R}_{n+1}(t+1)\left(\left(\begin{array}{c} I \\ \hline A_{n-1}(t) \\ \hline 0 \end{array}\right) - \left(\begin{array}{c} 0 \\ \hline B_{n-1}(t-1) \\ \hline I \end{array}\right)[R^r_{n-1}(t-1)]^{-1}F_{n-1}(t)\right)
$$

$$
= \left(\begin{array}{c} R^e_{n-1}(t) \\ \hline 0 \\ \hline F_{n-1}(t) \end{array}\right) - \left(\begin{array}{c} F^T_{n-1}(t)[R^r_{n-1}(t-1)]^{-1}F_{n-1}(t) \\ \hline 0 \\ \hline R^r_{n-1}(t-1)[R^r_{n-1}(t-1)]^{-1}F_{n-1}(t) \end{array}\right)
$$

$$
= \left(\begin{array}{c} R^e_{n-1}(t) - F^T_{n-1}(t)[R^r_{n-1}(t-1)]^{-1}F_{n-1}(t) \\ \hline 0 \\ \hline 0 \end{array}\right).
$$

Comparing this to (6.C.2) shows that

$$A_n(t) = \left(\begin{array}{c} A_{n-1}(t) \\ 0 \end{array}\right) - \left(\begin{array}{c} B_{n-1}(t-1) \\ I \end{array}\right)K^T_{n-1}(t), \tag{6.C.18}$$

where we have introduced the notation

$$K_n(t) = F^T_n(t)[R^r_n(t-1)]^{-1}. \tag{6.C.19}^\dagger$$

It also follows that

$$R^e_n(t) = R^e_{n-1}(t) - F^T_{n-1}(t)K^T_{n-1}(t). \tag{6.C.20}$$

By an analogous argument applied to (6.C.16), we find that

$$B_n(t) = \left(\begin{array}{c} 0 \\ B_{n-1}(t) \end{array}\right) - \left(\begin{array}{c} I \\ A_{n-1}(t) \end{array}\right)[K^*_{n-1}(t)]^T \tag{6.C.21}^\dagger$$

and

$$R^r_n(t) = R^r_{n-1}(t-1) - F_{n-1}(t)[K^*_{n-1}(t)]^T, \tag{6.C.22}^\dagger$$

where

$$K_n^*(t) = F_n(t)[R_n^e(t)]^{-1}. \qquad (6.C.23)^\dagger$$

6.C.5 A Ladder Form

We shall now show how the forward and backward predictions and prediction errors, defined in (6.C.4)–(6.C.6), are related. The expressions (6.C.18) and (6.C.21) will be instrumental for this.

Transposing (6.C.18) and delaying it one time step gives

$$A_n^T(t-1) = (A_{n-1}^T(t-1) \,|\, 0) - K_{n-1}(t-1)(B_{n-1}^T(t-2) \,|\, I). \qquad (6.C.24)$$

We note that

$$A_n^T(t-1)\varphi_n(t) = -\hat{x}_n(t),$$

$$(A_{n-1}^T(t-1) \,|\, 0)\varphi_n(t) = A_{n-1}^T(t-1)\varphi_{n-1}(t) = -\hat{x}_{n-1}(t)$$

and

$$(B_{n-1}^T(t-2) \,|\, I)\varphi_n(t) = B_{n-1}^T(t-2)\varphi_{n-1}(t) + x(t-n) = r_{n-1}(t-1)$$

in the notation of (6.C.4)–(6.C.6). Hence multiplying (6.C.24) by $\varphi_n(t)$ gives

$$\hat{x}_n(t) = \hat{x}_{n-1}(t) + K_{n-1}(t-1)r_{n-1}(t-1). \qquad (6.C.25)^\dagger$$

By analogous calculations [transposing (6.C.21), shifting it one time step backward, multiplying by $\varphi_n(t+1)$, and subtracting $x(t-n)$] we find that

$$r_n(t) = r_{n-1}(t-1) - K_{n-1}^*(t-1)e_{n-1}(t), \qquad (6.C.26a)$$

$$e_{n-1}(t) = x(t) - \hat{x}_{n-1}(t). \qquad (6.C.26b)$$

With e_n, (6.C.25)–(6.C.26) can be rewritten as

$$e_n(t) = e_{n-1}(t) - K_{n-1}(t-1)r_{n-1}(t-1),$$

$$r_n(t) = r_{n-1}(t-1) - K_{n-1}^*(t-1)e_{n-1}(t), \qquad (6.C.27)$$

$$e_0(t) = r_0(t) = x(t).$$

The signal flow in this recursion is depicted in figure 6.1.

Notice that when the p first elements of $x(t)$ are equal to $-y(t)$, then the first p rows of (6.C.25) read

$$\hat{y}_n(t) = \hat{y}_{n-1}(t) + K^y_{n-1}(t-1)r_{n-1}(t-1),$$

where K^y_{n-1} are the first p rows of the $\alpha \times \alpha$-matrix $-K_{n-1}$. This relation can also be written

$$\hat{y}_n(t) = K^y_0(t-1)r_0(t-1) + K^y_1(t-1)r_1(t-1)$$

$$+ \cdots + K^y_{n-1}(t-1)r_{n-1}(t-1). \tag{6.C.28}$$

This is exactly of the form (6.88) that we are seeking. The recursions (6.C.25) and (6.C.26) therefore contain the desired ladder form representation of the prediction, and the variables K^y_i are consequently the reflection coefficients.

We now discuss how (6.C.25) and (6.C.26) are to be used in recursive estimation, and what additional information is required. Suppose that at time $t-1$, we have the following quantities available:

$$R^e_i(t-1), F_i(t-1), R^r_i(t-2), \quad i=0, \ldots, M-1, \tag{6.C.29}$$

$$r_i(t-1), \qquad\qquad\qquad i=0, \ldots, M-1. \tag{6.C.30}$$

We want to update these variables to time t.

1. From (6.C.29) we can compute

$$K_i(t-1) \quad \text{and} \quad K^*_i(t-1), \quad i=0, \ldots, M-1, \tag{6.C.31}$$

using (6.C.19) and (6.C.22), (6.C.23).

2. With $K_i(t-1)$ and $K^*_i(t-1)$ given, $\hat{x}_i(t)$ can be computed from $r_{i-1}(t)$ for $i=1, \ldots, M$ using (6.C.25).

3. At time t, the information $x(t)$ arrives.

4. We can now compute $e_i(t)$ and $r_i(t)$ using (6.C.26) for $i=0, \ldots, M-1$. Hence the information (6.C.30) has been updated.

5. To update $R^r_i(t-1)$ from $R^r_i(t-2)$ we need, according to (6.C.10) and (6.C.13), $r_i(t-1)$ and $\beta_i(t)$.

6. To update $R^e_i(t)$ from $R^e_i(t-1)$ we need, according to (6.C.9) and (6.C.11), $e_i(t)$ and $\beta_i(t)$.

To be able to update the information (6.C.29) we need consequently only discuss how to find $\beta_i(t)$ and how to update $F_i(t)$ from $F_i(t-1)$. This will be discussed now.

6.C.6 Computation of $\beta_i(t)$

The idea is to update $\beta_{n+1}(t)$ from $\beta_n(t)$ based on an update formula for $L_{n+1}(t)$ from $L_n(t)$. To find the latter, we note from (6.B.22b, c) that [recall that $\bar{R}^*(t) = \bar{R}_{n+1}(t + 1)$]

$$\bar{R}_{n+1}(t)\begin{pmatrix} L_n(t) \\ 0 \end{pmatrix} = \begin{pmatrix} \varphi_n(t) \\ \check{x}_n(t - n - 1) \end{pmatrix}, \tag{6.C.32}$$

where

$$\check{x}_n(t - n - 1) = -B_n^T(t - 1)\varphi_n(t) = -\bar{r}_n(t - 1) + x(t - 1 - n). \tag{6.C.33}$$

The last equality in (6.C.33) follows from (6.C.4)–(6.C.6). Consequently, using (6.C.1), we have

$$\bar{R}_{n+1}(t)\left(\begin{pmatrix} L_n(t) \\ 0 \end{pmatrix} + \begin{pmatrix} B_n(t - 1) \\ I \end{pmatrix}[R_n^r(t - 1)]^{-1}\bar{r}_n(t - 1)\right)$$

$$= \begin{pmatrix} \varphi_n(t) \\ \check{x}_n(t - n - 1) \end{pmatrix} + \begin{pmatrix} 0 \\ \bar{r}_n(t - 1) \end{pmatrix} = \varphi_{n+1}(t),$$

from which we see that

$$L_{n+1}(t) = \begin{pmatrix} L_n(t) \\ 0 \end{pmatrix} + \begin{pmatrix} B_n(t - 1) \\ I \end{pmatrix}[R_n^r(t - 1)]^{-1}\bar{r}_n(t - 1). \tag{6.C.34}$$

Multiplying this expression by $\varphi_{n+1}^T(t)$ now gives

$$\beta_{n+1}(t) = \beta_n(t) + \bar{r}_n^T(t - 1)[R_n^r(t - 1)]^{-1}\bar{r}_n(t - 1). \tag{6.C.35}$$

With (6.C.13), this can also be written

$$\beta_{n+1}(t) = \beta_n(t) + [1 - \beta_n(t)]^2 r_n^T(t - 1)[R_n^r(t - 1)]^{-1}r_n(t - 1). \tag{6.C.36}^{\dagger}$$

6.C.7 Updating $F_i(t)$

The relation (6.C.14) defines $F_{n-1}(t)$. The left-hand side of (6.C.14) can be written

$$\bar{R}_{n+1}(t + 1)\begin{pmatrix} I \\ \hline A_{n-1}(t) \\ \hline 0 \end{pmatrix}$$

$$= \bar{R}_{n+1}(t+1)\left(\left(\frac{I}{A_{n-1}(t-1)}\right) - \left(\frac{0}{L_{n-1}(t)e_{n-1}^T(t)}\right)\right)$$

$$= [\lambda\bar{R}_{n+1}(t) + \varphi_{n+1}(t+1)\varphi_{n+1}^T(t+1)] \times \left(\frac{I}{A_{n-1}(t-1)}\right)$$

$$- \bar{R}_{n+1}(t+1)\left(\frac{0}{L_{n-1}(t)}\right)e_{n-1}^T(t). \tag{6.C.37}$$

Here we used (6.59b) and (6.B.11), respectively. The first term on the right-hand side equals

$$\lambda\left(\frac{R_{n-1}^e(t-1)}{\frac{0}{F_{n-1}(t-1)}}\right) + \varphi_{n+1}(t+1)e_{n-1}^T(t),$$

in view of the definition of $e_{n-1}(t)$. For the second term we find

$$\left(\begin{matrix} ******* \\ R_n(t) \\ ***** \end{matrix}\right)\left(\frac{0}{L_{n-1}(t)}\right) = \left(\frac{*}{\frac{\varphi_{n-1}(t)}{\check{x}_{n-1}(t-n)}}\right),$$

using (6.91) and (6.C.32). Collecting these expressions into (6.C.37) and just reading the last α rows now gives

$$F_{n-1}(t) = \lambda F_{n-1}(t-1) + x(t-n)e_{n-1}^T(t) - \check{x}_{n-1}(t-n)e_{n-1}^T(t)$$

$$= \lambda F_{n-1}(t-1) + \bar{r}_{n-1}(t-1)e_{n-1}^T(t)$$

using (6.C.33). With (6.C.13) we finally obtain

$$F_{n-1}(t) = \lambda F_{n-1}(t-1) + [1 - \beta_{n-1}(t)]r_{n-1}(t-1)e_{n-1}^T(t). \tag{6.C.38}^\dagger$$

With this the derivation is finished. The equations marked with a dagger † give a complete set of relations for updating all necessary quantities, as summarized in (6.92).

Appendix 7.A Convergence Analysis of Recursive Identification Methods with Adaptive Input Generation

In this appendix we give a lemma about the convergence analysis of the adaptive regulator (7.10)–(7.11), used with the general recursive identification algorithm (4.16). The approach follows the heuristic outline in section 7.3.1. The lemma will be proved using theorem 4.2. This means that we have to make the restriction to linear dynamics as in (4.52b). In the present case, both the predictor and the closed loop system that generates the input output data must be linear in the data. This is, of course, a restrictive assumption. However, it is not a necessary one; a counterpart of theorem 4.1 with nonlinear dynamics in (4.52b) can also be given (see Ljung, 1975).

We thus assume that the system can be described by the linear structure

$$x(t + 1) = F_{\mathcal{S}}x(t) + G_{\mathcal{S}}u(t) + w(t),$$
$$y(t) = H_{\mathcal{S}}x(t) + e(t). \tag{7.A.1}$$

We also specialize to a linear feedback law,

$$u(t) = L_1(\hat{\rho}(t))\varphi_z(t) + L_2(\hat{\rho}(t))\varphi_r(t). \tag{7.A.2}$$

Here $\varphi_z(t)$ and $\varphi_r(t)$ are vectors, constructed from input output data $\{y(t), z(t-1), \ldots, z(t-n)\}$ and from the reference signal $\{r(t), r(t-1), \ldots, r(t-n)\}$, respectively. L_1 and L_2 are matrix functions of the current value of the regulator parameter ρ. Compared to (7.11) we have imposed a linear structure on the regulator.

The equations (7.A.1) and (7.A.2) can be combined to yield a description of the closed-loop system:

$$\tilde{x}(t + 1) = \tilde{F}(\hat{\rho}(t))\tilde{x}(t) + \tilde{G}(\hat{\rho}(t))v(t),$$
$$z(t) = \tilde{H}(\hat{\rho}(t))\tilde{x}(t). \tag{7.A.3}$$

Here we introduced the notation

$$v(t) = \begin{pmatrix} r(t) \\ w(t) \\ e(t + 1) \end{pmatrix}. \tag{7.A.4}$$

As before,

$$z(t) = \begin{pmatrix} y(t) \\ u(t) \end{pmatrix}.$$

The complete adaptive control algorithm now follows by combining (7.A.3) and (4.16) and (7.10):

$$\varepsilon(t) = y(t) - \hat{y}(t), \tag{7.A.5a}$$

$$R(t) = R(t-1) + \gamma(t)[\eta(t)\Lambda^{-1}(t)\eta^{\mathrm{T}}(t) - R(t-1)], \tag{7.A.5b}$$

$$\hat{\theta}(t) = \hat{\theta}(t-1) + \gamma(t)R^{-1}(t)\eta(t)\Lambda^{-1}(t)\varepsilon(t), \tag{7.A.5c}$$

$$\hat{\rho}(t) = k(\hat{\theta}(t)), \tag{7.A.5d}$$

$$\tilde{x}(t+1) = \tilde{F}(\hat{\rho}(t))\tilde{x}(t) + \tilde{G}(\hat{\rho}(t))v(t), \tag{7.A.5e}$$

$$z(t) = \tilde{H}(\hat{\rho}(t))x(t), \tag{7.A.5f}$$

$$\zeta(t+1) = A(\hat{\theta}(t))\zeta(t) + B(\hat{\theta}(t))z(t), \tag{7.A.5g}$$

$$\begin{pmatrix} \hat{y}(t+1) \\ \operatorname{col}\eta(t+1) \end{pmatrix} = C(\hat{\theta}(t))\zeta(t+1). \tag{7.A.5h}$$

The driving noise and reference signals $v(t)$ are independent of ρ. We can therefore write the (fictitious) input-output data $z(t, \rho)$, that we would have obtained for a constant feedback law ρ, as

$$\tilde{x}(t+1, \rho) = \tilde{F}(\rho)\tilde{x}(t, \rho) + \tilde{G}(\rho)v(t); \ \tilde{x}(0, \rho) = 0,$$
$$z(t, \rho) = \tilde{H}(\rho)\tilde{x}(t, \rho). \tag{7.A.6}$$

The variable $z(t, \rho)$ will be well-defined as $t \to \infty$ only when ρ is such that $\tilde{F}(\rho)$ is stable:

$$\rho \in C_s \Rightarrow \tilde{F}(\rho) \text{ has all eigenvalues inside the unit circle.}$$

We can also compute the prediction errors ε and gradient approximations η that would result if the data $\{z(t, \rho)\}$ is fed into (7.A.5g, h) for a constant value of θ:

$$\zeta(t+1, \theta, \rho) = A(\theta)\zeta(t, \theta, \rho) + B(\theta)z(t, \rho),$$

$$\begin{pmatrix} \hat{y}(t \,|\, \theta, \rho) \\ \operatorname{col}\eta(t, \theta, \rho) \end{pmatrix} = C(\theta)\zeta(t, \theta, \rho). \tag{7.A.7}$$

Let

$$\varepsilon(t, \theta, \rho) = y(t, \rho) - \hat{y}(t \,|\, \theta, \rho). \tag{7.A.8}$$

We now introduce the condition

A1′   The data sequence $\{v(t)\}$ is such that the following limits exist for all $\rho \in C_s$ and all $\theta \in D_{\mathcal{M}}$:

(a) $\displaystyle \lim_{N \to \infty} \frac{1}{N} \sum_{t=1}^{N} \eta(t, \theta, \rho) \Lambda^{-1}(t) \varepsilon(t, \theta, \rho) \triangleq \bar{f}(\theta, \rho),$

(b) $\displaystyle \lim_{N \to \infty} \frac{1}{N} \sum_{t=1}^{N} \eta(t, \theta, \rho) \Lambda^{-1}(t) \eta^{\mathrm{T}}(t, \theta, \rho) \triangleq \bar{G}(\theta, \rho),$

(c) $\displaystyle \lim_{N \to \infty} \sup \frac{1}{N} \sum_{t=1}^{N} (1 + |v(t)|)^3 < \infty,$

where $\eta(t, \theta, \rho)$ and $\varepsilon(t, \theta, \rho)$ are defined from v^t by (7.A.6)–(7.A.8).

Compare this to assumption A1 in section 4.3.4. Obviously we could introduce counterparts of assumptions A2 and A3 and infer that they imply A1′ under certain conditions on $\{v(t)\}$. This would be entirely analogous to lemma 4.1.

We also introduce the assumptions

Reg 1   The matrix functions $L_i(\rho)$, $i = 1, 2$, in (7.A.2) are continuously differentiable for $\rho \in C_s$.

D1   The regulator design mapping $\rho = k(\theta)$ in (7.A.5d) is continuously differentiable for $\theta \in D_{\mathcal{M}}$.

Let D_{sc} be the inverse image of C_s under the mapping $k(\theta)$:

$$D_{sc} = \{\theta \mid k(\theta) \in C_s\}. \tag{7.A.9}$$

Also, let

$$\bar{D} = D_{\mathcal{M}} \cap D_{sc}. \tag{7.A.10}$$

We can now formulate the basic lemma.

LEMMA 7.A.1   Consider the algorithm (7.A.5). Assume that conditions M1, M2, R1, and G1 of section 4.3.4, as well as assumption A1′, Reg 1, and D1 hold. Assume also that

$$\hat{\theta}(t_k) \in \bar{D} \text{ and } |\tilde{x}(t_k)| + |\zeta(t_k)| < C \text{ for an infinite subsequence } t_k. \tag{7.A.11}$$

Suppose that there exists a positive function $V(\theta, R)$ such that

$$\frac{d}{d\tau} V(\theta_D(\tau), R_D(\tau)) \le 0 \quad \text{for } \theta_D \in \bar{D}, \tag{7.A.12}$$

when evaluated along solutions to the d.e.

$$\frac{d}{d\tau} \theta_D(\tau) = R_D^{-1}(\tau) \bar{f}(\theta_D(\tau), k(\theta_D(\tau))),$$

$$\frac{d}{d\tau} R_D(\tau) = \bar{G}(\theta_D(\tau), k(\theta_D(\tau))) - R_D(\tau), \tag{7.A.13}$$

where $\bar{f}$ and $\bar{G}$ are defined by A1′. Let

$$D_c = \left\{ \theta, R \left| \frac{d}{d\tau} V(\theta_D(\tau), R_D(\tau)) = 0 \right. \right\}. \tag{7.A.14}$$

Then as $t \to \infty$ either $\{\hat{\theta}(t), R(t)\}$ tends to D_c or $\{\hat{\theta}(t)\}$ has a cluster point on the boundary of $\bar{D}$.

Proof Equations (7.A.5d–h) can be merged into

$$\tilde{\zeta}(t+1) = \tilde{A}(\hat{\theta}(t))\tilde{\zeta}(t) + \tilde{B}(\hat{\theta}(t))v(t),$$

$$\begin{pmatrix} \hat{y}(t+1) \\ \mathrm{col}\,\eta(t+1) \end{pmatrix} = \tilde{C}(\hat{\theta}(t))\tilde{\zeta}(t+1). \tag{7.A.15}$$

The algorithm (7.A.5a–c) + (7.A.15) now looks exactly like the algorithm (4.70) (with particular choices of h and H) for which theorem 4.2 was given.

Our choices of h and H clearly satisfy Cr1 and Cr2 of theorem 4.2. Moreover the matrices $\tilde{A}$, $\tilde{B}$, and $\tilde{C}$ satisfy M2 as a consequence of assumptions Reg 1, D1, and the fact that A, B and C satisfy M2. M1 holds for $D_M \supset \bar{D}$.

The variables obtained from (7.A.15) when keeping θ constant are

$$\tilde{\zeta}(t+1, \theta) = \tilde{A}(\theta)\tilde{\zeta}(t, \theta) + \tilde{B}(\theta)v(t),$$

$$\begin{pmatrix} \hat{y}^*(t+1 \mid \theta) \\ \mathrm{col}\,\eta^*(t+1, \theta) \end{pmatrix} = \tilde{C}(\theta)\tilde{\zeta}(t+1, \theta).$$

Comparing these to (7.A.6) and (7.A.7) shows that

$$\hat{y}^*(t+1 \mid \theta) = \hat{y}(t+1 \mid \theta, k(\theta)),$$

$$\eta^*(t+1, \theta) = \eta(t+1, \theta, k(\theta)).$$

Hence the counterpart of assumption A1 in theorem 4.2 follows from A1′. (In fact it is sufficient to require the limits in A1′ to exist for $\rho = k(\theta)$.)

The assumptions of our lemma thus imply that theorem 4.2 can be applied to (7.A.5a–c) + (7.A.15). This proves the lemma. ∎

References

M. A. Aizerman, E. M. Braverman, and L. I. Rozonoer (1970)
Metody potentsialnych funktsij v teorii obuchenija mashin. Nauka, Moscow.

H. Akaike (1972)
Information theory and an extension of the maximum likelihood principle. Proc. 2nd Int.
Symp. Information Theory, Supp. to Problems of Control and Information Theory, pp.
267–281.

H. Akaike (1981)
Modern development of statistical methods. In P. Eykhoff, ed.: Trends and Progress in
System Identification, Pergamon Press, New York.

A. E. Albert and L. A. Gardner (1967)
Stochastic Approximation and Nonlinear Regression. Research Monograph 42, MIT Press,
Cambridge, Mass.

J. van Amerongen (1981)
A model reference adaptive autopilot for ships—practical results. Proc. 8th IFAC World
Congress, Kyoto, Japan.

B. D. O. Anderson and J. B. Moore (1979)
Optimal Filtering. Prentice-Hall, Englewood Cliffs, N.J.

K. J. Åström (1968)
Lectures on the identification problem—the least squares method. Report 6806, Division
of Automatic Control, Lund Institute of Technology, Lund, Sweden.

K. J. Åström (1970)
Introduction to Stochastic Control Theory. Academic Press, New York.

K. J. Åström (1972)
Unpublished notes.

K. J. Åström (1980a)
Why use adaptive techniques for steering large tankers? International Journal of Control,
vol. 32, pp. 689–708.

K. J. Åström (1980b)
Maximum likelihood and prediction error methods. Automatica, vol. 16, pp. 551–574.

K. J. Åström and T. Bohlin (1965)
Numerical identification of linear dynamic systems from normal operating records. IFAC
Symposium on Self-Adaptive Systems, Teddington, England. Also in P. H. Hammond, ed.:
Theory of Self-Adaptive Control Systems, Plenum Press, New York.

K. J. Åström, U. Borisson, L. Ljung, and B. Wittenmark (1977)
Theory and applications of self-tuning regulators. Automatica, vol. 13, pp. 457–476.

K. J. Åström and P. Eykhoff (1971)
System identification—a survey. Automatica, vol. 7, pp. 123–167.

K. J. Åström and C. Källström (1973)
Application of system identification techniques to the determination of ship dynamics. Proc.
3rd IFAC Symposium on Identification and System Parameter Estimation, the Hague
(North-Holland).

K. J. Åström and T. Söderström (1974)
Uniqueness of the maximum likelihood estimates of the parameters of an ARMA model.
IEEE Transactions on Automatic Control, Vol. AC-19, pp. 769–773.

K. J. Åström and B. Wittenmark (1971)
Problems of identification and control. Journal of Mathematical Analysis and Applications,
vol. 34, pp. 90–113.

K. J. Åström B. Wittenmark (1973)
On self-tuning regulators. Automatica, vol. 9, pp. 185–199.

B. S. Atal and M. R. Schroeder (1970)
Adaptive predictive coding of speech signals. Bell Syst. Tech. Journal, vol. 49, pp. 1973–1986.

G. Banon and J. Aguilar-Martin (1972)
Estimation linéaire recurrente de paramètres des processus dynamiques soumis a des perturbations aleatoires. Revue du Cethedec, vol. 9, pp. 39–86.

B. Bauer and H. Unbehauen (1978)
On-line identification of a load-dependent heat exchanger in closed loop using a modified instrumental variable method. Proc. IFAC 7th World Congress, Helsinki (Pergamon Press).

P. B. Belanger (1974)
Estimation of noise covariance matrices for a linear time-varying stochastic process. Automatica, vol. 10, pp. 267–275.

G. Bethoux (1976)
Approche unitaire des méthodes d'identification et de commande adaptive des procédes dynamiques. Thèse de 3ème cycle, Université de Grenoble.

G. J. Bierman (1977)
Factorization Methods for Discrete Sequential Estimation. Academic Press, New York.

P. Billingsley (1961)
The Lindeberg-Levy theorem for martingales. Proc. Am. Math. Soc., vol 12, pp. 788–792.

R. R. Bitmead and B. D. O. Anderson (1980a)
Lyapunov techniques for the exponential stability of linear difference equations with random coefficients. IEEE Transactions on Automatic Control, vol. AC-25, pp. 782–787.

R. R. Bitmead and B. D. O. Anderson (1980b)
Performance of adaptive estimation algorithms in dependent random environments. IEEE Transactions on Automatic Control, vol. AC-25, pp. 788–794.

J. Blum (1954)
Multidimensional stochastic approximation methods. Ann. Math. Stat., vol. 25, pp. 737–744.

T. Bohlin (1970)
Information pattern for linear discrete time models with stochastic coefficients. IEEE Transactions on Automatic Control, vol. AC-15, pp. 104–106.

T. Bohlin (1971)
On the problem of ambiguites in maximum likelihood identification. Automatica, vol. 7, pp. 199–210.

T. Bohlin (1976)
Four cases of identification of changing systems. In R. K. Mehra and D. G. Lainiotis, eds.: System Identification—Advances and Case Studies. Academic Press, New York.

A. van den Bos (1971)
Alternative interpretation of maximum entropy spectral analysis. IEEE Transactions on Information Theory, vol. IT-17, pp. 493–494.

G. E. P. Box and G. M. Jenkins (1970)
Time Series Analysis, Forecasting and Control. Holden-Day, San Francisco.

B. M. Brown (1971)
Martingale central limit theorems. Ann. Math. Stat., vol. 42, pp. 59–66.

J. P. Burg (1967)
Maximum entropy spectral analysis. Paper presented at the 37th Annual International Meeting, Soc. of Explor. Geophys., Oklahoma City.

P. E. Caines (1978)
Stationary linear and nonlinear system identification and predictor set completeness. IEEE Transactions on Automatic Control, vol. AC-23, pp. 583–594.

P. E. Caines and L. Ljung (1976)
Asymptotic normality of prediction error estimators. Control Systems Report No 7602, Dept of Electrical Engineering, University of Toronto, Canada.

R. L. Carroll and D. P. Lindorff (1973)
An adaptive observer for single-input single-output linear systems. IEEE Transactions on Automatic Control, vol. AC-18, pp. 428–434.

K. L. Chung (1968)
A Course in Probability Theory. Harcourt, Brace and World, New York.

D. W. Clarke (1967)
Generalized least squares estimation of parameters of a dynamic model. 1st IFAC Symposium on Identification in Automatic Control Systems, Prague.

H. Cox (1964)
On the estimation of state variables and parameters for noisy dynamic systems. IEEE Transactions on Automatic Control, vol. AC-9, pp. 5–12.

H. Cramér (1946)
Mathematical Methods of Statistics. Princeton University Press, Princeton, N.J.

H. Cramér and M. R. Leadbetter (1967)
Stationary and Related Stochastic Processes. Wiley, New York.

W. D. T. Davies (1970)
System Identification for Self-Adaptive Control. Wiley Interscience, London.

L. Dugard and I. D. Landau (1980a)
Recursive output error identification algorithms. Automatica, vol. 16, pp. 443–462.

L. Dugard and I. D. Landau (1980b)
Stochastic model reference adaptive controllers Proc. 19th IEEE Conference on Decision and Control, Albuquerque, New Mexico.

L. Dugard, I. D. Landau, and H. M. Silveira (1980)
Adaptive state estimation using MRAS techiques—Convergence analysis and evaluation. IEEE Transactions on Automatic Control, vol. AC-25, pp. 1169–1181.

A. Dvoretsky (1956)
On stochastic approximation. Proc. 3rd Berkeley Symposium on Mathematical Statistics and Probability, vol. 1, pp. 35–56.

B. Egardt (1979a)
Stability of Adaptive Controllers (Lecture Notes in Control and Information Sciences, no. 20). Springer-Verlag, Berlin.

B. Egardt (1979b)
Unification of some continuous-time adaptive control schemes. IEEE Transactions on Automatic Control, vol. AC-24, pp. 588–592.

B. Egardt (1980a)
Unification of some discrete-time adaptive control schemes. IEEE Transactions on Automatic Control, vol. AC-25, pp. 693–697.

B. Egardt (1980b)
Stochastic convergence analysis of model reference adaptive controllers. Proc. IEEE Conference on Decision and Control, Albuquerque, New Mexico.

P. Eykhoff (1974)
System Identification. Wiley, London.

V. Fabian (1960)
Stochastic approximation methods. Czechoslovak Math. J., vol. 10, pp. 85.

V. Fabian (1968)
On asymptotic normality in stochastic approximation. Ann. Math. Stat., vol. 39, pp. 1327–1332.

D. Falconer and L. Ljung (1978)
Application of fast Kalman estimation to adaptive equalization. IEEE Transactions on Communication, vol. COM-26, pp. 1439–1446.

P. Feintuch (1976)
An adaptive recursive LMS filter. Proc. IEEE, vol. 64, pp. 1622–1624.

B. Finigan and I. H. Rowe (1974)
Strongly consistent parameter estimation by the introduction of strong instrumental variables. IEEE Transactions on Automatic Control, vol. AC-19, pp. 825–831.

B. Friedlander (1981)
Application of recursive parameter estimation algorithms to adaptive signal processing. Proc. Workshop on Applications of Adaptive Systems Theory, Yale University.

B. Friedlander (1982a)
A modified pre-filter for some recursive parameter estimation algorithms. IEEE Transactions on Automatic Control, vol. AC-27, pp. 232–235.

B. Friedlander (1982b)
System identification techniques for adaptive noise cancelling. IEEE Transactions on Acoustics, Speech and Signal Processing, vol. ASSP-30, pp. 699–708.

B. Friedlander, L. Ljung and M. Morf (1981)
Lattice implementation of the recursive maximum likelihood algorithm. Proc. 20th IEEE Conference on Decision and Control, San Diego, California.

J.-J. J. Fuchs (1980)
Discrete adaptive control: A sufficient condition for stability and applications. IEEE Transactions on Automatic Control, vol. AC-25, pp. 940–946.

B. P. Fuhrt (1973)
New estimator for the identification of dynamic processes. IBK Report, Institut Boris Kidrič Vinča, Belgrade, Yugoslavia.

K. F. Gauss (1809)
Theoria motus corporum coelestium. English translation: Theory of the Motion of the Heavenly Bodies. Dover, New York (1963).

A. Gauthier and I. D. Landau (1978)
On the recursive identification of multi-input, multi-output systems. Automatica, vol 14, pp. 609–614.

S. Gentil, J. P. Sandraz, and C. Foulard (1973)
Different methods for dynamic identification of an experimental paper machine. Proc. 3rd IFAC Symposium on Identification and System Parameter Estimation, the Hague (North-Holland).

D. A. George, R. R. Bowen, and J. R. Storey (1971)
An adaptive decision feedback equalizer. IEEE Transactions on Communications, vol.
COM-19, pp. 281–293.

A. Gersho (1967)
Automatic equalization of highly dispersive channels for data transmission. International
Symp. on Information Theory, San Remo, Italy.

J. Gertler and Cs. Bányász (1974)
A recursive (on-line) maximum likelihood identification method. IEEE Transactions on
Automatic Control, vol. AC-19, pp. 816–820.

M. Gevers and V. Werts (1983)
A recursive least squares d-step-ahead predictor in lattice and ladder form. IEEE Transactions on Automatic Control, vol. AC-28, no. 4.

D. Godard (1974)
Channel equalization using a Kalman filter for fast data transmission. IBM Journal of Research and Development, vol. 18, pp. 267–273.

G. C. Goodwin, H. B. Doan, and A. Cantoni (1980)
Application of ARMA models to automatic channel equalization. Information Sciences,
vol. 22, pp. 107–129.

G. C. Goodwin and R. L. Payne (1977)
Dynamic System Identification: Experiment Design and Data Analysis. Academic Press,
New York.

G. C. Goodwin, P. J. Ramadge, and P. E. Caines (1980)
Discrete time multi-variable adaptive control. IEEE Transactions on Automatic Control,
vol. AC-25, pp. 449–456.

G. C. Goodwin, P. J. Ramadge, and P. E. Caines (1981)
Discrete time stochastic adaptive control. SIAM Journal of Control and Optimization, vol.
19, pp. 829–853.

G. C. Goodwin and K. S. Sin (1983)
Adaptive Filtering, Prediction and Control. Prentice-Hall, Englewood Cliffs, N.J.

L. J. Griffiths (1975)
Rapid measurement of digital instantaneous frequency. IEEE Transactions on Acoustics,
Speech and Signal Processing, vol. ASSP-23, pp. 207–222.

L. J. Griffiths (1977)
A continuously adaptive filter implemented as a lattice structure. Proc. 1977 IEEE International Conference on Acoustics, Speech and Signal Processing.

P. C. Gupta and K. Yamada (1972)
Adaptive short-time forecasting of hourly load using weather information. IEEE Transactions on Power Apparatus and Systems, vol. PAS-91, pp. 2085–2095.

I. Gustavsson, L. Ljung, and T. Söderström (1977)
Identification of processes in closed loop—identifiability and accuracy aspects. Automatica, vol. 13, pp. 59–75.

I. Gustavsson, L. Ljung, and T. Söderström (1981)
Choice and effect of different feedback configurations. In P. Eykhoff, ed.: Trends and Progress in System Identification, Pergamon Press, Oxford.

P. Hagander and B. Wittenmark (1977)
A self-tuning filter for fixed lag smoothing. IEEE Transactions on Information Theory, vol.
IT-23, pp. 377–384.

W. Hahn (1967)
Stability of Motion. Springer-Verlag, Berlin.

E. J. Hannan (1980)
Recursive estimation based on ARMA models. Annals of Statistics, vol. 8, pp. 762–777.

R. Hastings-James and M. W. Sage (1969)
Recursive generalized least squares procedure for on-line identification of process parameters. IEE Proceedings, vol. 116, pp. 2057–2062.

Y. C. Ho (1963)
On the stochastic approximation method and optimal filtering theory. Journal of Mathematical Analysis and Applications, vol. 6, pp. 152–154.

J. Holst (1977)
Adaptive prediction and recursive estimation. Report TFRT-1013, Department of Automatic Control, Lund Institute of Technology, Lund, Sweden.

T. C. Hsia (1977)
Identification: Least Squares Methods. Lexington Books, Lexington, Mass.

P. J. Huber (1973)
Robust regression: asymptotics, conjectures and Monte Carlo. Annals of Statistics, vol. 1, pp. 799–821.

R. Isermann (1974)
Prozessidentifikation. Springer-Verlag, Berlin.

R. Isermann and U. Baur (1974)
Two-step process identification with correlation analysis and least squares parameter estimation. Transactions ASME, Journal of Dynamic Systems, Measurement, and Control (ser. G), vol. 96, pp. 425–432.

R. Isermann, U. Baur, W. Bamberger, P. Knepo, and H. Siebert (1974)
Comparison of six on-line identification and parameter estimation methods. Automatica, vol. 10, pp. 81–103.

F. Itakura and S. Saito (1971)
Digital filtering techniques for speech analysis and synthesis. Conference Record, 7th International Congr. Acoust., Budapest.

A. Jakeman and P. C. Young (1979)
Refined instrumental variable methods of recursive time-series analysis. Part II. Multivariable systems. International Journal of Control, vol. 29, pp. 621–644.

A. H. Jazwinski (1969)
Adaptive filtering. Automatica, vol. 5, pp. 475–485.

A. H. Jazwinski (1970)
Stochastic Processes and Filtering Theory. Academic Press, New York.

J. L. Jeanneau and Ph. de Larminat (1975)
Metodo de regulacion adaptativa para sistemas de "fase no minima." 3 Congreso Nacional Informatica y Automatica, Madrid.

G. M. Jenkins and D. G. Watts (1969)
Spectral Analysis and Its Applications. Holden-Day, San Francisco.

C. R. Johnson, Jr. (1979)
A convergence proof for a hyperstable adaptive filter. IEEE Transactions on Information Theory, vol. IT-25, pp. 745–749.

C. R. Johnson, Jr. (1982)
The common parameter estimation basis of adaptive filtering, identification and control. IEEE Transactions on Acoustics, Speech and Signal Processing, vol. ASSP-30, pp. 587–592.

C. R. Johnson, Jr., M. G. Larimore, J. R. Treichler, and B. D. O. Anderson (1981)
SHARF convergence properties. IEEE Transactions on Circuits and Systems, vol. CAS-28, pp. 499–510.

E. I. Jury (1974)
Inners and Stability of Dynamic Systems. Wiley, New York.

T. Kailath (1976)
Lectures on Linear Least-Squares Estimation. Springer-Verlag, Wien.

T. Kailath (1980)
Linear Systems. Prentice-Hall, Englewood Cliffs, New Jersey.

R. E. Kalman (1958)
Design of a self-optimizing control system. Transaction ASME, vol. 80, pp. 468–478.

R. E. Kalman and R. S. Bucy (1961)
New results in linear filtering and prediction theory. Transactions ASME, Journal of Basic Engineering (ser. D), vol. 83, pp. 95–108.

R. L. Kashyap (1974)
Estimation of parameters in a partially whitened representation of a stochastic process. IEEE Transactions on Automatic Control, vol. AC-19, pp. 13–21.

R. L. Kashyap and A. R. Rao (1976)
Dynamic Stochastic Models from Empirical Data. Academic Press, New York.

M. G. Kendall and A. Stuart (1961)
The Advanced Theory of Statistics, vol. 2. Griffin, London.

R. Z. Khasminskii (1966)
On stochastic processes defined by differential equations with small parameter. Theory of Probability and its Applications, vol. 11, pp. 211–228.

J. Kiefer and J. Wolfowitz (1952)
Stochastic estimation of the maximum of a regression function. Ann. Math. Stat., vol. 23, pp. 462–466.

R. E. Kopp and R. J. Orford (1963)
Linear regression applied to system identification for adaptive control systems. AIAA Journal, vol. 1, pp. 2300–2306.

V. Kucera (1979)
Discrete Linear Control. Wiley, Chichester.

R. Kumar and J. B. Moore (1980)
State inverse and decorrelated state stochastic approximation. Automatica, vol. 16, pp. 295–311.

H. J. Kushner and D. S. Clark (1978)
Stochastic Approximation Methods for Constrained and Unconstrained Systems. Springer-Verlag, New York.

H. J. Kushner and H. Huang (1979)
Rates of convergence for stochastic approximation type algorithms. SIAM Journal of Control and Optimization, vol. 17, pp. 607–617.

H. J. Kushner and H. Huang (1981)
Asymptotic properties of stochastic approximations with constant coefficients. SIAM Journal of Control and Optimization, vol. 19, pp. 87–105.

H. Kwakernaak and R. Sivan (1972)
Linear Optimal Control Systems. Wiley Interscience, New York.

I. D. Landau (1974)
A survey of model reference adaptive techniques—theory and applications. Automatica, vol. 10, pp. 353–379.

I. D. Landau (1976)
Unbiased recursive identification using model reference techniques. IEEE Transactions on Automatic Control, vol. AC-21, pp. 194–202. See also I. D. Landau: An addendum to "Unbiased recursive identification using model reference adaptive techniques", IEEE Transactions on Automatic Control, vol. AC-23, pp. 97–99, 1978.

I. D. Landau (1978)
Elimination of the real positivity condition in the design of parallel MRAS. IEEE Transactions on Automatic Control, vol. AC-23, pp. 1015–1020.

I. D. Landau (1979)
Adaptive Control. The Model Reference Approach. Dekker, New York.

Ph. de Larminat (1979)
On overall stability of certain adaptive control systems. Proc. 5th IFAC Symposium on Identification and System Parameter Estimation, Darmstadt, FRG (Pergamon Press).

D. T. Lee (1980)
Canonical ladder form realizations and fast estimation algorithms. PhD dissertation, Stanford University.

D. T. Lee and M. Morf (1980)
Recursive square-root ladder estimation algorithms. The 1980 IEEE International Conf. on Acoustics, Speech and Signal Processing, Denver, Colorado.

D. T. Lee, M. Morf, and B. Friedlander (1981)
Recursive least squares ladder estimation algorithms. IEEE Transactions on Acoustics, Speech and Signal Processing, vol. ASSP-29, pp. 627–641.

K. Levenberg (1944)
A method for the solution of certain nonlinear problems in least squares. Quart. Appl. Math., vol. 2, pp. 164–168.

L. Ljung (1971)
Characterization of the concept of "persistently exciting" in the frequency domain. Report 7119, Division of Automatic Control, Lund Institute of Technology, Lund, Sweden.

L. Ljung (1974)
Convergence of recursive stochastic algorithms. Report 7403, Department of Automatic Control, Lund Institute of Technology, Lund, Sweden.

L. Ljung (1975)
Theorems for the asymptotic analysis of recursive stochastic algorithms. Report 7522, Department of Automatic Control, Lund Institute of Technology, Lund, Sweden.

L. Ljung (1976a)
On the consistency of prediction error identification methods. In R. K. Mehra and D. G. Lainiotis, eds.: System Identification—Advances and Case Studies. Academic Press, New York.

L. Ljung (1976b)
Consistency of the least-squares identification method. IEEE Transactions on Automatic Control, vol. AC-21, pp. 779–781.

L. Ljung (1977a)
On positive real transfer functions and the convergence of some recursions. IEEE Transactions on Automatic Control, vol. AC-22, pp. 539–551.

L. Ljung (1977b)
Analysis of recursive stochastic algorithms. IEEE Transactions on Automatic Control, vol. AC-22, pp. 551–575.

L. Ljung (1977c)
Some limit results for functionals of stochastic processes. Report LiTH-ISY-I-0167, Department of Electrical Engineering, Linköping University, Linköping, Sweden.

L. Ljung (1978a)
Strong convergence of a stochastic approximation algorithm. Annals of Statistics, vol. 6, pp. 680–696.

L. Ljung (1978b)
Asymptotic theory of prediction error estimates for dynamical systems. Report LiTH-ISY-I-0220, Department of Electrical Engineering, Linköping University, Linköping, Sweden.

L. Ljung (1978c)
Convergence analysis of parametric identification methods. IEEE Transactions on Automatic Control, vol. AC-23, pp. 770–783.

L. Ljung (1978d)
On recursive prediction error identification algorithms. Report LiTH-ISY-I-0226, Department of Electrical Engineering, Linköping University, Linköping, Sweden.

L. Ljung (1978e)
Some basic ideas in recursive identification. Conference on the Analysis and Optimization of Stochastic Systems, Oxford, England.

L. Ljung (1978f)
Convergence of an adaptive filter algorithm. International Journal of Control, vol. 27, pp. 673–693.

L. Ljung (1979a)
Asymptotic behaviour of the extended Kalman filter as a parameter estimator for linear systems. IEEE Transactions on Automatic Control, vol. AC-24, pp. 36–50.

L. Ljung (1979b)
Convergence of recursive estimators. Proc. 5th IFAC Symposium on Identification and System Parameter Estimation, Darmstadt, FRG (Pergamon Press).

L. Ljung (1980a)
The ODE approach to the analysis of adaptive control systems—possibilities and limitations. Proc. Joint Automatic Control Conference, San Francisco.

L. Ljung (1980b)
Asymptotic gain and search direction for recursive identification algorithms. Proc. IEEE Conference on Decision and Control, Albuquerque, New Mexico.

L. Ljung (1981)
Analysis of a general recursive prediction error identification algorithm. Automatica, vol. 17, pp. 89–100.

L. Ljung and P. E. Caines (1979)
Asymptotic normality of prediction error estimation for approximate system models. Stochastics, vol. 3, pp. 29–46.

L. Ljung, M. Morf, and D. Falconer (1978)
Fast calculations of gain matrices for recursive estimation schemes. International Journal of Control, vol. 27, pp. 1–19.

L. Ljung, T. Söderström, and I. Gustavsson (1975)
Counterexamples to general convergence of a commonly used recursive identification method. IEEE Transactions on Automatic Control, vol. AC-20, pp. 643–652.

S. Ljung (1983)
Fast algorithms for integral equations and recursive identification. PhD thesis no. 93, Department of Electrical Engineering, Linköping University, Linköping, Sweden.

R. Lucky (1965)
Automatic equalization for digital communication. Bell Syst. Tech. Journal, vol. 44, pp. 547–588.

R. Lucky, J. Salz, and E. J. Weldon (1968)
Principles of Data Communication. McGraw-Hill, New York.

G. Lüders and K. S. Narendra (1974)
Stable adaptive schemes for state estimation and identification of linear systems. IEEE Transactions on Automatic Control, vol. AC-19, pp. 841–847.

D. G. Luenberger (1973)
Introduction to Linear and Nonlinear Programming. Addison-Wesley, Reading, Mass.

O. Macchi and E. Eweda (1983)
Second order convergence analysis of stochastic adaptive linear filtering. IEEE Transactions on Automatic Control, vol. AC-28, pp. 76–85.

J. Makhoul (1975)
Linear prediction: a tutorial review. Proceedings IEEE, vol. 63, pp. 561–580.

J. Makhoul (1977)
Stable and efficient lattice methods for linear prediction. IEEE Transactions on Acoustics, Speech and Signal Processing, vol. ASSP-25, pp. 423–428.

J. Makhoul (1978)
A class of all-zero lattice digital filters: properties and applications, IEEE Transactions on Acoustics, Speech and Signal Processing, vol. ASSP-23, pp. 304–314.

J. D. Markel and A. H. Gray (1976)
Linear Prediction of Speech. Springer, Berlin.

D. W. Marquardt (1963)
An algorithm for least-squares estimation of non-linear parameters. Journal SIAM, vol. 11, pp. 431–441.

D. Q. Mayne (1967)
A method for estimating discrete time transfer functions. Advances of Control, 2nd UKAC Control Convention, University of Bristol.

R. K. Mehra (1970)
On the identification of variances and adaptive Kalman filtering, IEEE Transactions on Automatic Control, vol. AC-15, pp. 175–184.

R. K. Mehra (1976)
Synthesis of optimal inputs for multiinput-multioutput systems with process noise. In R. K. Mehra and D. G. Lainiotis, eds.: System Identification—Advances and Case Studies. Academic Press, New York.

R. K. Mehra (1981)
Choice of input signals. In P. Eykhoff, ed.: Trends and Progress in System Identification. Pergamon Press, Oxford.

R. K. Mehra and J. S. Tyler (1973)
Case studies in aircraft parameter identification. Proc 3rd IFAC Symposium on Identification and System Parameter Estimation, the Hague (North-Holland).

J. M. Mendel (1973)
Discrete Techniques of Parameter Estimation: The Equation Error Formulation. Dekker, New York.

J. B. Moore (1978)
On strong consistency of least squares identification algorithms. Automatica, vol. 14, pp. 505–509.

J. B. Moore and G. Ledwich (1980)
Multivariable adaptive parameter and state estimators with convergence analysis. Journal of the Australian Mathematical Society, vol. 21, pp. 176–197.

J. B. Moore and H. Weiss (1979)
Recursive prediction error methods for adaptive estimation. IEEE Transactions on Systems, Man and Cybernetics, vol. SMC-9, pp. 197–205.

M. Morf (1974)
Fast algorithms for multivariable systems. PhD dissertation, Stanford University.

M. Morf (1977)
Ladder forms in estimation and system identification. Proc. 11th Asilomar Conference on Circuits, Systems and Computers, pp. 424–429.

M. Morf, B. Dickinson, T. Kailath, and A. Vieira (1977)
Efficient solution of covariance equations for linear prediction. IEEE Transactions on Acoustics, Speech and Signal Processing, vol. ASSP-25, pp. 429–433.

M. Morf and D. T. Lee (1978)
Fast algorithms for speech modeling. Technical report no M308-1, Information Systems Laboratory, Stanford University.

M. Morf, A. Vieira, and D. T. Lee (1977)
Ladder forms for identification and speech processing. Proc. 16th IEEE Conf. on Decision and Control, New Orleans.

K. S. Narendra and Y.-H. Lin (1980)
Stable discrete adaptive control, IEEE Transactions on Automatic Control, vol. AC-25, pp. 456–461.

K. S. Narendra and L. S. Valavani (1978)
Direct and indirect adaptive control. Proc. 7th IFAC World Congress, Helsinki (Pergamon Press).

L. W. Nelson and E. Stear (1976)
The simultaneous on-line estimation of parameters and states in linear systems. IEEE Transactions on Automatic Control, vol. AC-21, pp. 94–98.

M. B. Nevelson and R. Z. Khasminskii (1973)
Stochastic Approximation and Recursive Estimation. Translations of Mathematical Monographs, vol. 47, Am. Math. Soc., Providence, Rhode Island.

J. Neveu (1975)
Discrete Parameter Martingales. North-Holland, Amsterdam.

R. F. Ohap and A. R. Stubberud (1976)
Adaptive minimum variance estimation in discrete-time linear systems. In C. T. Leondes, ed.: Control and Dynamic Systems, vol. 12, pp. 583–642. Academic Press, New York.

A. V. Oppenheim and R. W. Schafer (1975)
Digital Signal Processing. Prentice-Hall, Englewood Cliffs, N.J.

V. Panuska (1968)
A stochastic approximation method for identification of linear systems using adaptive filtering. Joint Automatic Control Conference, Ann Arbor.

A. Papoulis (1965)
Probability, Random Variables and Stochastic Processes. McGraw-Hill, New York.

P. C. Parks (1966)
Lyapunov redesign of model reference adaptive control systems. IEEE Transactions on Automatic Control, vol. AC-11, pp. 362–367.

V. Peterka (1975)
A square root filter for real time multivariate regression. Kybernetika, vol. 11, pp. 53–67.

V. Peterka (1979)
Bayesian system identification. Proc. 5th IFAC Symposium on Identification and System Parameter Estimation, Darmstadt, FRG (Pergamon Press).

V. Peterka (1981)
Bayesian approach to system identification. In P. Eykhoff, ed.: Trends and Progress in System Identification. Pergamon Press, Oxford.

R. L. Plackett (1950)
Some theorems in least squares. Biometrika, vol. 37, p. 149.

B. T. Polyak and Ya. Z. Tsypkin (1979)
Adaptive estimation algorithms: convergence, optimality, robustness. Automation and Remote Control, vol. 3, pp. 71–84.

B. T. Polyak and Ya. Z. Tsypkin (1980)
Robust identification. Automatica, vol. 16, pp. 53–63.

J. E. Potter (1963)
New statistical formulas. Memo 40, Instrumentation Laboratory, Massachusetts Institute of Technology.

C. R. Rao (1973)
Linear Statistical Inference and its Applications. Wiley, New York.

V. U. Reddy, B. Egardt, and T. Kailath (1981)
Optimized lattice-form adaptive line enhancer for a sinusoidal signal in broad-band noise. IEEE Transactions on Acoustics, Speech and Signal Processing, vol. ASSP-29, pp. 702–710.

H. Robbins and S. Monro (1951)
A stochastic approximation method. Annals of Mathematical Statistics, vol. 22, pp. 400–407.

A. P. Sage and C. D. Wakefield (1972)
Maximum likelihood identification of time varying and random system parameters. International Journal of Control, vol. 16, pp. 81–100.

D. J. Sakrison (1967)
The use of stochastic approximation to solve the system identification problem. IEEE Transactions on Automatic Control, vol. AC-12, pp. 563–567.

G. N. Saridis (1974)
Comparison of six on-line identification algorithms. Automatica, vol. 10, pp. 69–79.

G. S. Saridis (1977)
Self-Organizing Control of Stochastic Systems. Dekker, New York.

G. N. Saridis and G. Stein (1968)
Stochastic approximation algorithms for linear discrete time system identification. IEEE Transactions on Automatic Control, vol. AC-13, pp. 515–523.

E. H. Satorius and J. D. Pack (1981)
Application of least squares lattice algorithms to adaptive equalization. IEEE Transactions on Communications, vol. COM-29, pp. 136–142.

S. C. Shah (1981)
Internal model adaptive control. PhD dissertation, Stanford University.

T. Söderström (1973a)
On the uniqueness of maximum likelihood identification for different structures. Report 7307, Department of Automatic Control, Lund Institute of Technology, Lund, Sweden.

T. Söderström (1973b)
An on-line algorithm for approximate maximum likelihood identification of linear dynamic systems. Report 7308, Department of Automatic Control, Lund Institute of Technology, Lund, Sweden.

T. Söderström (1974)
Convergence properties of the generalized least squares identification method. Automatica, vol. 10, pp. 617–626.

T. Söderström (1975)
On the uniqueness of maximum likelihood identification. Automatica, vol. 11, pp. 193–197.

T. Söderström (1977)
On model structure testing in system identification. International Journal of Control, vol. 26, pp. 1–18.

T. Söderström, L. Ljung, and I. Gustavsson (1974a)
A comparative study of recursive identification methods. Report 7427, Department of Automatic Control, Lund Institute of Technology, Lund, Sweden.

T. Söderström, L. Ljung, and I. Gustavsson (1974b)
On the accuracy of identification and the design of identification experiments. Report 7428, Department of Automatic Control, Lund Institute of Technology, Lund, Sweden.

T. Söderström, L. Ljung, and I. Gustavsson (1978)
A theoretical analysis of recursive identification methods. Automati a, vol. 14, pp. 231–244.

T. Söderström and P. Stoica (1978)
Comparison of instrumental variable methods—consistency and accuracy aspects. Report UPTEC 7888R, Institute of Technology, Uppsala Univers .y, Uppsala, Sweden.

T. Söderström and P. Stoica (1981)
Comparison of instrumental variable methods—consistency and accuracy aspects. Automatica, vol.17, pp. 101–115.

T. Söderström and P. Stoica (1983)
The Instrumental Variable Approach to System Identification. Springer, Berlin.

V. Solo (1978)
Time series recursions and stochastic approximation. PhD dissertation, The Australian National University, Canberra, Australia.

V. Solo (1979)
The convergence of AML. IEEE Transactions on Automatic Control, vol. AC-24, pp. 958–963.

V. Solo (1980)
Some aspects of recursive parameter estimation. International Journal of Control, vol. 32, pp. 395–410.

V. Solo (1981)
The second order properties of a time series recursion. Annals of Statistics, vol. 9, pp. 307–317.

N. M. Sondhi and P. A. Berkley (1980)
Silencing echoes on the telephone network. Proc. IEEE, vol. 68, pp. 948–963.

F. Soong (1981)
Fast least-squares estimation and its applications. PhD dissertation, Stanford University.

S. D. Stearns (1980)
Error surfaces of recursive adaptive filters. Sandia report, Albuquerque, New Mexico.

K. Steiglitz and L. E. McBride (1965)
A technique for the identification of linear systems. IEEE Transactions on Automatic Control, vol. AC-10, pp. 461–464.

J. Sternby (1977)
On consistency for the method of least squares using martingale theory. IEEE Transactions on Automatic Control, vol. AC-21, pp. 346–352.

G. W. Stewart (1970)
Introduction to Matrix Computation. Academic Press, New York.

P. Stoica, J. Holst and T. Söderström (1982)
Eigenvalue location of certain matrices arising in convergence analysis problems. Automatica, vol. 18, pp. 487–489.

P. Stoica and T. Söderström (1979)
Consistency properties on an identification method using the instrumental variable principle. Revue Roumaine des Sciences Techniques, Série Électrotechnique et Énergétique, vol. 24, pp. 289–293.

P. Stoica and T. Söderström (1981a)
Asymptotic behaviour of some bootstrap estimators. International Journal of Control, vol. 33, pp. 433–454.

P. Stoica and T. Söderström (1981b)
Instrumental variable methods for identification of linear and certain nonlinear systems. Report UPTEC 8168R. Institute of Technology, Uppsala University, Uppsala, Sweden

P. Stoica and T. Söderström (1982a)
Identification of multivariable systems using instrumental variable methods. Proc. 6th IFAC Symposium on Identification and System Parameter Estimation. Washington, D.C. (Pergamon Press).

P. Stoica and T. Söderström (1982b)
A note on the parsimony principle. International Journal of Control, vol. 36, pp. 409–418.

P. Stoica and T. Söderström (1983a)
Optimal instrumental variable estimation and approximate implementations. IEEE Transactions on Automatic Control, vol. AC-28.

P. Stoica and T. Söderström (1983b)
Optimal instrumental variable methods for the identification of multivariable linear systems. Automatica, vol. 19.

V. Strejc (1980)
Least squares parameter estimation. Automatica, vol. 16, pp. 535–550.

J. L. Talmon and A. J. W. van den Boom (1973)
On the estimation of transfer function parameters of process and noise dynamics using a single-stage estimator. Proc. 3rd IFAC Symposium on Identification and System Parameter Estimation, the Hague (North-Holland).

C. L. Thornton and G. J. Bierman (1977)
Gram-Schmidt algorithms for covariance propagation. International Journal of Control, vol. 25, pp. 243–260.

C. L. Thornton and G. J. Bierman (1980)
UDU^T covariance factorization for Kalman filtering. In C. T. Leondes, ed.: Control and Dynamic Systems, vol. 16, Academic Press, New York.

J. R. Treichler (1979)
Transient and convergent behavior of the adaptive line enhancer. IEEE Transactions on Acoustics, Speech and Signal Processing, vol. ASSP-27, pp. 53–62.

Ya. Z. Tsypkin (1971)
Adaption and Learning in Automatic Systems. Academic Press, New York.

Ya. Z. Tsypkin (1973)
Foundation of the Theory of Learning Systems. Academic Press, New York.

J. W. Tukey (1961)
Discussion, emphasizing the connection between analysis of variance and spectrum analysis. Technometrics, vol. 3, pp. 191–219.

H. Unbehauen, B. Göhring, and B. Bauer (1974)
Parameterschätzverfahren zur Systemidentifikation. R. Oldenbourg Verlag, Munich.

A. Wald (1949)
Note on the consistency of the maximum likelihood estimate. Ann. Math. Stat., vol. 20, pp. 595–601.

A. Weiss and P. Mitra (1979)
Digial adaptive filters: Conditions for convergence, rates of convergence, effect of noise, and errors arising from the implementation. IEEE Transactions on Information Theory, vol. IT-25, pp. 637–652.

P. D. Welch (1967)
The use of fast Fourier transform for the estimation of power spectra: a method based on time averaging over short, modified periodograms. IEEE Transactions Audio Electroacoust. (Special Issue on Fast Fourier Transform and its Applications to Digital Filtering and Spectral Analysis), vol. AU-15, pp. 70–73.

T. Westerlund and A. Tyssø (1980)
Remarks on "Asymptotic behavior of th. xtended Kalman filter as a parameter estimator for linear systems". IEEE Transactions on Automatic Control, vol. AC-25, pp. 1011–1012.

H. P. Whitaker, J. Yamron, and A. Kezer (1958)
Design of model-reference adaptive control systems for aircraft. Report R-164, Instrumentation Laboratory, Massachusetts Institute of Technology.

S. White (1975)
An adaptive recursive digital filter. 9th Annual Asilomar Conference on Circuits, Systems and Computers.

B. Widrow, J. R. Glover, Jr., et al. (1975)
Adaptive noise cancelling: principles and applications. Proc. IEEE, vol. 63, pp. 1692–1716.

B. Widrow and M. E. Hoff, Jr. (1960)
Adaptive switching circuits. IRE WESCON Convention Record, part 4, pp. 96–104.

A. Willsky (1979)
Digital Signal Processing and Control and Estimation Theory: Points of Tangency, Areas of Intersection, and Parallel Directions. MIT Press, Cambridge, Mass.

B. Wittenmark (1974)
A self-tuning predictor. IEEE Transactions on Automatic Control, vol. AC-19, pp. 848–851.

B. Wittenmark and Y. Bar-Shalom (1979)
Model validation from estimated closed loop performance. Proc. 5th IFAC Symposium on Identification and System Parameter Estimation, Darmstadt, FRG (Pergamon Press).

K. Y. Wong and E. Polak (1967)
Identification of linear discrete time systems using the instrumental variable method. IEEE Transactions on Automatic Control, vol. AC-12, pp. 707–718.

W. R. Wouters (1972)
On-line identification in an unknown stochastic environment. IEEE Transactions on Systems, Man and Cybernetics, vol. SMC-2, pp. 666–668.

P. C. Young (1965)
Process parameter estimation and self-adaptive control. Proc. IFAC Symposium on Self-Adaptive Systems, Teddington, England (Plenum Press).

P. C. Young (1968)
The use of linear regression and related procedures for the identification of dynamic processes. Proc. 7th IEEE Symposium on Adaptive Processes, UCLA.

P. C. Young (1970)
An instrumental variable method for real-time identification of a noisy process. Automatica, vol. 6, pp. 271–287.

P. C. Young (1976)
Some observations on instrumental variable methods of time-series analysis. International Journal of Control, vol. 23, pp. 593–612.

P. C. Young and A. Jakeman (1979)
Refined instrumental variable methods of recursive time-series analysis. Part I: Single input, single output systems. International Journal of Control, vol. 29, pp. 1–30.

M. B. Zarrop (1979)
Optimal Experiment Design for Dynamic System Identification. Springer, Berlin.

Author Index

Subject Index